Effects of Climate Change on Social and Economic Factors

Berfin Göksoy Sevinçli
Bitlis Eren Üniversitesi İİBF, Turkey

Dilek Alma Savaş
Bitlis Eren Üniversitesi İİBF, Turkey

IGI Global
Publishing Tomorrow's Research Today

Published in the United States of America by
IGI Global
701 E. Chocolate Avenue
Hershey PA, USA 17033
Tel: 717-533-8845
Fax: 717-533-8661
E-mail: cust@igi-global.com
Web site: https://www.igi-global.com

Copyright © 2025 by IGI Global. All rights reserved. No part of this publication may be reproduced, stored or distributed in any form or by any means, electronic or mechanical, including photocopying, without written permission from the publisher.
Product or company names used in this set are for identification purposes only. Inclusion of the names of the products or companies does not indicate a claim of ownership by IGI Global of the trademark or registered trademark.

Library of Congress Cataloging-in-Publication Data

CIP Data Pending
ISBN:979-8-3693-5792-7
eISBN:979-8-3693-5794-1

Vice President of Editorial: Melissa Wagner
Managing Editor of Acquisitions: Mikaela Felty
Managing Editor of Book Development: Jocelynn Hessler
Production Manager: Mike Brehm
Cover Design: Phillip Shickler

British Cataloguing in Publication Data
A Cataloguing in Publication record for this book is available from the British Library.

All work contributed to this book is new, previously-unpublished material.
The views expressed in this book are those of the authors, but not necessarily of the publisher.

Table of Contents

Foreword xvii

Preface xix

Acknowledgment xxvii

Section 1
Conceptual Framework and Visibility of the Problem

Chapter 1
Conceptual Framework and Problem Visibility: Conceptual Explanations for
the Economics of Climate Change 1
    Selvi Vural, Gümüşhane University, Turkey

Chapter 2
Climate Change and Global Inequality: How Does Climate Change
Exacerbate Existing Global Inequalities and Its Implications 21
    Mohit Yadav, O.P. Jindal Global University, India
    Ajay Chandel, Lovely Professional University, India
    Harshita Agrawal, National Post Graduate College, Lucknow, India
    Majdi Quttainah, Kuwait University, Kuwait

Chapter 3
Social and Economic Impacts of Climate-Induced Migration and
Displacement 49
    Shashank Mittal, O.P. JIndal Global University, India
    Ajay Chandel, Lovely Professional University, India
    Thi Mai Le, Vietnam National University, Hanoi, Vietnam

Chapter 4
Impact of Climate Change on Mental Health and Economic Consequences of
Climate-Related Mental Health Issues 83
    Preet Kanwal, Lovely Professional University, India
    Shashank Mittal, O.P. JIndal Global University, India
    Hewawasam P. G. D. Wijethilaka, University of Colombo, Sri Lanka

Chapter 5
Effect of Climate Change and Agricultural Factors on the Technical
Efficiency of the Industrial Sector Across Indian States 105
    Ajay Kumar Singh, Department of Humanities and Social Sciences,
        Graphic Era University (Deemed), Dehradun, India
    Bhim Jyoti, Department of Seed Science and Technology, College of
        Forestry, Veer Chandra Singh Garhwali Uttarakh, India

Section 2
Evaluation of Climate Change Through Economy

Chapter 6
The Economic Impacts of Climate Change 137
    Firat Cem Dogan, Hasan Kalyoncu University, Turkey

Chapter 7
Economic Costs of Climate Change and How These Costs Can Be Mitigated 163
    Shashank Mittal, O.P. JIndal Global University, India
    Preet Kanwal, Lovely Professional University, India
    Hewawasam P. G. D. Wijethilak, University of Colombo, Sri Lanka

Chapter 8
Impacts of Climate Change on the Tourism Sector 199
    Fatma Fehime Aydin, Van Yuzuncu Yil University, Turkey
    Cemalettin Levent, Independent Researcher, Turkey

Chapter 9
Economic-Financial Issues and Climate Change: A Sustainability-Based
Approach.. 215
    Güven Güney, Atatürk University, Turkey
    Eda Bozkurt, Atatürk University, Turkey

Chapter 10
Impacts of Climate Change on Employment: An Evaluation on COVID-19
and Remote Working as an Alternative Employment Method 239
    Berivan Özay Acar, Yüzüncü Yıl Üniversitesi, Turkey

Section 3
Fighting Climate Change

Chapter 11
Methods of Combating Against the Problem of Climate Change 255
    Enes Yalçın, İzmir Kâtip Çelebi University, Turkey

Chapter 12
Turkey's Policies to Combat Climate Change 277
 Hande Saraçoğlu, Van Yüzüncü Yıl Üniversitesi, Turkey
 Hande Saraçoğlu, Van Yüzüncü Yıl Üniversitesi, Turkey

Chapter 13
Policies of European Countries to Combat the Climate Crisis 301
 Ezgi Kovancı, Adıyaman University, Turkey

Chapter 14
Calibrating Climate Change Curriculum Coverage in Some Modules at St. Peter's University 329
 Sikhulile B. Msezane, University of South Africa, South Africa
 Nonkanyiso Pamella Shabalala, University of South Africa, South Africa

Chapter 15
Utilizing OpenStreetMap Data for Local Climate Change Assessment and Policy Formulation 367
 Munir Ahmad, Survey of Pakistan, Pakistan
 Miguel Angel Osorio Rivera, Escuela Superior Politécnica de
  Chimborazo, Ecuador
 William Estuardo Carrillo Barahona, Escuela Superior Politécnica de
  Chimborazo, Ecuador
 Amara Nisar, University of the Punjab, Lahore, Pakistan
 Noor Ul Safa, GC Women University, Sialkot, Pakistan

Chapter 16
Climate Change and Sustainable Development: How Can Climate Change Be Addressed Within the Framework of Sustainable Development Goals? 387
 Ajay Chandel, Lovely Professional University, India
 Mohit Yadav, O.P. Jindal Global University, India
 Phuong Mai Nguyen, Vietnam National University, Hanoi, Vietnam

Chapter 17
Innovation and Technology to Address the Challenges of Climate Change 423
 Thi Minh Ngoc Luu, Vietnam National University, Hanoi, Vietnam
 Mohit Yadav, O.P. Jindal Global University, India
 Anugamini Srivastava, Symbiosis International University (Deemed),
  India
 Krishan Gopal, Lovely Professional University, India

Compilation of References 465

About the Contributors 525

Index 533

Detailed Table of Contents

Foreword xvii

Preface xix

Acknowledgment xxvii

Section 1
Conceptual Framework and Visibility of the Problem

Chapter 1
Conceptual Framework and Problem Visibility: Conceptual Explanations for the Economics of Climate Change 1
Selvi Vural, Gümüşhane University, Turkey

In the realm of economics, conceptual explanations serve as the foundation for understanding the complexities of market dynamics, resource allocation, and socio-economic phenomena. These explanations elucidate the theoretical frameworks that underpin economic models, guiding policymakers, analysts, and individuals in comprehending the intricate interplay of factors shaping economic outcomes. Conceptual explanations delve into fundamental concepts such as supply and demand, elasticity, production functions, utility theory, and market structures, offering insights into how these principles interact to influence economic behavior and outcomes. By providing a conceptual roadmap, these explanations facilitate informed decision-making, foster critical thinking, and contribute to the advancement of economic theory and practice.

Chapter 2
Climate Change and Global Inequality: How Does Climate Change
Exacerbate Existing Global Inequalities and Its Implications 21
    Mohit Yadav, O.P. Jindal Global University, India
    Ajay Chandel, Lovely Professional University, India
    Harshita Agrawal, National Post Graduate College, Lucknow, India
    Majdi Quttainah, Kuwait University, Kuwait

This chapter explores the intersection of climate change and global inequality, highlighting how climate impacts exacerbate existing disparities and challenge social justice. By examining case studies from Bangladesh, Puerto Rico, the Sahel region, and Pacific Island nations, it reveals the disproportionate effects of climate change on vulnerable populations. The analysis underscores the urgent need for integrated climate justice frameworks, community-based adaptation strategies, and sustainable development initiatives. Key global responses, including the Paris Agreement and international migration compacts, are assessed for their effectiveness in addressing these challenges. Future directions emphasize enhancing international cooperation, leveraging technological innovations, and improving monitoring and accountability. The chapter aims to provide a comprehensive understanding of how climate change intensifies inequality and offers actionable insights for building a more equitable and resilient future.

Chapter 3
Social and Economic Impacts of Climate-Induced Migration and
Displacement 49
    Shashank Mittal, O.P. JIndal Global University, India
    Ajay Chandel, Lovely Professional University, India
    Thi Mai Le, Vietnam National University, Hanoi, Vietnam

Climate-induced migration and displacement are emerging as critical issues in the context of global climate change. This chapter explores the drivers, social, and economic impacts of climate-induced migration, providing future projections and scenarios to understand the scale and nature of displacement. Key drivers include extreme weather events, sea-level rise, desertification, and water scarcity. Social impacts involve disruptions to communities, changes in demographics, and health challenges, while economic impacts affect labor markets, infrastructure, and resource allocation. Case studies illustrate real-world examples and policy responses, while resilience and adaptation strategies offer insights into managing these challenges. The chapter concludes with future projections highlighting the need for comprehensive planning and international cooperation to address the anticipated increase in climate-induced displacement.

Chapter 4
Impact of Climate Change on Mental Health and Economic Consequences of
Climate-Related Mental Health Issues 83
    Preet Kanwal, Lovely Professional University, India
    Shashank Mittal, O.P. JIndal Global University, India
    Hewawasam P. G. D. Wijethilaka, University of Colombo, Sri Lanka

Climate change poses significant challenges to mental health, impacting individuals and communities through extreme weather events, gradual environmental changes, and socio-economic disruptions. This chapter explores the complex relationship between climate change and mental health, highlighting direct and indirect pathways through which environmental factors exacerbate psychological distress. Case studies from various regions illustrate the widespread nature of these impacts, including increased rates of anxiety, depression, and PTSD. The chapter discusses practical, managerial, ethical, and societal implications, emphasizing the need for integrated strategies that include mental health services in disaster response and climate adaptation efforts. Future directions involve addressing research gaps, leveraging technological innovations, and fostering cross-sector collaboration. By understanding and addressing these mental health impacts, we can enhance resilience and support affected communities in the face of climate change.

Chapter 5
Effect of Climate Change and Agricultural Factors on the Technical
Efficiency of the Industrial Sector Across Indian States 105
    Ajay Kumar Singh, Department of Humanities and Social Sciences,
        Graphic Era University (Deemed), Dehradun, India
    Bhim Jyoti, Department of Seed Science and Technology, College of
        Forestry, Veer Chandra Singh Garhwali Uttarakh, India

This chapter estimates the TE of the industrial sector across Indian states using a stochastic frontier analysis. It also observes the impact of climatic and agricultural factors on gross value added (GVA) and TE of the industrial sector using a log-linear regression model. Annual average values of maximum and minimum temperature, precipitation and actual rainfall are considered as climatic factors, and irrigated area, cropping intensity, gross sown area, and credit deposit ratio are used as control variables in the empirical model. It compiles state-wise panel data of mentioned variables during 1991 – 2021. The results reveal that there is significant diversity in TE across states. TE and GVA are negatively impacted due to climate change. Gross irrigated area, cropping intensity, and gross sown showed a significant impact on TE and GVA. It provides policy proposals to reduce diversity in TE of industries across Indian states.

Section 2

Evaluation of Climate Change Through Economy

Chapter 6
The Economic Impacts of Climate Change 137
Firat Cem Dogan, Hasan Kalyoncu University, Turkey

Climate change profoundly impacts global economies, particularly the agricultural sector, through altered temperature and precipitation patterns, threatening food security and economic stability. This research examines climate change's effects on agriculture and economic growth, proposing mitigation policies. Methodologically, it analyzes the climate-economy relationship theoretically and empirically. Findings reveal declining agricultural productivity, disrupted supply chains, and increased natural disasters, necessitating a transition to sustainable energy and agriculture. Global collaboration and steadfast policies are crucial for addressing climate change, ensuring both economic prosperity and environmental sustainability.

Chapter 7
Economic Costs of Climate Change and How These Costs Can Be Mitigated 163
Shashank Mittal, O.P. JIndal Global University, India
Preet Kanwal, Lovely Professional University, India
Hewawasam P. G. D. Wijethilak, University of Colombo, Sri Lanka

This chapter explores the economic costs of climate change and strategies for mitigating these impacts. It examines the diverse and significant financial burdens associated with climate-related disasters, disruptions, and long-term environmental changes. The chapter outlines key areas of concern, including regional and sectoral variations in economic costs, and the use of economic models and methods to assess these impacts. It discusses various mitigation strategies, such as transitioning to a low-carbon economy and enhancing climate resilience and emphasizes the importance of international cooperation in addressing global climate challenges. The chapter concludes with an outlook on future developments and the implications for practical, managerial, ethical, and societal dimensions. Effective climate action requires comprehensive, collaborative approaches and a commitment to sustainability and equity.

Chapter 8
Impacts of Climate Change on the Tourism Sector 199
    Fatma Fehime Aydin, Van Yuzuncu Yil University, Turkey
    Cemalettin Levent, Independent Researcher, Turkey

Climate change, one of the most prominent problems all over the world in recent years, has significant impacts on many sectors as well as being an environmental problem. The tourism sector is one of the sectors most affected by climate change. The main objective of this study is to investigate the impact of climate change on tourism in selected countries of the world using panel data analysis based on the data set for the period 2004-2019. Based on the findings of this study, countries should adopt practices that will encourage tourist inflow to the country to increase tourism revenues, but at the same time, they should bring environmentally sensitive practices to the agenda so that both future generations and current generations can live in healthier environments.

Chapter 9
Economic-Financial Issues and Climate Change: A Sustainability-Based
Approach 215
    Güven Güney, Atatürk University, Turkey
    Eda Bozkurt, Atatürk University, Turkey

The situation in which the interaction of economic-financial and social structure with the environment is not evaluated as a whole causes present and future generations not to benefit equally from the opportunities brought by development. For this reason, it is very important for policy development to determine the relationship between climate change and economic-financial issues by considering the issue of climate change from a sustainable development perspective. Identifying which economic-financial events fuel climate change will help policymakers to develop measures and solutions in that area. For this purpose, in this study, considering the availability of the data set, estimations were made based on panel quantile regression for the countries in the World Bank database. Thus, with a new econometric method, the causes of climate change have been revealed from an economic and financial framework for the countries that make up each quantile.

Chapter 10
Impacts of Climate Change on Employment: An Evaluation on COVID-19
and Remote Working as an Alternative Employment Method 239
Berivan Özay Acar, Yüzüncü Yıl Üniversitesi, Turkey

Climate change is a growing concern around the world. This crisis causes significant impacts in various sectors such as agriculture, tourism, energy, and construction. Risks due to climate events are increasing, job losses are occurring, and workers' health is negatively affected. Especially as the frequency and severity of natural disasters increases, employment security decreases in these sectors. Recently, the COVID-19 pandemic has rapidly changed the business world and ways of doing business. Remote working has emerged as a result of this change and has been adopted as an alternative employment method. Unlike traditional office environments, the remote working model reduces greenhouse gas emissions, reduces carbon emissions caused by traffic congestion, and enables more efficient use of resources by optimizing energy consumption. Therefore, this study considers evaluating the effects of climate change on employment through the COVID-19 pandemic and the remote working model as an eco-strategy that will contribute to environmental sustainability.

Section 3
Fighting Climate Change

Chapter 11
Methods of Combating Against the Problem of Climate Change 255
Enes Yalçın, İzmir Kâtip Çelebi University, Turkey

This study addresses the critical issue of climate change, examining the contributing factors and the resulting problems. Key drivers of climate change, including greenhouse gas emissions, deforestation, and industrial pollution, are analyzed. The study also explores the multifaceted impacts of climate change, such as extreme weather events, rising sea levels, and biodiversity loss. Emphasis is placed on the role of international platforms and agreements at the nation-state level, such as the Paris Agreement, in combating climate change. Additionally, the study highlights effective sub-national initiatives and practices, showcasing local government actions and community-based projects aimed at reducing carbon footprints and enhancing climate resilience. Through comprehensive analysis, this research underscores the importance of coordinated efforts at both global and local levels to mitigate the adverse effects of climate change and promote sustainable development.

Chapter 12
Turkey's Policies to Combat Climate Change 277
    Hande Saraçoğlu, Van Yüzüncü Yıl Üniversitesi, Turkey
    Hande Saraçoğlu, Van Yüzüncü Yıl Üniversitesi, Turkey

Climate change, the most important environmental problem that concerns the whole world, is one of the biggest common problems facing humanity today. In this chapter, Turkey's policies to combat climate change will be examined in depth, and the country's trends in greenhouse gas emissions, efforts towards energy transformation, renewable energy investments, policies on forestry, and approaches to climate justice will be discussed. In addition, the challenges Turkey faces in combating climate change and suggestions for solutions to these challenges will be discussed. Thus, Turkey's current situation and future goals in combating climate change will be better understood.

Chapter 13
Policies of European Countries to Combat the Climate Crisis 301
    Ezgi Kovancı, Adıyaman University, Turkey

The climate crisis is profoundly impacting Europe, one of the most vulnerable regions to climate change globally. Increasing temperatures, erratic rainfall patterns, and rising sea levels are intensifying, threatening human life, economies, and ecosystems across the continent. European nations are actively implementing diverse policies and initiatives, such as the Paris Agreement and the European Green Deal, and leading in renewable energy adoption and sustainable transportation. In this study, the geography of Europe, which includes industrially advanced countries, has been examined within the context of the European Union's climate policies.

Chapter 14
Calibrating Climate Change Curriculum Coverage in Some Modules at St. Peter's University 329
    Sikhulile B. Msezane, University of South Africa, South Africa
    Nonkanyiso Pamella Shabalala, University of South Africa, South Africa

In this qualitative research approach study, the coverage of climate change education content at St. Peters University was investigated utilising a case study research design and document analysis. This study gathered pertinent information from three modules that covered education for sustainable development using the convenience sampling method. Realist social theory was employed as an analytical and theoretical framework. The study's conclusions showed that the modules' coverage of climate change risks, hazards, mitigation, resilience, and adaptation measures is lacking. Even though climate change was discussed inequitably, the implication is that students would not be able to cascade climate change education content to citizens during teaching practice, inhibiting awareness, and acting toward a behavioural change that encourages risk identification, climate change injustices, mitigation, adaptation, and resilience.

Chapter 15
Utilizing OpenStreetMap Data for Local Climate Change Assessment and Policy Formulation 367
    Munir Ahmad, Survey of Pakistan, Pakistan
    Miguel Angel Osorio Rivera, Escuela Superior Politécnica de Chimborazo, Ecuador
    William Estuardo Carrillo Barahona, Escuela Superior Politécnica de Chimborazo, Ecuador
    Amara Nisar, University of the Punjab, Lahore, Pakistan
    Noor Ul Safa, GC Women University, Sialkot, Pakistan

This chapter has delved into the significant role that OSM data can play in local climate change assessments and policy development. Utilizing the open, comprehensive, and continually updated geospatial information available in OSM, researchers and policymakers can equip themselves with essential tools to understand and tackle climate issues at the local scale. OSM offers numerous benefits for climate research, such as open access, extensive spatial coverage, and regular updates. Nevertheless, it is important to recognize the limitations, including potential data inconsistencies and gaps in certain regions. Implementing data validation techniques is crucial to ensure accurate analysis. OSM's success heavily depends on the active engagement of the global community. Ongoing contributions and efforts to enhance data quality are essential to maintain the platform's reliability and usefulness for climate research.

Chapter 16

Climate Change and Sustainable Development: How Can Climate Change Be Addressed Within the Framework of Sustainable Development Goals? 387

    Ajay Chandel, Lovely Professional University, India
    Mohit Yadav, O.P. Jindal Global University, India
    Phuong Mai Nguyen, Vietnam National University, Hanoi, Vietnam

Addressing climate change within the framework of the Sustainable Development Goals (SDGs) is crucial for achieving global sustainability. This chapter explores how integrating climate action with the SDGs can drive transformative change across sectors. It examines key areas such as mitigation and adaptation strategies, the role of finance and technology, and the need for effective policy integration. Highlighting challenges including policy coherence, financial constraints, and technological gaps, the chapter also identifies opportunities through innovation, global cooperation, and inclusive transitions. By fostering resilient cities and leveraging nature-based solutions, countries can enhance their climate resilience while advancing sustainable development. This comprehensive analysis underscores the importance of a coordinated, ambitious approach to climate action to ensure a sustainable and equitable future.

Chapter 17
Innovation and Technology to Address the Challenges of Climate Change 423

Thi Minh Ngoc Luu, Vietnam National University, Hanoi, Vietnam
Mohit Yadav, O.P. Jindal Global University, India
Anugamini Srivastava, Symbiosis International University (Deemed), India
Krishan Gopal, Lovely Professional University, India

This chapter explores the pivotal role of innovation and technology in addressing the multifaceted challenges of climate change. It examines advancements in renewable energy, digital technologies, circular economy models, and climate resilience strategies, highlighting their potential to drive significant progress in climate action. The chapter also discusses the barriers to innovation adoption, including financial constraints, regulatory hurdles, and social resistance. By analyzing case studies of successful climate innovations and considering future trends, it provides a comprehensive overview of how technological and strategic innovations can mitigate and adapt to climate impacts. The chapter emphasizes the importance of collaborative efforts and inclusive approaches in fostering effective climate solutions and advancing towards a sustainable, low-carbon future.

Compilation of References 465

About the Contributors 525

Index 533

Foreword

Humanity has faced a new existential crisis due to environmental problems arising from industrialization and urbanization. Environmental problems, which have been one of the most important agenda items of the international public opinion since the 1970s, have brought along the search for solutions at the global level. The Conference on the Human Environment was organized for the first time in Stockholm in 1972 under the leadership of the United Nations in order to intervene in the inevitable end of the world. In 1992, global warming and the resulting climate change became one of the priority topics and the UN Framework Convention on Climate Change was opened for signature. After 30 years since the convention entered into force in 1994, steps have been taken to take innovative and comprehensive measures and develop policies. It must be admitted that the economic development model of the industrial revolution is far from being an environmentally friendly development model. Therefore, a new economic development model needed to be built first. The most innovative model in this regard has found meaning in the concept of sustainable development. In resource utilization, the balance between economy and ecology must be observed and the right of future generations to live in a healthy and balanced environment must be protected, in other words, intergenerational justice must be ensured. This is the only way to develop strategies to adapt to and combat climate change. Because climate change has a profound impact on social and economic factors. Increasing temperatures and irregular precipitation in the world bring along many problems such as decreasing production, decreasing yields, spread of diseases, disruption of social order, displacement of the population, and employment constraints. Especially in recent years, we have witnessed an increase in the impact, prevalence and frequency of disasters caused by climate change.

Since it is located on the Mediterranean coast, Türkiye also experiences problems arising from climate change. Fires, floods, avalanches, landslides and landslides, drought and desertification can be given as examples of disasters experienced in Türkiye. Although its contribution to climate change is negligible, Türkiye is taking the necessary steps in this regard. In this context, President Recep Tayyip Erdoğan announced at the UN General Assembly that Türkiye will become a party to and ratify the Paris Climate Agreement prepared as a requirement of the UN Framework Convention on Climate Change. Türkiye also announced its 2053 Net Zero Emission Target and Green Development Revolution vision. In line with this vision, Türkiye has increased its national reduction target, which was announced in 2015 as up to 21% reduction compared to the Reference Scenario, to 41% in order to achieve net zero emissions by 2053. Türkiye has prepared the Green Deal Action Plan to align with the European Green Deal Action Plan. An indication of Türkiye's sincere and constructive contribution on this issue was the renaming of the Ministry of Environment and Urbanization as the Ministry of Environment, Urbanization and Climate Change and the establishment of the Climate Change Presidency within the Ministry. A new environmental movement against climate change was launched with the Zero Waste and Zero Waste Blue Project, which was patronized by the First Lady Emine Erdoğan and implemented by the Ministry of Environment, Urbanization and Climate Change. Since 2017, the project, which has been recognized by the UN and received awards, has been spreading as a model. As a matter of fact, the UN declared March 30 as International Zero Waste Day. The Climate Ambassadors Project, which was also initiated to improve climate sensitivity among young people, identified Climate Ambassadors in 208 universities in Türkiye.

It is pleasing that scientific studies on climate change, which threatens humanity, destroys biodiversity and will make our planet uninhabitable if no measures are taken, are becoming widespread. This book, prepared by academics as one of these studies, will fill an important gap in the field. In this book, the effects of climate change on social and economic factors such as health, education, tourism, agriculture, industry, employment, sustainable development, innovation-technology and global inequality are analyzed. The impacts of climate change on countries and policies to combat these impacts constitute the scope of this book. The book will provide a basis for studies on climate change and will also provide data on the socioeconomic impacts of climate change on an international scale. The target audience is researchers, students, scientists, policy makers and practitioners working on climate change. Since climate change is a multidimensional, multi-actor issue that requires a high level of cooperation, I wish success to my colleagues who contributed to the study, which was written with the aim of revealing the problem from different perspectives and developing solutions.

A. Menaf Turan

Ministry of Environment Urbanization and Climate Change of the Republic of Türkiye, Turkey & Department of Public Administration, Urbanization, and Environmental Problems, Faculty of Economics and Administrative Sciences, Van Yüzüncü Yıl University, Turkey

Preface

Climate change has emerged as one of the most urgent challenges of the 21st century, fundamentally altering the structure of social and economic systems worldwide. As global temperatures rise and extreme weather events become increasingly frequent, the impacts of climate change on societies have become profound and multifaceted. It is therefore important to examine the complex interactions between climate change and its impacts on a variety of social factors, such as population displacement and migration, resource scarcity, and public health. While the displacement of communities due to environmental degradation raises critical questions about the social impacts of climate-induced migration, scarcity of basic resources such as water and arable land exacerbates tensions and inequalities across regions. Economically, the consequences of climate change are alarming. It significantly impacts agricultural productivity and food security, which are vital for the livelihoods of populations globally. These disruptions threaten labor markets and employment rates, especially in vulnerable regions that are heavily dependent on climate-sensitive industries. Moreover, climate change increases economic inequality, creating a gap between those who can adapt and those who cannot. To address these challenges, it is imperative to examine effective policy responses that can mitigate the social and economic impacts of climate change. This includes exploring innovative strategies to reduce social vulnerability and policies to promote collaborative efforts to combat climate change. In this context, the book aims to highlight the urgent need for integrated approaches that address both the social and economic dimensions of climate change and ultimately contribute to the development of resilient communities and sustainable economies.

Climate change is increasingly recognized as a major driver of population displacement and migration, with a multitude of interconnected factors contributing to this phenomenon. Droughts, crop failures, and rising sea levels are some of the critical environmental stressors that are forcing people to leave their homes in search of more sustainable living conditions. These adverse effects of climate change are not limited to a single region; they are a global problem that leads to both internal and cross-border migration. Climate change mitigation efforts, combined with strategies to manage migration and displacement, are crucial to preventing a full-fledged humanitarian crisis.

The social impacts of climate-induced resource scarcity extend beyond migration, deeply affecting public health and social stability. Environmental pollution caused by climate change is degrading sea and air quality, which has significant social consequences for societies that depend on these resources for their livelihoods and well-being. For example, in Turkey, the compound effects of resource scarcity not only strain the economy, but also exacerbate social inequalities and stress social life. In addition, waste and garbage problems are increasing in many areas, contributing to public health crises and reducing the overall quality of life of residents. Degradation of water resources is another critical issue, as changes in the water cycle caused by climate change are leading to more frequent and severe droughts and floods. These extreme weather events disrupt social structures and livelihoods, affecting community stability and social cohesion. In addition, salinization of estuaries and coastal groundwater due to rising sea levels is reducing access to freshwater and further threatening the livelihoods and food security of coastal communities. These interconnected issues highlight the need for comprehensive interventions that address not only environmental sustainability but also social resilience to mitigate the far-reaching impacts of climate-induced resource scarcity.

The multifaceted impact of climate change on public health and public well-being extends beyond environmental factors to include social and economic dimensions. The study highlights that climate change poses a significant social challenge and requires a comprehensive approach to public health and public well-being. One critical intersection is the impact on labor productivity and employment. As extreme weather and rising temperatures become more frequent, they not only directly impact worker health and safety, but also lead to reduced labor productivity and economic instability. This economic downturn can exacerbate existing social inequalities, further straining community resources and support systems. As a result, there is an urgent need for strong social policies that address the health impacts of climate change. The study suggests that not only reducing environmental impacts but also protecting public health advocates policies that also aim to increase the resilience of populations and communities. These policies should be designed to support vulnerable populations and thus create a more equitable and healthy society in the face of climate challenges. Effective measures such as developing heat action plans, improving infrastructure that is resilient to extreme weather conditions, and ensuring access to health and social services are essential to protect community well-being in the face of the increasing threats of climate change.

Climate change profoundly affects agricultural productivity and food security, particularly through its effects on livestock production systems. Climate factors such as temperature and precipitation patterns are critical in determining the availability of pasture and other natural resources for livestock throughout the year. These resources are essential for maintaining the health and productivity of animals, which directly affects food security. As climate change disrupts these patterns, it increases the vulnerability of livestock systems, making them more susceptible to stresses such as drought and extreme weather events. This susceptibility can lead to significant declines in livestock productivity, further exacerbating food security problems. The major transformation expected in the global livestock sector, driven by the increasing demand for meat and milk, especially in developing countries, highlights the urgency of addressing the impacts of climate change on agricultural productivity. As a result, there is a critical need for adaptive strategies and interventions to mitigate these impacts and maintain global food security by ensuring the sustainability of animal production systems.

The impacts of climate change on labor markets extend beyond migration, significantly affecting economic structures and employment rates in various sectors. For example, the agricultural sector, which forms the backbone of many economies, is particularly vulnerable to climate-induced disruptions. Changes in agricultural yields will not only affect farmers, but will also have a cascading effect on the food processing industry, demonstrating the interdependence of different sectors. As agricultural productivity declines due to extreme weather events and water scarcity, there will be a direct reduction in labor demand in this sector, which will then affect related industries such as food production and distribution. Furthermore, the increase in food prices triggered by reduced agricultural output will force consumers to reallocate their spending, which in turn will reduce demand for other goods and services, affecting overall economic activity and employment. This scenario shows how climate change can have a ripple effect, disrupting economic balance and causing job losses across multiple sectors. Addressing the multifaceted impacts of climate change on labor markets therefore requires comprehensive interventions that include not only environmental policies but also economic and social strategies to mitigate the negative impacts on employment and livelihoods.

The economic consequences of climate change exacerbate regional inequalities as the impacts are not distributed equally across areas. Regions that are already economically disadvantaged often suffer the most and are increasingly vulnerable to climate-induced disruptions to agriculture, infrastructure and employment opportunities. For example, rural areas are particularly affected by climate change, facing significant shifts in employment and increasing levels of poverty. This deepens existing economic inequalities and can lead to cycles of deprivation that are difficult to break. In addition, the geopolitical consequences of climate change, such as resource conflicts and forced migration, put additional pressure on economically weaker regions, intensifying their struggles and widening the inequality gap. As a result, the impacts of climate change on economic inequality are multifaceted and far-reaching, affecting both local and global economic stability. Addressing these inequalities requires targeted interventions that take into account the unique vulnerabilities of different regions, aiming to mitigate the adverse effects of climate change and promote sustainable and inclusive economic growth.

Effective policies to mitigate the social impact of climate change should prioritize the integration of strong social protection mechanisms. Social protection serves as a critical buffer that helps individuals and communities manage and adapt to the unpredictable challenges posed by a changing climate. By ensuring that everyone has access to adequate social protection, policymakers can reduce the risk of exacerbating inequalities and prevent the exclusion of vulnerable populations during climate-related disruptions. Moreover, implementing universal social protection is not only an ethical imperative; it is also a It is a practical necessity to maintain social well-being and stability in the face of adversity. This comprehensive approach will support a just transition to a sustainable future by ensuring that no one is left behind as economies shift towards greener and lower-carbon energy sources. Embedding social protection in climate policies is therefore essential to foster public acceptance and ensure the long-term success of climate change mitigation and adaptation efforts.

To address the multifaceted challenges posed by climate change, economic policies must be carefully tailored to integrate ecological balance and sustainable development strategies. One of the primary considerations is to prioritize poverty eradication, as poor populations are disproportionately affected by climate-related crises and lack the resources to adapt effectively. This means that combating poverty is not only a moral imperative, but also a practical necessity to create resilient communities that can withstand environmental shocks. At the same time, balancing climate and population should be at the heart of economic development efforts, ensuring that growth does not come at the expense of environmental degradation. This approach would require a shift toward policies that support sustainable living practices and population management, perhaps through incentives for green technologies and family planning programs. In addition, integrating strategies to restore ecosystems into economic policies would help maintain the ecological balance that is crucial to mitigating the negative impacts of climate change. For example, investments in reforestation, wetland restoration, and sustainable agricultural practices can provide both economic and environmental benefits by creating jobs while preserving natural resources. The development and deployment of advanced technology and scientific research should also be used to restructure the energy economy, promote the transition to renewable energy sources, and reduce greenhouse gas emissions. By embedding these initiatives in economic policies, we can address the security, economic, and climate crises in a holistic way, prevent further ecological degradation, and promote global climate stability. Finally, these policies should include sustainable development principles that address both environmental and climate justice on a global scale, ensuring that the benefits of economic growth are distributed equitably and support long-term ecological health.

International agreements play an important role in reducing the social and economic impacts of climate change by ensuring that the needs and rights of marginalized groups are prioritized. These agreements serve as critical tools not only for achieving environmental goals but also for promoting social justice and equity. Policymakers can develop more effective, sustainable, and justice-oriented climate strategies by drawing on the experiences and knowledge of vulnerable groups. For example, indigenous communities, which often have extensive knowledge of sustainable natural resource management, can provide valuable insights that increase the effectiveness of climate policies. Furthermore, the inclusiveness of international climate agreements ensures that marginalized communities, such as economically and politically marginalized groups, are not left behind in the global effort to combat climate change. This inclusiveness is important to address the disproportionate impacts of climate change on these communities and thus to promote a more equitable distribution of resources and support. Finally, by examining the impacts of these agreements on marginalized groups, it becomes clear that justice and inclusiveness are crucial to effectively mitigating the social and economic impacts of climate change. As a result, the international community should prioritize listening to the voices of those most affected by climate change in order to support the effective implementation of these agreements.

As can be seen, climate change is one of the most complex and multidimensional crises of our time. This major environmental change caused by human activities has profound effects not only on ecosystems but also on global economic systems and social structures. This book examines the socio-economic effects of climate change in a comprehensive manner, emphasizing that this global problem must be addressed in economic, political and social dimensions. This study examines climate change in a wide range from agriculture to industry, from migration to health, from education to innovation, and discusses the dimensions of the problem and possible solutions in depth under each heading.

Chapter 1, "Conceptual Framework and Visibility of the Problem," forms the starting point of the book, examining the economic effects of climate change conceptual perspective. Declines in agricultural productivity, disruption of supply chains and increased natural disasters stand out as major threats to economic stability. The chapter emphasizes the necessity of a transition to sustainable energy and sustainable agriculture, arguing that global cooperation and decisive policies will play a key role in mitigating the effects of climate change.

The second chapter focuses on "The Impacts of Climate Change on Global Inequalities." Climate change is deepening existing inequalities by making the most vulnerable communities more vulnerable. These inequalities are illustrated with examples ranging from sea level rise in Bangladesh to economic hardship in Puerto Rico. The book emphasizes the need to strengthen climate justice frameworks and highlights the importance of community-based adaptation efforts and international cooperation.

The third chapter examines the social and economic impacts of "Climate Change-Related Migration and Displacement." Extreme weather events, sea level rise and resource scarcity are increasing forced migration, putting pressure on both migrant communities and infrastructures in their destinations. This chapter highlights the importance of resilient infrastructures and technological innovations, while also highlighting the need for integrating national and international policies.

The fourth chapter explores the relationship between "Climate Change and Mental Health." The chapter argues that climate change is causing mental health problems such as anxiety, depression, and post-traumatic stress disorder, and that a multi-sectoral approach should be adopted to address these problems. Climate change-related mental health problems should be addressed with solutions that promote broader resilience at both the individual and societal levels.

"Technical Efficiency Impacts of Climate Change on the Industrial Sector" is examined through the industrial sector in India. The fifth chapter focuses on the negative impacts of climate change on agriculture and industry, and argues that this has led to a decline in economic output. It emphasizes the need for technological developments and climate-friendly policies to increase productivity in the sector.

The sixth and seventh chapters of the book examine the economic costs of climate change and how these costs can be reduced. While economic costs range from the degradation of agriculture to the loss of biodiversity, the critical role of global cooperation and technological innovations in reducing these costs is discussed.

"The Effects of Climate Change on The Tourism Sector" is discussed in the eighth chapter. Climate change negatively affects tourism and reduces economic income, but it also reveals the need to develop environmentally sensitive tourism practices. The book draws attention to the importance of promoting sustainable tourism policies.

The ninth chapter focuses on the economic and financial issues related to climate change, emphasizing the importance of transitioning to environmentally friendly product production and shifting to renewable energy. It focuses on pricing policies, technological transfers and financial incentives for this transition to be successful.

The tenth chapter discusses "The Effects of Climate Change on Employment". In particular, green economy and remote working models stand out as effective tools in combating climate change. The book argues that the expansion of remote working policies can play an important role in reducing the carbon footprint.

"Methods to Combat Climate Change" are examined in detail in the third part of the book. The eleventh chapter emphasizes the role of international platforms and local initiatives in combating climate change. In addition to global efforts such as the Paris Agreement, the importance of local governments and community-based projects is discussed.

"Turkey's Climate Change Policies" is addressed in the twelfth chapter. Turkey's efforts to transition to renewable energy, energy efficiency strategies and sustainable land use policies are detailed. The importance of international cooperation and Turkey's participation in global climate initiatives is emphasized.

"European Countries' Policies to Combat the Climate Crisis" is also covered extensively in the thirteenth chapter. While the European Green Deal and other environmental policies of the European Union are examined, Europe's leadership role in this area is highlighted.

The inclusion of climate change awareness in education is discussed in the fourteenth chapter, using the example of St. Peters University. It is emphasized that the curriculum on climate change is inadequate, and it is stated that a more comprehensive approach should be adopted in education.

The fifteenth chapter, "The Role of OpenStreetMap Data in Climate Change Assessments and Policy-Making Processes," examines how open data sources can be used to respond to climate challenges at the local level.

The following sections focus on "Sustainable Development and Innovation". It emphasizes that technology and innovation are crucial in combating climate change. Renewable energy technologies, digital technologies and circular economy models are cited as key elements that will play a key role in coping with climate change in the future.

The findings of this study illuminate the profound and multifaceted impact of climate change on social and economic factors, especially through the lens of climate-induced migration. Environmental stressors such as drought, crop failure and rising sea levels due to climate change can increase migration, not only exacerbating existing vulnerabilities but also fueling fierce competition for dwindling resources, leading to increased social tensions and potential conflicts. The impacts on public health and social stability are alarming, as extreme weather events create a ripple effect that disrupts social structures and livelihoods, threatening social cohesion. The study highlights the urgent need for comprehensive international policies that address the root causes of climate change while providing robust support to affected populations. By bringing together vulnerable groups, particularly indigenous communities, with invaluable knowledge on sustainable resource management, policymakers can develop more effective and equitable climate strategies. Furthermore, integrating social protection mechanisms into climate policies is crucial to mitigating the negative impacts of climate change on public health and public well-being. However, the research also highlights the limitations of current frameworks that often ignore the voices of marginalized communities. Future research should prioritize these perspectives to ensure that international agreements reflect the needs and rights of those most affected. Moving forward, it is imperative that economic policies are designed to align ecological balance with sustainable development, thereby addressing the interconnected crises of security, economy, and climate. This holistic approach is crucial not only to prevent further ecological degradation, but also to promote global climate stability and ultimately increase resilience in the face of impending challenges.

In conclusion, this book provides a broad perspective on the socioeconomic impacts of climate change, providing a rich guide to the steps that need to be taken to combat this global challenge. This work is a valuable resource for policymakers, academics, and all interested parties.

Acknowledgment

We present this book to all children in the world and especially to Mir Agah and Ahi Doğa with the wish that they live in a healthy and sustainable environment.

We present this book, which addresses the effects of climate change on socio-economic factors, to our families who supported us when the idea of this work emerged, and to Müjde, whose death deeply saddened us while the work on the book was still ongoing.

We would like to thank our valuable authors who meticulously shared their knowledge and experience while writing the book chapters, and the relevant employees of the publishing house.

We received referee support from our expert professors while the book chapters were being evaluated. We would also like to thank all our referees.

Arbitration Board

Ahsen Saçli

Ajay Kumar Singh

Amit Yadav

Anish Kumar

Bahriye Eseler

Buket Aydin

Canan Kişlalioğlu

Celal Ince

Çağlar Karamaşa

Çağri Sürek

Eda Bozkurt

Enes Yalçin

Ezgi Kovanci

Firat Doğan

Haktan Sevinç

Hidayet Beyhan

Hikmet Selahattin Gezici

Ibrahim Aytekin

Ilke Bezen Tozkoporan

Kivanç Demirci

Mehmet Siddik Vangölü

Merter Akinci

Mohit Yadav

Özlem Topçuoğlu

Rahman Aydin

Veysel Erat

Yunus Savaş

Section 1
Conceptual Framework and Visibility of the Problem

Chapter 1
Conceptual Framework and Problem Visibility:
Conceptual Explanations for the Economics of Climate Change

Selvi Vural
https://orcid.org/0000-0002-3245-8599
Gümüşhane University, Turkey

ABSTRACT

In the realm of economics, conceptual explanations serve as the foundation for understanding the complexities of market dynamics, resource allocation, and socio-economic phenomena. These explanations elucidate the theoretical frameworks that underpin economic models, guiding policymakers, analysts, and individuals in comprehending the intricate interplay of factors shaping economic outcomes. Conceptual explanations delve into fundamental concepts such as supply and demand, elasticity, production functions, utility theory, and market structures, offering insights into how these principles interact to influence economic behavior and outcomes. By providing a conceptual roadmap, these explanations facilitate informed decision-making, foster critical thinking, and contribute to the advancement of economic theory and practice.

INTRODUCTION

Aside from the ever-increasing demands, needs and expectations, the scarcity of goods or services that can meet them is an undeniable fact. However, scarcity does not mean absence but limitation. In this context, economics is the branch of science that analyses how unlimited wants, needs or expectations can be met with

DOI: 10.4018/979-8-3693-5792-7.ch001

limited or scarce resources. Although it is possible to meet these wants, needs or expectations in the individual sense, addressing the demands and expectations of everyone at the social level to the same extent confronts us with the inadequacy of goods or services. At the same time, this situation reveals the imbalance between goods and services and some needs, and this imbalance is explained by the law of scarcity (Redish, 1984).

In the literature, there are four basic elements such as capital, natural resources, labour force and entrepreneurship, which constitute scarce resources or, in other words, factors of production. In this context, both financial (such as money, stocks) and production (machinery, building and stock capital) instruments that are produced and used in the production process have the characteristics of *capital*. *Natural resources* are underground or above-ground resources (such as air, water, oil, minerals, wind, etc.) that exist and are available in nature. *Labour force refers to* qualified and unqualified or physical and mental manpower, which is the main actor helping to carry out the production process. *Entrepreneurship is a* real or legal person who brings together capital, natural resources and labour force to produce goods or services and who can take high risks (Barnett and Morse, 2013).

On the other hand, the output obtained by the use of all these resources in economic terms reveals an income; capital is paid with interest, natural resources with rent, labour with wages and entrepreneurs with profit. Ultimately, in order to speak of an economy and an economic activity, these elements must come together and the income obtained contains an economic value. However, the economic organisation, production technologies, factors of production and their quantity differ in each society (Papava, 1994). Accordingly, the increase in the capacity of societies for these elements in the current conditions is considered to be an indicator of economic growth in a narrow sense and economic development as an end result of the process that follows it more broadly (Van den Berg, 2016).

Today, the increasing importance of economic concepts such as economic growth, economic development, sustainable economy or green growth and their close relationship with each other reveal the necessity of addressing and explaining the subject. The main problem and purpose of this study is to clearly reveal the relationship between concepts. In economics, conceptual explanations provide the foundation for understanding the complexities of market dynamics, resource allocation, and socioeconomic phenomena. In addition, these explanations clarify the theoretical frameworks that underlie economic models and guide policy makers, analysts, and individuals in understanding the complex interactions of factors that shape economic outcomes.

These explanations provide a conceptual roadmap, facilitating informed decision making, encouraging critical thinking, and advancing economic theory and practice. Economic growth is a tool or indicator of economic development and is also the basis of sustainable economy (King and Levine, 1994). On the other hand, in order to be able to talk about the existence of a sustainable economy, it is necessary to have environmental awareness or, in other words, to adopt environmentally friendly approaches, which in a sense indicates the existence of a sustainability policy on the axis of green growth (Kasztelan, 2017).

1. ECONOMIC GROWTH

It is seen that growth has gained a significant place in the literature, especially in the 1960s, but the inability to explain the developments in the world in the following period with the existing theories and the inability to develop new ones shows that the interest in the growth phenomenon has decreased. However, recent developments in the field of economics, the theories developed or seminal new concepts have brought a renewed popularity to the growth issue and brought the importance of the research area to the forefront. Economic growth is, in fact, an important objective and part of the macro-level economic policies of all countries (Easterly, 2005).

Economic growth, in its simplest form, refers to the quantitative increase in the production capacity of a country or a society over a certain period of time. This time period usually covers a period of one year. In other words, economic growth can be defined as the increase in the total or per capita amount of goods or services in a year. This increase is parallel to the increase in both national income and total production (Nurudeen and Usman, 2010). In addition, it is stated that when measuring economic growth, the rise and fall in production or investments and technological advances in a country or society in a certain period of time are taken into account. Since not all countries have the same economic organisation, production technology, factors or quantity, this suggests that the course of economic growth and sustainable growth practices may differ (Bilbao-Osorio and Rodríguez-Pose, 2004).

The fact that an increase qualifies as growth depends on its continuity. Otherwise, temporary or periodic increases do not reflect a sustainable growth. On the basis of sustainability, a continuous increase requires the expansion of the production potential or volume in a country in the long term or the effectiveness and efficiency in the utilisation of all these. In this framework, economic growth is handled with a visionary approach and is based on measuring its determination. The measurement of economic growth is largely based on the Gross National Product (GNP) as it reflects all the values produced by citizens in a country. In other words, the measure of economic growth is based on the amount of goods or services produced within those borders rather than the citizenship of the country (Aydın and Çetintaş, 2022).

1.1. Factors Affecting Economic Growth

According to Barro (2003), investments and technological progress are the main factors affecting economic growth. In particular, the most important indicator of the poverty or wealth of the countries after the war in 1945 is economic growth. Both investments and technological progress lead to an increase in per capita capital levels and productivity. Accordingly, the continuity of economic growth increases the requirements for new capital investments and technologies. Apart from these requirements, there are other factors affecting economic growth that reflect different methodological and conceptual perspectives and that differ from country to country or from one period to another (Boldeanu and Constantinescu, 2015).

Factors affecting economic growth are widely recognised in the literature. However, in general terms, *human and physical capital, structural and institutional policies, natural resources, technology, innovation and research and development, economic performance, trade openness, foreign direct investment, socio-cultural factors and demographic structure* are the main factors affecting economic growth. The effectiveness, functionality or management of these factors reveal the success of a country's economic policies. Therefore, this success also shows that each of these factors is a pillar and carrier of economic growth. Over time, various theories have been developed on the basis of these factors, which are the determinants of economic growth (Upreti, 2015).

1.2. Approaches to Explain Economic Growth

Economic growth, which shows the increase in production capacity and gross national product in a country in a one-year period, is a subject that economists emphasise and study. In this context, researchers determine the factors affecting growth by considering the conditions of the current period and the events experienced and interpret the subject with different approaches. Accordingly, a number of strategies and models are developed and even these can be changed or reshaped by taking into account the requirements of the current period in the following process. These models, in fact, show that the approaches to the subject have enriched or diversified (Quinn and Shapiro, 1991).

According to Aghion (1999), theories containing different perspectives of researchers are developed depending on the economy of the period. In the literature, these theories to explain economic growth are categorised under three main headings: *traditional, exogenous and endogenous. Traditional growth* theory covers the 17th and 18th centuries and is divided into two as mercantilism and physiocracy. *Exogenous growth theories* reflect a period that started in the 1970s and includes classical, Marxist, Keynesian, Harrod-Domar and neoclassical growth theories.

Endogenous growth theories, on the other hand, include new growth approaches, especially towards the end of the 1980s, when technological progress and developments were experienced. However, it is stated that these growth theories create a distinction on the basis of economic growth and economic development depending on the level of development of countries (Pietak, 2014).

As stated by Lewis (2013), while developed countries apply economic growth theories in order to maintain their current situation, developing countries tend to apply development theories in order to achieve growth to the extent necessary. On the other hand, while economic growth reflects the increase in a country's production capacity, national income or output in a given year, economic development reflects the economic or socio-cultural change and development of a country along with the increase in its national income. In the light of this information, economic growth is only quantitative, while economic development is an indicator of a qualitative and quantitative holistic change. Moreover, qualitative changes in a country's economy, industry, urbanisation, population growth rate, literacy level, medical facilities and artistic activities to improve the quality of life are the basis of economic development.

2. ECONOMIC DEVELOPMENT

The concept of development has been on the agenda of countries both economically and socially, especially in recent years as a result of globalisation. One of the reasons for this is that both the development plans of states are now seen within economic policies and researchers continue to maintain the dynamics of the field with different perspectives and the theories they have developed. In addition to all these, it is stated that while development before the Second World War was a protectionist argument in the name of industrialisation in the economic sense, the change/transformation in development efforts with the growth of global trade and economy afterwards caused countries to meet the phenomenon of economic development (Galbraith, 1964).

In those years, factors such as technology, foreign trade and human capital were seen as important determinants of the economic development process, and investments in health and education, which are particularly related to human capital, were considered to be a lever of economic development. However, it is observed that the term "progress" has also been frequently used in the literature to correspond to the concept of development in the past years (Singer, 1949; Anand and Sen, 2000). The first use of the term in its current meaning is found in a phrase mentioned by the United Nations in the 1947 development plan, and it is generally understood that development has a socio-cultural content rather than an economic meaning (Browne, 2012).

Economic development is seen both as an end result of the process following growth and as a part of the sustainability of economic growth (Arslan et al., 2022). Economic development is sometimes referred to as a sub-branch of economics and analyses the socio-economic structure in developing countries, which is considered to be changing or open to change in line with growth. In other words, economic development involves a socio-cultural development process in the long term. Regardless of the conceptualisation, dynamism is at the core of the concept of development. In this context, the principle of continuous innovation and improvement is of great importance for both developed and developing countries today (Læsse, 2010).

2.1. Determinants and Basic Indicators of Economic Development

According to Peri (2004), there are many determinants of economic development including *socio-cultural, administrative, geographical and economic*. These determinants are of great importance for countries or societies to pass through some critical stages and maintain their development. Accordingly, it is thought that the improvements and developments in the economic development policies of developing countries are basically ensured by the continuity of the increase in the amount and capacity of their production, the awareness and reality of the importance of the issue.

Researchers such as Moosa and Smith (2004) or Ledoux et al. (2002) point out that there are basically eight indicators of economic development and that some economists sometimes use different expressions to mean the same thing, simply by conceptualising the issue differently. However, *economic growth, natural resources, technological development, foreign trade, population growth, human capital, income distribution and gross national product per capita are among the* main indicators of economic development. These indicators play an important role in the economic development policy of countries and the success of the development policy reflects the effectiveness in the management of indicators. Of course, this situation is understood as a result of the measurement of economic development.

2.2. How is Economic Development Measured?

To be able to talk about the existence of economic development at some point, it is necessary to understand, comprehend and measure it correctly. In the literature, it is seen that a number of different criteria are taken into account in the measurement of economic development or the elements that researchers accept as criteria may vary. In this context, it is understood that GNP per capita or GNP in general is an accepted basic criterion for measuring economic development. On the other hand, changes in the efficiency and quantity of factors of production, increase in the share of industry in national income, increase in life expectancy, increase in literacy rate and improvement in health conditions are also among the basic criteria of economic development (England, 1998).

While gross national product is an indicator of economic growth on its own, it is among the important measures that help explain economic development, but it does not have the power to explain this situation alone. In other words, while the increase in gross national product affects growth to the same extent, it is considered that it does not have a significant effect on economic development. Accordingly, other criteria such as per capita income level, physical quality of life and human development index can be utilised in measuring development (Larson and Wilford, 1979; Morris, 1982).

2.3. Theoretical Approaches to Economic Development

There are three types of approaches to economic development: balanced growth, unbalanced growth and polar growth theories (Dang et al., 2015). However, each theory brings different perspectives to the issue by considering completely different factors related to economic development, being influenced by the conditions of the period, adhering to certain principles and taking both developed and developing countries as a basis. Therefore, in order to better explain and understand the subject, each theory should be briefly mentioned.

2.3.1. Balanced Growth Theory

Balanced growth theory focuses on the sustainability and stability of a country's long-term economic growth. However, the theory emphasises the interaction of various factors on the basis of efficient use of resources, capital accumulation, technological progress, income distribution, human capital, population growth and sustainability. The interaction of these basic factors creates a balanced growth pattern in the economy. This situation helps to reduce poverty, improve living standards and increase the general welfare of the society by making economic growth sustainable. At the same time, balanced growth theory argues that economic policies should be determined by considering long-term goals, resource management and social justice (Nath, 1962).

2.3.2. Unbalanced Growth Theory

Unbalanced growth theory tries to explain the situations where economic growth is not balanced and certain sectors grow faster than others or lag behind. Contrary to the balanced growth theory, this theory argues that economic growth does not occur at the same rate in all sectors and that certain sectors gain advantageous positions over others (Hansen, 1965). Unbalanced diffusion theory is based on factors such as clustering and spillover effects, technological progress and sectoral change, income effects or investment and finance.

On the other hand, this theory recognises the existence of imbalances in the economy and challenges the assumption of balanced growth theory that all sectors grow equally. In fact, the phenomenon of unbalanced growth is an important concept for understanding the economic development processes of countries and designing economic policies. Ultimately, it argues that economic policies should promote economic growth by supporting certain sectors and keeping others in balance (Saliminezhad and Lisaniler, 2018).

2.3.3. Polar Growth Theory

The theory of pole growth is widely used in the economic development strategies and local development planning of countries. According to this theory, focusing on the development of a particular economic, social or industrial area in a region or city encourages and accelerates development in a wider region or city. These areas are defined as poles in the theoretical framework and have the ability to influence other surrounding regions (Darwent, 1969). However, identifying the pole area, developing infrastructure, increasing economic diversity, attraction effect and employment creation are the basic elements of the pole growth theory. In this context, pole areas play an important role in reducing economic imbalances between regions, while increasing the development potential of the regions around them.

The effectiveness of pole growth strategies (at the planning-implementation stages) requires consideration of various factors such as the state of socioeconomic infrastructure, the attractiveness of investment, the capabilities of the local labour force and policy regulations (Polenske, 2017). In the light of this information, the impact of economic growth on the economic development process is taken into consideration and the role and importance of economic development in achieving sustainable economic growth is largely understood. Therefore, it is believed that sustainable economy should also be explained after mentioning the concepts of economic growth and development.

3. SUSTAINABLE ECONOMY

In these days of increasing environmental destruction or degradation both in our country and worldwide, the issue of sustainability is seen as one of the most discussed topics by governments, non-governmental organisations (NGOs), politicians and many researchers from various fields. Concepts such as economy, environment and energy are at the centre of research on sustainability. However, the concept of sustainable economy is fuelled by economic development and environmental awareness, and it is emphasised that development is an extension of economic growth. On the basis of sustainability, setting some targets by considering the economy, environment and energy resources and following policies to realise them indicate the existence of sustainable economic management (Chouinard et al., 2011).

According to Barrier (2017), sustainable economy, in its simplest form, is an economic model that aims to meet the basic needs and increase the welfare of both present and future generations, while protecting natural resources and the environment and ensuring their continuity. Unlike economic growth approaches, sustainable economy considers resource utilisation, environmental impacts and welfare of the society in a balanced manner. *Efficient use of natural resources, preference for renewable resources, waste and pollution reduction policies, ensuring social justice and equality, long-term planning, investment in green technologies and innovations reveal the* characteristics of sustainable economy.

Sustainable economy emphasises that economic growth and prosperity should be considered within a sustainable ecological (environmental) and social (social) framework. This approach argues that economic, ecological and social dimensions should be examined in a holistic approach, focusing on preventing the depletion of resources or ensuring their continuity on the one hand, reducing environmental pollution or increasing social welfare on the other. Only in this way, it is believed that the goals of creating a world that can meet the needs of not only current but also future generations can be achieved (Barbier, 2011).

3.1. Importance of Sustainable Economy

According to Chouinard et al. (2011), the importance of sustainable economy is increasing day by day due to a number of benefits it provides for both present societies and future generations. In this context, there are a number of elements that reveal the increasing importance of sustainable economy and are the focus of many studies. These elements and related explanations are as follows (Smol, 2019):

Conservation of natural resources: A sustainable economy helps to reduce the negative impacts on natural resources and the environment by promoting the efficient and balanced use of natural resources. Conservation of natural resources increases the chances of future generations to have these resources.

Improving the Health of the Environment and Ecosystems: A sustainable economy requires the adoption of environmentally sound policies and practices. In this way, the health and diversity of ecosystems can be maintained, biodiversity can be supported and natural balances can be preserved.

Climate Change and Reducing Carbon Emissions: Sustainable economy encourages the reduction of fossil fuel use and the transition to renewable energy sources. This means reducing greenhouse gas emissions and contributing to the fight against climate change.

Economic Resilience and Security: A sustainable economy increases economic resilience in the face of natural disasters, economic fluctuations and other crises. The use of renewable resources leads to energy security by reducing energy dependence.

Improving Social Justice and Welfare: Sustainable economy supports reducing income inequality and ensuring social justice. It increases access to education, health, housing and basic human rights and improves social welfare and quality of life.

Promoting Innovation and Green Technologies: Sustainable economy advocates investment in green technologies and innovation. Green technologies can offer environmentally friendly production methods, efficient energy use and solutions to reduce negative impacts on the environment.

Protecting the Rights of Future Generations: Sustainable economy aims to protect the rights of future generations while meeting the needs of current generations. Therefore, it secures the quality of life and well-being of future generations by using resources in a balanced and responsible manner (Smol, 2019).

3.2. Dimensions of Sustainable Economy

The main dimensions of sustainable economy are divided into three: economic, ecological (environmental) and social. These dimensions are used to assess and manage the sustainability of the economy. However, it is seen that each dimension interacts with each other in order to achieve sustainable economic growth targets. The main dimensions of sustainable economy and the issues addressed within this scope complement each other to a significant extent and a successful assessment or management approach is based on holism (Kristensen and Mosgaard, 2020).

According to Harris (2001), the *economic dimension* actually includes economic growth, productivity and sustainability of economic activities. Sustainable economy supports economic growth, employment and welfare on the one hand, while emphasising the need for efficient and effective use of natural resources on the other. Sustainable production and consumption models, green businesses and green economy, promotion of resource efficiency and environmentally friendly technologies, long-term economic planning and decision-making, and green financing and investments are some of the issues included in the economic dimension.

The ecological (environmental) dimension focuses on the effects of economic activities on the natural environment and the need to protect natural resources. This dimension aims to increase or secure the possibility of future generations to benefit from natural resources by preserving ecological balance and diversity. This dimension of sustainable economy includes reducing carbon emissions and combating climate change, protecting water and air quality, preserving biodiversity and improving ecosystem health, using renewable energy sources, waste management and recycling systems (Yáñez-Arancibia et al., 2014).

According to Dempsey (2011), the *social dimension* of sustainable economy is largely centred on the themes of people's quality of life, equality and social justice. Sustainable economy supports the process of securing access to basic needs for all people and increasing social welfare. On the other hand, improving access to education and health services, gender equality and women's participation in the economy, reducing poverty and preventing income inequality, protecting human rights and fundamental freedoms, and promoting a participatory and inclusive society are some of the issues analysed under the social dimension.

Sustainable economy aims to establish a balanced relationship between economic growth, environmental protection and social welfare by addressing these three main dimensions together. The interaction and harmony between economic, environmental and social dimensions is of great importance in achieving sustainable development goals. Only in this way, it is believed that while meeting the current needs of people, the goal of leaving a more livable world to future generations can be realised.

3.3. Key Indicators of a Sustainable Economy

The basic indicators of sustainable economy include criteria for assessing the sustainability of the economy by taking into account economic, environmental and social dimensions. In the literature, it is noteworthy that the dimensions and indicators of sustainable economy are generally grouped under the same theme. These dimensions are ecologically, economically and socially oriented, and indicators are examined under three main headings: economic growth, environmental protection, social welfare and justice (Martinho, 2021).

The first one is the *ecological (environmental) indicators* of sustainable economy. In this context, the focus is on issues such as *carbon footprint, share of renewable energy, water and energy efficiency or biodiversity conservation*. Carbon footprint is a measure of greenhouse gas emissions of a country, organisation or individual. A low carbon footprint is an indicator of less damage to the environment.

Renewable energy share refers to the ratio of renewable energy to total energy consumption. A high share of renewable energy means an energy system that is less dependent on fossil fuels. Water and energy efficiency measures the capacity to efficiently utilise the amount of water and energy used in production and consumption processes. Biodiversity conservation includes assessments of a country's or region's efforts to conserve and sustainably use its biodiversity (Vera and Langlois, 2007).

The economic indicators of sustainable economy largely focus on *green investment and employment, ecological productivity, green trade and economic inequality*. Green investment and employment measures investments in green sectors and environmentally friendly enterprises and the amount of employment in these areas. Ecological efficiency helps to assess the efficiency between the natural resources and energy used in production processes and the output obtained. Green trade includes the trade of environmentally friendly products and services and the exports and imports made in this field. Economic inequality measures inequality in income and wealth distribution and sustainable economy aims to reduce economic inequality by promoting social justice (Joung et al., 2013).

Social indicators are the last indicators of sustainable economy. As stated by Henderdon (1994), the *human development index (HDI)* addresses mostly social issues such as *access to education and health services, gender inequality and poverty rate*. However, the human development index (HDI) shows the level of human development of a country in terms of education, health and living standards. Access to education and health services includes an assessment of the population's access to education and health services in a society. Gender equality measures the level of egalitarianism, women's participation in the economy and their representation in leadership positions. The poverty rate shows the proportion of the population living below the poverty line and the issues of reducing poverty or increasing social welfare have an important place in sustainable economy goals (Boulanger, 2007).

Ultimately, these indicators are used to measure and evaluate different dimensions of a sustainable economy. It is the monitoring and guidance of these indicators that makes it possible to have a sustainable economy in which economic, environmental and social dimensions are handled in a balanced manner. On the other hand, the Report of the World Commission on Environment and Development entitled Our Common Future mentions that the principles of environmental integrity, social justice and economic welfare are the sine qua non of a sustainable economy (Imperatives, 1987). In fact, these issues, which are at the centre of sustainable economy, should also be addressed in a wide spectrum extending to the phenomena of green growth and green economy.

4. GREEN GROWTH

According to Jacobs (2013), green growth is a development model that minimises environmental impacts in the process of economic growth and emphasises the need for sustainable use of natural resources. This concept aims to achieve a balance between economic growth and environmental sustainability. Green growth aside, traditional economic growth models are based on rapid industrialisation and consumption-oriented development. Therefore, this situation consumes natural resources and may damage the environment. Therefore, increasing environmental problems, climate change and the threat of depletion of natural resources lead to the need for green growth.

Green growth basically encourages the use of environmentally friendly technologies and cleaner production methods. Environmentally sensitive or environmentally friendly practices such as switching to renewable energy sources, increasing energy efficiency, waste management and recycling are the main structural elements of green growth. These elements reveal the existence of a model that is needed not only for countries and societies but also for the whole world. This is because this model aims to achieve a harmony between the environment and the economy on the one hand, and on the other hand, it attaches great importance to social justice and human welfare.

Green growth aims to achieve sustainable growth or development goals, maintain the continuity of natural resources and reduce environmental impacts, while at the same time improving people's quality of life (Lyytimäki et al., 2018).

As a result, green growth prioritises not only the sustainability or stability of economic growth, but also the protection of the natural environment and the benefit of future generations from these natural resources. Therefore, many countries, societies and organisations are nowadays adopting green growth strategies, taking into account the compatibility between economic growth and environmental sustainability (Jänicke, 2012).

4.1. Green Growth and Sustainable Economy

As stated by Lorek and Spangenberg (2014), sustainable economy and green growth are two closely related concepts, but there are a number of features that fundamentally differentiate these two concepts. In this context, sustainable economy essentially aims to ensure long-term prosperity and sustainability of the planet by combining economic growth, environmental and social objectives. Green growth, on the other hand, offers a growth model that focuses on the efficient use of natural resources by minimising environmental impacts along with economic growth.

On the other hand, it would be appropriate to provide an explanation for the fact that the concepts of sustainable economy and green growth are closely related to each other. Green growth is a part of sustainable economy. However, sustainable economy is seen as an economic system in which green growth can be achieved. In addition, while sustainable economy balances the environmental and social dimensions of economic growth, green growth aims to achieve economic growth by minimising environmental impacts (Bartelmus, 2013).

According to Vazquez-Brust and Sarkis (2012), the reason why sustainable economy is seen as a precursor of green growth is the fact
that economic growth leads to unsustainable resource consumption and environmental pollution in countries or societies facing increasing environmental problems. However, green growth aims to ensure sustainability in economic growth by reducing environmental impacts and utilising natural resources more efficiently. In this context, strategies such as the development of green technologies, energy efficiency and the dissemination of cleaner production methods are the carrier elements of green growth.

According to Abid et al. (2022), the fact that green growth is seen as a sufficient solution in itself is sometimes criticised by researchers. At this point, it is argued that consumption-based economic growth is fundamentally unsustainable and that more fundamental changes are necessary for sustainability. Such criticisms are voiced by those who believe that focusing solely on green growth is inconsistent with the fundamental structural problems and the concept of growth in terms of sustainability.

4.2. Advantages and Disadvantages of Green Growth

After mentioning the importance and requirements of green growth, it is thought that it would be appropriate to mention its advantages and disadvantages. In this context; environmental protection, combating climate change, energy security, economic efficiency, innovation and employment opportunities are among the advantages of green growth. Cost, transition difficulties, limited resources, social impact and international inequality are seen as the disadvantages of green growth (Iravani et al., 2017).

In this framework, green growth is an important approach that combines environmental sustainability and economic growth. However, during its implementation, a balance must be achieved and its advantages as well as disadvantages must be taken into consideration. With the participation of policy makers and other segments of society, it is possible to maximise the advantages, minimise the disadvantages and move towards a sustainable future.

5. CONCLUSION

This study examines the relationship between the concepts of economic growth, economic development and sustainable economy and thoroughly examines the theoretical approaches in these areas. While economic growth generally refers to the quantitative increase in production capacity, economic development also covers qualitative changes that include social, cultural and environmental elements in a broader framework. Sustainable economy stands out as a model that aims to use natural resources effectively by considering the welfare of both current and future generations.

As revealed in the study, in order for economic growth and development to be sustainable, environmentally friendly policies should be adopted, technological advances should be encouraged and social justice should be ensured. In this context, green growth strategies are considered as an important tool that encourages economic growth while minimizing environmental impacts. However, it is emphasized that green growth alone is not a sufficient solution and structural changes are also necessary.

As a result, ensuring sustainable economic growth is an inevitable requirement for countries to achieve their long-term development goals. In order to achieve this goal, it is important to address economic, environmental and social dimensions together to build a more livable future.

REFERENCES

Abid, N., Ceci, F., & Ikram, M. (2022). Green Growth and Sustainable Development: Dynamic Linkage Between Technological Innovation, ISO 14001, and Environmental Challenges. *Environmental Science and Pollution Research International*, 29(17), 1–20. DOI: 10.1007/s11356-021-17518-y PMID: 34843051

Aghion, P., Caroli, E., & García-Peñalosa, C. (1999). Inequality and Economic Growth: The Perspective of The New Growth Theories. *Journal of Economic Literature*, 37(4), 1615–1660. DOI: 10.1257/jel.37.4.1615

Anand, S., & Sen, A. (2000). Human Development and Economic Sustainability. *World Development*, 28(12), 2029–2049. DOI: 10.1016/S0305-750X(00)00071-1

Arslan, H. M., Khan, I., Latif, M. I., Komal, B., & Chen, S. (2022). Understanding The Dynamics of Natural Resources Rents, Environmental Sustainability, and Sustainable Economic Growth: New Insights From China. *Environmental Science and Pollution Research International*, 29(39), 58746–58761. DOI: 10.1007/s11356-022-19952-y PMID: 35368236

Aydin, C., & Cetintas, Y. (2022). Does The Level of Renewable Energy Matter in The Effect of Economic Growth on Environmental Pollution? New Evidence From PSTR Analysis. *Environmental Science and Pollution Research International*, 29(54), 81624–81635. DOI: 10.1007/s11356-022-21516-z PMID: 35739444

Barbier, E. (2011, August). The Policy Challenges for Green Economy and Sustainable Economic Development. *Natural Resources Forum*, 35(3), 233–245. DOI: 10.1111/j.1477-8947.2011.01397.x

Barnett, H. J., & Morse, C. (2013). *Scarcity and Growth: The Economics of Natural Resource Availability*. Routledge. DOI: 10.4324/9781315064185

Barrier, E. B. (2017). The Concept of Sustainable Economic Development. In *The Economics of Sustainability* (pp. 87–96). Routledge. DOI: 10.4324/9781315240084-7

Barro, R. J. (2003). Determinants of Economic Growth in A Panel of Countries. *Annals of Economics and Finance*, 4, 231–274.

Bartelmus, P. (2013). The Future We Want: Green Growth or Sustainable Development? *Environmental Development*, 7, 165–170. DOI: 10.1016/j.envdev.2013.04.001

Bilbao-Osorio, B., & Rodríguez-Pose, A. (2004). From R&D to Innovation and Economic Growth in The EU. *Growth and Change*, 35(4), 434–455. DOI: 10.1111/j.1468-2257.2004.00256.x

Boldeanu, F. T., & Constantinescu, L. (2015). The Main Determinants Affecting Economic Growth. *Bulletin of the Transilvania University of Brasov. Series V, Economic Sciences*, 329–338.

Boulanger, P. M. (2007). Political Uses of Social Indicators: Overview and Application to Sustainable Development Indicators. *International Journal of Sustainable Development*, 10(1-2), 14–32. DOI: 10.1504/IJSD.2007.014411

Browne, S. (2012). *United Nations Development Programme and System (UNDP)*. Routledge. DOI: 10.4324/9780203806852

Chouinard, Y., Ellison, J., & Ridgeway, R. (2011). The Sustainable Economy. *Harvard Business Review*, 89(10), 52–62.

Dang, G., & Sui Pheng, L. (2015). Theories of Economic Development. *Infrastructure Investments in Developing Economies: The Case of Vietnam*, 11-26.

Darwent, D. F. (1969). Growth Poles and Growth Centers in Regional Planning-A Review. *Environment & Planning A*, 1(1), 5–31. DOI: 10.1068/a010005

Dempsey, N., Bramley, G., Power, S., & Brown, C. (2011). The Social Dimension of Sustainable Development: Defining Urban Social Sustainability. *Sustainable Development (Bradford)*, 19(5), 289–300. DOI: 10.1002/sd.417

Easterly, W. (2005), National Policies and Economic Growth: A Reappraisal. *Handbook of Economic Growth, 1*, 1015-1059.

England, R. W. (1998). Measurement of Social Well-Being: Alternatives to Gross Domestic Product. *Ecological Economics*, 25(1), 89–103. DOI: 10.1016/S0921-8009(97)00098-0

Galbraith, J. K. (1964). *Economic Development*. Harvard University Press. DOI: 10.4159/harvard.9780674333062

Hansen, N. M. (1965). Unbalanced Growth and Regional Development. *Economic Inquiry*, 4(1), 3–14. DOI: 10.1111/j.1465-7295.1965.tb00931.x

Harris, J. (Ed.). (2001). *A survey of Sustainable Development: Social and Economic Dimensions* (Vol. 6). Island Press.

Henderson, H. (1994). Paths to Sustainable Development: The Role of Social Indicators. *Futures*, 26(2), 125–137. DOI: 10.1016/0016-3287(94)90102-3

Imperatives, S. (1987), Report of The World Commission on Environment and Development: Our Common Future.

Iravani, A., Akbari, M. H., & Zohoori, M. (2017). Advantages and Disadvantages of Green Technology; Goals, Challenges and Strengths. *Int J Sci Eng Appl*, 6(9), 272–284. DOI: 10.7753/IJSEA0609.1005

Jacobs, M. (2013), Green Growth. *The Handbook of Global Climate and Environment Policy*, 197-214.

Jänicke, M. (2012). "Green Growth": From A Growing Eco-Industry to Economic Sustainability. *Energy Policy*, 48, 13–21. DOI: 10.1016/j.enpol.2012.04.045

Joung, C. B., Carrell, J., Sarkar, P., & Feng, S. C. (2013). Categorisation of Indicators for Sustainable Manufacturing. *Ecological Indicators*, 24, 148–157. DOI: 10.1016/j.ecolind.2012.05.030

Kasztelan, A. (2017). Green Growth, Green Economy and Sustainable Development: Terminological and Relational Discourse. *Prague Economic Papers*, 26(4), 487–499. DOI: 10.18267/j.pep.626

King, R. G., & Levine, R. (1994, June). Capital Fundamentalism, Economic Development, and Economic Growth. In Carnegie-Rochester Conference Series on Public Policy. North-Holland. DOI: 10.1016/0167-2231(94)90011-6

Kristensen, H. S., & Mosgaard, M. A. (2020). A Review of Micro Level Indicators for A Circular Economy-Moving Away From The Three Dimensions of Sustainability? *Journal of Cleaner Production*, 243, 118531. DOI: 10.1016/j.jclepro.2019.118531

Læsse, J. (2010). Education for Sustainable Development, Participation and Socio-Cultural Change. *Environmental Education Research*, 16(1), 39–57. DOI: 10.1080/13504620903504016

Larson, D. A., & Wilford, W. T. (1979). The Physical Quality of Life Index: A Useful Social Indicator? *World Development*, 7(6), 581–584. DOI: 10.1016/0305-750X(79)90094-9

Ledoux, L., Mertens, R., & Wolff, P. (2005, November). EU Sustainable Development Indicators: An Overview. *Natural Resources Forum*, 29(4), 392–403. DOI: 10.1111/j.1477-8947.2005.00149.x

Lewis, W. A. (2013). *Theory of Economic Growth*. Routledge. DOI: 10.4324/9780203709665

Lorek, S., & Spangenberg, J. H. (2014). Sustainable Consumption Within A Sustainable Economy-Beyond Green Growth and Green Economies. *Journal of Cleaner Production*, 63, 33–44. DOI: 10.1016/j.jclepro.2013.08.045

Lyyytimäki, J., Antikainen, R., Hokkanen, J., Koskela, S., Kurppa, S., Känkänen, R., & Seppälä, J. (2018). Developing Key Indicators of Green Growth. *Sustainable Development (Bradford)*, 26(1), 51–64. DOI: 10.1002/sd.1690

Martinho, V. J. P. D. (2021). Insights Into Circular Economy Indicators: Emphasising Dimensions of Sustainability. *Environmental and Sustainability Indicators*, 10, 100119. DOI: 10.1016/j.indic.2021.100119

Moosa, I. A., & Smith, L. (2004). Economic Development Indicators as Determinants of Medal Winning At The Sydney Olympics: An Extreme Bounds Analysis. *Australian Economic Papers*, 43(3), 288–301. DOI: 10.1111/j.1467-8454.2004.00231.x

Morris, M. D. (1978). A Physical Quality of Life Index. *Urban Ecology*, 3(3), 225–240. DOI: 10.1016/0304-4009(78)90015-3

Nath, S. K. (1962). The Theory of Balanced Growth. *Oxford Economic Papers*, 14(2), 138–153. DOI: 10.1093/oxfordjournals.oep.a040893

Nurudeen, A., & Usman, A. (2010). Government Expenditure and Economic Growth in Nigeria, 1970-2008: A disaggregated Analysis. *Business and Economics Journal*, 4(1), 1–11.

Papava, V. (1994). The Role of The State in The Modern Economic System. *Problems of Economic Transition*, 37(5), 35–48. DOI: 10.2753/PET1061-1991370535

Peri, G. (2004). Socio-Cultural Variables and Economic Success: Evidence From Italian Provinces 1951-1991. *Contributions in Macroeconomics*, 4(1), 20121025. DOI: 10.2202/1534-5998.1218

Piętak, Ł. (2014). Review of Theories and Models of Economic Growth. *Comparative Economic Research.Central and Eastern Europe*, 17(1), 45–60.

Polenske, K. R. (2017), Growth Pole Theory and strategy Reconsidered: Domination, Linkages, and Distribution. *Regional Economic Development*, 91-111.

Quinn, D. P., & Shapiro, R. Y. (1991). Economic Growth Strategies: The Effects of Ideological Partisanship on Interest Rates and Business Taxation in The United States. *American Journal of Political Science*, 35(3), 656–685. DOI: 10.2307/2111560

Redish, A. (1984). Why Was Specie Scarce in Colonial Economies? An Analysis of The Canadian Currency, 1796-1830. *The Journal of Economic History*, 44(3), 713–728. DOI: 10.1017/S0022050700032332

Saliminezhad, A., & Lisaniler, F. G. (2018). Validity of Unbalanced Growth Theory and Sectoral Investment Priorities in Indonesia: Application of Feature Ranking Methods. *The Journal of International Trade & Economic Development*, 27(5), 521–540. DOI: 10.1080/09638199.2017.1398270

Singer, H. W. (1949). Economic Progress in Underdeveloped Countries. *Social Research*, 1–11.

Smol, M. (2019). The Importance of Sustainable Phosphorus Management in The Circular Economy (CE) Model: The Polish Case Study. *Journal of Material Cycles and Waste Management*, 21(2), 227–238. DOI: 10.1007/s10163-018-0794-6

Upreti, P. (2015). Factors Affecting Economic Growth in Developing Countries. *Major Themes in Economics*, 17(1), 37–54.

Van Den Berg, H. (2016). *Economic Growth and Development*. World Scientific Publishing Company. DOI: 10.1142/9058

Vazquez-Brust, D. A., & Sarkis, J. (2012). *Green Growth: Managing The Transition to Sustainable Economies*. Springer Netherlands. DOI: 10.1007/978-94-007-4417-2

Vera, I., & Langlois, L. (2007). Energy Indicators for Sustainable Development. *Energy*, 32(6), 875–882. DOI: 10.1016/j.energy.2006.08.006

Yáñez-Arancibia, A., Dávalos-Sotelo, R., & Day, J. W. (2014). *Ecological Dimensions for Sustainable Socio Economic Development* (Vol. 64). WIT Press.

Chapter 2
Climate Change and Global Inequality:
How Does Climate Change Exacerbate Existing Global Inequalities and Its Implications

Mohit Yadav
https://orcid.org/0000-0002-9341-2527
O.P. Jindal Global University, India

Ajay Chandel
https://orcid.org/0000-0002-4585-6406
Lovely Professional University, India

Harshita Agrawal
National Post Graduate College, Lucknow, India

Majdi Quttainah
https://orcid.org/0000-0002-6280-1060
Kuwait University, Kuwait

ABSTRACT

This chapter explores the intersection of climate change and global inequality, highlighting how climate impacts exacerbate existing disparities and challenge social justice. By examining case studies from Bangladesh, Puerto Rico, the Sahel region, and Pacific Island nations, it reveals the disproportionate effects of climate change on vulnerable populations. The analysis underscores the urgent need for integrated climate justice frameworks, community-based adaptation strategies, and sustainable development initiatives. Key global responses, including the Paris

DOI: 10.4018/979-8-3693-5792-7.ch002

Agreement and international migration compacts, are assessed for their effectiveness in addressing these challenges. Future directions emphasize enhancing international cooperation, leveraging technological innovations, and improving monitoring and accountability. The chapter aims to provide a comprehensive understanding of how climate change intensifies inequality and offers actionable insights for building a more equitable and resilient future.

INTRODUCTION

The introduction gives a broader perspective into arguably one of the biggest challenges facing human civilization: climate change and global inequality-how does it heighten existing global inequalities, and what does this mean in terms of social justice? Climate change is not just an environmental issue but also a deep social and economic crisis that multiplies existing global inequalities, raising questions of justice and equity (Ali et al., 2016). As climate change worsens, it is the world's most vulnerable populations—those who have contributed the least to the problem—who are bearing the disproportionate brunt of its consequences. This harsh fact brings to the fore once again the topicality of treating climate change in relation to global inequality, and it needs to be tackled in the context of social justice (Cevik & Jalles, 2022).

Historically, the greenhouse gas emissions that drive the climate crisis have been produced first and foremost by the industrialized nations of the Global North. In contrast, the developing countries in the Global South have contributed the least to these emissions but are ironically suffering the severest of consequences. These impacts don't simply involve environmental degradation but also economic instability, food and water insecurity, and increased health risks. This is all a function of global inequality, where the most impacted communities are usually the least capable of coping with or adapting to climate change (Chen, 2024). The linkage between climate change and global inequality is rather multi-layer wise and complex. While inequalities exist between countries, they are also reflected within countries, where the most marginalized communities-the low-income households, women, children, and indigenous peoples-are the most vulnerable to such calamities. These are often disadvantaged in terms of the resources and the infrastructure that allows them to protect themselves from climate-related risks, such as extreme weather events, rising sea levels, and extended periods of drought. Mechanisms of inequality in social and economic relations ensure that they have minimal involvement in the decision-making related to their lives, entrenching their vulnerability (Chisadza et al., 2023).

Emphasizing social justice in an approach to deal with climate change calls on proper perspectives over the historical and current injustices that have characterized the modern world system. This calls for a critical assessment of policies and practices so formulated to cater to economic growth and development in the North at the expense of environmental sustainability and social equity in the South. This, in turn, requires the rethinking of climate action as centered on the needs and voices of the most impacted communities-those who must be not just protected but empowered with agency at the forefront of the transition to a sustainable future (Ciplet et al., 2015)

In more recent years, climate justice has emerged as a way of framing the glaring inequities in the distribution of climate impacts and as a call for restitution by those most to blame for the crisis. It befits broader principles of social justice that argue for all people being treated in a fair and humane manner, regardless of socio-economic standing or geographical location. A claim for a just and fair response as the climate crisis unravels (Emmerling et al., 2024).

This chapter will consider the multi-dimensional nature of the relationship between climate change and global inequality, including how climate change is perceived to magnify existing disparities and what this might mean for social justice at a global level. In this chapter, we will explore in detail the geographic, socioeconomic, and political factors that ultimately determine vulnerability to climate change and reflect on the consequences of these dynamics for human rights and international development. Through this chapter, we aim to underline one important point: the role that social justice needs to play in climate action and to provide some answers to how far we might travel towards an equable and sustainable future for humankind.

The Interplay Between Climate Change and Global Inequality

Climate change and global inequality are deeply embedded in a complex relational nexus and symbolize different burdens of the highly vulnerable populations of the world. This, in essence, deals with how climate change is becoming the greater intensifier of the existing inequalities, rendering the already vulnerable more vulnerable. Further, climate change actually discovered and aggravated the structural inequity that was prevalent among and within nations, communities, and individuals for a considerably long time, and, therefore, it is evident that the issue of climate change involves not only an environmental challenge but a deeply social and economic crisis at the heart (Gordon, 2007).

Undue Burden on Developing Countries and Marginalized Communities

One of the most prominent examples of the interplay of climate change and global inequality is how developing countries suffer the most. These nations are geographically located so that, due to their climate-sensitive dependency like agriculture and poor resources, they are very vulnerable to the adverse changes in climate. Be it nations in Sub-Saharan Africa, those in Southeast Asia, or the small island nations in the Pacific, they are being taken into account because extreme weather events—droughts, floods, hurricanes, along with others—shall recur much often and with total intensity through a change in climate (Green & Healy, 2022).

In other words, these areas are not only exposed to all the physical elements of climate change but also lack financial, technological, and infrastructural wherewithal to respond effectively. The richer nations will be able to put in place advanced technologies, infrastructure, and adaptive measures, thus reducing the impacts of climate change. The developing nations can hardly manage to get the capital and technical help required to ensure safety for their people. This reflects an overall injustice: those least responsible for the climate change are often those facing its worst impacts (Harlan et al., 2015).

Economic Disparities and Climate Vulnerability

Climate change further increases economic inequality between and within countries. Most developing countries depend on agriculture to make a living. This climatic effect on weather patterns makes them miss the set normal weather conditions and subjects them to huge uncertainties and risks. Prolonged droughts, unstable rainfall, and extremely high temperatures can destroy crops and threaten food insecurity, loss of income, and sometimes even famine. That's vicious cycle turns because of all these, and poor families are driven towards further poor conditions because of climate change impact (Hubacek et al., 2017).

At a country level, the impact of climate crisis falls especially hard on poor households, women, and women of endemic origin. These poor people often belong to areas having the most significant vulnerability to the effects of climate change, such as floodplains, coastal zones, or drought areas. They also have fewer resources with which to adapt, through relocation, rebuilding, or investment in protective measures (Islam & Winkel, 2017). Women also take a disproportionate share of the burden in terms of maintaining food, water, and energy supplies for their families—all components threatened by climate change. In addition to this, the indigenous groups that have all along been taking care of their ecosystems are now facing degradation of land and natural resources, a situation that aggravates the loss of culture and a specific way of life (Kenner, 2019).

Barriers to Adaptation and Participation in Decision-Making

The extent of which climate change will be adapted to is influenced by the possibility of access to resources, knowledge, and decision-making. Yet, with the global inequality, there at least seems to be a lack of voice from these very communities that are bound to suffer most due to climate change. International climate negotiations and national policies often tend toward domination by the interests of wealthier nations and powerful corporations, making the voices of those in the Global South and from all corners of marginalized communities seem glaringly quiet. Such exclusion also leads to solutions that may seem ineffective or may be even harmful to those at grave risk.

Take, for example, large projects for climatic mitigation, such as the construction of dams or the expansion of renewable energy infrastructure. Such large projects often result in the displacement of communities and the disturbance of their livelihoods, usually without compensation or other means of subsistence being put in place. This includes other market-based solutions, such as carbon trading, where profit is required above all. This allows the wealthy nations and corporations to continue polluting while the poor communities bear the burden of the impacts of environmental and social cost (Lomborg, 2020).

The Role of Historical Injustices

The interaction between climate change and inequality cannot be divorced from the historical context of global inequality. It is industrialization-the legacy of colonialism, exploitation, and resource extraction from the Global South by the Global North-that has driven much of the carbon emissions contributing to climate change. This historical injustice means that many developing countries have weak economies, fragile infrastructures, and a limited capacity to cope with the impacts of climate change (Millward-Hopkins et al., 2023.

In this light, the climate crisis is continuous with many more historical patterns of exploitation and inequality. It is the nations and communities that have been left out and exploited in the colonial times that are now left the most vulnerable to climate change and the well-to-do in better positions to protect themselves. This legacy of injustice underlines the need for climate action, which is not only directed at the environmental question of the crisis but seeks also to right some deep-seated social and economic inequalities that underpin it (Paglialunga et al., 2022).

The Need for an Equitable Response

Effectiveness in handling the interplay between climate change and global inequality shall be adopting an approach that stresses equity and justice. This will call for ensuring that in climate policy and action, vulnerable communities are put at the center with regard to needs and perspectives. This means giving strong financial and technical support to developing countries in adaptation to climate impacts and the transition to sustainable economies (Parks & Roberts, 2013).

Success in this effort, however, must be built upon international cooperation and solidarity. Richer countries, most responsible for climate change-induced disasters, must be in solidarity with—most importantly at the front—providing support for the Global South through building resiliency and pursuing sustainable development. This can be facilitated through mechanisms on climate finance, technology transfer, and capacity-building programs that enable developing countries to aggressively act on climate change (Parsons et al., 2024).

In short, the interaction of climate change with global inequality is a stark reminder that the climate crisis is not an environmental problem but a serious concern socially and economically. Fighting the changing climate is taking up the mantle to ensure that justice and equity will not leave behind those who need it most. By rooting out the causes of these inequalities and giving space to the voices of most marginalized, we can work towards a future where climate action will be both effective and just-inclusive (Pérez-Peña et al., 2021).

Geographic and Socioeconomic Dimensions of Vulnerability

These interlinked geographic and socioeconomic dimensions of vulnerability reveal that the place of somewhere and social factors together determine the degree to which a population is exposed to the impacts of climate and able to cope with it (Rao & Min, 2018). Both dimensions emphasize that vulnerability does not remain constant; rather, it may change markedly with geographical location and access to various resources, infrastructures, and social support systems. Being sensitive to these factors would help in the formulation of targeted and robust policies to reduce the negative effects of climate change on the most affected communities (Redclift & Sage, 1998).

Regional Disparities: Coastal Zones, Small Island Developing States, and Arid Lands

Geography plays a very crucial role in determining the rates of exposure various communities have towards climate change. Coastal zones, small island developing states and also arid lands are vulnerable due to their unique geographical settings. Take the increasing threat of sea level rise in coastal areas from melted polar ice caps and the thermal expansion of sea water due to warming (Roberts, 2001). Coastal zones, frequented by high population densities and major economic infrastructure like ports, tourism industries, and fisheries, will all be affected. Submergence of coastal zones can displace millions of people, a major loss in livelihood opportunities, and cause immense economic damages. Already, one can see the beginnings of the appearance of "climate refugees" as communities in low-lying coastal areas are forced to resettle because of the advancing sea (Roberts & Parks, 2006).

Threats from rising sea levels obsess small island nations in the Pacific and Caribbean. These countries are subject to devastating effects of a slight sea-level rise, including but not limited to a total loss in land of the islands, polluted fresh water, and loss of agricultural land. Commonly, these states have relatively low financial budgets and depend heavily on very small sectors, like tourism and fishing, which are themselves susceptible to changes in climate. Geographically, these nations are islands or located in insular regions, which only further worsens their plight in dealing with the impacts and recovery from all climate-triggered risks (Singer, 2018).

In arid and semi-arid areas, notably in Africa and parts of the Middle East, climate change is increasing the aridity and leading to increased durations of drought and desertification, putting an additional pressure on the scarce water resources. This spells food insecurity because cropping patterns in these very areas are mainly rain-fed and extremely susceptible to variation in precipitation patterns. These crop failures translate into food insecurity, malnutrition, and economic instability, which deepen poverty within the already shattered contexts. Furthermore, the competition for the increasingly scarce water resource heightens conflicts where the declining resources play among communities and nations in a fight for control (Sverdrup, 2019).

Socioeconomic Status and Climate Vulnerability: Case Studies of Low-Income Communities

Socioeconomic status is one of the critical determinants of climate vulnerability, as it varies in both the level of exposure to climate risks and the ability to adapt to these risks. Whether in the developing world or the developed, low-income communities bear the disproportionate impact in terms of climate change because of limited access to resources, poor infrastructure, and social marginalization (Gordon,

2007). Very often these neighborhoods sit in hazardous locations—very flood-prone plains, urban slums, or areas prone to stronger winds and rain; a majority of those houses are weakly built and lack environmental hazard resiliency. For instance, the informal settlements in most cities in the South are usually found in peripheral areas of the towns facing flooding areas on the lowlands or on unstable hillsides. Generally, populations in such areas have very basic access to services like clean water, sanitation, and healthcare. This can make them more susceptible to health consequences related to climate change, such as the spread of waterborne diseases following floods. In addition, these economically precarious communities are so due to the paucity of capacity in them for rebuilding after a disaster has occurred. An increased vulnerability to a cycle of impoverishment becomes a reality in the wake of climate stressors (Eghdami, 2023).

Even in the richer countries, climate change impacts are harshened in low-income communities. In America, for example, low-income and minority groups more often reside in areas with higher pollution levels, near hazardous waste sites, and with poorly maintained housing, which disposes them to greater vulnerability from events such as hurricanes and extreme heat waves. Many of these communities happen to be those without financial ability for evacuation at the time of any disaster, accessing insurance to cover damages, and relocating to safer areas (Jimoh, 2021).

The Role of Infrastructure and Access to Resources in Climate Resilience

Infrastructure and access to resources are very necessary to a community's ability to be resilient and to recover from climate hazards. The strong infrastructure in a country—good transport systems and good drainage systems as integral parts of the social system, and strong buildings and houses—lowers vulnerability to damage from extreme weather events. However, more often than not, infrastructure in developing countries and marginalized parts of the developed world is either weak or in a state of disrepair and thus loses its resilience to climate-related disasters (Chen, 2024).

Besides that, other key resources are financial resources, technologies, and information. Ease of access to early warnings, climate-smart agricultural practices, and strategies of reducing disaster risks assigns a community with a better stand to keep abreast of the climate impacts: to predict, prepare, and respond to them efficiently (Chisadza et al., 2023). The rest will have it hard. For instance, one group included farmers from developed countries who could predict and offset the effects of drought because of crop insurance, irrigation systems, and improved weather forecasting. In contrast, smallholder farmers in developing countries could be able to rely on traditional rain-fed agriculture, with no resources to invest in irrigation or other adaptive technologies that may potentially mitigate their vulnerability to crop failure and food insecurity.

Besides the accumulated physical infrastructure and resources, social infrastructures such as community systems and networks, education, and healthcare systems are critical aspects of climate resilience development. Elaborating, the existence of strong social networks enables vulnerable communities to receive support in times of crises; access to education and healthcare services results in extensive knowledge that communities can use to understand and manage climate risks. In most of the highest priority areas, however, social infrastructure is weak or absent as well, a factor exacerbating the impacts of climate change (Cevik & Jalles, 2022).

In conclusion, geographical and socio-economic dimensions of vulnerability to climate change underline uneven developments in climate risks and the role that location, economic status, and access to resources play in determining community ability to cope with and adapt to a changing climate. These vulnerabilities indeed demand a targeted response looking at both the physical and social drivers of climate risk, ensuring that the most vulnerable are not left behind in the global response to climate change (Ciplet et al., 2015).

Climate-Induced Migration and Displacement

Climate-induced migration and displacement are rapidly emerging as critical global challenges with profound implications for the affected individuals, communities, and nations. Accelerated climate change and associated impacts with altered frequencies and magnitudes of extreme weather events, rising sea levels, and environmental degradation are pushing millions of people into forced migration—be it temporary or permanent. It disrupts not only lives but also aggravates prevalent social, economic, and political tensions in the regions from which the people flee and the areas to which they come. Therefore, understanding drivers, patterns, and consequences of climate-induced migration is critical in the formulation of effective policies and assistive systems to address this increasing crisis (Ali et al., 2016).

Drivers of Climate-Induced Migration

Many of the major drivers behind climate-induced migration are many and often interlinked, reflecting the complex nature of these relationships. They could be broadly generalized as due to sudden-onset events, for example, hurricanes and flooding and fires, or as a result of slow-onset processes, for example, desertification, rising sea levels, and changing patterns in agriculture (Moore & Wesselbaum, 2023).

In general, events with sudden onset also led to sudden displacement. For example, hurricanes and floods destroy the shelters of individuals, while at the same time they damage or destroy infrastructure in any given community, leaving community members no other option but to flee from their residential areas, permanently in some instances. Destruction of livelihoods in agricultural-based rural areas may also create the need for migration to seek other economic opportunities. Large parts of communities have been displaced due to the increase in hurricane and typhoon occurrences and intensifications in places like Southeast Asia and the Caribbean; thus, they have become climate refugees (Torres et al., 2020).

Where slow-onset processes are the case, pressure on communities is longer-lasting and tends to generate more gradual migration. Sea levels are, for example, encroaching on coastlines, polluting freshwater supplies, and eroding land used both for agriculture and habitation. Countries like Bangladesh, with a large part of its population on low-lying coasts, might see millions displaced over the next few decades. In return, desertification and prolonged droughts are making big parts of Africa and the Middle East uninhabitable, moving rural people either to cities or across borders in search of better living conditions (Tarraga et al., 2020).

Patterns of Climate-Induced Migration

However, in developing a theory on the pattern of climate-induced migration, it has been found that such patterns are multifaceted, where several thrusting factors include the magnitude and period for which the particular climate change has hit, the state of preparedness, and the sociopolitical environment. The internal migration may be within the national border or across borders, where from one country people migrate to another (Reichman, 2022).

The extent of internal migration is higher in larger countries, where people can simply move to regions which may be less affected by climate change. For example, in the United States, hundreds of thousands of people were displaced by Hurricane Katrina and its aftermath in 2005, many moving to neighboring states. In China, desertification and lack of water in northern provinces have caused major internal migration to the more prosperous southern parts of the country with less impact due to climate change.

Cross-border migration often tends to be more complex and challenging, as it usually involves international borders but sometimes even legal and political environments that are often hostile. Climate impact migrants may not fall into the category of 'refugee' under international law, in its traditional definition of those fleeing persecution based on race, religion, nationality, or political opinion. This protection gap, at a legal level, means that many climate migrants do not have the right to protection and support, in contrast to refugees, making their resettlement or integration into the host country more complex.

Another emerging pattern relates to seasonal or temporary migration impelled by climate change. Seasonal migrations could turn to be one of the coping strategies for many in the face of climatic changes, as during the dry season, agricultural work may no longer be viable. These temporary movements may blossom into permanent relocations if environmental conditions do not improve or if opportunities present themselves as more sustainable in the destination area (Golf, 2012).

The Impact on Host Countries and Communities

Climate-induced migration can have a significant social, economic, and even political impact on the host countries and communities. In general, its impact depends on the scale of migration, receiving areas' capacity to absorb the new population, and pre-existing socio-economic conditions (Ahsan, 2014).

Sometimes, the inflow of migrants puts a serious burden on the local resources, infrastructure, and services, especially in those areas where the local people are barely surviving under poverty or an absence of development. For example, urban centers of developing countries that have unprecedented in-migration of rural people may lead to disorders in housing, health care, education, and employment. These pressures can heighten competition for resources, social tensions, and even, in extreme cases, conflict between migrant and host communities.

In other cases, however, migration can also lead to gains for the host communities. For example, immigrants can help local economies not just by labor, but also through entrepreneurship and the sharing of culture. In regions where there is a population decline, the influx of new residents can revitalize the economy and therefore help to mitigate some of the effects of demographic issues such as aging populations. Whether or not this migration burden or benefit depends squarely on the policies and attitudes of the host community and the support availed to both the migrants and the populations that host them (Ahsan, 2019).

Legal and Ethical Considerations in Supporting Displaced Populations

Legal and ethical issues of climate-induced migration are very complicated and multilevel. Under the presently existing international law-in particular, the 1951 Refugee Convention-there is no clear recognition of the status of persons displaced by climate change as refugees, and this puts them in a state of legal limbo. This lack of recognition translates into the fact that very often, climate migrants do not have access to the same rights and protection as traditional refugees, such as the right to asylum or protection against forcible return to their home countries (Khoshnood, 2018).

Efforts are underway to address this gap. Some countries and international organizations are starting to explore ways in which the definition of refugees can be extended to include persons displaced by climate change, or develop new legal frameworks that take consideration of the needs of migrants induced by climate change. For example, the UN's Global Compact for Migration adopted in 2018 includes provisions acknowledging the connection between climate change and migration and the international community's increased protection of climate migrants. Ethical Issues The ethical issues involved in climate-induced migration are questions of responsibility and justice (Jacobsen, 2003). Those countries that have been the biggest contributors to climate change-mostly because of their historical carbon emissions-are ironically the least affected by its impacts. While the contribution by every country and community has been unequal, those that contribute least to climate change are generally more vulnerable to its hazards. The international imbalance demands a response that encompasses not only humanitarian but also mitigation or compensatory support at the very roots for the community.

Climate-induced migration and displacement are pressing issues that require coordinated action at the global level. As climate change continues to reshape the environmental and social landscape, an immediate call would be to formulate and implement policies and practices aiming at protection and support interests of those forced to flee/displace, while looking into the greater issues of inequality and injustice undergirding this phenomenon (Gordon, 2007). Understanding complex drivers and patterns of climate-induced migration also enables the international community to understand legal and ethical gaps in existing frameworks. It is then that the displaced populations' needs can be responded to appropriately to guarantee a more just and sustainable future.

Implications for Social Justice and Human Rights

The intersection of climate change and global inequality carries major implications for social justice and human rights. More obvious manifestations of climate change continue to further deepen the current social and economic inequalities that threaten the rights and wellbeing of poor populations around the world. An understanding of the impediment brought forth by these facets in dealing with climate change requires a subtle analysis of how climate change affects social justice and human rights and then is committed to the integration of these themes into climate action and policy (Gordon, 2007).

Exacerbation of Inequalities

Climate change at present is disproportionally affecting the marginalized and disadvantaged while deepening existing social and economic divides. The low-income populations, the indigenous peoples, and other marginalized groups are more than likely to reside in those most climate-susceptible areas: floodplains, arid regions, or any area prone to extreme weather. These particular populations often have fewer resources and less access to services to help them adapt or recover from such climate-related shocks (Green & Healy, 2022).

Indeed, most people in developing countries who are in the most precarious situations are found in informal settlements or areas where infrastructure is at a bare minimum. In these cases, strong winds and other severe meteorological conditions, such as hurricanes or flooding, hit the hardest because their homes are inadequately built and their passage to emergency facilities is hindered (Harlan et al., 2015). The devastation of people's homes and sources of livelihood not only deepens poverty but also diminishes the potential for recovery and rebuilding of those communities, further entrenching the cycles of disadvantage.

It also increases the vulnerability of marginalized groups in more developed countries. For example, in the United States, low-income and minority communities are often sited in areas with higher levels of pollution and are less likely to have resources that mitigate the consequences of climate change, such as air conditioning during heat waves or flood insurance. Deep-seated inequalities are underpinned by disparities in the distribution of environmental hazards and access to protective measures (Hubacek et al., 2017).

Human Rights Consequences

Climate change impacts have serious human rights consequences, from the right to life and health, through to food, water, and shelter. As climate change intensifies, so does the threat from the exercise of these rights by individuals and communities.

Right to Life and Health: Increasingly violent weather events, rising temperatures, and natural disasters more frequently hit the very existence of human lives and health. These are the elderly, children, and those with other health conditions. For example, heatwaves result in heatstroke and exaggerate chronic diseases, while floods raise the risk for waterborne diseases. Inability to access proper healthcare during such occurrences only adds to these risks, further resulting in unnecessary deaths and health complications (Kenner, 2019).

Right to Food and Water: Agricultural productivity will decrease and there will be changes in water availability due to the effects of climate change. This would eventually pose a threat to the right of a person to adequate food and clean water. In drought or desertified regions, crop failures and reduced water supplies are associated with food insecurity and malnutrition. Those who depend on subsistence agriculture or fishing have heightened vulnerability to climate hazards because their livelihoods depend directly on climate conditions. Accordingly, the right to safe drinking water is further threatened by both sea-level rise and different types of contamination that may affect freshwater supplies (Islam & Winkel, 2017).

Right to Adequate Housing: Climate-related disasters destroy homes and infrastructure, a violation of the right to adequate housing. Displacement caused by either extreme weather events or sea-level rise can make individuals and families homeless or relegated to poor living conditions. In many instances, destroyed housing further exacerbates the social and economic disparities of the affected population due to their lack of access to available and safe shelter in the aftermath of the catastrophe (Lomborg, 2020).

Legal and Policy Responses

There is an intersection of climate change and social justice that firmly calls for legal and policy interventions for addressing causes that lead to inequity and the protection of the rights of vulnerable populations. Existing international mechanisms-while recognizing that climate action should support the most impacted communities, as reflected in the Paris Agreement-do very little to articulate the distinct needs and rights of those communities (Millward-Hopkins & Oswald, 2023).

Climate justice frameworks incorporate social justice into climate action in ways that elevate the needs of the most vulnerable and ensure full voice and agency in the processes of decision-making. Climate justice approaches work toward addressing issues of historical and systemic injustices at the roots of the present inequalities and toward ensuring equity and inclusiveness in all climate policies and interventions. This also involves financial and technical support to developing countries, community-led adaptation and resilience-building processes, and accountability to the needs of marginalized groups (Paglialunga, Coveri, & Zanfei, 2022).

Protection of Human Rights: Strengthening the protection of human rights in relation to climate change involves recognizing and protecting the rights of those most affected by the impacts of climate change. This includes ensuring that the implementation of climate policies will not infringe on basic rights, such as the right to food, water, and shelter, while taking adequate measures to support affected populations. Legal frameworks should be developed or improved with regard to the human rights dimensions of climate change, which may include measures regarding the protection of climate-induced migrants and internally displaced persons (Parks & Roberts, 2013).

It calls for **international cooperation and solidarity,** rooted in the social justice and human rights of climate change. Those developed nations that are mostly historically responsible for the particular problem of climate change have an added responsibility to support less developed ones with both mitigation and adaptation to climate impacts. Support should be provided through financial means, technology transfer, or capacity-building initiatives that would enable already vulnerable communities to become more resilient and adapt to the changing conditions (Parsons et al., 2024).

This will lead to the conclusion that from the standpoint of social justice and human rights, the ramifications caused by climate change are great and far-reaching. With the continuous and growing effect of climate change, the inequalities and challenges of human rights that come with it need to be attended to in order to ensure that climate action is truly equitable and inclusive. This can be done through the adoption of frameworks that genuinely put the concerns of the most vulnerable first, enhance the protection of human rights, and enjoin international cooperation in reaching a more equitable and resilient future for all (Pérez-Peña et al., 2021).

Global Responses and Policy Frameworks

The multi-dimensional impacts of climate change with its interlinkages to global inequality have led to the formation of several international and national policy frameworks and initiatives. These responses have attempted to mitigate the consequences of climate change, build resilience, and strive for social justice. As discussed in this section, key global responses and policy frameworks are dealt with based on their intent, successes, and gaps that remain (Rao & Min, 2018).

International Climate Agreements and Initiatives

1. **The Paris Agreement:** The Paris Agreement, adopted in 2015, represents an unprecedented effort by the international community to address the adverse alterations of the climate and to limit the increase in the global average temperature to well below 2°C above pre-industrial levels while pursuing efforts for its limitation to 1.5°C. In light of the latest scientific findings and the challenges thus posed, the Agreement makes an urgent call for a common effort toward the reduction of greenhouse gas emissions and toward supporting countries most in need through finance and technology. The principle of "common but differentiated responsibilities" is one of its salient features, recognizing that all countries are responsible for addressing climate change but with different contributions and capabilities. It also provides for climate adaptation under

support to developing countries and the creation of the Green Climate Fund for financing climate action (Redclift & Sage, 1998).
2. **The Global Compact for Migration:** Global Compact for Safe, Orderly and Regular Migration, adopted in 2018, truly recognizes the interlink between climate and migration-a call for holistic management of migration in a manner respectful of human rights and projects addressing the root causes of displacement. It contains commitments to improve international cooperation on migration, protect migrants' rights, and integrate migration into climate policies. Although non-legally binding, the Compact provides a framework for countries to develop policies and practices accounting for the impact of climate change on migration and displacement (Roberts, 2001).
3. **The Sendai Framework for Disaster Risk Reduction**: Adopted in 2015, the Sendai Framework aims to reduce disaster risk and further increase resilience, focusing attention on the root causes of risk, which include climate change. It recognizes that the guiding principles of risk-informed development, disaster preparedness, and community participation are indispensable in disaster risk reduction work. It shifts the focus to the need for holistic approaches to manage climate risks and build resilience, particularly in the most vulnerable communities (Roberts & Parks, 2006).

National Policies and Strategies

1. **Nationally Determined Contributions (NDCs):** The Paris Agreement requires countries to submit Nationally Determined Contributions, which detail their intents and actions in the way of greenhouse gas reductions and adaptation to impacts due to climate change. It reflects national priorities and capacities, and its implementation is vital with respect to the global climate goals. Most countries have integrated considerations about social justice and inequality within their NDCs; however, this ranges from low to high regarding the level of ambition and effectiveness (Singer, 2018).
2. **Adaptation Plans and Strategies**: Adaptation plans and strategies taken up by countries aim to address specific impacts or enhance resilience. These plans more often include vulnerable population protection, infrastructural improvement, and sustainable development. For instance, Bangladesh has an elaborative National Adaptation Plan that addresses community-based adaptation projects, improved flood management practices, and capacity-building activities for support of vulnerable communities (Sverdrup, 2019).

3. **Climate Action Plans:** In addition to adaptation plans at the national level, several countries have also implemented more general climate action plans, bringing mitigation and adaptation under one process. The policies and programs in this area typically include efforts to lower greenhouse gas emissions, develop renewable energy sources, and create more climate-resilient communities. The European Union's Green Deal, for example, involves a long list of actions toward achieving its goal of climate neutrality by 2050, including measures to protect vulnerable communities and further social equity (Torres et al., 2020).

Challenges and Opportunities

1. **Financing and Resource Allocation:** Indeed, financing and resource allocation continue to be some of the most critical challenges to the implementation of both global and national climate policies. While initiatives like the Green Climate Fund have been set up to help raise finances for developing nations, needs far outstrip the funds that are available. Much has to be done in terms of ensuring financial resources are well allocated and reach the most vulnerable communities if goals on climate and social justice are to be met (Ali et al., 2016).
2. **Integration of Climate and Social Justice Objectives:** The integration of social justice into the policy frameworks is of utmost importance, wherein any effective climate action is concerned. This requires ensuring that the various elements of climate policies are able to address the needs of the most marginalized and deprived sections of people and propagate equitable outcomes. While basic elements of social justice exist within different international and national frameworks, much more attention would be required on issues related to inequalities and ensuring that climate action does not result in the building of disparities (Cevik & Jalles, 2022).
3. **Enhancing International Cooperation:** Climate change is a kind of global problem that requires international cooperation and solidarity. Given the complexity and interconnectedness of the impacts, there is an urgent need to reinforce international cooperation in the areas of climate action, migration, and adaptation (Chen, 2024). The coordination between the countries, international organizations, and civil society should become better to ensure coherence and efficiency in acting upon climate change and its social and economic implications.
4. **Monitoring and Accountability**: Climate policy, in fact, requires rigorous monitoring and accountability for tracking progress, evaluating performance, and compliance. Transparency of reporting and periodic reviews with the help of independent assessments would bring about awareness of gaps and challenges

faced during the implementation process to continuously improve it (Ciplet et al., 2015).

Any meaningful attempt at dealing with both the challenges of climate change and working for social justice rests on global responses and policy frameworks. International agreements, such as the Paris Agreement and the Global Compact for Migration, provide critical frameworks through which collective action can take place. National policies and strategies pave the way toward implementations of climate and adaptation actions. These no doubt pose challenges in financing, social justice integration, and international cooperation along with monitoring. In overcoming these challenges and using the opportunities for improvement, the international community is able to move on toward a future that may be more just, resilient, and capable of enduring the impact of climate change (Chisadza et al., 2023).

Case Studies: Climate Change and Inequality in Action

Considering the relationship between climate change and global inequality, there is a great variation in deepening the clarity with respect to the impact of climate change and the various responses to such inequalities. Understanding real-world case studies will be greatly instrumental in this regard. This chapter consequently reviews case studies from a wide range of regions and contexts, establishing tangible effects of climate change on vulnerable communities and measures to mitigate such impacts.

1. Case Study: Bangladesh - Rising Sea Levels and Flooding

Background: Bangladesh is one of the most climate-vulnerable countries, with a densely populated coastal region that is extremely vulnerable to sea-level rise and flooding. With its mostly low-lying landscape, it is very much prone to all sorts of perils of climate change, such as increased frequency and intensity of cyclones and storm surges, riverbank erosion, and erratic rainfall distribution (Ahsan, 2019).

These are the impacts: Rising sea levels and increased flooding have immensely hit Bangladesh's coastal communities. Farmland, flooded by saltwater, has been associated with huge crop losses, affecting food security and people's livelihood. On one side, frequent flooding affects homes and infrastructures, displacing several thousand people yearly. Impacts are disproportionately felt by the poorest communities that have the least resources to adapt or recover from these disasters.

Answers: Various adaptive measures have been taken by Bangladesh as a response to the challenges. The investment has gone into building flood-resistant infrastructure, like raised embankments, cyclone shelters, and other things. Community-based adaptation programs in the construction of flood-resilient homes, early warning systems, and various initiatives that aim at enhancing local resilience. In addition, the Bangladesh Climate Change Strategy and Action Plan has outlined measures in detail in a comprehensive manner for the growth of actions in terms of climate impacts and support to the vulnerable (Ahsan et al., 2014).

2. Case Study-Puerto Rico-Hurricane Maria and Economic Disparities

Background: The U.S. unincorporated territory, Puerto Rico, has been profoundly impacted by Hurricane Maria in September 2017. The hurricane caused extensive destruction to the island, along with power outages and flooding, which were deeply heightened by pre-existing economic and social disparities.

Impacts: Hurricane Maria hit the poor and the most vulnerable sections of the population in Puerto Rico squarely. Poor, vulnerable, and in areas lacking in proper infrastructure, most of them became highly susceptible to the aftereffects brought about by the hurricane. The massive destruction of homes, schools, and health care facilities escalated differences that had existed in relation to equal access to basic services and economic opportunities. The prolonged recovery process sealed the deal on disparate resource allocation and show-cased how marginally positioned communities faced a choppier stretch of water than did others as they strove to overcome what was lost (Reyes Cruz, 2023).

Multiple recoveries have taken place since Hurricane María both on the island and abroad. Money was appropriated by the federal government for disaster relief and recovery, but this money often came with very slow and unorganized dissemination. Grassroots and community organizations gave much-needed immediate relief and are still fighting for just and equitable recovery efforts. Each of them has followed the track of doing the best for the populations most in need, by creating a path to long-term resilience through community-based initiatives (Peri et al., 2022).

3. Case Study: Sahel Region - Desertification and Migration

Introduction: The Sahel region in Africa covers several countries in West and Central Africa, including Mali, Niger, and Chad. It is seriously threatened by desertification and environmental degradation arising from the effects of climate change. Continuous and persistent droughts with decreasing rainfall in this semi-arid area have resulted in loss of arable land and water bodies.

Impacts: Desertification in the Sahel has had a deep impact on agriculture, which forms the main livelihood for so many people in this region. Due to this, agricultural productivity is being reduced, further contributing to food insecurity and economic hardship, thus forcing people to migrate in search of improved living standards. This environmental degradation leading to displacement has continued to spur increased migration into urban centers and across borders, normally associated with strained resources and heightened tensions within host communities (Pearson & Niaufre, 2013).

Desertification and all the accompanying repercussions have inspired a series of interventions in the Sahel. The Great Green Wall initiative, through the African Union, has been the subject of restorative land and sustainable land management practices to stem the tide of desertification. Other initiatives devised by native organizations include community-based activities aimed at enhancing agricultural techniques, augmenting water management, and easing migration management. This is meant to help build resilience in addition to providing alternative livelihoods for communities affected by environmental degradation.

4. Case Study: Pacific Island Nations - Climate-Induced Displacement

Background: Small Island states in the Pacific, such as Tuvalu, Kiribati, and the Marshall Islands, are facing existential threats due to rise in sea levels and higher incidence of extreme weather conditions. These islands, which number among the most vulnerable to the impacts of climate change, have a small land area and also limited resources to adapt to those impacts.

Impacts: Loss of land from sea-level rise and increased storm intensity has brought saltwater intrusion and destruction of key infrastructure to Pacific Island countries. The environmental impact already poses an existential threat, as the very states themselves are in question, with possible climate-induced displacements and loss of heritage. Communities find increased levels of salinization in both water sources and agricultural land, which affects food security and livelihoods (Thomas & Benjamin, 2018).

Answers: The Pacific Island nations have been quite active in pursuing international support and calls for action on the climate issue. Many have aggressively pursued climate adaptation projects, including building seawalls, elevating infrastructure, and protecting coastlines. In addition, the nations have pursued international diplomacy in an attempt to attract international climate finance and support options for relocation if it would be necessary. Another issue that has also come to the fore in international forums is climate-induced displacement, underpinning the need for global solidarity and action on behalf of these vulnerable nations (Green & Healy, 2022).

CONCLUSION

These diverse case studies detail how climate change affects inequality on a global spectrum. They epitomize exactly how these challenges brought on by climate change increase existing disparities, along with a myriad of responses to mitigate these issues. These lessons, drawn from processes of local adaptation to global advocacy, signal the need to embed social justice concerns within climate action. The case studies provide an important learning opportunity for policy makers and other stakeholders in their quest to devise more effective and just strategies that can help reduce the impacts of climate change while supporting vulnerable communities worldwide.

Future Directions: Toward a Just and Equitable Climate Future

While the impacts of climate change are worsening, the need for paying attention to a just and equitable approach towards climate actions is only growing bigger. The interconnected crises of climate change and global inequality will require innovative approaches, collaborative action, and a commitment to embedding issues of social justice within climate policy. This section considers possible future directions in pursuit of a more equitable climate future, informed by consideration of key areas of action and opportunity.

1. **Enhancing Climate Justice Frameworks Integrating Social Justice into Climate Policies:** For climate policies to be responsive to rising inequality, social justice concerns must be embedded in every level of climate action-from identification of specific vulnerabilities, participation, and prioritization of equitable outcomes in decision-making and policy design to actual implementation. It is necessary to develop or improve the frameworks of climate justice for articulating their needs and ensuring that climate action does not result in heightened inequality (Harlan et al., 2015). **Promoting Inclusive Policy Development:** The approach should focus on the development of inclusive policy, where diverse stakeholders in climate policies include the integration of marginalized communities into the design and implementation of climate policy. This participatory approach ensures that the voices of those most affected by climate change are heard and their views considered in the decision-making process. Community-led initiatives and grassroots organizations need to be at the forefront of pushing for the needs of the most vulnerable populations. They, thus, need to be supported and empowered to contribute at the policy development level (Hubacek et al., 2017).

2. **Enhance Resilience to Climate Change Invest in Community-Based Adaptation:** The community-based adaptation programs are very necessary in developing resilient ecosystems for vulnerable groups. These may include infrastructure projects, enhancing preparedness for disasters, and livelihoods that are sustainable. Investment in the local solution where the communities are leading processes for adaptation may result in enhanced resilience and reduced impacts of climate change. In addition, support to the community-based projects needs scaling up and should, if possible, focus on resources reaching the neediest (Islam & Winkel, 2017). **Improvement in Infrastructure and Services Resilience:** Building and adapting infrastructure to be climate-proof remains the core of every positive resilience story. This ranges from upgrading house, road, and health infrastructures to make them easily adaptable against extreme variability of weather and other hazards of climate change. Investment in resilient infrastructure should focus on the most vulnerable areas to climatic impacts and can be designed in cooperation with the affected communities in order to meet their needs (Kenner, 2019).
3. **Empowering Sustainable Development and Equity Empowerment for Sustainable Livelihoods**: The search for sustainable livelihoods involves the development of economic growth that is both ecologically sustainable and fair. This ranges from creating green employment and pursuing sustainable agriculture to investing in renewable energy ventures that provide real benefits to communities of the poor. The fact that all economic development initiatives are inclusive and nondiscriminatory will help a great deal in reducing inequalities and building resilience over the longer term. **Addressing Systemic Inequalities:** Address basic conditions of systemic inequalities to realize a just climate future. This requires root causes of vulnerability and exclusion, which are deeply entrenched in social, economic, and political factors that should be addressed. Policies and programs should attempt to reduce income, educational, and resource access gaps, while fostering social inclusion and the empowerment of marginal groups (Lomborg, 2020).
4. **Increased International Cooperation and Solidarity Improving International Climate Agreements**: Conventions like the Paris Agreement provide a framework for collective action; these need to be strengthened to ensure an adequate response to the needs of the most vulnerable. This may be done by increasing commitments with regard to financial support, technology transfer, and capacity building of developing countries in global agreements. Global agreements must zero in on issues of inequity and support equity in climate action (Parks & Roberts, 2013). **Improved Support for Climate-Induced Migration:** Climate-induced migration ripples into s plethora of challenges not only for the persons affected but also for the recipient communities. International cooperation is,

therefore, important in developing supportive policies and frameworks for migrants and ensuring that their rights are protected. This may include financial and technical support to manage migration, integration, and social cohesion in host communities, as well as to address the root causes of their displacement (Millward-Hopkins & Oswald, 2023).

5. **Tapping into Innovation and Technology Using Technology for Equity:** Technological changes can be major game-changers as people try to grapple with the vagaries of climate change and work toward equity. For example, advances in climate modeling, early warning systems, and renewable energy technologies enhance resilience and benefit the vulnerable. True as this may be, all technological benefits must be equitably apportioned; access to such innovations should be accorded to the marginalized (Parsons et al., 2024). **Supporting Just Transition Approaches:** Just transition defines the pathway to a low-carbon economy in a manner sensitive to workers and communities who may suffer from the transition process. This means training and support for workers whose industries are transitioning out of business, while creating new job opportunities in the emerging green sectors. Just transition strategies must center the needs and concerns of those most impacted by the transition and provide a seat at the table for them in shaping the future of work (Pérez-Peña et al., 2021).

6. **Enhancing Monitoring and Accountability Improving Transparency and Reporting:** Improved transparency and reporting can help in the monitoring of progress and accountability for ensuring that the policy is implemented as intended. The transparent reporting of climate actions and impacts on the vulnerable population provides a chance to find the lacunae and deficiencies in the implementations. Independent assessment and evaluation will facilitate letting the climate policy meet its aim with proper utilization of resources (Rao & Min, 2018). **Strengthening Accountability Mechanisms:** These are mechanisms for ensuring equity in climate actions and addressing the needs of marginalized sections, while holding governments, corporations, and other stakeholders responsible for their commitments and actions. Give support and amplify civil society organizations and advocacy groups, who have a significant role in monitoring and bringing actors to account (Redclift & Sage, 1998).

From fully incorporating social justice into climate action, developing resilience, and adaptation to sustainable development and international cooperation, it is all included. Community-based solutions, investing in system change and innovation, and technology can help drive a climate future that is inclusive and supportive of all communities. The way forward involves addressing the existing climate change impacts while, more so, building on the disparities that underpin laying liability to vulnerability and impeding progress. With proper efforts and equity, a more resilient and just world can be built for current and future generations.

CONCLUSION

The space that climate change inhabits is increasingly complex and evolving; thus, its impacts are only inextricably linked with global inequalities. Because of this fact, the effects of climate change striking the most vulnerable populations hard exacerbate current disparities and seriously threaten basic human rights. What is required is that these challenges be tackled with a holistic and equitable approach to include social justice within climate action, at the same time as acknowledging varied needs for different communities affected.

These case studies have demonstrated that responses to climate change need to be specific and all-inclusive. From the devastating impact of sea-level rise in Bangladesh to the economic hardship Puerto Rican communities have faced following Hurricane Maria, such examples have shown how climate change is magnifying existing inequalities and disrupting life. They also heighten community-led adaptation efforts, international cooperation, and robust policy frameworks that are highly needed to cushion such impacts and support vulnerable populations.

Beyond this, reinforcing climate justice frameworks, investment in community-based adaptation, and pursuit of sustainable development continue to have high relevance for building resilience and reducing disparities. Reinforced international cooperation and leveraging of technological innovations will introduce new tools that can serve to tackle climate change; these undertakings, however, have to be made with the active pursuit of equity and inclusion. Monitoring and accountability mechanisms will be indispensable for ensuring climate policies are enacted, hence making sure resources are placed where they need to be.

Ultimately, it is not possible to develop a just and equitable climate future without a response to current climate change impacts and to the root, systemic drivers of inequality giving rise to vulnerability. Welcoming such an approach includes social justice in depth and works toward a world where all communities can thrive but get a chance to do so, whatever their socioeconomic status or geographic location. A climate-resilient future with equity takes will and sustained commitment; it is attainable and necessary for the well-being of present and future generations.

REFERENCES

Ahsan, R. (2019). Climate-induced migration: Impacts on social structures and justice in Bangladesh. *South Asia Research*, 39(2), 184–201. DOI: 10.1177/0262728019842968

Ahsan, R., Kellett, J., & Karuppannan, S. (2014). *Climate induced migration: Lessons from Bangladesh* (Doctoral dissertation, Common Ground Publishing).

Ali, H., Dumbuya, B., Hynie, M., Idahosa, P., Keil, R., & Perkins, P. (2016). The social and political dimensions of the Ebola response: Global inequality, climate change, and infectious disease. *Climate change and health: improving resilience and reducing Risks*, 151-169.

Cevik, M. S., & Jalles, J. T. (2022). *For whom the bell tolls: climate change and inequality*. International Monetary Fund.

Chen, Q. (2024). The Intersection of Global Inequality and Climate Change. *International Journal of Education and Humanities*, 15(2), 378–381. DOI: 10.54097/3s08zx60

Chisadza, C., Clance, M., Sheng, X., & Gupta, R. (2023). Climate change and inequality: Evidence from the United States. *Sustainability (Basel)*, 15(6), 5322. DOI: 10.3390/su15065322

Ciplet, D., Roberts, J. T., & Khan, M. R. (2015). *Power in a warming world: The new global politics of climate change and the remaking of environmental inequality*. Mit Press. DOI: 10.7551/mitpress/9780262029612.001.0001

Eghdami, S., Scheld, A. M., & Louis, G. (2023). Socioeconomic vulnerability and climate risk in coastal Virginia. *Climate Risk Management*, 39, 100475. DOI: 10.1016/j.crm.2023.100475

Emmerling, J., Andreoni, P., & Tavoni, M. (2024). Global inequality consequences of climate policies when accounting for avoided climate impacts. *Cell Reports Sustainability, 1*(1).

Goff, L., Zarin, H., & Goodman, S. (2012). Climate-induced migration from Northern Africa to Europe: Security challenges and opportunities. *The Brown Journal of World Affairs*, 18(2), 195–213.

Gordon, R. (2007). Climate change and the poorest nations: Further reflections on global inequality. *U. Colo. L. Rev.*, 78, 1559.

Green, F., & Healy, N. (2022). How inequality fuels climate change: The climate case for a Green New Deal. *One Earth*, 5(6), 635–649. DOI: 10.1016/j.oneear.2022.05.005

Harlan, S. L., Pellow, D. N., Roberts, J. T., Bell, S. E., Holt, W. G., Nagel, J., & Brulle, R. J. (2015). Climate justice and inequality. *Climate change and society. Sociological Perspectives*, 2015, 127–163.

Hubacek, K., Baiocchi, G., Feng, K., Muñoz Castillo, R., Sun, L., & Xue, J. (2017). Global carbon inequality. *Energy, Ecology & Environment*, 2(6), 361–369. DOI: 10.1007/s40974-017-0072-9

Islam, N., & Winkel, J. (2017). Climate change and social inequality.

Jacobsen, K., & Landau, L. B. (2003). The dual imperative in refugee research: Some methodological and ethical considerations in social science research on forced migration. *Disasters*, 27(3), 185–206. DOI: 10.1111/1467-7717.00228 PMID: 14524045

Jimoh, M. Y., Bikam, P., & Chikoore, H. (2021). The influence of socioeconomic factors on households' vulnerability to climate change in semiarid towns of Mopani, South Africa. *Climate (Basel)*, 9(1), 13. DOI: 10.3390/cli9010013

Kenner, D. (2019). *Carbon inequality: The role of the richest in climate change.* Routledge. DOI: 10.4324/9781351171328

Khoshnood, K. (2018). Methodological and ethical challenges in research with forcibly displaced populations. *The Health of Refugees: Public Health Perspectives from Crisis to Settlement*, 209.

Lomborg, B. (2020). Welfare in the 21st century: Increasing development, reducing inequality, the impact of climate change, and the cost of climate policies. *Technological Forecasting and Social Change*, 156, 119981. DOI: 10.1016/j.techfore.2020.119981

Millward-Hopkins, J., & Oswald, Y. (2023). Reducing global inequality to secure human wellbeing and climate safety: A modelling study. *The Lancet. Planetary Health*, 7(2), e147–e154. DOI: 10.1016/S2542-5196(23)00004-9 PMID: 36754470

Moore, M., & Wesselbaum, D. (2023). Climatic factors as drivers of migration: A review. *Environment, Development and Sustainability*, 25(4), 2955–2975. DOI: 10.1007/s10668-022-02191-z

Paglialunga, E., Coveri, A., & Zanfei, A. (2022). Climate change and within-country inequality: New evidence from a global perspective. *World Development*, 159, 106030. DOI: 10.1016/j.worlddev.2022.106030

Parks, B. C., & Roberts, J. T. (2013). Inequality and the global climate regime: breaking the north-south impasse. In *The Politics of Climate Change* (pp. 164–191). Routledge.

Parsons, E. S., Jowell, A., Veidis, E., Barry, M., & Israni, S. T. (2024). Climate change and inequality. *Pediatric Research*, 1–8.

Pearson, N., & Niaufre, C. (2013). Desertification and drought related migrations in the Sahel–the cases of Mali and Burkina Faso. *The State of Environmental Migration, 3*.

Pérez-Peña, M. D. C., Jiménez-García, M., Ruiz-Chico, J., & Peña-Sánchez, A. R. (2021). Analysis of research on the SDGs: The relationship between climate change, poverty and inequality. *Applied Sciences (Basel, Switzerland)*, 11(19), 8947. DOI: 10.3390/app11198947

Peri, G., Rury, D., & Wiltshire, J. C. (2022). The economic impact of migrants from hurricane maria. *The Journal of Human Resources*, 0521-11655R1. DOI: 10.3368/jhr.0521-11655R1

Rao, N. D., & Min, J. (2018). Less global inequality can improve climate outcomes. *Wiley Interdisciplinary Reviews: Climate Change*, 9(2), e513. DOI: 10.1002/wcc.513

Redclift, M., & Sage, C. (1998). Global environmental change and global inequality: North/South perspectives. *International Sociology*, 13(4), 499–516. DOI: 10.1177/026858098013004005

Reichman, D. R. (2022). Putting climate-induced migration in context: The case of Honduran migration to the USA. *Regional Environmental Change*, 22(3), 91. DOI: 10.1007/s10113-022-01946-8 PMID: 35814810

Reyes Cruz, C. A. (2023). *Causes of displacement: a look into the state of Puerto Rico's housing crisis before, during, and after Hurricane Maria* (Doctoral dissertation).

Roberts, J. T. (2001). Global inequality and climate change. *Society & Natural Resources*, 14(6), 501–509. DOI: 10.1080/08941920118490

Roberts, J. T., & Parks, B. (2006). *A climate of injustice: Global inequality, north-south politics, and climate policy*. MIT press.

Singer, M. (2018). *Climate change and social inequality: The health and social costs of global warming*. Routledge. DOI: 10.4324/9781315103358

Sverdrup, H. U. (2019). The global sustainability challenges in the future: the energy use, materials supply, pollution, climate change and inequality nexus. In *What Next for Sustainable Development?* (pp. 49–75). Edward Elgar Publishing. DOI: 10.4337/9781788975209.00013

Tarraga, J. M., Piles, M., & Camps-Valls, G. (2020). Learning drivers of climate-induced human migrations with Gaussian processes. *arXiv preprint arXiv:2011.08901*.

Thomas, A., & Benjamin, L. (2018). Policies and mechanisms to address climate-induced migration and displacement in Pacific and Caribbean small island developing states. *International Journal of Climate Change Strategies and Management*, 10(1), 86–104. DOI: 10.1108/IJCCSM-03-2017-0055

Torres, P. H. C., Leonel, A. L., Pires de Araújo, G., & Jacobi, P. R. (2020). Is the brazilian national climate change adaptation plan addressing inequality? Climate and environmental justice in a global south perspective. *Environmental Justice*, 13(2), 42–46. DOI: 10.1089/env.2019.0043

Chapter 3
Social and Economic Impacts of Climate-Induced Migration and Displacement

Shashank Mittal
O.P. JIndal Global University, India

Ajay Chandel
https://orcid.org/0000-0002-4585-6406
Lovely Professional University, India

Thi Mai Le
https://orcid.org/0000-0001-9720-308X
Vietnam National University, Hanoi, Vietnam

ABSTRACT

Climate-induced migration and displacement are emerging as critical issues in the context of global climate change. This chapter explores the drivers, social, and economic impacts of climate-induced migration, providing future projections and scenarios to understand the scale and nature of displacement. Key drivers include extreme weather events, sea-level rise, desertification, and water scarcity. Social impacts involve disruptions to communities, changes in demographics, and health challenges, while economic impacts affect labor markets, infrastructure, and resource allocation. Case studies illustrate real-world examples and policy responses, while resilience and adaptation strategies offer insights into managing these challenges. The chapter concludes with future projections highlighting the need for comprehensive planning and international cooperation to address the anticipated increase in

DOI: 10.4018/979-8-3693-5792-7.ch003

climate-induced displacement.

INTRODUCTION

Climate-induced migration and displacement have emerged as one of the critical concerns of the modern world that bears significant social, economic, and political repercussions. As climate change is gaining pace, so is its prominence through more frequency and intensity of various natural catastrophes-tornadoes, flooding, droughts, and wildfires-alongside slow-onset phenomena such as sea-level rise, desertification, and soil degradation. These changes perpetuate disruption in local ecosystems and livelihoods alike, forcing communities to seek life elsewhere in their quest for safety, stability, and better living standards. Whereas migration has been a human response to environmental change throughout history, today's scale and complexity of climate-induced migration pose new challenges to the individual, communities, and nations (Ahsan, 2019).

Climate-induced migration is the movement of people that is mainly being caused by environmental changes and degradation emanating from climate change. The movement could be either temporary or permanent, voluntary or forced, within national borders or across international boundaries (Ahsan et al., 2014). A forced displacement is generally an event under the broad description of climate-induced migration, involving sudden and unplanned movements due to acute environmental events like hurricanes or floods. On the other hand, migration can be a more gradual process where people relocate pre-emptively to avoid risks in the future, given the slow-onset climate impacts of rising sea levels or prolonged droughts. These distinctions are important to understand while developing appropriate policy responses and support mechanisms for the affected populations (Aleinikoff, 2024). The drivers behind climate-induced migration range from environmental to socio-economic. Environmental drivers refer to natural disasters, the loss of habitable land, degradation of ecosystems that directly threatens people's safety, homes, and livelihoods. Such environmental change often reinforces prevailing socio-economic vulnerabilities-like poverty, resource inaccessibility, and political instability-and may thus also force the individuals and groups to move. For instance, in areas where agriculture is the main livelihood activity, prolonged drought reduces crops due to this livelihood activity, culminating in food insecurity and economic hardship, hence compelling one to shift the migration search for opportunities elsewhere. In coastlines liable to increase in sea levels, for instance, settlements can shift into other areas as the houses and infrastructure become increasingly vulnerable to flooding and erosion (Almulhim et al., 2024).

The social impacts of climate-induced migration go so deep that they do not only hurt the migrants themselves, but also the host communities to which they migrate. Migrants and displaced people many a time find it hard to get access to basic needs, such as shelter, food, or healthcare, let alone finding jobs or settling into new social and cultural milieus. These are also issues that may be exacerbated by discrimination, legal barriers, and lack of support networks, further adding to their vulnerability and marginalisation (Askland et al., 2022). The receiving communities may also be affected by changes in demographic composition, cultural dynamics, and social cohesion, aside from additional pressures on public services and infrastructure. The sheer size of migrant influx sometimes creates social tensions and conflicts, particularly when such resource competition falls in a resource-scarce setting, or if the cultural and political gap is huge between migrants and hosts (Bose & Lunstrum, 2012).

The economic impacts of climate-induced migration can be both positive and negative. While on one hand, migrants contribute to the local economy through filling labor shortages, business creation, and bringing diverse competencies and experiences in, on the other hand, large-scale displacement can overstress public resources, increase unemployment, and reduce wages in host communities-particularly in developing countries with low capacity to absorb large numbers of newcomers. Besides, out-migration might cause the most vulnerable regions to lose human capital, which will affect local development and increase poverty and inequality (Choudhury & Shahi, 2024). Therefore, economic impacts are a concern that should be viewed as an issue of careful planning and coordination by the governments, international organizations, and local communities in managing migration to attain maximum benefit with minimum cost (Draper, 2020).

In sum, climate-induced migration and displacement are complex phenomena interfacing with a wide range of social, economic, and environmental issues. The unfolding dynamics of climate change will make the comprehension and addressing of these challenges highly critical in ensuring well-being for populations at risk and furthering sustainable development globally. This chapter reviews the various dimensions of climate-induced migration in terms of its causes, social and economic implications, as well as necessary policy responses to resiliently adapt to an unpredictable future (Draper & McKinnon, 2018).

Drivers of Climate-Induced Migration

Climate-induced migration is the complex process caused by environmental and socio-economic drivers that make it impossible for individuals and communities to continue living in their traditional habitats. Understanding the various drivers of climate-induced migration brings out the full scope of the phenomenon for both understanding and responding to it appropriately. Generally, drivers of climate-induced migration can be summarized as environmental factors, socio-economic factors, and the interplay between these elements (Ferreira, 2017).

Environmental Factors

The immediate causes of climate-induced migration are environmental in nature. These range from sudden-onset events to slow-onset events due to climate change. Events in the former category are those that are acute, dramatic environmental changes, such as hurricanes, floods, and wildfires, which force immediate population displacement. For instance, hurricanes and typhoons have devastated communities living around coastlines, destroying homes and buildings and consequently making residents move elsewhere. Likewise, wildfire incidents, which sometimes spread by successive drought and rise in temperature, may lead to the complete devastation of entire areas with high casualties and loss of property that could further result in mass evacuations (Francis, 2019).

The slow-onset events refer to those processes of environmental change that have a gradual effect on the onset of displacement over a period of time. In this category fall sea level rise, desertification, soil degradation, and changes in rainfall. For example, the rise in sea levels is of concern for low-lying coastal areas and island nations because of the submergence of the land and contamination of freshwater supplies through the gradual intrusion of seawater (Gemenne, 2011). This gradual but eventually unstoppable process could leave communities with no other choice but to seek long-term adaptation strategies that involve resettling elsewhere. On the other hand, desertification and soil degradation result in uninhabitable land, as the agriculture around which most people base their livelihoods in rural communities is lowered in productivity. As these lands become ever more desolate, their inhabitants are left with little choice but to move in search of richer soil and basic livelihoods (Islam & Hasan, 2016).

Socio-economic Factors

While environmental factors provide the proximate causes for climate-induced migration, socioeconomic factors significantly affect who moves and how well equipped the person is to adapt. The principal socioeconomic driver of vulnerability to climate change is poverty. Generally, the investment required to prepare for and respond to environmental disasters is beyond the means of poor communities, so although poor people are more likely to be displaced by climate induced events, they are less able to safely and sustainably relocate (Islam & Shamsuddoha, 2017).

Climate induced migration is also driven by lack of access to essential resources such as water, food, and energy. This competition for scarce resources is heightened by the alteration in availability and distribution, which might then lead to conflict and displacement. In this regard, communities could conflict over the limited supplies of water because of increased evaporation and reduced rainfall in some regions, hence prompting migration into other areas with plentiful resources. In addition, food insecurity, wherein agricultural yields are decreasing and competition for arable land is increasing, is often a driver of human migration as people seek out better opportunities (Jayawardhan, 2017).

Moreover, political instability and issues of governance can aggravate the impacts caused by climate-induced migration. Where governments are weak or corrupt, effective policies and programs are absent to support communities in adapting to the changing environmental conditions and managing displacement. Lack of governance makes people more vulnerable and thus is likely to lead to migration (Kälin, 2010). Additionally, in politically unstable regions, climate-induced migration may lead to increased tension due to additional needs on basic resources and social amenities, hence leading to conflicts and displacements.

Intersection of Environmental and Socio-Economic

Socio-environment interface knits a network of complex drivers in which climate-induced migration is going to have. Very often, that interface gives birth to vulnerability and environmental changes affect a huge number of people most negatively who are socio-economically most fragile (Lyster & Burkett, 2018). For example, in an agricultural-based economy, if drought gets prolonged, it tends to reduce the yield of crop besides unfavorable effects on economic well-being and food security while the migration pressure increases. Similarly, coastal communities reliant on fishing could incur not only environmental but also economic damages due to the increase in sea level and changes in the pattern of the migration of fish (Mayer, 2011).

The adaptive capacity-which is defined by the ability of individuals and communities to adjust to potential damages, utilize opportunities, or respond to the consequences-is also molded by the interaction between environmental and socio-economic factors. The more access a community has to resources, education, technology, and institutional support, the more adaptive it would be to environmental change and less likely to be displaced. That is to say, risk of climate-induced displacement is higher among communities that lack these endowments. For example, somebody in an urban area with different job opportunities and social service access is likely to be more resilient to the negative impact of climate than a rural farmer who depends solely on the land (McMichael et al., 2010).

In sum, the drivers of climate-induced migration are multi-dimensional, encompassing several environmental and socio-economic factors that interact in complex ways. While, therefore, environmental changes are the immediate triggers for migration, it is socio-economic conditions that underpin individuals' vulnerability and their adaptive capacity. Understanding these drivers and the way in which they intersect is crucial to devising comprehensive strategies to respond to climate-induced migration and to support the affected populations. As these changes continue to worsen, response strategies will be paramount in the process of mitigating climate-induced migration impacts that address environmental and socio-economic drivers to ensure sustainable and resilient communities (Mehedi et al., 2010).

Social Impacts of Climate-Induced Migration

The effects of climate-induced migration are devastating from a social perspective for migrants and host communities in receiving regions, while it also caused problems for local communities left behind. These impacts are multi-dimensional and complex and touch social structures, relationships, and processes at very different and multiple levels. The social consequences of climate-induced migration can only be ascertained in such a way that it will be possible to develop policies and interventions in providing protection to the affected vulnerable population and allowing healthy integration for the host community. This section discusses different social impacts that movers and host communities bear as a result of climate-induced migration (Mostaque & Hasan, 2015).

Impacts on Movers and Host Communities

For many migrants and displaced persons, the process of moving away from a certain place is often highly cumbersome in nature, as a result of the effects of climate change. The most immediate and serious effects which people experience when they are displaced involve health and psychological issues. Forced migration, when caused by a sudden-onset event such as floods, hurricanes, or wildfires, may be traumatizing and a very stressful experience (Naser & Afroz, 2009). Long-term loss of livelihood and community can be associated with grief, anxiety, and a sense of hopelessness. Physical conditions in the processes of displacement, including but not limited to shelter conditions with temporary overcrowding, poor sanitation, and limited health care, can accelerate health problems and add to vulnerability from infectious diseases (Oluwaseyi & Stilinski, 2024).

Not only health issues but significant social integration issues also prevail for migrants and displaced persons in their new settings. These may include issues of language barriers, cultural differences, and discrimination that could pose a challenge to social cohesion and the building of new social networks. For example, displaced people may face rejection from the host communities among which they might be perceived as strangers or competitors for meager resources. This may be seen to breed conflict, isolation, and even violence between migrants and people in the community hosting them. Legal restrictions, such as failure to confer refugee status or limited access to employment, education, and public services, further contribute to this social exclusion (Poddar, 2024).

Gender-specific impacts and vulnerabilities also play a critical role in shaping the experiences of climate-induced migrants and displaced populations. Women and girls are particularly at risk for violence, exploitation, and human trafficking resulting from displacement and in temporary settlement settings. They often have to assume the responsibility of securing water, food, and fuel for their families, which is increasingly difficult and dangerous to provide in new, often insecure environments. The reduced level of access women and girls have to information, resources, and decision-making processes that are integral to effective adaptation and recovery further increases their vulnerability (Salik et al., 2020).

Impacts on Host Communities

These communities receiving climate-induced migrants also display a range of social impacts, both positive and negatives. The most marked change will be seen in the demographic composition and cultural dynamics that take place in the host communities. A large-scale influx of migrants often leads to the social composition of the receiving areas being changed, new languages, and new customs, and traditions. In fact, this diversity has the potential to contribute to the enriching of cultural life, besides helping in fostering mutual understanding. Simultaneously, however, it can create tensions-especially in those situations where a host community might feel their cultural identity threatened by newcomers, or when resources and opportunities are competed for (Schwan & Yu, 2018).

Other significant social impacts of climate-induced migration upon host communities involve the strain created upon social services and infrastructural capabilities. Outflow of migrants can strain meager resources such as housing, health, and education facilities, and public utilities. The same may also take many forms, like enlargement of schools and hospitals, longer times before being attended by a doctor, and greater demands from the public mass transit. In developing countries or areas with fragile infrastructure, these pressures often exceed local capacities, further eroding service quality and fueling social tensions between migrants and host communities (Teye & Nikoi, 2022).

Other critical issues related to social cohesion and the potential for conflict also arise in host communities experiencing significant climate-induced migration. Such sudden and often unplanned mass arrivals of migrants tend to dissipate established social networks and create competition for available jobs, land, and other resources of the host community, therefore potentially leading to conflicts and social unrest. In some instances, such tensions might be exacerbated by preexisting ethnic, religious, or political divisions within the host community. This would mean the negative effects that could arise in terms of misunderstanding, prejudice, and discrimination, further deteriorating social cohesion if compared with the host population (Thomas & Benjamin, 2018).

While it is true that climate-induced migration has caused so much damage socially, this kind of migration can also have a number of positive social effects on host communities. For instance, migrants offer a bundle of different skills, knowledge, and experiences to share for application in the socio-economic development of the receiving area. These might include filling labor shortages, creating businesses, or introducing new farming or technological practices that improve the economy in a locality. In addition, the fact that migrants are present affords an opportunity for intercultural exchange and mutual learning, enabling social inclusion and resilience in a community hosting them (Tremblay & Trudel, 2013).

Impacts on Communities of Origin

The social impacts of climate-induced migration do not only affect the migrants and host communities but also the communities of origin left behind. Changes in family structures and social networks are among the most serious impacts on these communities. When people or families move, they usually leave their aging parents, children, or other dependents, who may then rely on remittances or the support from those who have left. This may generate new kinds of vulnerabilities and social dependence, in cases where migrants are unable to send regular support or if the remittances are inadequate to meet those needs.

The second important impact on transforming communities is linked to the outflow of human capital. Youth and skills out-migration of educated people often means an improvement in the opportunities to be availed, but a sort of "brain drain" where valuable human resources useful for local development are lost to the community. Also, this talent and labor could be the loss of general resilience and adaptability for a community striving to escape environmental shocks, bouncing back after them, and building long-term sustainability.

In sum, the social effects of climate-induced migration are far-reaching and many-sided, with migrants, host communities, and communities of origin facing quite different challenges. Such impacts, therefore, need an integrated response that takes into consideration the different needs and different levels of vulnerability of various populations, while also enhancing social inclusion and social cohesion, and seizing the potential contribution of migration to durable development. Understanding and accepting of the social dimensions of climate-induced migration will, therefore, enable policy and practice in appropriately serving these displaced populations to make more resilient and adaptive communities.

Economic Impacts of Climate-Induced Migration

The economic impacts of climate-induced migration are multifaceted and cut across both the moving migrants and the communities they leave or go to. These take the form of various dimensions such as employment, stabilization of income, deployment of resources, and growth and development in general. Understanding the economic implications of climate-induced migration is crucial for the development of strategies that aim at minimizing negative impacts while optimizing any potential opportunities. In this chapter, the approach to economic impacts takes a look at those experienced by migrants or forcibly displaced people and host communities, as well as origin communities.

Impact on Migrants and Displaced People

Climate-induced migration is, in most instances, a cycle of economic challenges and opportunities to both the migrant and displaced. Foremost among these are employment and livelihood. Many times, migrants struggle to get into the labor markets at their destinations, especially if they lack relevant skills and certifications or even important languages. This is further exacerbated by areas with higher rates of unemployment, or wherever, migrants are seen to compete with the available opportunities. This means that the majority of migrants end up in low-wage, informal, or precarious work, which may precipitate precarious economic situations and heightened vulnerability (Jayawardhan, 2017).

Other major concerns of climate-induced migrants pertain to unstable income situations. The process of displacement generally disrupts regular income streams and savings, hence creating financial insecurity. As a matter of fact, the migrant may incur many costs: the cost of transport, housing costs, and costs incurred due to the loss of income during the transition period. It results in debt accumulation and more economic hardships as some would sell out their personal belongings to settle the costs of movement or even borrow money at very high interest rates. Additionally, many migrants, therefore, may not have adequate social support to help meet subsistence basics related to food, healthcare, and education, compounding their economic vulnerabilities.

Nonetheless, amongst all these adversities, migration may also provide some economic opportunities for displaced people. Some may have access to new types of jobs or better pay and steadier conditions once a move has been made. Workers who have highly valued skills or knowledge in their new home may also find employment that greatly improves their economic circumstances. Additionally, migration can create entrepreneurial opportunities, such as opening new businesses or trade that contributes to their economic resilience and potential integration into the host community (Almulhim et al., 2024).

Impacts on Host Communities

The host communities to which climate-induced migrants move also incur a number of economic impacts-both positive and negative. The most immediate of these impacts is the increased demand for goods and services. Upon their arrival in large numbers, migrants can boost the local economy through increased demand for housing, food, and clothing, among other basic needs. This could be the short-term boom in an economic trend, driving huge benefits from the perspective of local businessmen and service providers. Besides, migrants can bring in new skills and knowledge, as well as cultural practices that may enhance economic diversity and innovation in the host community.

While these changes lead to economic benefits for migrants, they often breed competition for resources and employment. In cases where economic opportunities or even basic resources are scarce, such an influx can put undue additional stress on housing, health, education, and social services. This inflation in demand is believed to often lead to the rising costs of living, such as through increased rent and higher prices of goods and services, which pose a problem for the budgets of migrants and natives alike. This could sometimes generate competition that builds resentment in the minds of the locals-perhaps a perception that foreigners are being favored with preference while the locals continue to suffer from loss of job opportunities.

Climate-induced migration also significantly affects public expenditure and infrastructure. Local governments could be faced with increased costs in servicing a growing population that would require an expansion of health, education, and social welfare programs or an upgrading of infrastructure to cater for newly arriving residents. Such a situation can further strain public budgets, especially in developing countries or regions that have low fiscal capacity. In some instances, the resources have to be used to cope with their situation, thereby denting the governments' ability to invest in infrastructure development, economic diversification, and long-term social programs.

Impacts on Communities of Origin

The economics implications do not stop there where the migrants and host communities are affected, they go to the Communities of Origin. Among the economic impacts that are most important to any community experiencing out-migration, therefore, would be that loss of labor or human capital. That is, if skilled, educated, and youthful people leave in search of better opportunities, it is a brain drain that will strip the community of resources that are needed most in local development and economic growth. Productive loss can affect the community in that it is not able to go through innovation, stay afloat in productivity, and adaptation to environmental changes, therefore widening its economic vulnerabilities.

Migration can also have positive economic impacts towards origin communities in the form of remittances. Money is sent by migrants to their families and communities, which is very crucial since it acts as substantial income to the people left behind. These remittances can help reduce poverty, raise living standards, and finance education, health, and small business development. In some instances, remittances pave the way for local economic development through the demand for goods and services, job creation, and investment in infrastructure and community projects.

Yet, the economic benefits of remittances often remain uneven and do not blunt the negatives of labor loss. This overdependence on remittances may breed economic dependency, where people develop no interest in taking care of their local development and diversification. This makes communities more vulnerable to changes in migration policies or economic recessions in host countries. Consequently, those are factors that slow down the inflow of remittances and upset local economies (Jayawardhan, 2017).

Broader Economic Impacts

Beyond the immediate impacts on migrants, host communities, and communities of origin, climate-induced migration has broader economic implications that reverberate throughout the economic systems at national and global levels. The economic effects of climate-induced migration may further exacerbate existing gaps between economic classes, as people from low-income regions are generally found to be the most risk-sensitive, least able to cope with environmental changes, and least able to adapt or migrate. This will create disparities between and within regions and countries, where areas with a better economy will be able to attract skilled labor and invest in adaptive capacity, while poor are being enforced with less adaptive capacity and greater vulnerability (Draper, 2020).

Economic resilience and adaptation are very crucial aspects in light of this climatic-induced migration. It is also expected that countries and regions which manage migration effectively will mingle the immigrants with their economies, and better economic results will be observed. Effective policies and strategies are required for the economic integration, social protection, and sustainable development of the host communities and communities of origin. The aspect of economic implications, in terms of climate-induced migration, will, at the global level, remediate the economic impacts through international cooperation, investment in climate adaptation, economic development, and migration management.

Overall, the economic effects of climate-induced migration are several and in multiple dimensions, affecting both individual and community large economic systems and areas. While migration might cause significant and diverse economic woes, it is equally laden with opportunities for increased economic growth, development, and resilience. There is, therefore, a need to understand these economic drivers in making policies and strategies to be put in place for the mitigation of these negative outcomes, to leverage potential benefits, and to support implementation of sustainable and inclusive economic development in the face of a changing climate.

Case Studies of Climate-Induced Migration and Displacement

Presenting cases regarding specific conditions of climate-induced migration and displacement widens our perspective toward various impacts and challenges being faced. The case studies are important in the attempt to demonstrate the many dimensions of the migration process in association with environmental change, and for calling attention to responses by individuals, communities, and states. The cases stress full, context-specific management of climate-induced migration. This section we will look at different case studies from different parts of the world that tells about different perpetrators, impacts, and reactions to climate-change-induced migration and displacement.

1. River Erosion and Sea-Level Rise: Bangladesh

Not only is Bangladesh highly impoverished in the South Asian region, but it is also one of the most risk-prone regions for climate change. Blades of natural disasters like cyclones and floods, incidents of river erosion, and so on hit this country from time and again. River erosion is yet another major driver of displacement in the country. The rivers, mainly Brahmaputra, Ganges, and to a lesser extent Meghna, have the habit of inundation every year and thereby bring about massive erosion

(Draper, 2020). People lose their dwellings, agricultural lands, and infrastructure; hence the consequent population displacement is huge. Thus, this displaces thousands of people into other territories each year, including crowded metropolitan cities like Dhaka, which exerts pressures onto other facilities and services. Sea level rise allows saltwater intrusion into agricultural lands and freshwater bodies, adding to the increased devastation of agricultural lands and freshwater bodies, further dwindling food security and livelihoods of millions of people. Many coastal communities are forced to move further inland, adding to the already existing trend of rural-urban migration. These environment changes altogether contributed to enormous economic losses, increased poverty, and social instability. Actions taken by the Bangladeshi government and international agencies have included the construction of embankments, promoting climate-resilient agriculture, and developing early warning systems. These continue to be useful measures, but in most cases, the measures in place do not match the magnitude of the displacement that looms large, so there has been a kind of call for a more holistic approach to arrive at solutions (Jayawardhan, 2017).

2. The Pacific Islands: Sea-Level Rise and Loss of Territory

The Pacific includes such nations as Kiribati, Tuvalu, and the Marshall Islands, where sea-level rise has given rise to the most extreme cases of migration that are settled under duress. Most are small island states with low elevations, rendering them highly prone to rising sea levels, as well as its concomitant danger—storm surges. Many cases have the reduction of habitable space or the whole community going under due to coastal erosion, saltwater intrusion, and loss of freshwater sources. For instance, in Kiribati, several villages have already been moved due to flooding, and its government has already bought land in Fiji for potential relocation of its citizens. The prospect of turning into climate refugees presents unique legal, social, and cultural challenges to Pacific Islanders. In such cases, the process of displacement due to climate change is equated with the very survival of a nation, threatening to erase its sovereignty, culture, and identity from the world map. These latter states have appealed to the international community for financial support for adaptation measures such as building seawalls and enhancing water management, and ways for migration according to international law. And all of these efforts combined, don't really make it clear if such island nations are going to exist or not (Almulhim et al., 2024).

3. The Sahel Region of Africa: Desertification and Resource Scarcity

The Sahel, the semi-arid expanse, locates indirectly in countries that border the southern edge of the Sahara, primarily in Mali, Niger, Chad, and Burkina Faso. It experiences lots of climate-induced migration that results from desertification and a scarcity of resources. The Sahel is already semi-arid, known to be erratic in rainfall, recurrently having prolonged periods of drought, leading to desertification and deterioration of the land. Such shifts in the environment due to climate change have severely reduced the productivity of agriculture and threatened the livelihoods of millions of people who engage in farming and pastoral labor.

This has made many Sahelians migrate in search of a good life. Indeed, most of them move from rural to urban areas and even cross borders to other neighboring counties. This has made the migration of rural people into cities to be at an unprecedented rate, due to rapid urbanization, and puts lots of stress on infrastructure and services that are already overstretched in those urban centres. It has also exacerbated a dispute between communities over such precious resources as water and arable land, increasing interests in vying further, mainly in the case of farmers versus herders, leading to further displacements.

A number of national and regional-level initiatives are already being taken, including sustainable land management, livelihood diversification, and better regional cooperation in order to manage shared resources. However, the presence of conflict, political instability, and scarce resources exacerbate the challenges of addressing climate change impacts and providing support to the affected population (Draper, 2020).

4. The United States: Hurricanes and Wildfires

Over the last couple of years, in the US, there has been an event in which people were forced to migrate by the force of climate; natural catastrophes, such as hurricanes or fires, seem to become more frequent and intense. Hurricanes such as Katrina in 2005, Harvey in 2017, and Maria in 2017 have caused considerable damage and displacement, especially in coastal states like Louisiana, Texas, and Florida and Puerto Rico. All of these have created displacement of hundreds of thousands of people, many of whom can't go back home or build lives again.

Also, a state such as California, Oregon, Colorado has remained prone to more frequent and devastating wildfires due to prolonged droughts, higher temperatures, and a changed pattern of precipitation. For instance, in 2018, the Camp Fire in California razed down Paradise, displacing nearly 50,000 people. The economic impacts of such disasters range from property damages, loss of livelihoods, and increased insurance costs. On a social scale, people suffer trauma, community loss, and long-term displacement.

In response to those threats, the federal, state, and local governments of the United States have adopted disaster preparedness and response measures including early warning systems and investment in resilient infrastructure. However, the scale and frequency of climate-induced disasters continues to outpace existing systems, making the call for more robust adaptation and mitigation strategies towards addressing the increasing impacts of climate change (Draper & McKinnon, 2018).

5. Syria: Drought, Conflict, and Complex Migration Dynamics

While the political instability and subsequent civil war predominantly explain the situation in Syria, there are also climate-induced factors, particularly protracted drought, which has greatly determined the dynamics of mass migration in this region. Between 2006 and 2010, Syria suffered the most severe drought in its history, with irreparable loss to the agricultural sector. Apart from this, it resulted in widespread crop failures, loss of livestock, and rising food prices; taken together, these factors compelled an estimated 1.5 million Syrians from rural areas into cities searching for livelihoods other than farming.

This huge internal displacement increased existing social and economic pressure within the cities, contributing to heightened poverty, unemployment, and social unrest. This mix of environmental stressors, socio-economic challenges, and political instability provided a fertile ground for conflict, which further intensified displacement within and across the borders. For instance, the case of Syria highlights how climate-induced and other migration drivers interact, showing that environmental changes can be a threat multiplier in regions already facing many challenges (Jayawardhan, 2017).

International responses to the crisis in Syria have, therefore, focused on emergency assistance as far as civilian migration to other regions is concerned, where some on-grid cogeneration facilities have continued to operate even inside refugee' camps. However, there is a realization of the need to address environmental and economic drivers of displacement to work upon climate resilience and sustainable development in the region (Draper, 2020).

The case studies have been thus used to demonstrate the multi-faceted and complex nature of climate-induced human migration and displacement in various regions and contexts. Though every one of these cases has its particular challenges and drivers, common themes emerge relating to the necessity of implementing sound comprehensive adaptation strategies, satisfactory governance, and international cooperation on how best to deal with the effect of climate change on migration. These case studies are helpful in having a clearly established idea about the complexity of climate-induced migration and are important in devising the appropriate, tailored responses in context to support affected populations and nurture sustainable and resilient communities.

Policy Responses and Legal Frameworks

The responding policy to this enormous challenge in the migration and displacement of people, due to climate issues, has to be apt at the local, national, regional, and international levels. This means bringing into balance humanitarian needs and development imperatives, thus combining shorter-term relief and longer-term strategies toward resilience and adaptation. The section further discusses a variety of implemented or proposed policy responses and legal frameworks on climate-induced migration, with special reference to the role of governments, international organizations, and corporative response in managing the impacts of climate change and human mobility.

1. National and Local Policy Responses

National and sub-national authorities have played an important role in the development of policies and, more importantly, implementation in managing climate-induced migration and displacement. The latter policy content often embraces subscription to adaptation modalities, measures to reduce disaster risks, as well as social protection mechanisms to cushion the affected populations (Jayawardhan, 2017).

Adaptation and Resilience Building: Most of the countries prone to climate-induced migration have outlined national adaptation plans (NAP) as well as strategies to create resilience against climatic change. These usually comprise infrastructure improvements, the creation of flood defences, building retrofits designed for severe weather events, and expenditures on water-wise urban infrastructure. Meanwhile, in areas such as Bangladesh, a range of adaptation measures—for example, cyclone shelters and an early warning system—has been undertaken to reduce the instances of displacement arising from natural hazards (Draper, 2020).

Disaster risk reduction (DRR): It is only through effective disaster risk reduction that displacement by such abrupt-onset events as hurricanes, floods, and wild fires can be reduced. Most governments have an DRR framework aiming to be prepared for, respond to, and recover from disasters. Some of the elements that can be included in these frameworks are land-use planning, policies, building codes, and evacuation plans. For example, Japan has an integrated DRR strategy oriented toward community-based disaster preparedness programs and drills, ensuring preparedness to handle natural calamities.

Social Protection and Livelihood Support: Social protection programs include cash transfers, unemployment benefits, and food assistance as part of the vital support for communities due to climate-induced displacement. This class of program contributes to income stabilization and forms a safety net that could help the displaced persons in building back their lives and livelihoods. Social protection schemes have been adopted in countries like Ethiopia within the Productive Safety Net Programme to improve the resilience and livelihood of the vulnerable population in regions with high food insecurity and low agricultural production to prevent distress migration.

2. Community and Regional Policy Frameworks

Besides national importance, regional organizations and frameworks assume high significance when it comes to addressing climate-induced migration and displacement, specifically in the regions where environmental changes cut across national boundaries or affect multiple countries. Enhanced regional cooperation would only leverage greater efficiency in the national policies and a knowledge, resource, and best practices sharing platform (Draper & McKinnon, 2018).

Regional Agreements and Action Plans: There are several agreements and action plans as developed by various regional organizations to address climate-induced migration and displacement. Also, the African Union has created the African Union Climate Change Strategy, which has components on managing migration and displacement because of climate change issues. Similarly, in this part of the world, the Pacific Islands Forum has developed the Framework for Resilient Development in the Pacific to enhance regional cooperation on issues relating to climate change adaptation, disaster risk management, and human mobility (Almulhim et al., 2024).

Regional frameworks always allow room for **cross-border collaborations** on issues of climate-induced migrations. There are established mechanisms for countries to collaborate in areas like data sharing, early warning alert systems, and response mechanisms. A regional example for dealing with the question of regional displacement of people due to violence, including other climate impacts in Central America, is the Comprehensive Regional Protection and Solutions Framework—MIRPS. It offers the possibility of facilitating collaboration across countries in the region for enhanced protection and solutions for the displaced people (Jayawardhan, 2017).

Regional Capacity Building: Regional institutions also have a mandate to build the capacity of the member states in the management of climate-induced migration. Activities could be in the form of technical assistance, training, and micro-funding to enable the national or county governments to come up with effective policies, and finally, to formulate and implement the said policies. For example, regarding adaptation, the Caribbean Community has established the Caribbean Climate Change Centre, through which member states receive the appropriate technical support and training to strengthen their adaptive capacity toward changing climate and migration and displacement.

3. International Policy Responses

On the international scene, a number of policy frameworks and initiatives have been developed to address the global dimensions of climate-induced migration and displacement. Large parts of these frameworks primarily focus on increasing international cooperation, legal protections, and financial support for adaptation and resilience-building efforts.

The Global Compact for Migration and the Global Compact on Refugees: In 2018, the United Nations adopted the Global Compact for Safe, Orderly and Regular Migration (GCM) and the Global Compact on Refugees (GCR). Whilst not legally binding, either instrument establishes a multilateral framework for co-operation among countries regarding the phenomenon of migration and cases of displacement, including those induced by climate change. For example, the GCM states that it, " recognizes climate change as one of the drivers of migration and inspired us to intensify international cooperation to address the impacts of climate change, including through the facilitation of "development, issuing of licences to other forms of legal migration, provision of humanitarian assistance to affected persons, and sustainable return to their countries of origin".

The United Nations Framework Convention on Climate Change (UNFCCC) – While UNFCCC has gradually taken cognizance of addressing human mobility in climate change. The Paris Agreement 2015 integrates the measures to be taken in order to enhance the adaptive capacity and strengthen resilience to climate change, potentially reducing the numbers of people being forced to leave the area where they reside. The UNFCCC's taskforce on Displacement, through the Warsaw International Mechanism on Loss and damage, is designed in such a way that it will develop recommendations on integrated approaches to avert, minimize and address displacement related to the adverse impacts of climate change.

International Funding Mechanisms: International funding channels such as the Green Climate Fund and the Adaptation Fund are very critical in financing projects that enhance resilience to climate change and support communities affected by climate-induced migration. Such funds often prioritize projects in developing countries that are most vulnerable to climate change, including infrastructure improvement, building adaptive capacity, and supporting livelihoods that are truly sustainable.

4. Legal Frameworks and Protections

Legal frameworks protect the rights of those displaced by climate change and guarantee assistance and support. International legal frameworks generally either neglect to address climate-induced displacement explicitly or are sparse, thus offering insufficient protection and support to these populations.

International Refugee Law: Protection for refugees-that is, people forced to flee their country owing to a well-founded fear of persecution-is afforded through the 1951 Refugee Convention and its 1967 Protocol. People who were displaced due to environmental causes or climate change do not fall under this definition; therefore, they are not accorded any formal legal status or protection under international refugee law (Draper & McKinnon, 2018).

Human Rights Frameworks: Human rights law itself provides a broader framework for the protection of persons affected by climate-induced displacement. Fundamental international human rights instruments, such as the Universal Declaration of Human Rights and the International Covenant on Economic, Social and Cultural Rights provide life, security, and an adequate standard of living that may be in jeopardy throughout the process of climate-induced displacement. Regional human rights bodies, such as the Inter-American Court of Human Rights have also articulated this connection, articulating that states have an obligation to protect the rights of people displaced due to climate change (Almulhim et al., 2024).

Creation of New Legal Instruments: Deficiencies in existing legal regimes have generated different proposals for new legal instruments which will grant protection to people displaced by climatic change. The most notable include demands by some scholars and policy thinkers for a new international convention on climate-induced displacement with special legal protection and rights of the concerned populations. Others have suggested the expansion of the definition of "refugee" in the 1951 Refugee Convention to include those people displaced by environmental factors, although this way is fraught with political and legal hurdles (Ahsan, 2019).

5. Civil Society and Non-Governmental Organizations

These CSOs and NGOs are of immense contribution, especially in the case of climate-induced migration and displacement, for direct support of the affected population and for policy change.

Advocacy and Awareness-raising: Many CSOs and NGOs work to raise awareness about the impacts of climate change on migration and displacement, advocating for stronger legal protections and policy responses at national and international levels. Such organizations conduct research and documentation, bringing into light experiences and needs of the displaced population, besides calling for their inclusion in policy discussions and decision-making processes.

Direct Aid and Assistance: Communities that are actually displaced or induced to displace due to climate change often require the essential assistance of NGOs in the form of emergency relief, legal assistance, and livelihood support. Organizations such as the International Organization for Migration and the Norwegian Refugee Council provide shelter, food, water, and health care to the displaced persons while supporting the rebuilding and recovery efforts of those who were displaced.

Community-Based Adaptation and Resilience Building: In addition to such policy-driven and local rapid response mechanisms, many NGOs also undertake community-based adaptation and resilience-building work, supporting vulnerable groups in developing locally led responses. Often, this involves sustainable livelihoods development, enhanced disaster preparedness, and community building for social cohesion that aids community adjustments to an altered environment, thus reducing their need for such relocation.

Effective policy responses and legal frameworks are needed to address these increasingly complex challenges. Despite progress at the national, regional, and international levels, there are still gaps in the provision of legal protection and policy responses for displaced persons. Further development of such frameworks and deeper cooperation among governments, international organizations, and civil society would further support the affected communities in ensuring that the migration and displacement arising from the impacts of climate change are managed in a way that builds resilience and sustainability with human rights (Ahsan, 2019).

Therefore, with rising climate-induced migration and displacement, increasing resilience and adaptive capacity becomes two of the major strategies that help develop a mitigative approach toward climate change impacts on communities worldwide (Jayawardhan, 2017).

Resilience and Adaptation Strategies

Resilience and adaptation are broad terms that cover measures, infrastructural enhancement, sustainable resource management, community-based initiatives, and policy interventions. These strategies are designed to reduce vulnerability, enhance the ability to cope at a community level, and ensure that the solutions are practically sound and feasible; hence, they prevent the worst forms of forced migration. This section explores diverse resilience and adaptation strategies; it also points out how such approaches are to be integrated into broader development and climate-action frameworks (Draper & McKinnon, 2018).

1. Community-Based Adaptation

Community-based adaptation empowers local communities to design and undertake adaptation strategies in their context and culture, according to their specific needs. Community-based adaptation gives recognition to the fact that communities have unique knowledge and skills relevant for conceiving and designing effective and sustainable measures of adaptation.

Local Knowledge and Practices: Local tales, knowledge, and practices vary depending on the available knowledge use and honing through generations on how to cope with changes in surroundings. For instance, across much of the Pacific Islands, traditional farming practices are resilient to extreme events such as agroforestry raised-bed agriculture. Such traditional practices can be integrated into modern adaptation strategies and further help communities in building resilience to climate change, conserving cultural heritage.

Participatory Planning and Decision-Making: Community participatory planning entails the identification of areas of vulnerability, the setting of priorities, and the design of adaptation measures. An effective CBA has to make sure that the proposed adaptation strategies are locally relevant, culturally appropriate, and well-accepted by the community. For example, in Bangladesh, it has developed community-led processes to construct houses that can resist floods, and locally-developed early warning of cyclones (Almulhim et al., 2024).

This will be achieved through the strengthening of social networks and institutions. Building social capital and strengthening community institutions are core features of CBA. Strong social networks and institutions facilitate improved collective action, information sharing, and mutual help in times of crises. In Kenya, for example, community-based organizations have played an important role in the promotion of sustainable practices in water management and the provision of support to households affected by drought, enhancing resilience to climate-induced displacement.

Infrastructure Development and Technological Innovations

Resilient investment in infrastructure coupled with the adoption of technological innovations holds a primary key towards the lowering of vulnerability to climate-induced displacement and a boost in adaptive capacity. This will help the communities in being safe from extreme weather conditions, improved access to basic amenities and resources, as well as sustainable development.

Resilient Infrastructure: Building resilient infrastructure—such as flood defenses, stormwater management systems, and climate-resilient housing—is a key effort that is needed to protect communities from the impacts of climate change. In the coastal city of New Orleans, the main adaptation measure is through the use of levees and walls to bar flooding from hurricanes and additional storm surges. In addition, in Netherland, dikes, storm surge barriers, and water storage systems have also been developed to protect low-lying areas from additional levels of sea-level rise and flooding.

Sustainable Water and Agricultural Management: Sustainable water and agricultural management as well as water use practices are among the most important ways of increasing coping capabilities due to improving both rural livelihoods and economies of regions mainly vulnerable to droughts and water scarcity. Examples of these techniques are rainwater harvesting, drip irrigation, and varietal crops resistant to drought. For example, in Ethiopia, due to terracing and soil conservation, much erosion of soil has been reduced, and water percolation in agriculture has improved, therefore supporting other livelihoods and reducing the level of migration.

Technological Innovations: Such technological improvement innovations as early-warning systems and climate-smart agriculture and renewable energy solutions help improve adaptive capacity and, thus, reduce vulnerability to populations vulnerable to the effects of climate change. Early warning systems on natural disasters such as tsunamis, hurricanes, and floods can provide essential information to such populations at risk, ensuring that they evacuate in time and lessen the possibility of suffering displacement. Countries are now implementing advanced meteorological systems and mobile-based alert services for disaster preparedness and response, saving lives and reducing displacement in places such as India (Jayawardhan, 2017).

3. Sustainable Livelihoods and Diversification

This is through sustainable livelihood promotion and income diversification among many other tools of adaptation. This significantly reduces climate change-induced human mobility. All these kinds of livelihood diversification may provide better means to cope with environmental shocks and stresses at the community level, enhancing the community capacity to be better able to withstand new conditions.

Sustainable agriculture and fisheries practices will help uphold the practice of maintaining and averting food security and livelihoods within communities pertaining to climate change. For instance, in a coastal area, there can be the practice of aquaculture and mariculture adopted to offer alternative income to communities that are affected by dwindling fish stocks because of warming and acidification of the ocean. In Vietnam, the promotion of integrated rice-fish farming systems contributes to farmers' adapting to new water regimes and thereby enhancing food security with less stress on marine resources (Almulhim et al., 2024).

Nonagricultural Livelihoods: Promotion of non-agricultural options of livelihood in areas of small-scale manufacturing, tourism, services work in reducing dependence on climatic-sensitive sectors and improves economic resilience. Some countries have designed programs providing training and support for alternative livelihoods, for example, those in handicrafts, eco-tourism, and digital services. In Morocco, for example, co-operatives of women have been made to produce and market the argan oil, which is an alternative source of income in drought-affected and land degraded rural communities (Draper & McKinnon, 2018).

Financial Services Access: Access to microfinance, insurance, and savings programs is important in supporting sustainable livelihoods and enhancing adaptive capacity. Financial inclusion helps households invest in adaptation measures, such as climate-resilient seeds and irrigation systems, and acts as a safety net in times of crisis. For instance, in Bangladesh, as an effect of this financial access, through microfinance programs, smallholder farmers have afforded climate-resilient practices and made investments in other income-generating activities, in turn reducing vulnerability to displacement due to climate change (Draper, 2020).

4. Policy Integration and Governance

In other words, climate adaptation and resilience strategies integrated into larger development policies could act as an enabler of governance supporting every possible effective response to the changes in climate during human migration and displacement. Coherent policies and solid institutions will support the actual implementation of adaptation measures, coordination, and involvement of the most vulnerable populations.

Mainstreaming Climate Adaptation into Development Planning: This process helps ensure that climate adaptation is mainstreamed into national and local development planning, whereby climate resilience is a consideration in all policy decision areas and resource allocations. This will go a long way in ensuring an integrated response to climate change through addressing the root causes of vulnerability, thus reducing the risks of displacement. The issue in point is, for example, the Rwandan national development strategy that incorporates climate resilience into the major priorities for economic growth and poverty reduction, thereby focusing on the sustainable land management, renewable energy, and disaster risk reduction subsectors (Ahsan, 2019).

He also said that the pathways of institutional capacity enhancement would lead to the development of the appropriate institutions at the different layers of government and thus enable it to deal with climate adaptation in a very effective manner to build resilience. Higher institutional capacity for coordination across sectors, ensuring implementation of adaptation measures, and mobilizing resources is used for taking climate action. For example, the establishment of the Climate Change Commission has significantly strengthened the country's ability to coordinate climate policy and implement adaptation initiatives in the Philippines throughout various government agencies and sectors (Jayawardhan, 2017).

Inclusiveness in governance and stakeholder engagement are fundamentally considered in developing adaptation measures and delivering social cohesion. Broad stakeholder engagement, from local communities to civil society organizations, and from the private sector to groups of beneficiaries, increases the legitimacy of adaptation measures. In South Africa, for instance, the government has developed multi-stakeholder platforms for climate adaptation planning that include voices from the level of local communities, indigenous people, and women's groups (Draper, 2020).

5. International Cooperation and Support

International cooperation and support constitute integral components in addressing the global dimensions of climate-induced migration and displacement and improving resilience and adaptation efforts in more vulnerable countries. Adaptation initiatives are supported by global frameworks, funding mechanisms, and partnerships through much-needed resources, knowledge, and technical assistance (Almulhim et al., 2024).

Global Frameworks and Agreements: International agreements that underpin collaboration in adapting to climate and building resilience include the Paris Agreement and the Sendai Framework for Disaster Risk Reduction. These frameworks call for an integrated approach to tackling climate change, disaster risks, and sustainable development, thus calling for coherent and coordinated actions among countries and regions (Draper & McKinnon, 2018).

Financial Support and Technology Transfer: Financial support and the transfer of technology from developed to developing countries are quite significant in enabling vulnerable countries to adapt effectively. International financial backing mechanisms, such as the Green Climate Fund and the Global Environment Facility, have funded financial resources for projects that boost climate resilience and encourage sustainable growth. Moreover, technology transfer schemes such as the Climate Technology Centre and Network have been established to disseminate innovative solutions and best practices for adaptation.

Other elements of international cooperation include activities such as capacity building and knowledge sharing, leading to the better preparation and ability to develop and implement adaptation strategies within countries. Indeed, local capacity can be enhanced by training programs, technical assistance, and collaborative research on adaptation to climate change and resiliency. For instance, the Global Adaptation Network (GAN) is one such global platform that involves sharing good practices around climate adaptation knowledge among policymakers, practitioners, and researchers globally (Ahsan, 2019).

CONCLUSION

Both resilience and adaptation have to play a significant part in forming the entire approach to managing impacts from climate-induced displacement/migration. Investment in development of community-driven adaptation, resilient infrastructure, and sustainable livelihoods, together with policy integration and international cooperation by governments, organizations, and communities, would enhance the ability of the people to cope with the challenges of climate change and thereby reduce displacement. These strategies have to be made population specific; therefore, they have to be inclusive, sustainable, and in line with broader development and climate action goals (Draper, 2020).

Future Projections and Scenarios

Policymakers have to understand future projections and scenarios of climate-induced migration and displacement so that they prepare in their respective nations and work closely with researchers and humanitarian organizations to avert the climate risk impacts. These projections are prepared based on different factors—that is, climate models, socio-economic trends, and patterns of migration. In considering potential future scenarios, stakeholders can anticipate the magnitude and character of displacement, identify the region and population at risk, and elaborate strategies required for the management of foreseen changes. The section goes on to describe current drivers of climate-induced migration and displacement, envisions possible outcomes and implications for global governance, and points to future projections and scenarios.

1. Climate Change and Migration Projections

Future projections of climate-induced migration and displacement are inherently uncertain, given that they depend on a set of factors including the magnitude of climate change, the effectiveness of adaptation measures, and socioeconomic conditions. Nonetheless, climate models and research provide valuable insights into possible future trends and patterns of migration and displacement (Almulhim et al., 2024).

Increased Temperatures and Events of Extreme Weather: With increasing temperatures throughout the world, extreme weather events in the form of hurricanes, floods, droughts, or heatwaves are likely to be experienced more frequently and intensely. These are bound to make natural vulnerabilities more profound and result in frequent and intense displacement. In fact, the studies carried out by the World Bank take into consideration that climate change may displace more than 216 million people internally across six major regions by 2050 due to climate change.

The region includes Sub-Saharan Africa, South Asia, and Latin America (Draper & McKinnon, 2018).

Sea Level Rise and Coastal Erosion: SLR is amongst the greatest contributors to long-term displacement, especially for low-lying coasts and small island states. According to the Intergovernmental Panel on Climate Change projections for high-emission scenarios, global sea levels could rise up to 1 meter by the end of the century. Such a rise in sea level will displace many people who live in low-lying areas within coastal cities, towns, and countryside with densely inhabited and economically fragile regions like South and Southeast Asia, the Pacific Islands, and areas of Africa. In fact, by 2100, it is said that there could be over 150 million displaced people because of rising sea levels alone (Ahsan, 2019).

Desertification and Land Degradation: Desertification and land degradation, to be driven by higher temperatures and changing precipitation patterns, are projected to impact agricultural productivity and food security, hence heightened migration from the affected regions. Places that are already facing desertification, such as the Sahel region in Africa, will probably see continued higher levels of migration as communities move in search of arable land, water, and livelihoods. According to the United Nations Convention to Combat Desertification, up to 135 million people could be displaced by desertification in 2045 if current trends continue.

Water scarcity and resource conflicts: With climate change, the water scarcity situation is likely to increase, especially in agriculture-based and countries with low water resources. These conditions will create competition for the available water resources, leading to conflicts such as what is currently being experienced in MENA, South Asia, and Sub-Saharan Africa, which could also be a cause of forced migration. According to the World Resources Institute, by 2040, 33 countries might face extremely high water stress and, consequently, increased migration pressures. 2. Socioeconomic Scenarios and Migration Outcomes The impacts of climate change on migration and displacement are not uniform and will differ depending on the socioeconomic context, governance, and directions of development. A number of scenarios illustrate how different drivers may interact at various levels in influencing future migration patterns and consequences (Draper, 2020).

High Adaptation, Low Emission Scenario: This would involve a very low level of emissions with strong adaptation measures that could possibly minimize impacts related to migration and displacement due to climate change. Successful adaptation strategies, which include improved infrastructure, better management of resources, and social protection programs, could enhance resilience and reduce the need for migration. The migration, in this regard, would be much more controlled and planned, with proactive movements being a part of adaptation strategies rather than flights compelled by extreme events or resource scarcity (Almulhim et al., 2024).

Low Adaptation, High Emissions Scenario: On the contrary, a high-emission scenario with low adaptations could be visualized as much more associated with unmanaged displacement in a wide sense. Climate change impacts would be stronger, disasters more frequent and intense, and sea levels would be high under this scenario. This would also make the communities' adaptive capacity low, leading to displacement amidst weak governance, poor conditions of infrastructure, and poverty. In this scenario, migration is likely to become more chaotic, unplanned, and with high humanitarian and social challenges (Draper & McKinnon, 2018).

The key driver of migration outcomes is socio-economic inequality. This means that in a future characterized by prevailing or increased inequality, the populations that are most vulnerable to climate-induced displacement-low-income communities, women, children, and marginalized groups among others-would be hardest hit. These are the populations that do not have substantial resources to safely adapt or migrate, hence being pushed towards exposure to forced displacement and its risks. On the other hand, in wealthier communities, there is the possibility of adaptation and voluntary migration, including alternatives to migration, such as relocation to less vulnerable zones or investing in protection measures (Ahsan, 2019).

Urbanization and Internal Displacement: A second structural driver that will shape future migration flows is rapid urbanization. While cities are already growing and will continue to be a destination for migrants, they are also found to be in areas that are prone to climate impacts. The migrating people would add to their load, apart from the vulnerabilities of cities to climate change, such as flooding, heat stress, and shortage of infrastructure. In other probable and future scenarios, internal displacement within countries could rise, or people might migrate from rural to urban areas, or within a city to its parts due to the impacts of climate change (Jayawardhan, 2017).

3. Regional Hotspots and Migration Corridors

Indeed, some regions and areas of passage can easily be especially hotspots developed under a mix of environmental, socio-economic, and political impetus to climate-induced migration and displacement. It is through the identification of such hotspots and their respective corridors that potential migration may flow that targeted policy responses and resource allocation can be guided.

South Asia: A large part of its population lives in low-lying coastal areas and river deltas, as well as in arid regions, making it one of the potential hotspots regarding climate-induced displacement. In Bangladesh, India, and Pakistan, for instance, increased flooding and heat stress due to sea-level rise could intensify both internal and cross-border migration. The tendency of migration from rural to urban areas will be more pronounced, like in Dhaka, Bangladesh, and Mumbai, India, as people seek safety and livelihoods in cities.

Sub-Saharan Africa: The next region that may experience significant migration on the grounds of climatic changes is Sub-Saharan Africa. This, in turn, will be the result of increased desertification and serious water shortage extreme weather events. In particular, the Sahel region is likely to see increased migration, with pastoralists and farmers moving in search of grazing land, water, and livelihoods. Rural-to-urban and cross-border migration corridors may become more pronounced, with destinations such as Lagos in Nigeria and Addis Ababa in Ethiopia receiving a number of displaced populations.

The **Pacific Islands and the Small Island Developing States (SIDS)** include some of the most vulnerable nations to climate-induced displacement because of the rise in sea levels, storm surges, and lack of available land. Countries such as Kiribati, Tuvalu, and the Maldives are also so vulnerable that in some cases, whole populations may have to be moved if current trends are allowed to continue. The migration corridors from these islands to larger, more stable countries like New Zealand and Australia might become increasingly relevant in search of safety and stability.

Middle East and North Africa (MENA): MENA's region already shows high levels of water scarcity and political instability; this area is envisaged to have additional pressure for migration due to climate change. With increasingly scarce water resources and decreased agricultural productivity, increased migration in Syria, Yemen, and Egypt could also be escalated. Rural-to-urban and a cross-border migration corridors to neighbors and Europe could be huge in the coming decades (Draper & McKinnon, 2018).

4. Implications for Global Governance and Policy

Future projections and scenarios of climate-induced migration and displacement create a wide range of implications for global governance and policy. Reckoning with these budding challenges requires coordination at all levels-local, national, regional, and international-to build resilience, achieve sustainable development, and realize human rights.

This means viewing climate-induced migration within a global framework by strengthening international cooperation. While the GCM and the Paris Agreement develop some sort of intergovernmental framework for action at the global level, this needs to be more robust in light of the challenges thrown up by climate-induced displacement. Strengthening cooperation on data sharing, research, and policy development will provide countries with a deeper understanding of migration dynamics.

Legal protection needs to be increased for displaced persons and the enforcement of human rights due to climate change. The already existing international legal frameworks, including the 1951 Refugee Convention, do not explicitly refer to displacement caused by climate change-that is why many displaced persons do not get proper protection. The development of new legal instruments or the expansion of existing ones regarding climate-displaced persons would provide necessary protection and rights for people affected by this disaster (Ahsan, 2019).

Promoting Sustainable Development and Resilience Building: Addressing climate-induced migration and displacement at the source requires focusing on sustainable development and building resilience. Adaptation, disaster risk reduction, and livelihood investment reduce vulnerability by building the resilience of communities to resist climate impacts. Climate resilience will have to be built into national development plans, infrastructure projects, and social protection programs if long-term benefits are to be achieved with reduced potential for forced migration (Almulhim et al., 2024).

Preparing for Future Displacement Scenarios: With the ever-changing dynamics of displacement scenarios, policymakers and humanitarian organizations have to be prepared with contingency plans, early warning system enhancements, and response capacity expansion. This is inclusive of both short-term, sudden-onset displacement due to extreme weather events and long-term, gradual displacement due to slow-onset changes like sea-level rise. Anticipatory planning and preparedness can reduce the humanitarian and social impacts of displacement; it would ensure that affected populations receive timely support and resources commensurate with their needs (Draper & McKinnon, 2018).

Future projections and scenarios of climate-induced migration and displacement show the urgent need to adopt proactive and comprehensive approaches in dealing with climate change challenges. Only by having a vision into how things may look in the future can policymakers, researchers, and humanitarian organizations develop targeted strategies to protect vulnerable populations in ways that foster resilience and assure sustainable development in a changing climate (Ahsan, 2019).

CONCLUSION

Climate change is a factor both in migration and displacement; these are complex issues that need urgent, multi-faceted measures. Drivers of migration-extreme weather events, sea-level rise, desertification, and resource scarcity-are likely to further increase in the future with continued climate change, along with associated important social and economic consequences for affected populations. The projected future scenarios discussed underscore how important this understanding is to preparation for the efficient averting of climate-induced displacement impacts. Evidence shows that the scale and scope of climate-induced migration are bound to expand if the present trends continue unabated. The projected increases in extreme weather events, rising sea levels, and resource scarcity make the needs for robust adaptation and resilience strategies an urgent call. Climate change is likely to displace people in astronomical numbers in the absence of vigorous measures to reduce gas emissions and build any noticeable adaptive capacity. Increased vulnerability and strain on resources and infrastructure will be felt in areas of origin and destination alike. The challenge posed by climate migration is multi-dimensional. Among these, community-based adaptation initiatives take a vital place, since they ensure that responses originate from and are led at the level of local populations with meaningful participation to develop and implement contextual and culturally appropriate strategies. Resilient infrastructure and technological innovations are also crucial to provide the wherewithal both physically and technically to resist climate impacts and sustain communities affected. Further, sustainable livelihoods ensure lesser dependence on climate-sensitive sectors due to enhanced economic resilience, hence reducing the need for forced migration.

Policy integration and effective governance are, therefore, at the heart of addressing climate-induced displacement. Developing planners at the national and sub-national level can harness a more coherent and sustainable response to the challenges posed by climate change. Building institutional capacity and supporting inclusive governance help ensure that adaptation is effectively undertaken and all stakeholders' needs, especially the vulnerable population, are met.

International cooperation and support are also exceedingly important in addressing the global dimensions of climate-induced migration. Global frameworks, such as the Paris Agreement and the Global Compact for Migration, provide a foundation for coordinated action that needs strengthening through increased financial support, technology transfer, and capacity building. In this way, countries will be in a position to share knowledge, resources, and best practices that will help solve the problem of climate-induced displacement and support the populations concerned.

The one thing that becomes clear, looking into the future, is planning and preparedness. Thoughtful consideration and action with regards to future scenarios of climate-induced migration will have to be made in order to reduce humanitarian and social impacts associated with displacement. For any management of anticipated increases in displacement, developing comprehensive contingency plans, improving early warning systems, and building response capacities will have to become critical elements in ensuring support for those affected.

Conclusively, climate-induced migration and displacement remain the immediate concern of humanity, to be engaged by various streams and levels of governance simultaneously. The integration of resilience and adaptation strategies, enhancement of policy frameworks, and multinational cooperation in the face of climate change impacts will make better preparedness and response more possible with a view toward supporting vulnerable communities and embarking on a more sustainable and equitable future for all.

REFERENCES

Ahsan, R. (2019). Climate-induced migration: Impacts on social structures and justice in Bangladesh. *South Asia Research*, 39(2), 184–201. DOI: 10.1177/0262728019842968

Ahsan, R., Kellett, J., & Karuppannan, S. (2014). *Climate induced migration: Lessons from Bangladesh* (Doctoral dissertation, Common Ground Publishing).

Aleinikoff, T. A. (2024). Climate-Induced Displacement and the International Protection of Forced Migrants. *Social Research*, 91(2), 421–444. DOI: 10.1353/sor.2024.a930749

Almulhim, A. I., Alverio, G. N., Sharifi, A., Shaw, R., Huq, S., Mahmud, M. J., ... Abubakar, I. R. (2024). Climate-induced migration in the Global South: an in depth analysis. *NPJ Climate Action, 3*(1), 47.

Askland, H. H., Shannon, B., Chiong, R., Lockart, N., Maguire, A., Rich, J., & Groizard, J. (2022). Beyond migration: A critical review of climate change induced displacement. *Environmental Sociology*, 8(3), 267–278. DOI: 10.1080/23251042.2022.2042888

Bose, P., & Lunstrum, E. (2012). Environmentally induced displacement and forced migration. *Refuge: Canada's Periodical on Refugees*, 29(2), 5–10. DOI: 10.25071/1920-7336.38163

Choudhury, A., & Shahi, S. K. (2024, June). Climate-Induced Displacement and Sustainable Development. In *NDIEAS-2024 International Symposium on New Dimensions and Ideas in Environmental Anthropology-2024 (NDIEAS 2024)* (pp. 51-64). Atlantis Press. DOI: 10.2991/978-2-38476-255-2_5

Draper, J. (2020). *Justice in Climate-Induced Migration and Displacement* (Doctoral dissertation, University of Reading).

Draper, J., & McKinnon, C. (2018). The ethics of climate-induced community displacement and resettlement. *Wiley Interdisciplinary Reviews: Climate Change*, 9(3), e519. DOI: 10.1002/wcc.519

Ferreira, V. (2017). Climate induced migration: Legal challenges. *Intergenerational responsibility in the 21st century*, 107-121.

Francis, A. (2019). Climate-induced migration & free movement agreements. *Journal of International Affairs*, 73(1), 123–134.

Gemenne, F. (2011). Climate-induced population displacements in a 4 C+ world. *Philosophical Transactions of the Royal Society A: Mathematical, Physical and Engineering Sciences, 369*(1934), 182-195.

Islam, M. R., & Hasan, M. (2016). Climate-induced human displacement: A case study of Cyclone Aila in the south-west coastal region of Bangladesh. *Natural Hazards*, 81(2), 1051–1071. DOI: 10.1007/s11069-015-2119-6

Islam, M. R., & Shamsuddoha, M. (2017). Socioeconomic consequences of climate induced human displacement and migration in Bangladesh. *International Sociology*, 32(3), 277–298. DOI: 10.1177/0268580917693173

Jayawardhan, S. (2017). Vulnerability and climate change induced human displacement. *Consilience*, (17), 103–142.

Kälin, W. (2010). Conceptualising climate-induced displacement. *Climate change and displacement: Multidisciplinary perspectives, 81*, 102.

Lyster, R., & Burkett, M. (2018). Climate-induced displacement and climate disaster law: Barriers and opportunities. In *Research handbook on climate disaster law* (pp. 97–114). Edward Elgar Publishing. DOI: 10.4337/9781786430038.00012

Mayer, B. (2011). The international legal challenges of climate-induced migration: Proposal for an international legal framework. *Colo. J. Int'l Envtl. L. & Pol'y*, 22, 357.

McMichael, A., McMichael, C., Berry, H., & Bowen, K. (2010). *Climate-related displacement: health risks and responses. Climate Change and Population Displacement: Multidisciplinary Perspectives*. Hart Publishing Ltd.

Mehedi, H., Nag, A. K., & Farhana, S. (2010). *Climate Induced Displacement*. Case Study of Cyclone Aila in the Southwest Coastal Region of Bangladesh.

Mostaque, L. Y., & Hasan, S. (2015). Climate Change Induced Migration: Impact on Slumaisation. *Participatory Community Assessment for Priority Problem Diagnosis in Bajura District, Nepal: What Matters Most–Poverty or Climate Change?* 223.

Naser, M. M., & Afroz, T. (2009). Human rights implications of climate change induced displacement. *Bond L. Rev.*, 21(3), i. DOI: 10.53300/001c.5543

Oluwaseyi, J., & Stilinski, D. (2024). *The Impact of Climate Change on International Migration: Analyzing the Social*. Political, and Economic Consequences of Climate-Induced Displacement.

Poddar, A. K. (2024). Climate Change and Migration: Developing Policies to Address the Growing Challenge of Climate-Induced Displacement. *The International Journal of Climate Change*, 16(1), 149–170. DOI: 10.18848/1835-7156/CGP/v16i01/149-170

Salik, K. M., Shabbir, M., & Naeem, K. (2020). Climate-induced displacement and migration in Pakistan: Insights from Muzaffargarh and Tharparkar Districts.

Schwan, S., & Yu, X. (2018). Social protection as a strategy to address climate-induced migration. *International Journal of Climate Change Strategies and Management*, 10(1), 43–64. DOI: 10.1108/IJCCSM-01-2017-0019

Teye, J. K., & Nikoi, E. G. (2022). Climate-induced migration in West Africa. In *Migration in West Africa: IMISCOE regional reader* (pp. 79–105). Springer International Publishing. DOI: 10.1007/978-3-030-97322-3_5

Thomas, A., & Benjamin, L. (2018). Policies and mechanisms to address climate-induced migration and displacement in Pacific and Caribbean small island developing states. *International Journal of Climate Change Strategies and Management*, 10(1), 86–104. DOI: 10.1108/IJCCSM-03-2017-0055

Tremblay, M., & Trudel, M. È. S. O. (2013). The Climate-Induced Migration: What Protection for Displaced People? *The International Journal of Climate Change*, 4(4), 67–81. DOI: 10.18848/1835-7156/CGP/v04i04/57870

Chapter 4
Impact of Climate Change on Mental Health and Economic Consequences of Climate-Related Mental Health Issues

Preet Kanwal
https://orcid.org/0009-0006-5114-8381
Lovely Professional University, India

Shashank Mittal
O.P. JIndal Global University, India

Hewawasam P. G. D. Wijethilaka
https://orcid.org/0009-0006-9611-5735
University of Colombo, Sri Lanka

ABSTRACT

Climate change poses significant challenges to mental health, impacting individuals and communities through extreme weather events, gradual environmental changes, and socio-economic disruptions. This chapter explores the complex relationship between climate change and mental health, highlighting direct and indirect pathways through which environmental factors exacerbate psychological distress. Case studies from various regions illustrate the widespread nature of these impacts, including increased rates of anxiety, depression, and PTSD. The chapter discusses practical, managerial, ethical, and societal implications, emphasizing the need for integrated strategies that include mental health services in disaster response and climate

DOI: 10.4018/979-8-3693-5792-7.ch004

adaptation efforts. Future directions involve addressing research gaps, leveraging technological innovations, and fostering cross-sector collaboration. By understanding and addressing these mental health impacts, we can enhance resilience and support affected communities in the face of climate change.

INTRODUCTION

Climate change is one of the ultimate global challenges of our century, given that its impacts have gone beyond mere physical environmental change to a level of human mental and emotional psychology. Whereas the environmental impacts of climate change, namely the rising temperatures, extreme events, shifting ecosystems, among other similar factors, have received an appropriate level of awareness, how the phenomenon relates to mental health remains an important area where much needs to be understood. Knowledge about how climate change impacts mental health is essential to establish comprehensive strategies pertaining to the immediate and long-term effects on individuals and societies.

Climate change is multi-dimensional in its impact on mental health and serious. Severe weather events such as hurricanes, flooding, and wildfires can cause immediate psychological trauma and long-lasting stress. Events of this kind can also leave survivors with PTSD, anxiety, and depression in the aftermath of natural catastrophes as they try to cope with lost homes and livelihoods. Chronic stress resulting from active environmental changes-such as gradual landscape degradation and the threat of future disasters-may be part of a pervasive sense of anxiety and hopelessness. This results in uneasiness, 'eco-anxiety' or 'climate anxiety,' which increasingly is recognized as a valid psychological condition with grave implications for mental health.

Apart from that, the altered climate aggravates prevailing social and economic inequalities, which in turn impact mental health effects. The dynamics of climate change make vulnerable those low-income groups, indigenous peoples, and individuals of developing lands. These groups are routinely subjected to much more environmental damage and have fewer subsistence resources and avenues of recovery because of the additional stress that climate change places on them. As a result, they are more prone to mental health issues. This exacerbates inequality at its roots and thus heightens the total impact of climate change (Charlson et al., 2021; Baudon & Jachens, 2021).

Economic costs due to climate-related mental health are large and thus need consideration. These mental health disorders result in higher treatment costs and costs for conditions related to the disorder. Besides this fact, these mental problems may affect the productive workforce by impairing their work capacity; hence, productivity loss, absenteeism, and disability claims are incurred. These factors not only affect individuals but also extend into economic growth and stability. Understanding these economic implications informs the development of policies and interventions that address both the mental health and economic dimensions of climate change (Charlson et al., 2021; Clemens et al., 2020).

This chapter will review the interplay between climate change and mental health, looking at various ways through which the direct and indirect pathways of environmental changes affect psychological well-being. We will also look at the economic repercussions of mental health complications stemming from climate change in order to outline the requirement for effective policy responses and interventions. Drawing on this comprehensive approach, our aim is to add to the growing complexity of understanding how climate change influences mental health and to help further strategies that reduce these impacts and help affected populations cope.

The Link Between Climate Change and Mental Health

Climate change strongly influences mental health through various direct and indirect means. Understanding these linkages is important in addressing the complete spectrum of impacts that environmental changes have on psychological well-being.

Direct Impacts: Extreme Weather Events and Trauma

The immediate way that climate change would affect mental health is through extreme weather events. According to recent changes in climate, all natural disasters like hurricanes, floods, wildfires, and heatwaves have increased in severity and frequency. These events can immediately cause acute psychological trauma among the affected persons. Common manifestations of mental health issues among survivors are post-traumatic stress disorder, anxiety neurosis, and depression. The trauma of having lost homes, possessions, and loved ones can be so long-lasting, disrupting the lives of individuals and building up a feeling of instability and loss ideation (Shoib, 2023; Nicholas et al., 2020; Yellowlees, 2022; Schwartz et al., 2022).

Other factors contributing to the state of stress include the intensity and frequency of these extreme events. The constant threat of these disasters continuing into the future can build chronic anxiety and a feeling of vulnerability across the board. This continuous exposure to environmental threats further worsens mental health issues and contributes to a state of psychological distress that is hard to overcome (Nicholas et al., 2020; Schwartz et al., 2022; Obradovich et al., 2018).

Indirect Impacts: Anxiety, Depression, and Stress Related to Climate Change

Beyond the immediate effects of natural disasters, however, climate change impacts mental health through more oblique channels. The incremental and cumulative environmental degradation-in the forms, for example, of steadily increasing temperatures, rising sea levels, and biodiversity loss-serve as an all-pervading source of unease and feelings of helplessness, a feeling that is quite colloquially called "eco-anxiety" or "climate anxiety." It involves dread, despair, and concern about the future of the planet and all life on Earth.

Anxiety about ecologies manifests on a wide spectrum, from low-level concern to psychological anguish. Citizens concerned about climate change present symptoms similar to anxiety and depression clinical states, such as chronic concerns, disturbed sleep, and concentration problems. Such feelings of powerlessness over what is happening with the environment in most parts of Earth engender hopelessness and despair, which in turn contributes to the worsening of mental health conditions (Goudet, 2024; Seritan & Seritan, 2020).

Vulnerable Populations: Disproportionate Effects on Specific Groups

The consequences of climate change to mental health are not spread evenly across populations. Vulnerable populations-low-income communities, indigenous peoples, elderly individuals, and those with prior existing mental health conditions-endure a disproportionate share of the climate-related stresses. These populations often face an increased exposure to environmental hazards and have fewer resources with which to manage and recover from climate-related impacts.

Low-income communities could live in areas more prone to environmental hazards and with less mental health service and support available. In Indigenous peoples, who have grown mostly with cultural and spiritual links to their lands, the degradation and displacement caused by environmental degradation can give rise to profound psychological distress. Older adults may experience further hardship as they try to cope with extreme weather events and emergent health burdens. Interventions and support such as these will, therefore, go a long way toward addressing their needs and hence mitigating the above-mentioned adverse mental health impacts of climate change on these vulnerable populations.

Conclusion The interplay of climate change with mental health is indeed complex since the points of interaction are both direct and indirect. Whereas extreme weather variability leads to immediate psychological trauma, continuous environmental changes make people continuously experience feelings of eco-anxiety and distress.

Vulnerable populations suffer the most, and a need exists for comprehensive and inclusive approaches that address mental health in view of climate change. Recognizing such links, taking them into consideration is very important in preparing effective strategies for safeguarding mental well-being in an increasingly uncertain environmental future.

Mechanisms of Impact

The pathways through which climate change influences mental health are multifaceted, being both physiologically, psychologically, and socioeconomically complex. Identification of these mechanisms will be instructive in attempting to devise strategies to mitigate the adverse impacts on mental health from climate change and in providing support for those affected.

Physiological Pathways: Stress Responses and Mental Health

Climate change has the potential to trigger physiological stress responses that may have direct effects on mental health. Extreme heat, air pollution, and natural disasters cause the stress response in the body to affect psychological well-being. For example, long-term exposure to high temperatures has been linked to increased cortisol levels, a stress hormone triggered by extreme temperature, which may be associated with disturbances in mood and mental health.

Heat waves and poor air quality could also contribute to increasing symptoms of pre-existing conditions such as cardiovascular and respiratory diseases. Research also documents that the physiological stress from these events also disrupts sleep patterns, adds to fatigue, and raises the vulnerability to mental health problems, including anxiety and depression.

Psychological Pathways: Fear, Uncertainty, and Eco-Anxiety

The psychosocial effects of climate change go well beyond the immediate impacts of weather extremes. Ongoing environmental changes, coupled with uncertainty about the future, create an environment in which a suite of psychological problems can occur. Perhaps most representative is eco-anxiety: the chronic concern and fear for the degradation of the environment and its consequences on future generations.

It may be brought about by a plethora of factors that include knowledge about environmental problems, media information on this issue, and involvement in climate catastrophes. Symptoms for eco-anxiety are similar to clinical anxiety disorders and may include chronic concern, attacks of panic, and helplessness. These feelings can further be exacerbated by such factors as uncertainty about the future and lack of control over events that happen with the environmental processes, which only adds to psychological distress (Hayes & Poland, 2018; Heeren et al., 2021).

Besides anxiety related to environmental aspects, a general overwhelming sense of loss and grief regarding the deterioration of nature can also influence mental health. It is the loss of natural landscapes, wildlife, and even cultural heritage that evokes deep feelings of sadness and mourning, hence influencing overall well-being and mental health (Schwartz et al., 2022; Gunasiri et al., 2022).

Socioeconomic Pathways: Displacement, Loss, and Inequality

Climate change enhances the current social and economic disparities, furthering the inequity in mental health. Some of the socioeconomic pathways through which the mental health of an individual could be affected by climate change include displacement, loss of livelihood, and increased inequality.

Displacement due to climate-related events, either through rising sea levels or extreme weather, is usually characterized by intense psychological distress. The loss of home and community from displacement disrupts social networks and support systems; feelings of isolation and anxiety result. Economic consequences of climate change, including job losses and reduction in income opportunities, further increase stress levels and heighten vulnerability to mental health issues (Charlson et al., 2021; Opoku et al., 2021; Hayes & Poland, 2018).

These also include socioeconomic consequences that are more devastating to the most vulnerable populations, particularly low-income communities and marginalized groups. These are usually more exposed to various environmental hazards and have limited capacity to resist and recover from weather-related stresses, exacerbating the inequalities presently experienced and increasing cases of mental health disorders in these groups.

Overall, the mechanisms of impact by which climate change affects mental health have proven to be varied and complexly entwined. Physiological stress responses, psychological factors such as eco-anxiety, and socioeconomic challenges all contribute to shaping mental health outcomes. Coming to understand these mechanisms is important in the development of targeted interventions and policies that will address the full spectrum of mental health impacts associated with climate change.

CASE STUDIES AND EVIDENCE

Caselet 1: Hurricane Katrina and Mental Health in New Orleans, USA

The Hurricane Katrina that hit New Orleans in August 2005 forms one of the important caselets toward understanding the impacts of severe weather on mental health. The hurricane caused massive destruction, forcing the displacement of thousands of residents, rendered them houseless, jobless, and bereft of all essential services. A lot of research that followed showed a phenomenal increase in the rates of post-traumatic stress disorder (PTSD), anxiety, and depression among the survivors. Prolonged recovery and displacement following the trauma of the disaster brought to the fore the psychological toll of such disasters.

Caselet 2: The 2019-2020 Australian Bushfires

The bushfires of Australia, from the end of 2019 through the early parts of 2020, are yet another powerful example of how climate change can have deep impacts on mental health. These fires engulfed wide areas, burned homes and wildlife habitats, and critical infrastructure. Research conducted in the affected regions depicts a significant rise in cases of psychological distress, as most individuals received symptoms of PTSD, anxiety, and depression. This was in addition to the long-lasting consequences of the fires themselves, including ongoing uncertainty regarding future fire hazard and recovery challenges, which exacerbated mental health issues for the afflicted. Caselet 3: Heatwaves and Mental Health in Athens, Greece A severe heatwave struck Athens, Greece during summer 2007. The heatwave introduced a new dimension of heat-related sicknesses and deaths. Research carried out on the psychological impact that this heatwave had on its residents revealed increased levels of stress and anxiety. Added to the extreme heat, health hazards, and disruption in life's routine, there was seen a surge in mental health issues. Thus, the study called for public health strategies that can focus on physical and psychological impact during such extreme temperatures.

Caselet 4: Flooding and Mental Health in Bangladesh

Bangladesh, due to its high vulnerable flood hazard, is really a very important case study in respect of the psychological impacts of climate-induced natural disasters. Annual monsoon floods displace communities, disrupt livelihoods, and destroy homes. Research in the flood-prone areas of Bangladesh has documented the higher prevalence rates of depression, anxiety, and PTSD among the affected populations.

This repetitive cycle of flooding and recovery consolidates chronic psychological stress, which itself has been calling for integrated mental health support as an intrinsic part of disaster response frameworks.

Caselet 5: Rising Sea Levels and Mental Health in Kiribati

Kiribati is a small, low-lying island nation in the Pacific that faces serious threats from rising sea levels associated with climate change. As seawater has gradually encroached on this land, communities have been displaced and arable land has been destroyed. In Kiribati, studies on the psychological effects have reported high levels of anxiety and despair among its residents due to having to contemplate losing their homes and ways of life. The slow yet inevitable rise in sea level maintains ongoing psychological stress and a great degree of powerlessness about the future.

Joint Analysis

These case studies represent a wide variety of mental health impacts from climate change, each with its specific environmental and socioeconomic contexts. There are similarities running across these caselets-that report the wide prevalence of PTSD, anxiety, and depression following extreme weather events, and the deep psychological toll of displacement and loss.

Examples include Hurricane Katrina in New Orleans, where the acute trauma was followed through with prolonged recovery and displacement, hence an acute and chronic combination of stressors resulting in poor mental health. Similarly, the bushfires in Australia and the heatwave in Greece both contributed to considerable mental health consequences due to extreme weather events, as evidenced by increased rates of psychological distress in each case. The floods in Bangladesh and sea-level rise in Kiribati offer insight into chronic psychological trauma associated with incremental environmental events that are recurring and gradual in nature. These mental health impacts detail the need for long-term consideration.

Taken together, these cases underscore the importance of integrated, multimodal mental health into disaster response and climate adaptation strategies. Undeniably, it will be important to address acute psychological impacts associated with extreme weather. Equally critical, however, are the ongoing needs of persons coping with continued environmental change and displacement. Analysis points out that successful mental health interventions take into consideration acute and chronic stressors while being tailored to specific needs and contexts in affected communities.

These diverse examples help create a deeper understanding of how climate change impacts mental health and develops necessary focused strategies to support the people affected. This joint analysis also underlines a holistic approach that addresses the environmental and psychological dimensions of climate change, ensuring adequate attention is given to mental health in climate adaptation and disaster responses (Gunasiri et al., 2022; Kumar et al., 2023).

Climate-Related Mental Health Issues: Economic Consequences

The economic consequences of climate-related mental health issues involve key and multi-faceted ramifications that affect the individuals themselves, the healthcare system, and wider economies. These range from direct to indirect costs, further emphasizing the urgent need for integrated approaches in addressing the mental health impacts of climate change.

Direct Costs: Healthcare Expenses and Treatment

The most significant economic impacts related to climate-related mental health are those that derive from increased demand for healthcare services. In many instances, when natural disasters associated with climate variability or environmental change occur, psychological diagnosis, counseling, and treatment are necessary. This adds to healthcare systems, forcing administrators to shift resources and focus on an increased demand for mental health services.

The treatments involved in the forms of therapy and medication besides hospitalization can make anxiety, depression, and PTSD costly. For example, studies have indicated that in the wake of disasters, expenditure as a result of mental health problems escalates. These associated individual and family financial burdens, further coupled with the burden to the public health systems, make a strong case for the adequate funding and resource allocation that are needed to provide mental health services in disaster-prone areas.

Indirect Costs: Loss of Productivity, Absenteeism, and Disability

Climate-related mental health problems also burden the economy indirectly through lost productivity and absenteeism. Individuals with mental health disorders may have difficulty fulfilling their job responsibilities or may generally be less productive at work. Anxiety, depression, and post-traumatic stress disorder symptoms can impact a worker's concentration, decision-making capability, and total work capacity, thus reducing efficiency and output.

Absenteeism is yet another major problem whereby persons suffering from severe mental health problems may take more frequent or longer leaves from work, reducing their personal incomes while employers and, eventually, the general economy are being dealt a blow. The many lost work hours and the need for temporary replacements add to the absenteeism costs that contribute to the general economic loss.

Besides productivity losses, mental health conditions result in increased claims for disability. Such severe cases of mental health disorders render a person disabled, hence commanding extra costs from social security systems and insurance providers. The chronicity of some of the mental health conditions may result in prolonged disability, thus increasing the economic burden.

Long-term Economic Consequences: Effects on Workforce and Economic Growth

The long-term economic consequences of climate-related mental health problems go far beyond the front-line health care costs and lost productivity. As repeatedly stated, prolonged states of mental health will influence the capability to enter the workforce fully and hence reduce career advancement and earning potential. This has wider ramifications for economic development generally, since a reduced workforce and lower productivity may impede economic growth and innovation.

It is also important to note that psychological effects of climate change can drive further socioeconomic inequality. Generally, low-income populations and those among marginal groupings tend to be more vulnerable to psychosocial stressors induced by climate change and have additional challenges in seeking services for mental health, further heightening economic inequalities-forcing the dint of economic growth and development on less than even terms (Xue, 2024; Liu et al., 2020).

These long-term economic impacts will have to be addressed in a multi-faceted approach, including investment in mental health support, promotion of resilience, and integrating mental health considerations into climate adaptation and disaster response strategies. If policymakers and stakeholders recognize and address such economic consequences, then better strategies to reduce the vulnerability of mental health to climate change may be developed that will help support overall economic stability and growth.

In all, the economic consequences of climate-related mental health issues involve individual, health system, and broad economywide direct and indirect costs. Appreciating these costs is quite critical in devising holistic strategies that support mental health and build economic resilience in the face of climate change (Nicholas et al., 2020; Hayes & Poland, 2018; Palinkas et al., 2020)

Policy and Interventions

Mental health impacts of climate change call for concerted, multi-modal solutions, ranging from policy formulation to interventions. Such strategies should integrate mental health concerns into climate adaptation and disaster response mechanisms for holistic consideration of the affected people's psychological needs. The section explores current policies, recommendations for improvement in policies, and programs that served as successful interventions.

Current Policies Addressing Mental Health and Climate Change

The many policies and frameworks that exist today address the issue of mental health in the context of climate change in fragmentary ways. Various countries have disaster response policies with embedded mental health components that take into account the psychological toll taken by natural disasters. The World Health Organization has developed guidelines for mental health and psychosocial support in emergencies-the framework on the issue of how to handle mental health needs during and after disasters.

Apart from that, some countries have integrated climate change into their public health policy. For instance, the Climate Change Act of the United Kingdom engages in adapting the country's public health system to address the health consequences related to mental health as far as climate change is concerned (Atta, 2024; Qin, 2024). In the same vein, the National Climate Resilience and Adaptation Strategy of Australia repeats this call in a dissemination plan which has given particular weight to the need for mental health support in the context of climate change and extreme weather events.

But even in this case, policies mostly do not detail mental health impacts and may be incomplete for vulnerable populations' needs. What is needed are more holistic and integrated approaches that link mental health support to climate adaptation and disaster preparedness (Xue, 2024; Liu et al., 2020).

Recommendations for Policy Improvement

A few important recommendations might help in formulating and improving policies that effectively address climate-related issues of mental health.

1. Include Mental Health in Climate Adaptation Plans: Mental health should be foregrounded in all climate adaptation and disaster response plans. This will involve the formulation of strategies that cover the issues of mental health responses to extreme weather events, altered environments, and displacement.

2. Scale Up Funding and Resources to Mental Health Services: Scaling up mental health services and support requires adequate funding and resources. This means that governments must commit specific funds to mental health initiatives due to climate change, in addition to ensuring that the resources reach the communities concerned.
3. Intersectoral Collaboration: The mental health effects of global climate change require a collaborative intersectoral response among the public health, climate science, disaster management, and mental health sectors. There are interdisciplinary approaches to provide more comprehensive solutions and help in giving a much-needed boost in enhancing the effectiveness of policies.
4. Develop Targeted Support for Vulnerable Populations: Policies should focus on the needs of vulnerable populations, including low-income communities, indigenous peoples, and people with pre-existing conditions of mental health. Specific interventions can provide substantial assistance in solving these problems in subgroups.
5. Improve Data Collection and Research: Improvements in data collection and research will provide fuller insights into the mental health impacts of climate change, apart from assessing the efficacy of interventions. This will be done through longitudinal studies that show evolving trends in mental health responses and determine the best practices for support.

Examples of Successful Interventions and Programs

The following interventions and programs serve as illustrations of effective support for mental health in relation to climate change:

1. The Australian Bushfires Mental Health Response: A full-scale mental health response was created as a result of the 2019-2020 bushfires in Australia, with increased funding for mental health services, opening of crisis support lines, and counseling and psychological support provided to the communities concerned. It also involved the training of local practitioners in dealing with disaster trauma.

2. The WHO's Mental Health and Psychosocial Support in Emergencies (MHPSS) Framework: The MHPSS framework by WHO provides guidelines for addressing mental health needs during emergencies, including those arising from climate-related disasters. The framework emphasizes community-based support, the integration of mental health into disaster response efforts, and training for healthcare workers.

3. The Climate and Health Alliance's (CAHA) Mental Health and Climate Change Network: Mental Health and Climate Change Network-CAHA provides a network for advocacy, research, and policy development relating to the mental health consequences of climate change. It also serves as a forum for raising awareness and promoting best practices to support the integration of mental health in climate and health policies.

4. Resilience and Wellbeing Programme in Bangladesh: The country of Bangladesh, with its recurrent flooding, has devised a program on resilience and wellbeing that offers mentally supportive service delivery in the communities affected. It incorporates community-based counseling, psycho-education, support of social networking, and livelihood development.

5. Island Resilience Initiative in Kiribati: Kiribati has implemented the Island Resilience Initiative to address mental health concerns that could emanate from the increase in sea levels. Some of these initiatives include, but are not limited to, mental health support services, community-based activities, and coping strategies for displacement and environmental change.

In a nutshell, what is needed is a multi-faceted solution that looks into the impacts of climate change on mental health by incorporating both climate adaptation and disaster response policies. Emendation of the available policies, resource allocation, encouragement of collaboration, and actual implementation of successful interventions are urgently called for to better support people and communities affected by the repercussions of climate-related mental health consequences.

Future Directions

Addressing the Mental Health Impacts of Climate Change: Forward-looking strategies and innovation in approaches are needed to better comprehend and mitigate the effects of climate change. Just as climate change continues to evolve, so too must the methods to support mental health in affected populations. The following are key future directions for research, innovation, and action that can help us better address the mental health challenges posed by climate change.

Research Gaps and Opportunities

1. Longitudinal Studies on Mental Health Impacts: The majority of research into the mental health impacts of climate change needs to adopt a longitudinal approach in order to understand their development and modulation by continuous environmental alteration. In this light, deeper insight will be afforded into the time course of development and continuing evolution of mental health problems associated with such environmental events; it will also bring important lessons into the development of more effective interventions and support systems.

2. **Impact of Climate Change on Specific Mental Health Conditions:** Further attention is needed regarding the effects of climate change on specific forms of mental disorders such as PTSD, anxiety, depression, and eco-anxiety. In such instances, a deeper understanding of the way in which climate change perpetuates these conditions provides a premise for targeted treatments and preventions (Innocenti et al., 2023; Hickman et al., 2021).

3. **Effectiveness of Different Interventions:** more research is needed on the effectiveness of interventions, including for mental health in light of climate change. For instance, the role of community-based support, telehealth services, and the integration of mental health services into disaster response frameworks are critical and needs to be clearly evaluated. The research needs to focus more on finding best practices which can also be scalable (Hickman et al., 2021; Ediz, 2023).

4. **Cultural and contextual variations:** Future research needs to debate how cultural and contextual factors influence the presentation of mental health and treatment outcomes. Knowledge of such variations may help to tailor mental health support for specific populations and enhance broader efficacy.

Innovations in Addressing Climate-Related Mental Health Issues

1. **Technology-Driven Solutions:** Developments in technology have opened new avenues for giving mental health support in the context of climate change. Similarly, modern Web-based digital platforms, mobile apps, and virtual reality can provide accessible one-stop-shop mental health support services, including teletherapy, mindfulness training, and stress management tools. Further explorations into the capabilities of such technologies can improve access to mental health service delivery, especially within isolated or underserved communities.

2. **Community-Based Approaches**: Creative approaches at the community level that build upon local wisdom and resources will provide solutions in trying to stem the tide of climate-related mental health problems. Community-led initiatives, peer support networks, and local resilience programs all have the potential to foster social cohesion and culturally supportive mental health.

3. **Integration of Mental Health into Climate Policies**: In this respect, embedding mental health into climate policies and plans for adaptation is of critical essence. Further work must be developed on a framework which directly spells out mental health needs within strategies of climate adaptation, ensuring that response and disaster preparedness plans will definitely incorporate mental health support.

4. **Cross-Sector Collaboration:** Collaboration among public health, climate science, mental health professionals, and policy makers can render more inclusive solutions. Cross-sectoral collaboration will help in knowledge sharing, resource sharing, and the sharing of best practices to respond more effectively and help better coordinate responses related to climate change and mental health.

Role of Technology and Data in Mitigating Impacts

1. **Data Collection and Analysis:** There is a greater need to consider how to improve data collection and analysis in the understanding of the mental health impacts of climate change. With modern data analytics and artificial intelligence, the improvement in data trends, assessment of risk factors, and anticipation of future mental health needs would make identification easier. Robust data systems and leveraging big data could put more teeth into our response to emergent mental health challenges.

2. **Predictive Modeling and Risk Analysis:** The predictive modeling technique will help in making assessments of the potential mental health impacts brought about by future climate scenarios. By the simulation of the various climate change scenarios and the respective potential effects which such changes could bring to mental health, one is better prepared for the psychological consequence of these changes in the environment.

3. **Personalized Mental Health Interventions:** Technology can allow for interventions in mental health that are personalized to meet individual needs and circumstances. For example, AI-powered tools can provide personalized suggestions related to mental health support, considering one's risk factors and preference, which can make the intervention more relevant and effective.

4. **Public Awareness and Education:** Another direction in which technology may be instrumental is that of raising public awareness and educating communities about the impacts of climate change on mental health. Social media, online campaigns, and educational platforms are methods through which the public may be better informed about issues surrounding climate-related mental health, the building of resilience, and proactivity in mental health care.

In all, future directions towards addressing climate-related mental health features involve a combination of research, innovation, and collaboration. Finding the gaps in research studies, exploring newer technologies, integrating mental health into climate policies, and using data will help build a greater capacity to support individuals and communities battling with the psychological impacts of climate change. Efforts will be translated to more effective and sustainable solutions, which means that mental health support will go with the changing climate.

Implications

This has far-reaching and deep implications on a number of levels. These include practical, managerial, ethical, and societal dimensions, all of which lie at the core of framing responses to the mental health effects of climate change.

Practical Implications

At a practical level, responding to mental health concerns related to climate change involves revisiting disaster response strategies and climate adaptation policies. Integration of mental health into the emergency response system in a timely manner is absolutely essential to provide efficient support to the affected people. This calls for infrastructural development in mental health facilities in disaster-prone areas. A specialized training and capacity building in managing climate trauma needs to be evolved among the mental health professionals. Practical implications include the requirement of proper data collecting and monitoring mechanisms required to comprehend and deal with the needs related to mental health. This will also be realized through the adoption of technology-driven solutions, for instance, telehealth and mobile mental health applications, which will be intending to increase care and support access in those areas that are either geographically remote or underserved. These steps are important in setting the foundation for a robust, responsive mental health support system befitting the climate change challenge.

Managerial Implications

A major implication of climate-related mental health issues for managers and organizational leaders is in developing and implementing policies that address the psychological well-being of the working population in relation to climate change. This can include workplace programs that support mental health, manage stress, and build resilience. It is also crucial that organizations consider the possibility of lower productivity and increased absenteeism due to climate-related mental health issues and plan mechanisms for offsetting this effect. Organizational risk management and business continuity need, for this reason, be adjusted to incorporate considerations of mental health to retain a workforce that can continue to be healthy and productive. Also, managers are advised to develop a supportive working environment, which is adapted to the psychological realities of a changing climate. This will further contribute to well-being and resilience among employees.

Ethical Considerations

The ethics underlying in the response to climate-related mental health concern equal access to mental health care and support. Many vulnerable populations, including low-income communities and marginalized groups, are more often highly deprived in regard to services on mental health and may, therefore, be more at risk due to climate change. Ethically, interventions and policies need to be designed in a way that they reduce such disparities by enabling access to much-needed support. Another aspect of ethics involves privacy and data security with the use of technology in mental health interventions. Safeguarding the privacy of personal information and responsible use will be key in maintaining trust and ethical standards in mental health care. Ethically, decision-making processes should involve the communities concerned, and interventions must be sensitive and respectful to culture.

Societal Implications

Climate change-related mental health has a number of multi-dimensional, far-reaching impacts on society. Since the impact of climate change is increasingly occurring at a very large scale, there is an emerging need for awareness and action at the societal level. This kind of awareness can be brought about by public education campaigns through highlighting the impacts of climate change on mental health and offering some strategies to build resilience. Society has to work on attitudes towards mental health and climate change to make them more holistic and interrelated in their solution. Besides, overcoming this climate-related mental health challenge requires an effort by governments, non-governmental organizations, and communities as a whole. There has to be collaboration on all levels in the formulation and implementation of specific policies, support systems, and interventions. The wider social implications are also those of increased social cohesion and solidarity in the light of communities pulling together to deal with adverse impacts on mental health due to climate change, and to adapt to a shifting environment.

These implications of climate-related mental health issues cut across practical, managerial, ethical, and societal domains. In sum, these implications require a multilayered approach: integrating mental health support into disaster response and climate adaptation strategies; developing appropriate workplace policies; upholding ethical standards; and increasing awareness and action at the level of society. With this possible consideration of implications, various stakeholders will be able to give more support to persons and communities suffering from the psychological burdens of climate change and improve resilience and well-being in general.

CONCLUSION

Climate change is a deep and multi-dimensional challenge that cuts across different aspects of human wellbeing and societal functioning. With an accelerating climate change, the impacts on mental health are increasingly evident, and comprehensive integrated strategies addressing such impacts are urgently needed. In this regard, the chapter has discussed the linkage between climate change and mental health by analyzing the direct and indirect pathways through which environmental change influences psychological well-being.

All these climatic extreme events, gradual environmental change, and socioeconomic disruption contribute to a wide range of mental health disorders, from anxiety and depression to PTSD. These case studies and evidence will give a good example of the diversity and pervasiveness of the impacts, as climate change aggravates current mental health conditions and fosters new psychological challenges. From hurricanes to heatwaves, from rising sea levels to recurrent floods, the evidence indicates a similar pattern in the emergence of psychological distress related to climatic events (Hayes & Poland, 2018; Heeren et al., 2021).

Addressing these mental health impacts can best be done by a multisectoral approach that would consider practical managerial, ethical, and societal concerns. On a practical note, mental health services should be integrated into disaster response and climate adaptation strategies to ensure immediate and appropriate support. The managers and organizational leaders must form policies and programs about the psychological well-being of their employees in such a way that it influences productivity and absenteeism. Considering the ethical perspective, equal access to mental health care must focuses on disparities in care and protection of privacy. Socially, it is an awareness-building and creation of shared responsibility needed to build resilience and support affected communities.

Looking ahead, a number of priorities emerge that will be important in further developing our response to climate-related mental health impacts. Research gaps need to be filled through longitudinal studies and exploration of specific mental health conditions and interventions. Innovations in technologies and community-based approaches similarly hold promise for improving mental health. Collaboration across sectors and integration of mental health into climate policies will further help our ability to respond effectively. Moreover, data utilization and predictive modeling may also serve in preparing for and mitigating future mental health challenges.

Conclusion: Climate change and mental health are at an interface that conveys grave concern and necessitates urgent and sustained attention. An integrated approach-weaving practical interventions, managerial strategies, ethical considerations, and societal actions-could better equip individuals and communities to grapple with the psychological consequences of climate change. Addressing these challenges not only improves mental health outcomes but contributes to broader goals of resilience, equity, and sustainability in the face of a changing climate.

REFERENCES

Atta, M., Zoromba, M. A., El-Gazar, H. E., Loutfy, A., Elsheikh, M. A., El-ayari, O. S. M., Sehsah, I., & Elzohairy, N. W. (2024). Climate anxiety, environmental attitude, and job engagement among nursing university colleagues: A multicenter descriptive study. *BMC Nursing*, 23(1), 133. Advance online publication. DOI: 10.1186/s12912-024-01788-1 PMID: 38378543

Baudon, P., & Jachens, L. (2021). A scoping review of interventions for the treatment of eco-anxiety. *International Journal of Environmental Research and Public Health*, 18(18), 9636. DOI: 10.3390/ijerph18189636 PMID: 34574564

Charlson, F., Ali, S., Benmarhnia, T., Pearl, M., Massazza, A., Augustinavicius, J., & Scott, J. (2021). Climate change and mental health: A scoping review. *International Journal of Environmental Research and Public Health*, 18(9), 4486. DOI: 10.3390/ijerph18094486 PMID: 33922573

Clemens, V., Hirschhausen, E., & Fegert, J. (2020). Report of the intergovernmental panel on climate change: Implications for the mental health policy of children and adolescents in europe—a scoping review. *European Child & Adolescent Psychiatry*, 31(5), 701–713. DOI: 10.1007/s00787-020-01615-3 PMID: 32845381

Ediz, Ç., & Yanik, D. (2023). The effects of climate change awareness on mental health: Comparison of climate anxiety and hopelessness levels in turkish youth. *The International Journal of Social Psychiatry*, 69(8), 2157–2166. DOI: 10.1177/00207640231206060 PMID: 37874036

Goudet, J., Binte Arif, F., Owais, H., Uddin Ahmed, H., & Ridde, V. (2024). Climate change and women's mental health in two vulnerable communities of bangladesh: An ethnographic study. *PLOS Global Public Health*, 4(6), e0002080. DOI: 10.1371/journal.pgph.0002080 PMID: 38935627

Gunasiri, H., Wang, Y., Watkins, E., Capetola, T., Henderson-Wilson, C., & Patrick, R. (2022). Hope, coping and eco-anxiety: Young people's mental health in a climate-impacted australia. *International Journal of Environmental Research and Public Health*, 19(9), 5528. DOI: 10.3390/ijerph19095528 PMID: 35564923

Hayes, K., & Poland, B. (2018). Addressing mental health in a changing climate: Incorporating mental health indicators into climate change and health vulnerability and adaptation assessments. *International Journal of Environmental Research and Public Health*, 15(9), 1806. DOI: 10.3390/ijerph15091806 PMID: 30131478

Heeren, A., Mouguiama-Daouda, C., & Contreras, A. (2021). On climate anxiety and the threat it may pose to daily life functioning and adaptation: a study among european and african french-speaking participants. DOI: 10.31234/osf.io/a69wp

Hickman, C., Marks, E., Pihkala, P., Clayton, S., Lewandowski, R., Mayall, E., & Susteren, L. (2021). Climate anxiety in children and young people and their beliefs about government responses to climate change: A global survey. *The Lancet. Planetary Health*, 5(12), e863–e873. DOI: 10.1016/S2542-5196(21)00278-3 PMID: 34895496

Innocenti, M., Santarelli, G., Lombardi, G., Ciabini, L., Zjalic, D., Russo, M., & Cadeddu, C. (2023). How can climate change anxiety induce both pro-environmental behaviours and eco-paralysis? the mediating role of general self-efficacy. *International Journal of Environmental Research and Public Health*, 20(4), 3085. DOI: 10.3390/ijerph20043085 PMID: 36833780

Kumar, P., Brander, L., Kumar, M., & Cuijpers, P. (2023). Planetary health and mental health nexus: Benefit of environmental management. *Annals of Global Health*, 89(1), 49. DOI: 10.5334/aogh.4079 PMID: 37521755

Liu, J., Potter, T., & Zahner, S. (2020). Policy brief on climate change and mental health/well-being. *Nursing Outlook*, 68(4), 517–522. DOI: 10.1016/j.outlook.2020.06.003 PMID: 32896304

Nicholas, P., Breakey, S., White, B., Brown, M., Fanuele, J., Starodub, R., & Ros, A. (2020). Mental health impacts of climate change: Perspectives for the ed clinician. *Journal of Emergency Nursing: JEN*, 46(5), 590–599. DOI: 10.1016/j.jen.2020.05.014 PMID: 32828480

Obradovich, N., Migliorini, R., Paulus, M., & Rahwan, I. (2018). Empirical evidence of mental health risks posed by climate change. *Proceedings of the National Academy of Sciences of the United States of America*, 115(43), 10953–10958. DOI: 10.1073/pnas.1801528115 PMID: 30297424

Opoku, S., Filho, W., Fudjumdjum, H., & Adejumo, O. (2021). Climate change and health preparedness in africa: Analysing trends in six african countries. *International Journal of Environmental Research and Public Health*, 18(9), 4672. DOI: 10.3390/ijerph18094672 PMID: 33925753

Palinkas, L., O'Donnell, M., Lau, W., & Wong, M. (2020). Strategies for delivering mental health services in response to global climate change. DOI: 10.20944/preprints202010.0150.v1

Qin, Z. (2024). The relationship between climate change anxiety and pro-environmental behavior in adolescents: the mediating role of future self-continuity and the moderating role of green self-efficacy. DOI: 10.21203/rs.3.rs-3930493/v1

Schwartz, S., Benoit, L., Clayton, S., Parnes, M., Swenson, L., & Lowe, S. (2022). Climate change anxiety and mental health: Environmental activism as buffer. *Current Psychology (New Brunswick, N.J.)*, 42(20), 16708–16721. DOI: 10.1007/s12144-022-02735-6 PMID: 35250241

Seritan, A., & Seritan, I. (2020). The time is now: Climate change and mental health. *Academic Psychiatry*, 44(3), 373–374. DOI: 10.1007/s40596-020-01212-1 PMID: 32162168

Shoib, S., Hussaini, S. S., Armiya'u, A. Y., Saeed, F., Őri, D., Roza, T. H., Gürcan, A., Agrawal, A., Solerdelcoll, M., Lucero-Prisno, D. E.III, Nahidi, M., Swed, S., Ahmed, S., & Chandradasa, M. (2023). Prevention of suicides associated with global warming: Perspectives from early career psychiatrists. *Frontiers in Psychiatry*, 14, 1251630. Advance online publication. DOI: 10.3389/fpsyt.2023.1251630 PMID: 38045615

Xue, S., Massazza, A., Akhter-Khan, S. C., Wray, B., Husain, M. I., & Lawrance, E. L. (2024). Mental health and psychosocial interventions in the context of climate change: A scoping review. *Npj Mental Health Research*, 3(1), 10. Advance online publication. DOI: 10.1038/s44184-024-00054-1 PMID: 38609540

Yellowlees, P. (2022). Climate change impacts on mental health will lead to increased digitization of mental health care. *Current Psychiatry Reports*, 24(11), 723–730. DOI: 10.1007/s11920-022-01377-6 PMID: 36214930

Chapter 5
Effect of Climate Change and Agricultural Factors on the Technical Efficiency of the Industrial Sector Across Indian States

Ajay Kumar Singh
https://orcid.org/0000-0003-0429-0925
Department of Humanities and Social Sciences, Graphic Era University (Deemed), Dehradun, India

Bhim Jyoti
Department of Seed Science and Technology, College of Forestry, Veer Chandra Singh Garhwali Uttarakh, India

ABSTRACT

This chapter estimates the TE of the industrial sector across Indian states using a stochastic frontier analysis. It also observes the impact of climatic and agricultural factors on gross value added (GVA) and TE of the industrial sector using a log-linear regression model. Annual average values of maximum and minimum temperature, precipitation and actual rainfall are considered as climatic factors, and irrigated area, cropping intensity, gross sown area, and credit deposit ratio are used as control variables in the empirical model. It compiles state-wise panel data of mentioned variables during 1991 – 2021. The results reveal that there is significant diversity in TE across states. TE and GVA are negatively impacted due to

DOI: 10.4018/979-8-3693-5792-7.ch005

climate change. Gross irrigated area, cropping intensity, and gross sown showed a significant impact on TE and GVA. It provides policy proposals to reduce diversity in TE of industries across Indian states.

1. BACKGROUND

Scientific research concluded that climate is changing due to multiple reasons including human intervention and natural change (Hussain et al., 2023). GHGs and CO2 emissions are accountable for climate change and occurrence of natural disaster at global level (Galushko & Gamtessa, 2022). The quantity of GHGs emissions has increased at unprecedented rate due to urbanization and industrialization after the 1970s (Hussain et al., 2023) Climate change posed a negative and significant impact on most sectors across countries (Singh & Jyoti, 2019). Climate change shows the variation in weather indicators from their mean values in a long period of time. Most scientists have been proved that climatic factors are changing significantly worldwide (Singh & Jyoti, 2019). Existing researchers and international organizations like IPCC and UNFC already explained that climatic factors are significantly varied. It would create multiple problems of livelihood security in all sectors. Hence, climate change has become a global concern at present (Hussain et al., 2023).

Industrial development provides job opportunities to skills and unskilled workers. It is also a prime driver to develop multiple sources of energy for consumers (Xu & Zhang, 2024). It also meets the different requirements of other sectors like agricultural, transport, textile, etc. For example, industries manufacture chemicals, pesticides, seeds, machines, irrigation instruments, fertilizer, and other tools for the agricultural sector. Therefore, there is inter-dependency between the industrial and agricultural sector. Furthermore, industrial development depends on multiple indicators at macro and micro level (Kumar & Paul, 2019; Singh et al., 2019a; Singh & Kumar, 2022c). Tax regime, intensity of machine, labour intensity, labour law and technological advancement help to enhance industrial production (Soni et al., 2017). Foreign direct investment, exports and imports also develop a favourable ecosystem for industrial development (Maroof et al., 2019; Singh & Kumar, 2022b). Intellectual property rights (IPRs) provide the protection of intellectuals of scientists, researchers and industries (Singh et al., 2020). IPRs help to reduce the imitation rate of technology innovation and mechanical process in the manufacturing sector (Singh & Kumar, 2022b). Hence, IPRs is a great driver to increase industrial development. Technological development and exports and imports of high-tech goods and services also reflect industrial development (Singh et al., 2020). It is necessary to increase growth of high-tech industries to increase industrial growth (Aneja & Arjun, 2021).

The agricultural sector solves the food-grain problem of people and it provides employment to rural people (Ashraf & Singh, 2022). Industries also create jobs for rural and urban dwellers in India. This process creates physical assets, infrastructure and road connectivity in India. Therefore, the industrial sector of India is directly dependent on its agricultural sector. Although, extreme variability in climatic factors may not be supportive for the agricultural sector in India (Sudarshan et al., 2021). Also, this sector is in stress due to urban transformation, sustainable development of cities, over population pressure on geographical areas in India (Hussain et al., 2023). India has a large geographical area with high diversity in population size, social-economic structure of people that are caused to increase more climate vulnerability (Hussain et al., 2023). Therefore, it seems from the literature perspectives that industrial development is highly dependent on climate change in India (Singh & Jyoti, 2019). Subsequently, it can be accepted that industrial development is also negatively associated with climate change. Industrial development is accountable to increase environmental degradation, and depletion of green space and water bodies in most cities (Hussain et al., 2023).

However, very few studies could observe the impact of climate change on the industrial sector in India. Further, it cannot be denied that the performance of industrial development cannot be measured effectively due to its multiple associations with other sectors. Therefore, scientific research articles use TE of the industrial and manufacturing sectors to explain their performance. While few studies could estimate the TE of the industrial sector. This research, therefore, is proposed to assess the answer for following research questions:

· How can researchers measure the TE of the industrial sector at state level?

· What may be the further implications of TE of the industrial sector?

· How and why TE of the industrial sector varies across states and over a period of time?

· How technical efficiency can be useful to increase industrial development in Indian states?

· Can climate adaptation strategies be conducive to reduce the negative implications of climate change on TE of the industrial sector in India?

This chapter is achieved the following objectives:

· To measure the TE of industrial sectors across Indian states during 1991 – 2021 using stochastic frontier analysis.

· To assess the expected reasons that are responsible for increasing the variation in technical efficiency of industrial sectors across Indian states.

· To examine the impact of climatic and agricultural factors on GVA and TE of the industrial sector in Indian states using panel data analysis.

The chapter is divided in 9 broad sections: The 1st section explains the background of the chapter. The 2nd section provides the findings of previous studies which estimate the impact of climate change on industrial development in India and other countries. Also, how existing researchers used different indicators to examine the impact of climate change on industrial development in India? It also explores the importance of TE in the production sectors. It also presents the summary of related studies which already estimated the impact of explanatory variables on TE of the industrial and agricultural sector across countries. The 3rd section highlights the research methodology including study area, source of data, conceptual framework of stochastic frontier analysis, research design and formulation of regression equations. The 4th section provides the statistical interpretation of estimated TE of the industrial sector across Indian states. The correlation coefficient of GVA of the industrial sector with variables is given in the 5th section. The statistical inference of regression results is provided in the 5th section. The 6th section explains the conclusion. Policy implications are highlighted in the 7th section. The 8th section explores the innovation and contribution of this study. At the end, it acknowledged the limitations and suggested a reliable research gap to undertake further studies.

2. REVIEW OF LITERATURE

2.1. Climate Change and Social-Economic Activities

Existing literature focused their investigation for assessing the impact of climate change on the agricultural sector due to its high dependency on weather factors (Sudarshan et al., 2021). Climate change impact on food security, health security, and nutritional security is also investigated by the existing researchers (Singh et al., 2024). Climate change also has a negative impact on employment of agricultural workers, income of the community, and rural infrastructure. Children, women and weaker sections of the society become more vulnerable due to climate change and natural disasters (Jyoti & Singh, 2020). Climate change also hampers agricultural marketing (Singh & Singh, 2021). Several studies provided an empirical association of climate change with the agricultural sector, food security, hunger, poverty, employment, income inequality and health security in developed and developing countries including India (Singh & Jyoti, 2019). Therefore, climate change has an adverse impact on social and economic development, and social welfare of the farmers and common citizens in largely agricultural intensive countries.

2.2. Production, Gross Value Added, Gross Output, and TE of Industrial Sector

Measurement of TE is effective to observe the role of inputs in the industrial sector. Most scholars calculated the TE of different sectors in India. Rajesh (2007) observed the TE of the unorganized sector in Kerala (India). As per the finding of this research, credit facilities and labour force play a significant role to increase TE. Mazumdar et al. (2009); Debnath and Sabastian (2014) applied data envelopment analysis to estimate the TE of the pharmaceutical sector in India. It observed that R&D expenditure and imported technologies help to increase TE. Energy intensity has a significant contribution in the manufacturing sector (Sahu & Narayanan, 2011). Sahu (2015); Sahu and Narayanan (2015) estimated the TE of the manufacturing sector in India using stochastic frontier analysis. Thereupon, TE and gross output of industries depend on several variables including internal and external variables. FDI, exports and imports of goods and services also help to increase TE of industries (Goldar & Sharma, 2015).

FDI, equity openness and inflation also have a significant impact on the industrial sector (Maroof et al., 2019). Infrastructural development and ICT also create a favourable ecosystem in manufacturing firms (Mitra et al., 2016). Thampy and Tiwary (2021) noticed a positive impact of sector specific technologies in the manufacturing sector. R&D expenditure, export and imports, capital stock and labour force are also supportive to increase industrial production. Export intensity, marketing, R&D intensity and capita intensity also enhance production in the manufacturing sector (Tyagi & Nauryal, 2016). Information technology agreement is also crucial for increasing TE of firms (Chaudhuri, 2016). Size of firms, intensity of inputs and imported materials produce a positive impact on TFP of firms in India (Satpathy et al., 2017). Energy and its sources also help to increase the production of industries (Soni et al., 2017).

Sales growth, innovation, quality certification, R&D expertise and skilled labour are also significant in the manufacturing sector (Singh et al., 2019a). The bank size, GDP and contribution of per employees also enhance TE of industries (Singh & Malik, 2018). Moreover, better infrastructure, appropriate tax regime, labour laws and environmental standards are also vital in the manufacturing sector (Mehta & Johan, 2017). Availability of resources is also positive to increase sustainability of the manufacturing sector (Basu et al., 2018). Financial development and banking sector are also significant in the manufacturing sector (Thampy & Tiwary, 2021). Human and financial capital are also favourable in this sector (Thampy & Tiwary, 2021). Singh and Kumar (2022a) reported that literacy rate, credit facilities and total emolument showed a positive impact on GVA of industries in India. The production of industries is also determined by labour intensity, R&D expertise, skilled and unskilled labour force, and R&D investment (Singh & Kumar, 2022c). Bhat and Kaur (2024) applied SFA to examine the TE of the IT industry in India. Foreign capital and profits have a significant impact on TE. Xu and Zhang (2024) explore the role of capital, labour and sources of energy in industrial production in China using SFA.

2.3. Climate Change and Industrial Development

Industrial development directly leads to increased possibilities of climate change. The industrial development is caused to increase GHGs and CO_2 emission (Xu & Zhang, 2024). Thus, it is essential to abate CO2 emissions to increase the efficiency of inputs in the industrial sector (Xu & Zhang, 2024). Hence, climate change is increasing due to extensive industrialization (Hussain et al., 2023). For instance, India, China and Brazil are contributing to extensive CO2 emission due to agricultural and industrial development (Naudé, 2010). Extreme development of industries is leading to decrease in green space and depletion of environment and ecosystem services in India (Hussain et al., 2023). Hence, industrial development is a significant cause for climate change. On the other hand, industrial development and its growth are greatly dependent on climate change and natural disaster directly and indirectly. Therefore, the production of the industrial sector is also negatively affected due to climate change.

Industrial production is likely to decline due to decline in production of commercial crops (Kumar & Sharma, 2014; Sudarshan et al., 2021). Production of textile industries cannot be sustained without production of cotton (Nzuza, 2021). Sugarcane crop is highly sensitive due to climate change in India (Kumar et al., 2015). Also, sugarcane yield declines due to climate change in India (Jyoti & Singh, 2020). The production of other cash crops like cotton, sesame, soyabean etc. are also expected to decline due to change in climatic factors (Singh et al., 2017). Singh and Jyoti (2019) also reported the negative impact of climate on many commercial crops like potato, groundnut, sesame and cotton in India. These are major agro-based crops in India. Hence, climate change would reduce production of agro-based industries in India (Singh & Singh, 2021). Moreover, climate change also would be responsible for reducing labour productivity in the industrial sector (Sudarshan et al., 2021). Manufacturing sector also bears the cost of climate change (Nzuza, 2021).

2.4. Impact of Climate Change, Environmental and Inputs on TE of Agricultural Sector

Most scholars examined TE of the production sector in India (Singh et al., 2019b; Zahid & Shah, 2024). Mehta and Johan (2017) explored the environmental standard in the performance of the manufacturing sector in India. Singh et al. (2019b) explored the impact of climatic factors on TE of sugarcane farming using stochastic frontier analysis. TE sugarcane farming is negatively affected due to climate change in India. Most crops are negatively impacted due to climate change. Maity and Singh (2021) examined the TE of tea gardens in Assam and West Bengal. Gogoi et al. (2022) estimated the TE of organic tea growers in Assam (India). It indicates that organic fertilizer, machine and land area showed a significant impact on TE. Kandpal et al. (2022) calculated TE of wheat crop production in India. Liu et al. (2023) observed the impact of adaptation strategies on TE of rice growers in China. Irrigated area and arable land help to increase TE of rice farmers. Climate-smart agriculture would be useful to sustain the production of the farming sector.

2.5. Summary of Review of Literature

The preceding sub-sections reveal that climate change is caused to reduce agricultural production, food security and environmental development. Several studies also estimated the TE of different industries and crop production in the agricultural sector. Moreover, previous studies also assessed the factors enhancing TE and output of the industrial sector. Industrial sector also contributes to increasing CO_2 and GHGs emission in the environment. Hence, industrial development is also accountable to increase climate change. While several studies have estimated the impact of climate change on the TE of the farming sector in India. Further, it is true that production of the industrial sector may decline due to decrease in agricultural production. Hence, there is greater scope to estimate the impact of climatic and agricultural factors on TE and value added of the industrial sector in India. Therefore, this chapter makes a crucial contribution towards the existing literature as it assesses the impact of climatic and agricultural factors on TE and GVA of the industrial sector in India.

3. RESEARCH METHODS

3.1. Study Area and Source of Data

Industrial development is dependent on the agricultural sector. Hence, 20 agricultural intensive states are considered in this chapter. These states are selected as per their share in agricultural and industrial sectors of India (Table 1). These states contribute 95.77% of gross sown area and 96.17% of gross irrigated area in India. Also, these states have more than 90% of contribution in gross value added, value of gross output, total persons engaged, total emoluments and total inputs in the industrial sector of India (Table 1). Accordingly, the statistics of related variables are used during 1991 – 2021 in empirical investigations.

Table 1. % share of selected states in agricultural and industrial sectors of India (in 2019-20)

States/Indicators	Agricultural sector		Industrial sector				
	Gross sown area	Gross irrigated area	Gross value added	Value of gross output	Total persons engaged	Total emoluments	Total inputs
AP	3.63	3.47	3.48	4.46	3.99	3.63	4.65
Assam	1.99	0.47	1.18	0.86	1.58	0.77	0.80
Bihar	3.68	5.25	0.48	0.85	0.77	0.39	0.92
Chhattisgarh	2.79	1.92	1.71	1.84	1.38	1.52	1.87
Gujarat	6.08	7.22	15.85	18.14	12.44	12.95	18.59
Haryana	3.28	5.75	5.50	6.64	6.16	6.89	6.86
HP	0.45	0.19	2.18	1.30	1.37	1.45	1.13
J & K	0.56	0.47	0.49	0.33	0.40	0.32	0.30
Jharkhand	0.91	0.24	1.90	1.53	1.24	1.66	1.46
Karnataka	6.74	4.53	7.16	6.20	6.50	7.56	6.01
Kerala	1.28	0.49	1.52	2.39	2.05	1.81	2.57
MP	12.98	12.11	2.96	3.22	2.49	2.38	3.27
Maharashtra	11.03	4.39	14.53	13.80	12.26	16.85	13.66
Odisha	2.25	1.25	2.82	3.08	1.70	2.06	3.13
Punjab	3.90	7.39	2.30	2.38	3.98	2.75	2.39
Rajasthan	12.58	10.53	4.20	3.67	3.60	3.61	3.56
Tamil Nadu	2.82	3.04	11.04	10.32	16.02	13.80	10.18
UP	13.35	20.71	5.70	6.26	6.80	6.38	6.37
Uttarakhand	0.51	0.52	3.65	2.68	2.63	2.35	2.49
West Bengal	4.95	6.23	3.17	3.98	4.43	3.65	4.14
Total	95.77	96.17	91.83	93.94	91.80	92.79	94.36
Other States	4.23	3.83	8.17	6.06	8.20	7.21	5.64
All India	100.00	100.00	100.00	100.00	100.00	100.00	100.00

Source: RBI (GoI); Annual Survey of Industries (GoI).

However, the statistics of selected indicators are available during 1999 – 2021 for Jharkhand, Uttarakhand and Chhattisgarh. Hence, it compiles imbalanced panel data for 20 states to estimate the TE of the industrial sector, and to estimate the implications of industrial indicators on GVA. Accordingly, it detects the climatic and agricultural factors on GVA and TE of the industrial sector for the mentioned states in Table 1. The state-wise statistics of industries are derived from annual surveys of industries (GoI). The data of agricultural factors are derived from the official website of the Reserve Bank of India (RBI). Finally, the statistics of climatic factors are derived from the website of ICRISAT.

3.2. Descriptive Analysis

The graphical presentation of estimated TE is presented. The correlation coefficients among the variables are also estimated using Karl – Pearson correlation coefficient analytical technique.

3.3. Measurement of Technical Efficiency (TE)

Data Envelopment Analysis (DEA) and Stochastic Frontier Analysis (SFA) can be applied to estimate the TE of industrial and agricultural sectors (Sen & Das, 2016; Maity & Singh, 2021; Kathayat et al., 2021; Bhat & Kaur, 2024). The SFA is supportive of examining the role of individual indicators which are used to estimate the TE of a specific production sector (Pattnayak & Chadha, 2013). The technique is effective to measure the TE of the industrial sector in parametric and non-parametric conditions (Sen & Das, 2016; Aneja & Arjun, 2021). The SFA is developed to estimate the TE of industrial sectors across Indian states (Chaudhuri, 2016). The stochastic frontier analysis (SFA) is applied based on a linear regression model in this chapter. The functional form of this model is adopted from Singh et al. (2019b). The SFA includes the functional relationship of dependent and independent variables, and composite error-term. While composite error term has two components i.e., natural and technical error-term. The SFA model is specified as:

$$(O)st = \alpha + f(Xst, \beta) + \dot{\varepsilon}st \qquad (1)$$

$$(O)st = \alpha + f(Xst, \beta) + (vst - ust) \qquad (2)$$

Here, O is output of a specific state in a given year, X is the set of IVs, $\dot{\varepsilon}_{st}$ is the composite-error term; α is constant term; s is a respective state; and t is year in equation (1). While v_{st} is natural error-term and u_{st} is technical error-term in equation (2). The chapter used 20 states in a panel data and these states have high diversity in multiple indicators that are associated with industrial and agricultural development. Hence, the coefficients of IVs with output are estimated through a time varying model in this chapter (Gogoi et al., 2022). TE is the ratio of total output with total input. Thus, TE is observed as:

$$(TE)st = (O)st/[f(Xst, \beta) \exp(ust)] \qquad (3)$$

Here, *TE* is technical efficiency of a specific state in equation (3). The highest value of TE infer that a state achieved optimum output while its lowest value indicates that the states need to improve optimum output.

3.4. Description of Indicators for Measuring the TE of Industrial Sector

Earlier studies used different indicators like output, sales revenue, net profits, gross value added, annual turnover to estimate the TE of a specific industry (Mitra et al., 2016; Sen & Das, 2016; Kumar & Paul, 2019). Rajesh (2007) considered the gross value added of the unorganized sector to estimate its TE in Kerala (India). Kumar and Arora (2012) used gross output to estimate the TE of sugar industries in India. Sahu (2015); Chaudhuri (2016) also applied output of manufacturing industries to estimate their TE. Mahajan et al. (2014) applied net sales revenue as a response variable to examine the TE of pharmaceutical industries.

The chapter used the ratio of most indicators to maintain the standardized values in estimation of TE of the industrial sector. The ratio of GVA with total person engaged is used as a dependent variable. While ratio of total emoluments with total persons engaged, ratio of total persons engaged with number of factories, ratio of fixed capital with total persons engaged, ratio of total inputs with value of gross output, ratio of gross capital formation with value of gross output, ratio of invested capital with value of gross output, ratio of physical working capital with value of gross output, ratio of productive capital with value of gross output and ratio of working capital with value of gross output are comprised as independent variables in TE estimation (Singh & Kumar, 2022a). The brief summary of indicators included for estimation of TE is given in Table 2. The ratio of fixed capital with total person engaged shows the capital intensity of the industrial sector (Sahu & Narayanan, 2011). Labour and capital help to increase TE of manufacturing firms (Mahajan et al., 2014; Tyagi & Nauryal, 2016).

Table 2. Indicators included for TE estimation

Indicators	Symbol	Unit
Gross value added/total persons engaged	RVATPE	Ratio
Total emoluments/total persons engaged	RTETPE	
Total persons engaged/number of factories	RTPENF	
Fixed capital/total persons engaged	RFCTPE	
Total inputs/value of gross output	RTIVGO	
Gross capital formation/value of gross output	RGCFVGO	
Invested capital/value of gross output	RICVGO	
Physical working capital/value of gross output	RPWCVGO	
Productive capital/value of gross output	RPCVGO	
Working capital/value of gross output	RWCVGO	

Source: Rajesh (2007); Sahu and Narayanan (2011); Mitra et al. (2016); Singh and Malik (2018).

3.5. Explanation of Dependent and Independent Variables

This chapter is expected to examine the impact of climatic and agricultural factors on GVA and TE of industries in India. Rajesh (2007) also used GVA and TE of the unorganized sector as dependent variables in the empirical model. Mazumdar et al. (2009) applied output of the firms as a dependent variable for estimating TE of the pharmaceutical sector in India. Sahu and Narayanan (2015) also assessed the impact of specific characteristics of firms on TE in India. Singh and Kumar (2022a) analyse the impact of industrial indicators and literacy rate and financial development on gross value added of industrial development in India. Singh et al. (2019b) also estimated the impact of climatic factors on TE of sugarcane farming India using a log-linear regression model. Gross irrigated area, gross sown area and cropping intensity are used as agricultural factors in the regression model (Singh et al., 2019b). While annual actual rainfall, annual average precipitation, annual average maximum temperature and annual average minimum temperature are used as climatic factors in the statistical investigation (Singh et al., 2019b). Credit deposit ratio (CDR) helps to increase money supply in the financial market. Hence, it may be supportive to increase agricultural and industrial development. Thus, this variable is also used as an independent variable in the empirical model. Hence, in this chapter two functional forms are formulated as:

$$(GVA) = f\{CDRSCBU, CI, GIA, GSA, AARF, AAP, AAMAT, AAMIT\} \quad (4)$$

$$(TE) = f\{CDRSCBU, CI, GIA, GSA, AARF, AAP, AAMAT, AAMIT\} \quad (5)$$

The description of variables stated in equation (4) and (5) are provided in Table 3.

Table 3. Summary of dependent and independent variables

Indicators	Symbol	Unit
Gross value added	GVA	Million
Technical efficiency	TE	Number
CDR of scheduled commercial banks	CDRSCBU	Ratio
Cropping intensity	CI	%
Gross irrigated area	GIA	'000' Ha.
Gross sown area	GSA	'000' Ha.
Annual actual rainfall	AARF	mm
Annual average precipitation	AAP	mm
Annual average maximum temperature	AAMAT	^{0}C
Annual average minimum temperature	AAMIT	^{0}C

Source: Author's compilation based on existing studies like Rajesh (2007); Sahu and Narayanan (2015); Singh et al. (2019b).

3.6. Formulation of Empirical Models

The Cobb-Douglas production function model is formulated to examine the impact of climate change on gross value added and TE of the industrial sector (Sahu, 2015). Similar model is also adopted by Sahu and Narayanan (2015); Mitra et al. (2016); Kumar and Paul (2019) to examine the TE enhancing factors in manufacturing firms of India. While state-wise, panels of *DV*s and *IV*s are compiled to achieve the prescribed objectives of this chapter. For the empirical investigations, TE and gross value added of respective states are used as *DV*, and average maximum temperature, and average minimum temperature, average precipitation, average evapotranspiration and annual actual rainfall are concluded as *IV*s in the mentioned variables. Accordingly, total factors productivity of each state is also estimated to identify the overall impact of climate change on TE and gross value added of the industrial sector. Following regression equations are applied for assessing the impact of climatic and agricultural factors on the aforesaid *DV*s:

$log(GVA)_{st} = \beta_0 + \beta_1 log(CDRSCBU)_{st} + \beta_2 log(CI)_{st} + \beta_3 log(GIA)_{st} + \beta_4 log(GSA)_{st} + \beta_5 log(AARF)_{st} + \beta_6 log(AAP)_{st} + \beta_7 log(AAMAT)_{st} + \beta_8 log(AAMIT)_{st} + \phi_{st}$

(6)

$$log(TE)st = \xi 0 + \xi 1\ log(CDRSCBU)st + \xi 2\ log(CI)st + \xi 3\ log(GIA)st + \xi 4\ log(GSA)st$$
$$+ \xi 5\ log(AARF)st + \xi 6\ log(AAP)st + \xi 7\ log(AAMAT)st + \xi 8\ log(AAMIT)st + \lambda st \qquad (7)$$

Here, *log* is natural logarithm of corresponding variables; β_0 and ξ_0 are constant coefficients; $\beta_1, \beta_2, ... \beta_8$ and $\xi_1, \xi_2, ... \xi_8$ are the coefficients of associated *IV*s; and ϕ_{st} and λ_{st} are the error-terms in equation (6) and (7). The constant coefficients measure the total factor productivity (TFP); and regression coefficients measure the elasticities of corresponding variables in stated equations.

3.7. Application of Statistical Software and Estimation of Coefficients of *IV*s

SPSS software is used to examine the correlation coefficients among the indicators. While STATA statistical software is used to estimate the TE of industrial sector and to examine the coefficients of explanatory variables.

4. TECHNICAL EFFICIENCY (TE) OF INDUSTRIAL SECTOR IN INDIAN STATES

The trend in the estimated TE of the Industrial sector during 1991 – 2021 for the northern region is given in Figure 1. TE of the industrial sector in the eastern and north-eastern region is provided in Figure 2. Figure 3 provides the trend in TE of the industrial sector in the central region. Figure 4 shows the trend in TE of the industrial sector in the southern region. Figure 5 infer that trend in TE of industrial sector in western region. Northern states have diversity in TE of the industrial sector. Also, the TE of the industrial sector has an increasing trend in Haryana, Himachal Pradesh, Jammu and Kashmir, Punjab and Rajasthan. It is noticed that TE of the industrial sector has continuously increased in all Indian states after 1991. Himachal Pradesh has higher TE as compared to other states in the Northern region (Figure 1). Haryana has the lowest TE of industrial sector in the Northern region. Assam and Bihar have higher TE as compared to other states in eastern and north-eastern region (Figure 2).

Jharkhand has the lowest value of TE in this region. The TE of industrial sector for central region infer that Uttarakhand has higher TE with comparison to other states (Figure 3). Chhattisgarh has the lowest value of TE in this region. Uttar Pradesh and Madhya Pradesh have large population sizes and the population has high reliance on the agricultural sector. These states also have the lion share of many commercial crops like sugarcane and soybean (Kumar & Sharma, 2014). Uttar Pradesh and

Madhya Pradesh have scope to increase TE of the industrial sector in the central region. The TE of industrial sector for southern region specify that Andhra Pradesh and Tamil Nadu have higher TE as compared to Kerala (Figure 4). While Kerala has the lowest TE in this region. Maharashtra and Gujarat are the highly industrial states in western region. Despite that Maharashtra has lower TE as compared to Gujarat (Figure 5). Thus, both the states have scope to improve TE in the industrial sector.

Figure 1. TE of industrial sector in northern region

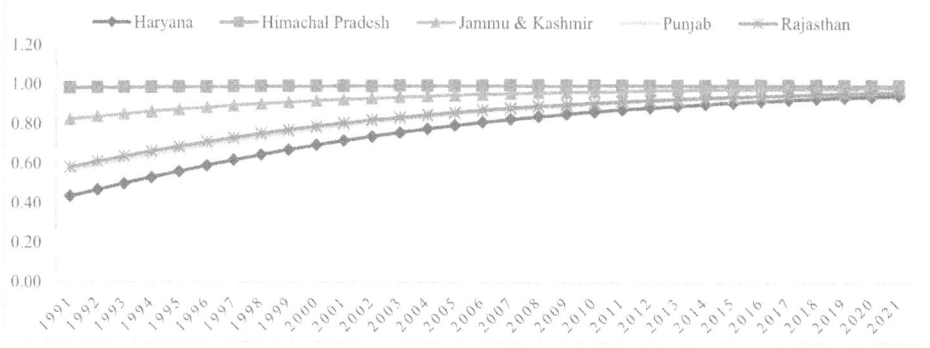

Source: *Author's estimation.*

Figure 2. TE of industrial sector in eastern and north-eastern region

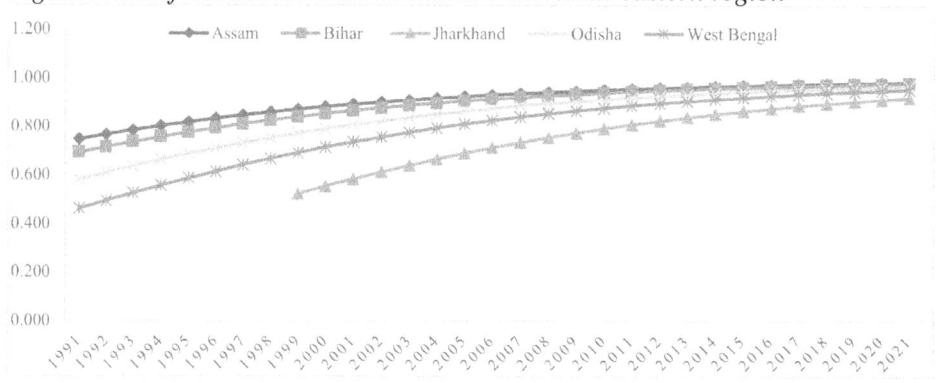

Source: *Author's estimation.*

Figure 3. TE of industrial sector in central region

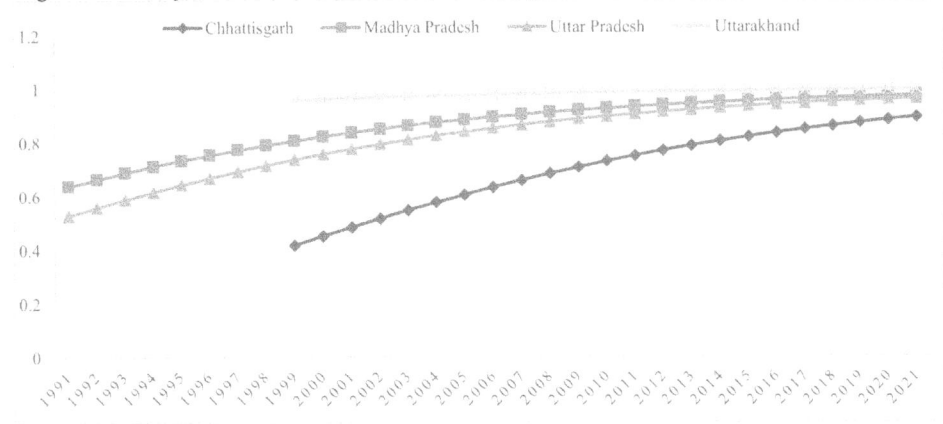

Source: *Author's estimation.*

Figure 4. TE of industrial sector in southern region

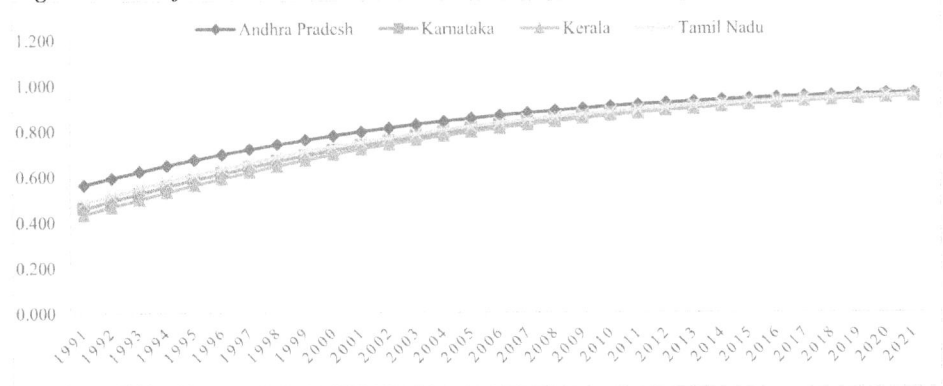

Source: *Author's estimation.*

Figure 5. TE of industrial sector in western region

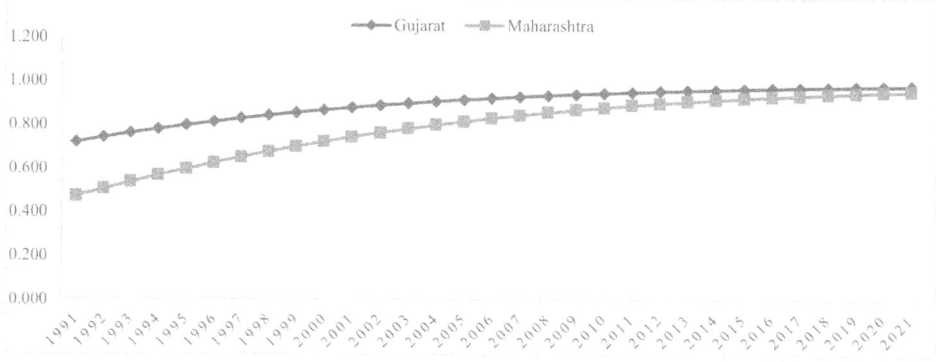

[Line chart showing Gujarat and Maharashtra technical efficiency from 1991 to 2021. Gujarat rises from ~0.75 to ~0.95; Maharashtra rises from ~0.48 to ~0.93.]

Source: *Author's estimation.*

5. STATISTICAL INFERENCE OF REGRESSION RESULTS

The correlation coefficient of gross value added per person engaged with industrial indicators like total emolument, total person engaged per factory, fixed capital formation per person engaged (capital intensity), ratio of total input with value of gross output, ratio of gross capital formation with value of gross output, ratio of invested capital with value of gross output, ratio of physical working capital with value of gross output, ratio of physical capita with value of gross output and ratio of working capital with value of gross output is presented in Table 4. The results infer that few variables have a statistically significant association with gross value added per person. Gross value added is positively associated with total emolument and fixed capital formation. Hence, total emolument and fixed capital would be helpful to increase gross value added on the industrial sector.

Table 4. Correlation coefficients among the indicators of industries

Indicators	GVATPE	TETPE	TPENF	FCTPE	TIVGO	GCFVGO	ICVGO	PWCVGO	PCVGO	WCVGO
GVATPE	1	0.884[b]	0.034	0.726[b]	0.012	-0.115[b]	-0.008	-0.153[b]	0.023	-0.149[b]
TETPE	0.884[b]	1	-0.046	0.736[b]	0.083[a]	-0.165[b]	-0.004	-0.157[b]	0.001	-0.202[b]
TPENF	0.034	-0.046	1	-0.004	-0.052	0.253[b]	0.364[b]	0.095[a]	0.377[b]	0.085[a]
FCTPE	0.726[b]	0.736[b]	-0.004	1	0.023	0.009	0.409[b]	-0.082[a]	0.380[b]	-0.263[b]
TIVGO	0.012	0.083[a]	-0.052	0.023	1	-0.078[a]	-0.111[b]	-0.067	-0.125[b]	-0.081[a]
GCFVGO	-0.115[b]	-0.165[b]	0.253[b]	0.009	-0.078[a]	1	0.486[b]	0.200[b]	0.501[b]	0.225[b]

continued on following page

Table 4. Continued

Indicators	GVATPE	TETPE	TPENF	FCTPE	TIVGO	GCFVGO	ICVGO	PWCVGO	PCVGO	WCVGO
ICVGO	-0.008	-0.004	0.364^b	0.409^b	-0.111^b	0.486^b	1	0.249^b	0.964^b	0.070^a
PWCVGO	-0.153^b	-0.157^b	0.095^a	-0.082^a	-0.067	0.207^b	0.249^b	1	0.230^b	0.184^b
PCVGO	0.023	0.001	0.377^b	0.380^b	-0.125^b	0.501^b	0.964^b	0.230^b	1	0.296^b
WCVGO	-0.149^b	-0.202^b	0.085^a	-0.263^b	-0.081^a	0.225^b	0.070^a	0.184^b	0.296^b	1

Source: Author's estimation. **Note:** ^{b} and ^{a} signify that correlation coefficients are significant at the 1% and 5% significant level, respectively.

Conversely, gross value added is negatively associated with physical working capital and gross capital formation. Total emolument is positively associated with fixed capital and total input. Fixed capital is also positively correlated with invested capital and total emolument. Gross capital formation is also expected to be improved as the number of factories, invested capital, physical working capital, physical capital and working capital increase. The invested capital is positively associated with fixed capital, physical working capital, physical capital and working capital. Moreover, the physical working capital is positively correlated with gross capital formation, invested capital, physical and working capital. Hence, there is a significant correlation among the gross capital formation, invested capital, physical working capital, physical capital and working capital in the industrial sector in India.

The GVA per person is negatively associated with gross irrigated area, gross sown area, annual actual rainfall, annual average precipitation, annual average maximum temperature and annual average minimum temperature (Table 5). Gross irrigated area is also negatively associated with annual actual rainfall and annual average precipitation. Gross sown area is also negatively correlated with cropping intensity, annual actual rainfall and annual average precipitation. Cropping intensity signifies the use of a specific arable land for growing multiple crops in a year. While cropping intensity is also negatively associated with gross sown area, annual average precipitation, annual average maximum temperature and annual average minimum temperature.

Table 5. Correlation coefficient of gross value added per person engaged with climatic and agricultural factors

Indicators	RGVATPE	CDRSCBU	GIA	GSA	CI	AARF	AAP	AAMAT	AAMIT
RGVATPE	1	0.024	-0.051	-0.05	0.014	-0.005	-0.014	-0.087^a	-0.061
CDRSCBU	0.024	1	0.119^b	0.274^b	-0.280^b	-0.036	-0.293^b	0.486^b	0.489^b
GIA	-0.051	0.119^b	1	0.772^b	0.220^b	-0.012	-0.421^b	0.367^b	0.186^b

continued on following page

Table 5. Continued

Indicators	RGVATPE	CDRSCBU	GIA	GSA	CI	AARF	AAP	AAMAT	AAMIT
GSA	-0.050	0.274[b]	0.772[b]	1	-0.137[b]	-0.022	-0.381[b]	0.548[b]	0.323[b]
CI	0.014	-0.280[b]	0.220[b]	-0.137[b]	1	0.025	-0.004	-0.445[b]	-0.479[b]
AARF	-0.005	-0.036	-0.012	-0.022	0.025	1	0.039	-0.023	0.002
AAP	-0.014	-0.293[b]	-0.421[b]	-0.381[b]	-0.004	0.039	1	-0.139[b]	0.143[b]
AAMAT	-0.087[a]	0.486[b]	0.367[b]	0.548[b]	-0.445[b]	-0.023	-0.139[b]	1	0.913[b]
AAMIT	-0.061	0.489[b]	0.186[b]	0.323[b]	-0.479[b]	0.002	0.143[b]	0.913[b]	1

Source: Author's estimation. **Note:** [b] and [a] signify that correlation coefficients are significant at the 1% and 5% significant level, respectively.

The impact of explanatory variables on gross value added is estimated through PCSE estimation (Table 6). The estimates reported that 37% variation in GVA of the industrial sector depends on undertaken climatic and agricultural factors. The agricultural sector is supportive to meet the requirement of raw material for the industrial sector. Thus, the impact of cropping intensity, gross irrigated area and gross sown area on gross value added is reported positive (Table 6). The results claimed that gross value added is to be increased by 0.43% and 0.17% as cropping intensity and gross irrigated area increase by 1%, respectively. While all climatic factors (except annual average minimum temperature) showed a negative impact on gross value added to the industrial sector. The results indicate that gross value added is projected to decrease by 0.31% and 8.70% due to 1% increase in annual average precipitation and annual average maximum temperature, respectively. Availability of finance is favourable for farmers and producers to increase their production scale. Hence, the impact of credit deposit ratio is noticed positively on the GVA of the industrial sector.

The results based on stochastic frontier analysis are provided in Table 7. The impact of total emolument, gross capital formation and physical capital on gross value added per person is detected positive (Table 7). Hence, these variables are found useful to increase GVA and TE of the industrial sector. Conversely, total inputs, invested capital, physical working capital and working capital produce a negative impact on GVA. Thus, these variables are not found significant for improving the gross value added and TE of the industrial sector.

Table 6. Impact of climatic and agricultural factors on gross value added

Number of obs.		595	R-squared		0.3717
Number of groups		20	Wald Chi²		600.87
Obs. per group: min		23	Prob > Chi²		0.0000
Obs. per group: Max		31	Estimated covariances		210.0000
Average		29.75	Estimated coefficients		9
log(GVA) =DV	Reg. Coef.	Std. Err.	z	P>|z|	[95% Conf. Interval]
log(CDRSCBU)	**1.2607**	**0.1672**	**7.54**	**0.000**	**0.9331**	**1.5884**
log(CI)	0.4320	0.2610	1.66	0.098	-0.0795	0.9436
log(GIA)	0.1726	0.0881	1.96	0.050	-0.0001	0.3453
log(GSA)	0.1035	0.1378	0.75	0.453	-0.1667	0.3736
log(AARF)	-0.1181	0.1079	-1.09	0.274	-0.3296	0.0934
log(AAP)	-0.3134	0.2231	-1.4	0.160	-0.7507	0.1239
log(AAMAT)	-8.6996	2.4250	-3.59	0.000	-13.4526	-3.9466
log(AAMIT)	6.0217	1.2798	4.71	0.000	3.5134	8.5300
Con. Coef.	15.5397	5.1677	3.01	0.003	5.4112	25.6682

Source: Author's estimation.

Table 7. Impact of indicators of industries on gross value added per person through stochastic frontier analysis

Number of obs.		596	Obs. per group: max		31
Number of groups		20	Wald Chi²		2209.96
Obs. per group: min		23	Prob > Chi²		0.000
Average		29.8	Log likelihood		371.8373
RGVATPE =DV	Reg. Coef.	Std. Err.	z	P>|z|	[95% Conf. Interval]
RTETPE	4.5329	0.1625	27.9	0.000	4.2145	4.8514
RTPENF	0.0005	0.0003	1.49	0.136	-0.0002	0.0011
RFCTPE	0.0061	0.0010	5.84	0.000	0.0041	0.0081
RTIVGO	-0.0366	0.0164	-2.24	0.025	-0.0687	-0.0045
RGCFVGO	0.3531	0.0792	4.46	0.000	0.1979	0.5084
RICVGO	-1.3726	0.1645	-8.34	0.000	-1.6949	-1.0502
RPWCVGO	-0.0278	0.0599	-0.46	0.642	-0.1452	0.0896
RPCVGO	1.1941	0.1806	6.61	0.000	0.8402	1.5481
RWCVGO	-0.7922	0.1752	-4.52	0.000	-1.1356	-0.4488
Con. Coef.	0.3202	0.0450	7.11	0.000	0.2320	0.4085
/mu	0.5213	0.0808	6.46	0.000	0.3630	0.6796

continued on following page

Table 7. Continued

Number of obs.		596	Obs. per group: max		31
Number of groups		20	Wald Chi²		2209.96
Obs. per group: min		23	Prob > Chi²		0.000
Average		29.8	Log likelihood		371.8373
RGVATPE =DV	Reg. Coef.	Std. Err.	z	P>|z|	[95% Conf. Interval]
RTETPE	**4.5329**	**0.1625**	**27.9**	**0.000**	**4.2145**	**4.8514**
/eta	-0.0924	0.0076	-12.2	0.000	-0.1072	-0.0776
/lnsigma2	-2.3862	0.3547	-6.73	0.000	-3.0814	-1.6911
/ilgtgamma	1.6377	0.4283	3.82	0.000	0.7982	2.4771
sigma2	0.0920	0.0326			0.0459	0.1843
gamma	0.8372	0.0584			0.6896	0.9225
sigma_u2	0.0770	0.0326			0.0131	0.1409
sigma_v2	0.0150	0.0009			0.0132	0.0167

Source: Author's estimation.

The association of TE of industries with climatic and agricultural factors are presented in Table 8. The results infer that credit deposit ratio, gross irrigated area and gross sown area are negatively correlated with TE (Table 8). Temperature is negatively associated with TE of industries. It means that agricultural and climatic factors have crucial associations with TE of the industrial sector in India.

Table 8. Correlation coefficient of TE of industries with climatic and agricultural factors

Indicators	TE	CDRSCBU	GIA	GSA	CI	AARF	AAP	AAMAT	AAMIT
TE	1	-0.386[b]	-0.212[b]	-0.137[b]	0.003	0.002	0.002	-0.386[b]	-0.433[b]
CDRSCBU	-0.386[b]	1	0.119[b]	0.274[b]	-0.280[b]	-0.036	-0.293[b]	0.486[b]	0.489[b]
GIA	-0.212[b]	0.119[b]	1	0.772[b]	0.220[b]	-0.012	-0.421[b]	0.367[b]	0.186[b]
GSA	-0.137[b]	0.274[b]	0.772[b]	1	-0.137[b]	-0.022	-0.381[b]	0.548[b]	0.323[b]
CI	0.003	-0.280[b]	0.220[b]	-0.137[b]	1	0.025	-0.004	-0.445[b]	-0.479[b]
AARF	0.002	-0.036	-0.012	-0.022	0.025	1	0.039	-0.023	0.002
AAP	0.002	-0.293[b]	-0.421[b]	-0.381[b]	-0.004	0.039	1	-0.139[b]	0.143[b]
AAMAT	-0.386[b]	0.486[b]	0.367[b]	0.548[b]	-0.445[b]	-0.023	-0.139[b]	1	0.913[b]
AAMIT	-0.433[b]	0.489[b]	0.186[b]	0.323[b]	-0.479[b]	0.002	0.143[b]	0.913[b]	1

Source: Author's estimation. **Note:** [b] signifies that correlation coefficient of corresponding variables are statistically significant at the 1% level.

The impact of climatic and agricultural factors on TE is examined through PCSE model (Table 9). The results infer that credit deposit ratio, capital intensity and gross irrigated area showed a negative impact on TE. TE is expected to decrease by 0.10%, 0.17% and 0.04% as 1% increase in credit deposit ratio, cropping intensity and gross irrigated area. However, gross sown area is found positive to increase TE of industries. The impact of credit deposit ratio on TE is also reported negative. All climatic factors produce a negative impact on TE. TE is likely to be decreased by 0.03% as 1% increase in annual average precipitation. While TE is to decrease by 0.50% and 0.13% due to 1% increase in annual average maximum temperature and annual average minimum temperature, respectively.

Table 9. Impact of climatic and agricultural factors on TE of industries

Number of obs.		595		R-squared		0.327
Number of groups		20		Wald Chi²		175.20
Obs. per group: Minimum		23		Prob > Chi²		0.000
Obs. per group: Maximum		31		Estimated covariances		210.00
Average		29.75		Estimated coefficients		9
TE	Reg. Coef.	Std. Err.	z	P>|z|	[95% Conf. Interval]	
log(CDRSCBU)	**-0.1028**	0.0176	-5.85	0.000	-0.1372	-0.0683
log(CI)	-0.1725	0.0275	-6.28	0.000	-0.2264	-0.1186
log(GIA)	-0.0436	0.0063	-6.96	0.000	-0.0558	-0.0313
log(GSA)	0.0922	0.0130	7.07	0.000	0.0667	0.1178
log(AARF)	-0.0080	0.0112	-0.71	0.475	-0.0301	0.0140
log(AAP)	-0.0291	0.0192	-1.51	0.130	-0.0667	0.0086
log(AAMAT)	-0.5064	0.2282	-2.22	0.026	-0.9537	-0.0591
log(AAMIT)	-0.1310	0.1134	-1.16	0.248	-0.3532	0.0912
Con. Coef.	3.8272	0.5434	7.04	0.000	2.7621	4.8923

Source: Author's estimation.

6. CONCLUSION

This chapter uses SFA to estimate the TE of the industrial sector for 20 Indian states during 1991-2021. There reported a significant diversity in TE of the industrial sector across Indian states. TE of the industrial sector show a consistently increasing trend in it after 1991. Also, TE of the industrial sector also varies across regions of India. The diversity in TE is due to variation in total emolument, fixed capital, total input, gross capital formation, and working capital. The TE of the industrial sector is also varied due to agricultural production enhancing factors like irrigated area, gross sown area and cropping intensity. Credit deposit ratio, infrastructure development, financial development, literacy rate, technical skills of people, and government policies have significant impact on TE of industrial sector in India. Himachal Pradesh has a high TE of Industrial sector in northern region of India. Assam and Bihar have high TE in eastern and north-eastern region of India. In the central region, Uttarakhand and Madhya Pradesh have the highest TE. All states in the southern region have a better TE of Industrial sector as compared to other regions. Gujarat has a higher TE of industrial sector as compared to Maharashtra in western region. The results suggested that Maharashtra, Kerala, Chhattisgarh, Uttar Pradesh, Jharkhand, West Bengal and Odisha, Haryana, Punjab, and Rajasthan have significant scope to improve TE of the industrial sector in India. These states have dominant positions in the agricultural sector. Hence, these have high potential to improve their TE by adopting more agro-industries.

Gross value added of the industrial sector is positively associated with total emolument and fixed capital formation. While it is negatively associated with gross capital formation, physical working capital and working capital. Furthermore, gross value added is also negatively associated with agricultural factors like gross irrigated area, gross sown area and cropping intensity. All climatic factors are also negatively associated with gross value added. Hence, the results provide a confirmation that gross value added to industrial decline due to variation in climatic factors. The correlation coefficient of TE of the industrial sector with climatic and agricultural factors produces statistically significant results. TE of the industrial sector is negatively associated with credit deposit ratio, gross irrigated area, gross sown area, and maximum and minimum temperature.

The regression results based on PCSE estimation infer that credit deposit ratio, cropping intensity, gross irrigated area and gross sown area showed a positive impact on GVA of the industrial sector in India. Hence, the empirical results indicate that agricultural factors also help to increase the gross value added of the industrial sector in India. While climatic factors like annual actual rainfall, annual average precipitation and annual average maximum temperature showed a negative impact on the GVA of the industrial sector in India. Hence, the results demonstrate that gross value added is significantly declined due to climate change in India.

The regression results based on SFA infer that GVA is expected to be improved as total emolument and gross capital formation increase. The results also claimed that TE of the industrial sector is expected to decline as credit deposit ratio, cropping intensity and gross irrigated area increase. Most specifically, climate change has a negative effect on TE of the industrial sector in India. Therefore, TFP of most states are also reported negative.

7. POLICY IMPLICATIONS

Technical efficiency (TE) helps to understand the importance of technological change in the production sectors. The estimated values of TE also provide how an individual input or different inputs can be useful to increase output in a specific sector? Industrial development is highly dependent on technological advancement. India is an emerging country and it is giving more importance to increasing industrial development. TE depends on many variables like technological change, innovation, advertisement, marketing, transport, foreign trade, availability of raw materials and water, transport, skills of workers, taxes, inflation, government policies, digital infrastructure, social media etc. Thereafter, TE of the industrial sector depends on multiple variables including agricultural output. Production activities of most industries depend on the agricultural sector directly and indirectly. Agricultural production activities are also hampering due to multiple reasons like climate change, decreasing ecosystem services, rising cost of cultivation, decreasing arable land, and use of arable land commercial purposes at global level (Galushko & Gamtessa, 2022).

Furthermore, to ensure the sustainability in the agricultural production sector and its dependent sectors are crucial agenda of sustainable development goals (SDGs). Subsequently, the above-mentioned activities are creating obstacles to attain the SD worldwide. Hence, the Indian government should focus on above mentioned indicators to increase industrial development and to increase the TE of the industrial sector. Social and economic development would be declined due to decrease in gross value added of industries in India. Hence, India should bring more climate resilience, adaptation mitigation practice to increase agricultural production, and gross value added and TE of the industrial sector (Singh & Jyoti, 2023). India should develop green technologies, environmental technologies, energy intensity technologies, digital technologies and green farm management practices for increasing the green growth of the industrial sector (Jyoti et al., 2023). Environmental sustainability would be highly effective to reduce climate change impacts in the industrial and agricultural sector in India. Adoption of climate smart technologies should be promoted to increase agricultural production (Zahid & Shah, 2024). The government and financial organization should provide green funds to the farmers and business community to develop green ventures to promote sustainability in business, industrial and agricultural sectors. Also, the policy makers must pay equal attention to increase agricultural and industrial development in India.

8. SIGNIFICANCE AND VALUE ADDITION OF THIS CHAPTER

The chapter explains the importance of industrial development in India as per the existing studies. The determinants of industrial development are also described in this chapter. The obstacles of industrial development are highlighted in this chapter It also provides the conceptual framework for assessing the association of industrial development with climate change. The chapter estimates the TE of the industrial sector for selected 20 states of India during 1990 – 2021. The SFA is applied to estimate the TE of the industrial sector. Accordingly, it explores the comparative performance of selected states in TE of the industrial sector. It applied correlation coefficient analysis to assess the correlation of gross value added with industrial indicators. The similar technique is also considered to observe the association of climatic and agricultural factors with GVA and TE of the industrial sector in India. It also uses a log-linear regression model to expose the impact of climatic and agricultural factors on GVA and TE of the industrial sector. The results found significant implications of climatic and agricultural factors on gross value added and TE of the industrial sector in India. Thus, the regression results increase the attention of policy makers to adopt effective climate action plans to increase industrial development in India.

9. LIMITATIONS AND FURTHER RESEARCH DIRECTION

The chapter explores the impact of climate change on TE of the industrial sector across India states. This chapter used only climatic and few agricultural factors in the empirical investigation. It found a negative impact of most climatic and agricultural factors on GVA and TE of the industrial sector in India. Also, the integrated impact of all climatic factors on TE and GVA of the industrial sector are also reported negative and statistically significant. However, the TE of the industrial sector depends on different non-climatic factors like technological upgradation, skilled workers, social media, digitalization, digital technology, transportation, infrastructure development, marketing, entrepreneurial attitude of the people, entrepreneurship ecosystem, government policies, etc. Also, the gross output, gross value added and TE of the industrial sector depend on foreign trade, FDI, tariff and exports and imports. While this chapter could not consider these variables in the regression analysis. Hence, further study can assess the impact of above-mentioned variables on TE of the industrial sector in India. Moreover, existing researchers can examine the interlinkages of climate change with the agricultural and industrial sector in further studies. The chapter also found a negative impact of total input, invested capital, physical working capital and working capital on gross value added. Hence, existing researchers can assess the reasons why these indicators produce a negative impact on GVA of the industrial sector in India in further study.

REFERENCES

Aneja, R., & Arjun, G. (2021). Estimating components of productivity growth of Indian high and medium-high technology industries: A non-parametric approach. *Social Sciences & Humanities Open*, 4(1), 1–10. DOI: 10.1016/j.ssaho.2021.100180

Ashraf, S. N., & Singh, A. K. (2022). Implications of appropriate technology and farm inputs in the agricultural sector of Gujarat: Empirical analysis based on primary data. *Agricultural Economics Research Review*, 35(2), 59–77. DOI: 10.5958/0974-0279.2022.00031.3

Basu, P., Ghosh, I., & Das, P. K. (2018). Using structural equation modelling to integrate human resources with internal practices for lean manufacturing implementation. *Management Science Letters*, 8(1), 51–68. DOI: 10.5267/j.msl.2017.10.001

Bhat, N. A., & Kaur, S. (2024). Technical Efficiency Analysis of Indian IT Industry: A Panel Data Stochastic Frontier Approach. *Millennial Asia*, 15(2), 327–348. DOI: 10.1177/09763996221082199

Chaudhuri, D. D. (2016). Impact of economic liberalization on technical efficiency of firms: Evidence from India's electronics industry. *Theoretical Economics Letters*, 6(1), 549–560. DOI: 10.4236/tel.2016.63061

Debnath, R. M., & Sabastian, V. J. (2014). Efficiency in the Indian iron and steel industry – an application of data envelopment analysis. *Journal of Advances in Management Research*, 11(1), 4–19. DOI: 10.1108/JAMR-01-2013-0005

Galushko, V., & Gamtessa, S. (2022). Impact of climate change on productivity and technical efficiency in Canadian crop production. *Sustainability (Basel)*, 14(1), 1–21. DOI: 10.3390/su14074241

Gogoi, M., Buragohain, P. P., & Gogoi, P. (2022). Technical efficiency of organic tea growers of assam, India: A study in Dibrugarh district. *Estudios de Economía Aplicada*, 40(2), 1–10. DOI: 10.25115/eea.v40i2.6438

Goldar, B., & Sharma, A. K. (2015). Foreign investment in Indian industrial firms and impact on firm performance. *The Journal of Industrial Statistics*, 4(1), 1–18.

Hussain, S., Hussain, E., Saxena, P., Sharma, A., Thathola, P., & Sonwani, S. (2023). Navigating the impact of climate change in India: A perspective on climate action (SDG13) and sustainable cities and communities (SDG11). *Frontier in Sustainable Cities*, 5(1), 1–22. DOI: 10.3389/frsc.2023.1308684

Jyoti, B., & Singh, A. K. (2020). Projected sugarcane yield in different climate change scenarios in Indian states: A state-wise panel data exploration. *International Journal of Food and Agricultural Economics*, 8(4), 343–365. https://www.foodandagriculturejournal.com/vol8.no4.pp343.pdf

Jyoti, B., Singh, A. K., & Ashraf, S. N. (2023). Appropriate technologies and their implications in the agricultural sector. In Alex Khang, P. H. (Ed.), *Advanced Technologies and AI-Equipped IoT Applications in High-Tech Agriculture* (pp. 65–87). IGI Global. DOI: 10.4018/978-1-6684-9231-4.ch004

Kandpal, A., Kumara, K., Sendhil, R., & Balaji, S. J. (2022). Technical efficiency in Indian wheat production: regional trends and way forward. In Kashyap, P. L., Gupta, V., Gupta, O. P., Sendhi, R., Gopalareddy, K., Jasrotia, P., & Singh, G. P. (Eds.), *New Horizons in Wheat and Barley* (pp. 475–490). Spinger Nature. DOI: 10.1007/978-981-16-4134-3_17

Kathayat, B., Dixit, A. K., & Chandel, B. S. (2021). Inter-state variation in technical efficiency and total factor productivity of India's livestock sector. *Agricultural Economics Research Review*, 34(3), 59–72. DOI: 10.5958/0974-0279.2021.00015.X

Kumar, A., & Sharma, P. (2014). Climate change and sugarcane productivity in India: An econometric analysis. *Journal of Social and Development Sciences*, 5(2), 111–122. DOI: 10.22610/jsds.v5i2.811

Kumar, A., Sharma, P., & Ambrammal, S. K. (2015). Climatic effects on sugarcane productivity in India: A stochastic production function application. *International Journal of Economics and Business Research*, 10(2), 179–203. DOI: 10.1504/IJEBR.2015.070984

Kumar, R. A., & Paul, M. (2019). *Industry level analysis of productivity growth under market imperfections*. Working Paper, 207, Institute for Studies in Industrial Development, New Delhi.

Kumar, S., & Arora, N. (2012). Evaluation of technical efficiency in Indian sugar industry: An application of full cumulative data envelopment analysis. *Eurasian Journal of Business and Economics*, 5(9), 57–78.

Liu, Y., Ruiz-Menjivar, J., Zavala, M., & Zhang, J. (2023). Examining the effects of climate change adaptation on technical efficiency of rice production. *Mitigation and Adaptation Strategies for Global Change*, 28(55), 1–17. DOI: 10.1007/s11027-023-10092-3

Mahajan, V., Nauriyal, D. K., & Singh, S. P. (2014). Efficiency and ranking of Indian pharmaceutical industry: Does type of ownership matter? *Eurasian Journal of Business and Economics*, 7(14), 29–50. DOI: 10.17015/ejbe.2014.014.02

Maity, S., & Singh, K. (2021). Frontier production functions, technical efficiency and panel data: With application to tea gardens in India. *International Journal of Business and Globalisation*, 27(4), 571–591. DOI: 10.1504/IJBG.2021.113797

Maroof, Z., Hussain, S., Jawad, M., & Naz, M. (2019). Determinants of industrial development: A panel analysis of South Asian economies. *Quality & Quantity*, 53(1), 391–1419. DOI: 10.1007/s11135-018-0820-8

Mazumdar, M., Rajeev, M., & Ray, S. C. (2009). Output and input efficiency of manufacturing firms in India: A case of the Indian pharmaceutical sector. Working Paper No. 219, The Institute for Social and Economic Change, Bangalore, India.

Mehta, Y., & Johan, R. A. (2017). Manufacturing sectors in India: Outlook and challenges. *Procedia Engineering*, 174(1), 90–104. DOI: 10.1016/j.proeng.2017.01.173

Mitra, A., Sharma, C., & Véganzonès-Varoudakis, M. A. (2016). Infrastructure, ICT and firms' productivity and efficiency: An application to the Indian manufacturing. In De Beule, F., & Narayanan, K. (Eds.), *Globalization of Indian Industries*. India Studies in Business and Economics. DOI: 10.1007/978-981-10-0083-6_2

Naudé, W. (2010). Climate change and industrial policy. *Sustainability (Basel)*, 3(7), 1003–1021. DOI: 10.3390/su3071003

Nzuza, Z. W. (2021). Effect of climate change on the manufacturing sector. In Olarewaju, O. M., & Ganiyu, I. O. (Eds.), *Handbook of Research on Climate Change and the Sustainable Financial Sector* (pp. 463–476). IGI Global. DOI: 10.4018/978-1-7998-7967-1.ch028

Pattnayak, S. S., & Chadha, A. (2013). Technical efficiency of Indian pharmaceutical firms: A stochastic frontier function approach. https://conference.iza.org/conference_files/pada2009/pattnayak_s5193.pdf

Rajesh, R. S. N. (2007). *Technical efficiency in the informal manufacturing enterprises: Firm level evidence from an Indian state*. MPRA Paper No. 7816. https://mpra.ub.uni-muenchen.de/7816/

Sahu, P. K. (2015). Technical efficiency of domestic and foreign firms in Indian manufacturing: A firm level panel analysis. *Arthshastra Indian Journal of Economic & Research*, 4(2), 7–21. DOI: 10.17010/aijer/2015/v4i2/65534

Sahu, S. K., & Narayanan, K. (2011). Determinants of energy intensity in Indian manufacturing industries: A firm level analysis. *Eurasian Journal of Business and Economics*, 4(8), 13–30.

Sahu, S. K., & Narayanan, K. (2015). Environmental certification and technical efficiency: A study of manufacturing firms in India. *Journal of Industry, Competition and Trade*, 16(2), 1–17. DOI: 10.1007/s10842-015-0213-9

Satpathy, L. D., Chatterjee, B., & Mahakud, J. (2017). Firm characteristics and total factor productivity: Evidence from Indian manufacturing firms. *Margin*, 11(1), 77–98. DOI: 10.1177/0973801016676013

Sen, J., & Das, D. (2016). Technical efficiency in India's unorganized manufacturing sector: A non-parametric analysis. *International Journal of Business and Management*, 4(4), 92–101.

Singh, A. K., Ashraf, S. N., & Arya, A. (2019a). Estimating factors affecting technical efficiency in Indian manufacturing sector. *Eurasian Journal of Business and Economics*, 12(24), 65–86. DOI: 10.17015/ejbe.2019.024.04

Singh, A. K., & Jyoti, B. (2019). Measuring the climate variability impact on cash crops farming in India: An Empirical Investigation. *Agriculture and Food Sciences Research*, 6(2), 155–164. DOI: 10.20448/journal.512.2019.62.155.165

Singh, A. K., & Jyoti, B. (2023). Appropriate technology and adaptation strategies mitigate the adverse impact of climate change on the agricultural sector: A case study in Gujarat, India. In Alex Khang, P. H. (Ed.), *Advanced Technologies and AI-Equipped IoT Applications in High-Tech Agriculture* (pp. 389–416). IGI Global. DOI: 10.4018/978-1-6684-9231-4.ch023

Singh, A. K., Jyoti, B., & Sankaranarayanan, K. G. (2024). A review for analyzing the impact of climate change on agriculture and food security in India. In Samanta, D., & Garg, M. (Eds.), *The Climate Change Crisis and Its Impact on Mental Health* (pp. 227–251). IGI Global. DOI: 10.4018/979-8-3693-3272-6.ch017

Singh, A. K., & Kumar, S. (2022a). Assessing the performance and factors affecting industrial development in Indian states: An empirical analysis. *Journal of Social Economic Research*, 8(2), 135–154. DOI: 10.18488/journal.35.2021.82.135.154

Singh, A. K., & Kumar, S. (2022b). Expert's perception on technology transfer and commercialization, and intellectual property rights in India: Evidence from selected research organizations. *Journal of Management, Economics, and Industrial Organization*, 6(1), 1–33. DOI: 10.31039/jomeino.2022.6.1.1

Singh, A. K., & Kumar, S. (2022c). Measuring the factors affecting annual turnover of the firms: A case study of selected manufacturing industries in India. *International Journal of Business Management and Finance Research*, 5(2), 33–45. DOI: 10.53935/26415313.v5i2.211

Singh, A. K., Narayanan, K. G. S., & Sharma, P. (2017). Effect of climatic factors on cash crop farming in India: An application of Cobb-Douglas production function model. *International Journal of Agricultural Resources, Governance and Ecology*, 13(2), 175–210. DOI: 10.1504/IJARGE.2017.086452

Singh, A. K., Narayanan, K. G. S., & Sharma, S. (2019b). Measurement of technical efficiency of climatic and non-climatic factors in sugarcane farming in Indian staes: Use of stochastic frontier production function approach. *Climatic Change*, 5(19), 150–166.

Singh, A. K., & Singh, B. J. (2021). Projected productivity of cash crops in different climate change scenarios in India: Use of marginal impact analysis technique. *Finance & Economics Review*, 3(1), 63–87. DOI: 10.38157/finance-economics-review.v3i1.281

Singh, A. K., Singh, B. J., & Ashraf, S. N. (2020). Implications of intellectual property protection, and science and technological development in the manufacturing sector in selected economies. *Journal of Advocacy. Research in Education*, 7(1), 16–35.

Singh, D., & Malik, G. (2018). Technical efficiency and its determinants: A panel data analysis of Indian public and private sector banks. *Asian Journal of Accounting Perspectives*, 11(1), 48–71. DOI: 10.22452/AJAP.vol11no1.3

Soni, A., Mittal, A., & Kapshe, M. (2017). Energy intensity analysis of Indian manufacturing industries. *Resource-Efficient Technologies*, 3(1), 353–357.

Sudarshan, A., Somanathan, E., Somanathan, R., & Tewari, M. (2021). Climate change may hurt Indian manufacturing due to heat stress on workers. https://epic.uchicago.in/climate-change-may-hurt-indian-manufacturing-due-to-heat-stress-on-workers/

Thampy, A., & Tiwary, M. K. (2021). Local banking and manufacturing growth: Evidence from India. *IIMB Management Review,* 33(2), 95-104. DOI: 10.1016/j.iimb.2021.03.013

Tyagi, S., & Nauryal, D. K. (2016). Profitability determinants in Indian drugs and pharmaceutical industry: An analysis of pre and post TRIPS period. *Eurasian Journal of Business and Economics*, 9(17), 1–21. DOI: 10.17015/ejbe.2016.017.01

Xu, G., & Zhang, W. (2024). Abating carbon emissions at negative costs: Optimal energy reallocation in China's industry. *Environmental Impact Assessment Review*, 105(1), 1–13. DOI: 10.1016/j.eiar.2023.107388

Zahid, K. B., & Shah, H. (2024). Impact of climate change on farm-level technical efficiency in Punjab, Pakistan. *Climate Research*, 92(1), 65–77. DOI: 10.3354/cr01727

Section 2
Evaluation of Climate Change Through Economy

Chapter 6
The Economic Impacts of Climate Change

Firat Cem Dogan
https://orcid.org/0000-0002-2398-1484
Hasan Kalyoncu University, Turkey

ABSTRACT

Climate change profoundly impacts global economies, particularly the agricultural sector, through altered temperature and precipitation patterns, threatening food security and economic stability. This research examines climate change's effects on agriculture and economic growth, proposing mitigation policies. Methodologically, it analyzes the climate-economy relationship theoretically and empirically. Findings reveal declining agricultural productivity, disrupted supply chains, and increased natural disasters, necessitating a transition to sustainable energy and agriculture. Global collaboration and steadfast policies are crucial for addressing climate change, ensuring both economic prosperity and environmental sustainability.

1. INTRODUCTION

Over the course of billions of years, the phenomenon of climate change, occurring within the natural balance of the Earth, has significantly veered towards adverse consequences with the escalating human activities accompanying the Industrial Revolution. The increase in human activities has led to a notable augmentation of greenhouse gas emissions into the environment, consequently accelerating global warming. This escalating global warming has heightened the frequency of meteo-

DOI: 10.4018/979-8-3693-5792-7.ch006

rological disasters such as droughts and floods, thereby exerting negative impacts on the economy.

As a consequence of the widespread utilization of fossil fuels, climate change has emerged as a prominent issue globally, elevating temperatures and giving rise to various problems. These impacts affect numerous areas such as water resources, forests and vegetation, biodiversity, agriculture, and human health. Among the increasing adverse effects in recent years are alterations in precipitation patterns, temperature rises, droughts, desertification, and natural disasters. These detrimental effects pose a significant threat to the economy by influencing agricultural productivity and growth rates. Climate change scenarios encompass higher temperatures, changes in precipitation patterns, and an increase in atmospheric carbon dioxide (CO_2) concentrations (Mahato, 2014: 1).

Following the Industrial Revolution, the increasing energy demand was met through the utilization of fossil fuels, significantly elevating greenhouse gas emissions in the atmosphere. Measurements indicate a marked increase in the concentrations of carbon dioxide (CO_2), methane, and nitrous oxide in the atmosphere since 1750 as a consequence of human activities. CO_2 stands out as the most significant anthropogenic greenhouse gas in this regard. Prior to the Industrial Revolution, the global atmospheric concentration of CO_2 was 280 parts per million (ppm), whereas by 2005, this value had risen to 379 ppm, surpassing the natural range of 180 to 300 ppm (IPCC, 2007: 2).

The factors such as climate change and global warming, driven by high greenhouse gas emissions, pose a serious threat to both developed and developing countries. Regular occurrences of floods, heatwaves, and storms due to climate change negatively impact agricultural production and the livestock sector, consequently harming the agricultural economy. Countries most affected by these effects experience significant economic losses due to floods and storms. Despite the alterations in the environmental ecosystem caused by the global temperature rise over the past decade, farmers in less developed countries often continue traditional farming methods. This situation leads to the inability of these countries to adopt modern agricultural technologies to adapt to temperature increases and enhance productivity. Consequently, these countries become vulnerable to losses caused by climate change. As a result, agricultural production decreases, leading to a reduction in the share of the agricultural sector in Gross Domestic Product (GDP). This adversely affects economic growth in these countries. Therefore, countries struggling to adopt adaptation and resilience strategies to climate change experience a decline in agricultural income and face difficulty in sustainably increasing their GDP (Khalid et al., 2016: 40).

Climate change has become a significant source of concern worldwide in contemporary times. The increase in greenhouse gases in the atmosphere, along with factors such as global temperature fluctuations and rising sea levels, also impacts economic systems. In this section, we will focus on the theoretical foundations of climate change and delve into the potential effects of these changes on economic structures. We will attempt to understand the economic ramifications of climate change from the economic consequences of global warming caused by greenhouse gases to the sectoral impacts of rising sea levels, considering a broad perspective. This analysis could aid in developing more effective strategies to combat climate change, taking into account not only economic aspects but also social and environmental dimensions.

In this framework, we can initially concentrate on the underlying causes of the accumulation of greenhouse gases in the atmosphere and the resultant global temperature rise. A comprehensive understanding of the greenhouse effect mechanism and the intricate interactions between ecosystems and economic activities can assist in better comprehending the economic implications of climate change. Subsequently, we will evaluate the sectoral-level impacts of climate change by examining various sectors such as agriculture, energy, water resources, and transportation. By delving into sector-specific risks, opportunities, and adaptation strategies, we aim to comprehend how each sector can effectively cope with climate change. Lastly, we can assess the effects of globally implemented policies and strategies concerning the economic consequences of climate change. Matters such as international cooperation, carbon reduction targets, and financing mechanisms play a pivotal role in shaping the economic effectiveness of global endeavors to combat climate change.

This section will establish a foundation towards understanding the complexity of climate change within an economic context and provide readers with a more informed perspective on coping with climate change.

2. CONCEPTUAL AND THEORETICAL FRAMEWORK

Climate is a dynamic system subject to natural processes that evolve over long periods of time. This process, where natural factors develop spontaneously and human factors are excluded, is referred to as climate variability. Climate change, on the other hand, denotes the alteration in climate resulting from human activities directly or indirectly modifying the composition of the atmosphere, in addition to the natural variability observed over comparable timeframes (Duffy, 2008: 545; IPCC, 1992: 3).

Climate change is a universal issue that profoundly impacts ecosystems, climate systems, and societies on our planet. This section aims to provide a broad theoretical perspective on the economic dimensions of climate change. It seeks to establish a theoretical framework to comprehend the complex effects of climate change on various sectors such as agriculture, energy, health, infrastructure, and others.

Economic growth and social progress have led to significant changes in human lifestyles worldwide. These changes have intensified competition between developed and developing countries. In this competitive environment, countries make considerable efforts to enhance their prosperity and develop their physical and human capital (Shahbaz et al., 2016).

In the last twenty years, particularly in developing countries, welfare levels have steadily risen, largely due to the impact of technological advancements (Sarkodie et al., 2019). However, this increase has been accompanied by a notable rise in energy consumption, concurrent with the expansion of economic activities and industrialization processes (Pata et al., 2016: 255).

As developing countries pursue their economic growth objectives, concerns about the environmental impacts and sustainability issues arising from the increase in energy consumption are becoming increasingly significant. This situation underscores efforts to ensure the sustainability of economic development and enhance energy efficiency solutions. High energy demand may necessitate effective management of environmental resources and expedite the transition to renewable energy sources.

Globalization, while promoting economic growth, is also one of the key factors bringing about environmental changes. While globalization undeniably supports development, it also brings significant issues such as environmental degradation and ecological pollution as negative externalities. In recent years, the environmental impacts of globalization and trade liberalization have become fundamental issues in international trade (Shahzadi et al., 2019).

Globalization, particularly with the increasing importation of resource-intensive products such as natural gas, minerals, petroleum, and coal, has been contributing to harmful effects on regional environments and escalating environmental pressure. This situation has led environmental issues to become not only national but also global concerns, as the accelerated transfer of resources from one country to another exacerbates environmental problems (Kan et al., 2019).

Country-level assessments indicate that the increased demand, production, and consumption pressures brought about by globalization's impact on the economy, particularly the release of toxic gases into the atmosphere, contribute to the deterioration of environmental quality (Kumar, 2019).

This situation has sparked a significant debate on how globalization should be balanced with environmental sustainability. Developing effective policies and strategies to manage the economic benefits brought about by globalization in harmony with environmental responsibility and sustainability has become a global imperative. In this context, it is critically important for both developed and developing countries to develop strategic policies to sustainably balance economic growth and welfare increases, minimize environmental impacts, and enhance energy efficiency. These efforts should aim to strike a balance not only for economic growth but also for environmental sustainability.

2.1 Historical Context of Climate Change

The discussion will cover the evolution of climate change science, including key milestones in scientific discovery and policy development, such as the establishment of the IPCC, the Kyoto Protocol, and the Paris Agreement.

2.1.1 Evolution of Climate Change Science

The scientific understanding of climate change began in the late 19th century with foundational work by Svante Arrhenius. In 1896, Arrhenius proposed that increasing concentrations of carbon dioxide (CO_2) in the atmosphere would enhance the greenhouse effect, leading to global warming (Arrhenius, 1896). His theory was among the first to link anthropogenic activities with climate changes, marking a significant departure from the then-prevailing climatic theories. Arrhenius's hypothesis was based on the idea that CO_2 traps heat radiating from the Earth's surface, thus increasing global temperatures.

2.1.2 Advancements in the Mid-20th Century

The 1950s brought substantial advancements in climate science, notably through the work of Charles David Keeling. Keeling's measurements of atmospheric CO_2 at Mauna Loa Observatory, beginning in 1958, revealed a consistent increase in CO_2 concentrations, now known as the Keeling Curve (Keeling et al., 1976). This empirical evidence supported the earlier theoretical predictions and underscored the impact of human activities on atmospheric composition. The Keeling Curve remains one of the most crucial datasets in climate science, illustrating the ongoing rise in greenhouse gases.

2.1.3 Formation of International Climate Bodies

The late 20th century saw the formalization of international efforts to address climate change. In 1988, the Intergovernmental Panel on Climate Change (IPCC) was established by the United Nations Environment Programme (UNEP) and the World Meteorological Organization (WMO) (IPCC, 1988). The IPCC's mission was to assess scientific research on climate change, its impacts, and potential adaptation and mitigation strategies. The IPCC's comprehensive reports have been instrumental in shaping global climate policy and increasing awareness of climate issues.

2.1.4 Key International Agreements

The establishment of the IPCC was followed by significant international agreements aimed at combating climate change. The 1992 Earth Summit in Rio de Janeiro resulted in the creation of the United Nations Framework Convention on Climate Change (UNFCCC), a foundational treaty that set the stage for future climate negotiations and actions (UNFCCC, 1992). The UNFCCC established a framework for international cooperation to address climate change, focusing on reducing greenhouse gas emissions and supporting climate resilience. In 1997, the Kyoto Protocol was adopted at the Third Conference of the Parties (COP3) to the UNFCCC. The Kyoto Protocol was a landmark agreement that set legally binding targets for developed countries to reduce their greenhouse gas emissions (UNFCCC, 1997). The protocol represented a significant step forward in international climate policy, establishing the principle of differentiated responsibilities between developed and developing countries. The Paris Agreement, adopted in 2015 at COP21 in Paris, marked a new era in international climate diplomacy. The agreement aims to limit global warming to well below 2°C above pre-industrial levels and to pursue efforts to limit the temperature increase to 1.5°C (UNFCCC, 2015). Unlike the Kyoto Protocol, which focused on specific emission reduction targets, the Paris Agreement emphasizes a universal approach, with each country setting its own climate action plans and reporting on progress. The Paris Agreement represents a collective commitment to addressing climate change and fostering global cooperation.

2.1.5 Recent Developments and Ongoing Challenges

In recent years, the focus has shifted towards implementing the commitments made under the Paris Agreement and enhancing climate resilience. The Intergovernmental Panel on Climate Change (IPCC) has continued to provide periodic assessments, reflecting the latest scientific knowledge and offering policy recommendations (IPCC, 2021). The latest reports emphasize the urgency of reducing emissions and adapting to the impacts of climate change, highlighting the critical role of national and international policies in achieving climate goals.

Despite these efforts, challenges remain in translating international agreements into effective action. Issues such as climate finance, equitable distribution of responsibilities, and technological innovation continue to influence the global response to climate change. Ongoing research and international negotiations aim to address these challenges and advance global climate policy.

3.1 In-Depth Analysis of Climate Change Impacts

The chapter provides a solid introduction to the core concepts of climate variability and climate change, distinguishing between natural and anthropogenic factors. This foundation is crucial for understanding the economic implications of climate change. However, a deeper exploration of specific problems such as extreme weather events, biodiversity loss, and health impacts is necessary for a comprehensive overview.

3.1.1 Extreme Weather Events

Climate change has significantly intensified the frequency and severity of extreme weather events, which have profound socio-economic impacts. For instance, heatwaves have become more common and severe due to rising global temperatures. Research indicates that heatwaves are linked to increased mortality rates and adverse health effects. The European heatwave of 2003, for example, resulted in approximately 70,000 excess deaths across the continent (Robine et al., 2008). Studies suggest that such extreme temperature events will become more frequent in the future due to ongoing global warming (Fischer & Schär, 2010).

Hurricanes have also become more intense as a result of warmer ocean temperatures. Emanuel (2005) shows that the energy available for hurricanes, known as Tropical Cyclone Heat Potential, has increased, leading to more powerful storms. The impact of hurricanes on infrastructure and economies has been severe, as evidenced by Hurricane Katrina in 2005 and Hurricane Maria in 2017, both of which caused significant damage and economic disruption (Klein & Nicholls, 2004; Boaden et al., 2020).

In addition to hurricanes, increased precipitation and extreme weather patterns have exacerbated flooding, while altered precipitation patterns contribute to prolonged droughts. Climate change has increased the likelihood of extreme precipitation events, resulting in more frequent and severe flooding (Pall et al., 2007). Conversely, rising temperatures and changing precipitation patterns are leading to more intense and prolonged droughts, affecting water availability and agricultural productivity (Trenberth et al., 2014).

3.1.2 Loss of Biodiversity

Climate change also poses a significant threat to global biodiversity. The shifting climate alters habitat conditions, impacting both flora and fauna. Many ecosystems, such as coral reefs and temperate forests, are experiencing significant stress due to rising temperatures and acidification. Coral reefs, in particular, face severe bleaching events, with approximately 75% of the world's coral reefs affected in the past decade (Hughes et al., 2017).

Changes in temperature and precipitation patterns disrupt species' life cycles and migration patterns, leading to an increased risk of extinction. Parmesan and Yohe (2003) document shifts in species distributions and changes in migration patterns, with many species struggling to adapt to rapidly changing environmental conditions. The loss of biodiversity undermines ecosystem services, which are crucial for human well-being.

3.1.3 Health Impacts

The health impacts of climate change are extensive and multifaceted. Increased temperatures contribute to heat-related illnesses such as heatstroke and dehydration. The World Health Organization (2021) highlights that heatwaves can exacerbate pre-existing health conditions and lead to increased mortality, particularly among vulnerable populations such as the elderly and those with chronic illnesses.

Climate change also influences the distribution of vector-borne diseases. Rising temperatures and changing precipitation patterns affect the habitats of vectors like mosquitoes and ticks, potentially expanding the range of diseases such as malaria and dengue fever (Ryan et al., 2019).

Moreover, climate change impacts mental health, with increased stress and anxiety related to extreme weather events and the perceived threat of future climate impacts. Research discusses how extreme weather events and environmental degradation contribute to mental health issues, including depression and anxiety (Cianconi et al., 2020).

In this subsection, first, the Environmental Kuznets Curve will be explained. Following that, indices measuring climate change and data related to climate change will be presented with figures.

4.1 Environmental Kuznets Curve

In a study conducted by Simon Kuznets in 1955, an important perspective on the relationship between economic development, income inequality, and economic growth was presented (Kuznets, 1955). This study suggested that at the onset of a country's development process, economic growth and income inequality would increase. According to Kuznets, as countries reach higher levels of development, there is a decrease in income inequality alongside the shift of labor from agriculture to industry, and the relationship between growth and inequality takes the form of an "Inverted U." At the beginning of development, income disparities arise due to higher incomes in the industrial and service sectors compared to agriculture. However, as the development process progresses, a decrease in income inequality is observed. Factors such as increased participation in education, balanced income distribution, and improvements in labor quality play a significant role in this decrease. These factors emerging alongside the development process play a key role in reducing income inequality. In this context, the relationship between economic growth and changes in income distribution takes shape according to the stages of development, and this situation can be considered an important guide in determining countries' development policies (Kuznets, 1955).

Simon Kuznets' Environmental Kuznets Curve graph is depicted in Figure 1.

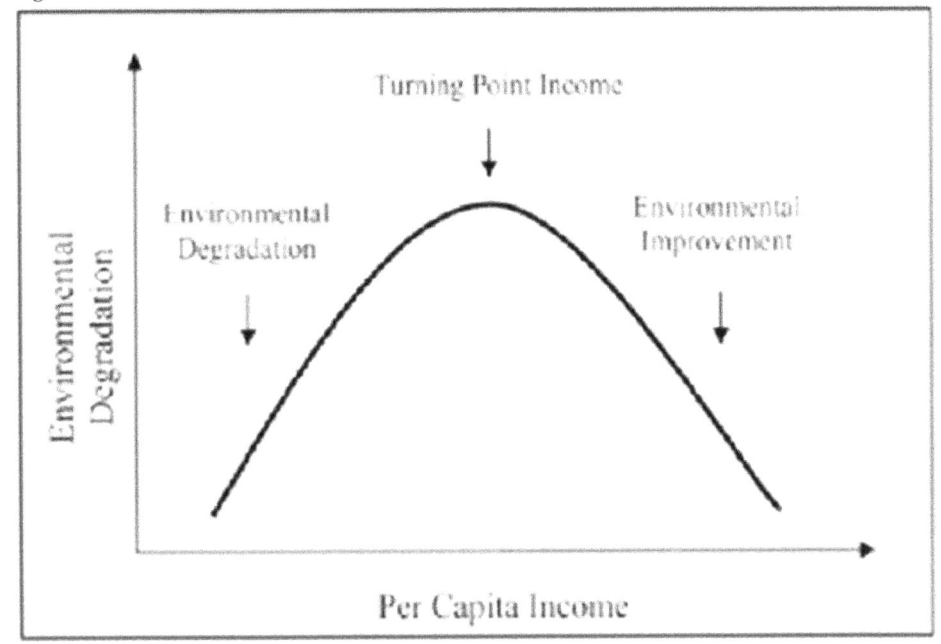

Figure 1. Environmental Kuznets Curve

Source: *Figure created by the author.*

Environmental degradation increases with increasing income up to a certain threshold level, beyond which environmental quality improves as income per capita rises. This relationship is illustrated by an inverted U-shaped curve, as shown in Figure 1. It is commonly referred to as the Environmental Kuznets Curve (EKC), which stems from the findings of Kuznets (1955). The EKC hypothesis aims to depict the long-term relationship between environmental impact and economic growth. As economies develop, particularly during the initial stages marked by agricultural intensification and resource extraction, the rate of resource depletion surpasses the rate of resource replenishment, leading to an increase in both the quantity and toxicity of waste generation. At more advanced stages of development, there is a structural shift towards information-intensive industries and services. This transition is accompanied by heightened environmental awareness, stricter enforcement of environmental regulations, advancements in technology, and increased investments in environmental protection. Consequently, environmental degradation stabilizes and gradually diminishes. As income surpasses the turning point of the EKC, it is assumed that the transition towards improved environmental quality commences. Thus, the EKC illustrates the natural progression of economic development, starting from a pristine agrarian economy, transitioning into a polluting industrial economy, and ultimately evolving into a cleaner service-based economy (Arrow et al., 1995).

This situation demonstrates that, as predicted by the Environmental Kuznets Curve hypothesis, the relationship between economic growth and environmental degradation can reverse after a certain point.

The explanation of the Environmental Kuznets Curve emphasizes three main factors: scale effect, composition effect, and technology/technique effect. The scale effect focuses on the relationship between economic growth and the increase in production volume. This increase leads to a higher utilization of natural resources. The increased use of natural resources, in turn, results in resource depletion and increased waste generation (Grossman and Krueger, 1991).

The composition effect focuses on the share of the industrial sector in economic activities. As income levels increase, over time, the share of industry in the total economy decreases while the share of the service sector and knowledge-based sectors increases. These sectors generally cause less harm to the environment compared to industry, and environmental degradation decreases after reaching a certain income level (Janicke et al., 1997). The composition effect encompasses the transition from an agricultural society to an industrial society, and then to a knowledge-based society (Dinda, 2004; Karaca, 2012).

The technological effect emphasizes that the use of outdated technologies leads to inefficient use of resources and increased waste in output-oriented production systems (Yandle, 2004). An increase in the country's income level generally leads to increased investment in research and development (R&D) activities. The new technologies emerging from R&D activities can be effective in reducing environmental damage (Grossman and Krueger, 1991).

The scale effect represents the increase in environmental degradation in the Environmental Kuznets Curve, while the technological effect represents the decrease in environmental degradation. The composition effect, on the other hand, is a factor that entirely influences the Environmental Kuznets Curve (Karaca, 2012). These three effects provide an important framework for understanding the complex relationships between environmental change and economic growth.

Grossman and Krueger (1991) laid the foundations of the Environmental Kuznets Curve (EKC) hypothesis, initiating the discussion on the dilemma between economic growth and environmental pollution in this study. The EKC hypothesis suggests that environmental pollution increases in the early stages of income growth, but decreases after income reaches a certain level. The EKC hypothesis, which includes a second-degree equation, provides researchers with important information such as the shape of the income-environmental quality relationship (if the EKC is valid, inverse-U shape; if not, U shape), the turning point of income, and whether the scale, structural, or technological effect is valid. During the period when the scale effect is valid, income increases lead to an increase in energy use. In the structural effect period, gains resulting from economic growth trigger structural transformation in the economy. In the final stage, the technological effect comes into play, and green technology is used to improve environmental quality while environmentally harmful energy technologies are abandoned. Thus, understanding the dynamics between environmental degradation and economic growth within the framework of the EKC hypothesis provides an important roadmap for shaping environmental policies and economic strategies at various stages (Çağlar, 2022: 914-915).

4.1.1 Global Climate Risk Index

Global Climate Risk Index determines how much countries are affected by extreme weather events. These events can be meteorological events (such as tropical storms or hurricanes), hydrological events (such as storm surges or flash floods), or climatological events (such as wildfires or droughts). Index scores are derived from rankings of countries in the following indicators and calculated by averaging them according to weights (Germanwatch, 2021).

Number of deaths - Weight: 1/6
Number of deaths per 100,000 people - Weight: 1/3
Total of losses in dollars based on purchasing power parity (PPP) - Weight: 1/6
Losses per capita Gross Domestic Product (GDP) - Weight: 1/3

The Global Climate Risk Index is depicted in Figure 2.

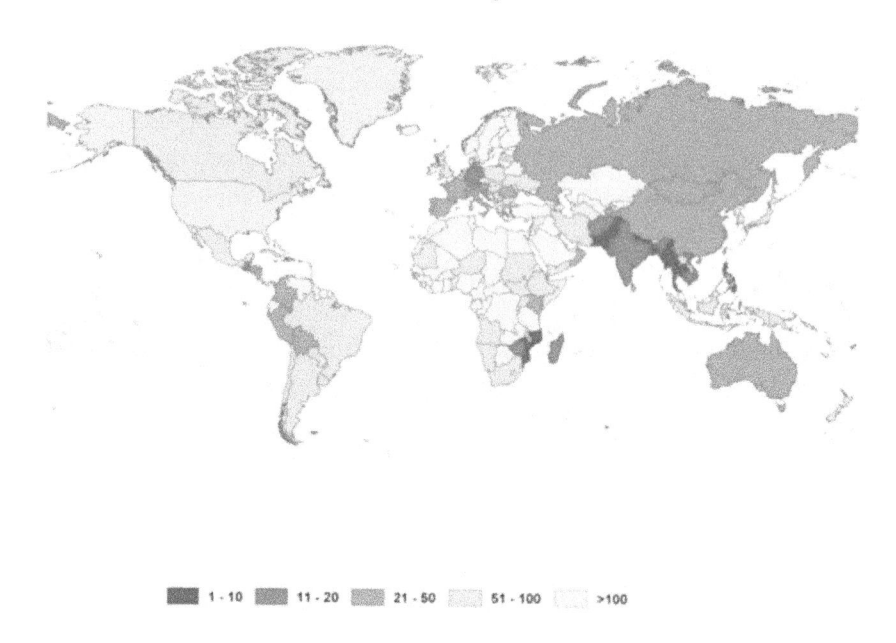

Figure 2. 2021 Global Climate Risk World Map

■ 1 - 10   ■ 11 - 20   ■ 21 - 50   ▨ 51 - 100   ▨ >100

Source: *Germanwatch 2021.*

According to Figure 2, countries most affected by weather events are observed particularly in countries located in Asia and the Pacific, as well as Australia, in the southern part of Europe, and in countries in the western part of South America. The least affected countries are those in North Africa, Northern Europe, and the Middle East.

4.1.2 Economic Impacts of Climate Change

Global warming and climate change have significant impacts on the economies of both developed and developing countries. If necessary emission reduction and adaptation measures are not taken, it appears inevitable that these impacts will reach significant proportions. Due to the greenhouse effect, it is estimated that a temperature increase of 1-2°C on the Earth's surface in the future will affect about 10% of ecosystem regions. While some forest ecosystems may expand, other regions may experience adverse effects such as increased forest fires and insect infestations. Additionally, marine life and coral reefs in the oceans will also be negatively affected by climate change. According to current calculations, the economic cost of just a 1°C increase in global warming is estimated to reach $2 trillion annually by 2050.

A study conducted in the EU suggests that the cumulative global economic cost of global warming could amount to 74 trillion Euros (Bağraç and Doğan, 2016).

To avoid the cost ranging between 5% and 20% of the global economy after 2050 due to global warming, it is emphasized that a global investment is required for reducing greenhouse gas emissions. According to the IPCC, carbon emissions can be reduced by 20% to 40% using current technology by the year 2020. While half of this emission reduction can be achieved without additional cost, it is estimated that the other half can be achieved with an approximate cost of 10 Euros per ton of CO_2 annually (IPCC, 2023: 1-8).

4.1.3 The Impact of Climate Change on Agriculture

Agriculture is a crucial economic activity tightly linked to climate and soil conditions. Therefore, any changes in climate will directly affect the agricultural sector. While agricultural activities contribute to climate change by emitting greenhouse gases, they are also influenced by changes in temperature and CO_2 levels. Generally, elevated CO_2 levels can positively impact crop growth, thereby increasing agricultural productivity. However, high temperatures may have differing effects, particularly contributing to the extension of the growing season in countries with cold climate conditions found at moderate to high latitudes, thus positively influencing the agricultural sector. Conversely, in countries with relatively warm climates at lower latitudes, high temperatures may have adverse effects.

Climate is a primary factor underlying agricultural production. In this context, changes in temperature, precipitation, atmospheric CO_2 levels, the frequency of extreme weather events, and rising sea levels affect the agricultural sector. These effects can be listed as follows (cited from Dellal and McCarl, 2007; Dellal, 2008: 105):

> Temperature, precipitation, CO_2 levels, and extreme weather events can significantly influence plant yield, harvesting time, and grazing efficiency. When droughts or heavy rainfall occur frequently and intensely, agricultural losses tend to increase. These situations can also affect agricultural product prices depending on changes in production quantity.
>
> For plant development, not only temperature and rainfall but also soil moisture level, moisture storage capacity, and soil fertility are critical. Losses in soil moisture are often associated with temperature increases. Irrigation may be applied to maintain soil moisture levels in response to temperature increases; however, excessive evaporation caused by high temperatures can complicate this balancing process.

Temperature increases can disrupt the balance between heat production in animals and the efficient use of heat. This can have negative effects on mortality rates, feed consumption rates, live weight gain, milk production, and pregnancy rates.

Temperature increases lead to increased evaporation, resulting in reduced irrigation water availability. Additionally, temperature increases affect the timing and duration of snowfall, reducing the amount of water needed during summer months.

Greenhouse gases are gases that absorb and re-emit long-wave infrared radiation in the atmosphere. These gases are natural components of the atmosphere and can also arise from human activities. Greenhouse gases are generally classified into two main groups: "natural greenhouse gases" and "indirect greenhouse gases." Water vapor (H_2O), carbon dioxide (CO_2), methane (CH_4), nitrous oxide (N_2O), and ozone (O_3) are among the natural greenhouse gases (UNFCCC, 1992: 3).

With the increase in production associated with globalization, greenhouse gas emissions have risen worldwide. This situation is depicted in Figure 3.

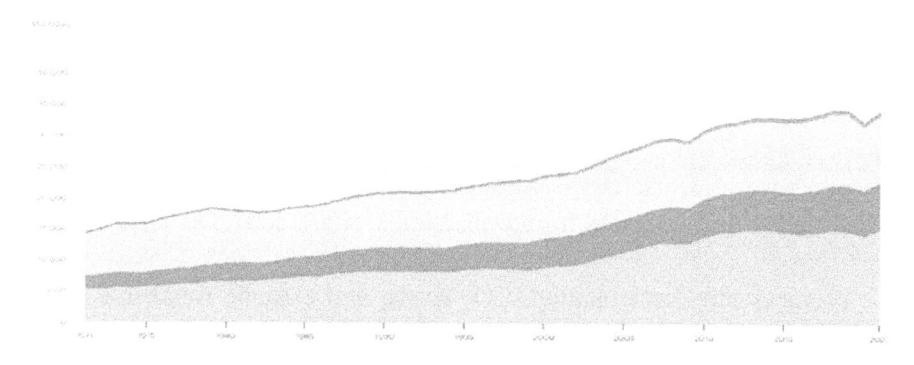

Figure 3. Greenhouse Gas Emissions from Fuels Worldwide, 1971-2021

Note: *Coal, oil shale, and oil are shown in turquoise, natural gas in navy blue, biofuels and waste in green.*
Source: *Compiled from the International Energy Agency's 2023 report.*

When examining Figure 3, it can be observed that greenhouse gas emissions from coal, oil shale, and oil were 5000 megatons in 1971, approximately doubling by 2021. Natural gas emissions, on the other hand, increased nearly threefold. There has not been a significant increase in greenhouse gas emissions from biofuels and waste.

Climate change also leads to rising sea levels, exposing coastal agricultural areas to flooding and saltwater intrusion. This situation can reduce productivity in these regions. Rising sea levels, accompanied by increased storm and flood events, can result in the erosion of arable land. These adverse effects can negatively impact groundwater resources. Additionally, high temperatures, drought, increased risk of forest fires, proliferation of pests, and the desertification process of agricultural areas can be triggered. This situation can also increase food prices worldwide (Reti, 2007: 54-55).

CO_2, emitted in any combustion process, is a gas released in higher quantities compared to other greenhouse gases. Its impact on global warming is associated with human activities due to its high volume, continuous emission, and long lifespan. Modern industrial societies contribute to climate change and global warming issues through the energy sources they utilize to sustain progress. While approximately 77% of greenhouse gases in the atmosphere consist of CO_2, the remaining 33% is attributed to CH_4, N_2O, and F-gases. About 75% of CO_2 emissions (57% of total greenhouse gases) in the atmosphere stem directly from the use of fossil fuels for energy production. The remaining emissions result from factors such as deforestation and improper land use (Uzmen, 2007: 54; Houghton, 2004: 29).

Especially since the Industrial Revolution, there has been a continuous increase in carbon dioxide emissions worldwide, parallel to the rising production levels. This trend is illustrated in Figure 4.

Figure 4. Trend of Carbon Dioxide Emissions (CO_2 ppm) in the World Between 1800-2020.

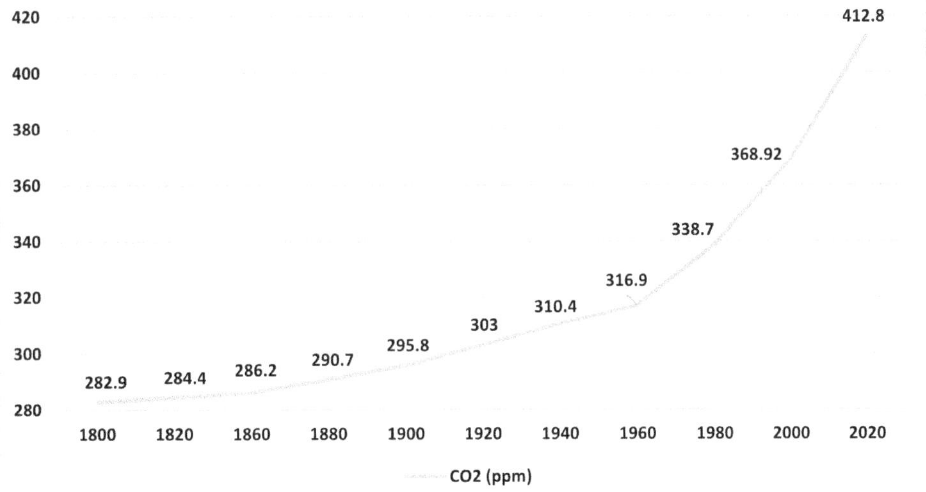

Year	CO2 (ppm)
1800	282.9
1820	284.4
1860	286.2
1880	290.7
1900	295.8
1920	303
1940	310.4
1960	316.9
1980	338.7
2000	368.92
2020	412.8

Source: *European Environment Agency (EEA) 2023.*

According to Figure 4, it is observed that after the Industrial Revolution, which began in England in 1760 and later spread to the United States and all of Europe, the continuous increase in carbon dioxide emissions in the world has been influenced by industrialization and mass production. Carbon dioxide emissions increased from 282.9 ppm in the 1800s to 303 ppm in 1920, and to 412.8 ppm in 2020.

The lifespan of a CO_2 molecule in the atmosphere is variable and typically ranges between 5 to 200 years. This implies that even if CO_2 emissions were to be halted, the Earth would continue to warm for a certain period. In order to maintain the Earth's former balance, the concentration of CO_2 in the atmosphere needs to be kept below 350 ppm. If the atmospheric CO_2 concentration cannot be reduced to the critical level of 350 ppm, the delicate balance and structure necessary for life on Earth will be completely lost (Hansen et al., 2008: 217).

Considering the lifespan of CO_2 in the atmosphere and the impact of intensive fossil fuel use, urgent measures are required to reduce CO_2 concentration and bring it down to 350 ppm. This is because any delay of 20 years in CO_2 reduction efforts would result in reaching the critical level in the 2300s, while a delay of 40 years would push it to the 3000s (Hansen et al., 2012: 8).

In light of this information, it is an undeniable fact that timely and effective intervention regarding carbon dioxide emissions holds crucial importance for long-term sustainability and the well-being of both humanity and the planet.

4.1.4 The Impact of Climate Change on the Economy

The increased temperatures associated with global warming, along with health issues caused by it, can lead to significant effects on labor productivity through heat stress, humidity, and other extreme weather events. Particularly in low- and middle-income countries, workers are more exposed to the effects of temperature due to inadequate ventilation conditions. The rise in daily temperatures and humidity, along with fluctuations in precipitation patterns due to global warming, may result in these effects being experienced more frequently and intensely. Heat stress can lead to a decrease in the physical and mental capacities of workers and consequently an increase in occupational accidents (Ramsey et al., 1983: 110).

This situation can lead to declines in labor productivity. Workers instinctively tend to reduce work intensity or increase the frequency of short rest breaks to cope with the heat. These behavioral changes can slow down workers' daily activities and tasks, thereby reducing output. Additionally, additional expenditures on occupational health services for workers can also increase costs (Kjellstrom et al., 2009: 217-218).

Climate change can potentially lead to decreases in agricultural productivity, resulting in price fluctuations that can challenge the central bank's capacity to combat inflation. This uncertainty may raise questions about the reliability of the central bank and its monetary policy in ensuring price stability (Başoğlu, 2014: 85).

To illustrate the global trend of price increases, *the Food and Agriculture Organization of the United Nations* (FAO) has developed monthly and annual food price indices. This index typically measures the average prices of food products during a specific period. It provides information to consumers and economic analysts about understanding and monitoring the general level of food prices in a particular geographic region or country. The food price index usually includes basic food items such as meat, dairy products, grains, vegetable oils, and sugar, which constitute a significant portion of consumer expenditures, reflecting the price changes of food products. Additionally, this index serves as an important indicator that highlights various economic and social factors such as food security, the effects of agricultural policies, and changes in consumer spending patterns (FAO, 2023).

Economic crises, climate change, and epidemic diseases trigger inflation worldwide, particularly increasing food prices. The food price index can be used as an indicator of inflation because increases in food prices often signify a rise in overall price levels. This situation is illustrated in Figure 5.

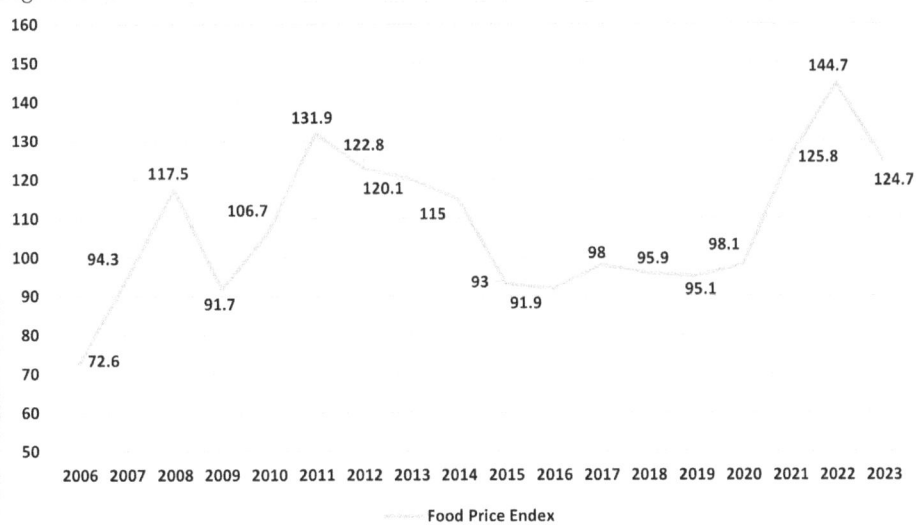

Figure 5. Food Price Index in the World Between 2006-2023 Years

Year	Food Price Index
2006	94.3
2007	117.5
2008	72.6
2009	106.7
2010	91.7
2011	131.9
2012	122.8
2013	120.1
2014	115
2015	93
2016	91.9
2017	98
2018	95.9
2019	95.1
2020	98.1
2021	125.8
2022	144.7
2023	124.7

Source: *Food and Agriculture Organization of the United Nations (FAO), 2023.*

Looking at Figure 5, it can be observed that following the global economic crisis originating in the United States in 2008 and subsequently spreading worldwide, the food price index rose from 91.7 to 131.9. Later, food prices returned to the 2009 level in 2016. Following the COVID-19 pandemic in 2019, the index surged from 95.1 to 144.7. In summary, economic crises, epidemic diseases, and climate change have increased the World Food Index.

Retrospective studies conducted by scientists reveal that carbon dioxide (CO2) levels have not risen to their current levels over the past 650,000 years (Walker and King, 2009: 34). It is known that this increase is largely attributed to human-induced causes. The issue of greenhouse gas emissions is particularly a problem associated with the contributions of highly industrialized developed countries (Climate Change and Power, 2002: 4-5).

Therefore, especially in highly developed countries with high per capita income, carbon dioxide and greenhouse gas emissions are higher compared to developing and underdeveloped countries. This situation is illustrated in Table 1.

Table 1. Per Capita Carbon Dioxide Emissions, Total Carbon Dioxide Emissions, and Per Capita Gross National Income Data in Developed, Developing, and Underdeveloped Countries as of 2023

	Country	Per Capita CO_2 Emissions (Tons)	CO_2 Emissions (Tons)	Per Capita Income ($)
Developed Countries	Norway	7,5	43,456,012	106.177,00
	Ireland	7,7	55,688,96	103.986,00
	Switzerland	3,6	39,666,930	93.259,00
	United States of America	14,9	5,011,686,600	76.329,00
	Iceland	9,5	38,684,15	73.446,00
	Denmark	4,9	39,689,77	67.790,00
	Australia	15,0	414,988,700	65.099,00
	Sweden	3,6	44,694,415	56.373,00
	Canada	14,2	675,918,610	54.917,00
	Germany	8,0	775,752,190	48.718,00
	England	4,7	367,860,350	46.125,00
	France	4,6	331,533,320	40.886,00
	Japan	8,5	1,239,592,060	33.823,00

	Country	Per Capita CO$_2$ Emissions (Tons)	CO$_2$ Emissions (Tons)	Per Capita Income ($)
Developing Countries	Russia	2,3	1,661,899,300	15.270,00
	Argentina	4,2	200,708,270	13.650,00
	China	8,0	10,432,751,400	12.720,00
	Mexico	4,0	441,412,750	11.496,00
	Türkiye	5,1	368,122,740	9.743,00
	Brazil	2,2	462,994,920	8.917,00
	South Africa	7,2	390,557,850	6.766,00
	Indonesia	2,2	530,035,650	4.788,00
	India	2,0	2,533,638,100	2.410,00
Underdeveloped Countries	Haiti	0,2	3,086,897	1.748,00
	Senegal	0,1	8,247,295	1.598,00
	Tanzania	0,1	9,731,560	1.192,00
	Ethiopia	0,2	10,438,855	1.027,00
	Rwanda	0,1	1,403,087	966,00
	Uganda	0,1	5,009,493	964,00
	Liberia	0,2	846,7	754,00
	Malawi	0,1	1,815,598	645,00
	Niger	0,1	2,088,475	585,00
	Central African Republic	0,1	1,526,878	427,00
	Sierra Leone	0,1	1,272,332	259,00

Source: Data compiled from the World Bank and Our World in Data.

According to Table 1, it can be observed that countries belonging to the group of developed and developing countries, such as the USA, China, Japan, Germany, India, and Russia, have significantly high total and per capita carbon dioxide emissions. On the other hand, it is evident that the emissions of carbon dioxide in underdeveloped countries are considerably low.

In order to assess countries' efforts and performance regarding climate change, the Climate Change Performance Index (CCPI) was developed by the non-profit organizations Germanwatch and NewClimate Institute in the city of Bonn, Germany. The CCPI is a measure that evaluates countries' performance in combating climate change worldwide. It assesses countries' environmental efficiency, energy use, greenhouse gas emissions, renewable energy use, and performance in climate policies based on various criteria. This index aims to identify countries that are successful in combating climate change as well as those that need to make more efforts.

The main criteria considered by the CCPI (Climate Change Performance Index) in 2024 are as follows:

Energy Use and Efficiency: Countries' performance in the energy sector, clean energy use, and energy efficiency.

Renewable Energy: The level of investment in renewable energy sources and the amount of energy generated from these sources.

Greenhouse Gas Emissions: Reductions in countries' greenhouse gas emissions and the policies they implement to control emissions.

Forest Protection and Use: Policies and practices aimed at conserving forests and managing them sustainably.

Climate Policies: Policies and strategies implemented by countries to combat climate change.

The performance of countries in combating climate change is shown in Figure 6.

The Climate Change Performance Index evaluates 60 countries and the European Union, which are responsible for more than 90% of global greenhouse gas emissions, based on four categories: greenhouse gas emissions, renewable energy, energy use, and climate policy. These evaluations consist of five different levels: 'very high', 'high', 'medium', 'low', and 'very low'. The primary objective of the index is to determine the environmental impacts of participating countries, observe the effectiveness of national climate policies, and promote sustainable energy use. It aims to encourage global environmental responsibility by promoting environmentally friendly policies at the national level and contributing to effective measures against global climate change (CCPI, 2024).

The performance of countries in combating climate change is shown in Figure 6.

Figure 6. Map of Countries' Climate Change Performance

Source: *Climate Change Performance Index (CCPI) 2023.*

According to the climate change performance map shown in Figure 6, countries in Northern Europe, Central and Western Europe, and South Asia particularly demonstrate very high and high scores in the climate change performance index. These regions play a leading role in environmentally friendly policies and sustainable energy practices. On the other hand, some countries such as Canada, the USA, Russia, Turkey, and certain countries in the Middle East have very low scores in the climate change performance index. This highlights the need for further development of environmental sustainability and climate policies in these regions. These countries may need to take more effective measures in areas such as reducing greenhouse gas emissions and transitioning to renewable energy sources. The map visually illustrates the different performances in combating climate change globally, indicating the need for greater collaboration and efforts within the international community on this issue.

5. CONCLUSION

Climate change significantly affects economic systems worldwide and poses significant challenges in various sectors. Losses in agricultural productivity increase concerns about food security, while supply chain disruptions in the industrial sector and the increasing frequency of natural disasters threaten economic stability. Additionally, dependence on fossil fuels in the energy sector complicates the transition to sustainable energy. In this context, developing and implementing effective solutions to combat climate change is of critical importance. Transitioning to clean energy, implementing sustainable agricultural practices, and setting carbon reduction targets are important steps in increasing economic resilience. However, this transformation process requires global cooperation and steadfast policies. Consequently, climate change has become not only an environmental issue but also an economic priority. Therefore, it is important for the international community to come together, adopt sustainability-focused policies, and make concerted efforts to combat climate change. Taking determined action to combat climate change is vital for ensuring both economic prosperity and environmental sustainability and leaving a healthy world for future generations.

REFERENCES

Arrhenius, S. (1896). On the Influence of Carbonic Acid in the Air upon the Temperature of the Ground. *The London, Edinburgh and Dublin Philosophical Magazine and Journal of Science*, 41(251), 237–276. DOI: 10.1080/14786449608620846

Arrow, K., Bolin, B., Costanza, R., Folke, C., Holling, C. S., Janson, B., Levin, S., Maler, K., Perrings, C., & Pimental, D. (1995). Economic growth, carrying capacity, and the environment. *Science*, 15, 91–95. PMID: 17756719

Başoğlu, A. (2014). An Attempted Model and Econometric Analysis on the Economic Effects of Global Climate Change. *Karadeniz Technical University, Institute of Social Sciences*, Department of Economics, Doctoral Program.

Boaden, P., Morrison, G., & Smith, L. (2020). Economic Impacts of Hurricane Maria on Puerto Rico. *Weather, Climate, and Society*, 12(3), 563–578.

Cianconi, P., Betrò, S., & Janiri, D. (2020). The Impact of Climate Change on Mental Health: A Systematic Review. *Frontiers in Psychiatry*, 4(1), 1–12. DOI: 10.3389/fpsyt.2020.00074 PMID: 32210846

Climate Change and Power. (2002). *Economic Instruments For European Elecetrıcıty* (Vrolijk, C., Ed.). The Royal Institute of International Affairs.

Dinda, S. (2004). Environmental Kuznets Curve Hypothesis: A Survey. *Ecological Economics*, 49(4), 431–455. DOI: 10.1016/j.ecolecon.2004.02.011

Emanuel, K. (2005). Increasing Destructiveness of Tropical Cyclones Over the Past 30 Years. *Nature*, 436(7051), 686–688. DOI: 10.1038/nature03906 PMID: 16056221

European Enviroment Agency. (n.d.). Carbon Emission Data. https://www.eea.europa.eu/data-and-maps/daviz/atmospheric-concentration-of-carbon-dioxide 5#tabchart_5_filters=%7B%22rowFilters%22%3A%7B%7D%3B%22column-Filters%22%3A%7B%22pre_config_polutant%22%3A%5B%22CO2%20 (ppm)%22%5D%7D%7D

Fischer, E. M., & Schär, C. (2010). Consistent Evidence of Enhanced Summer Heat in Europe. *Geophysical Research Letters*, 37(2), L20704.

Greiving, S. (2013). ESPON CLIMATE-Climate Change and Territorial Effects on Regions and Local Economies. Applied Research Project, 1(4). https://www.espon.eu/export/sites/default/Documents/Projects/AppliedResearch/CLIMATE/ESPON_Climate_Final_Report-Part_C-ScientificReport.pdf

Hansen, J., Sato, M., Kharecha, P., Beerling, D., Berner, R., Masson-Delmotte, V., Pagani, M., Raymo, M., Royer, D. L., & Zachos, J. C. (2008). Target Atmospheric CO2: Where Should Humanity Aim? *The Open Atmospheric Science Journal*, 2(1), 217–231. DOI: 10.2174/1874282300802010217

Hansen, J. (2012). Scientific Case for Avoiding Dangerous Climate Change to Protect Young People and Nature. Proceedings of the National Academy of Sciences (Submitted paper), http://arxiv.org/ftp/arxiv/papers/1110/1110.1365.pdf

Houghton, J. (2004). *Global Warming* (3rd ed.). Cambridge University Press. DOI: 10.1017/CBO9781139165044

Hughes, L., & Salinger, M. J. (2006). Climate Change and Heatwaves. *Australian Journal of Public Health*, 30(1), 1–8.

Hughes, T. P., Kerry, J. T., & Simpson, T. (2017). Global Warming and Reefs. *Science*, 359(6371), 158–160.

IPCC. (1988). *Report of the Intergovernmental Panel on Climate Change*. United Nations Environment Programme (UNEP) and World Meteorological Organization (WMO).

IPCC. (2007). Summary for Policymakers. A report of Working Group I of the Intergovernmental Panel on Climate Change, https://www.ipcc.ch/pdf/assessmentreport/ar4/wg1/ar4-wg1-spm.pdf

IPCC. (2021). *Climate Change 2021: The Physical Science Basis*. Intergovernmental Panel on Climate Change (IPCC).

IPCC. (2021). *The Physical Science Basis*. Intergovernmental Panel on Climate Change.

IPCC. (2023). https://www.ipcc.ch/report/ar6/syr/downloads/report/IPCC_AR6_SYR_LongerReport.pdf

Janicke, M., Binder, M., & Mönch, H. (1997). Dirty industries: Patterns of change in industrial countries. *Environmental and Resource Economics*, 9(4), 467–491. DOI: 10.1007/BF02441762

Kan, S., Chen, B., Meng, J., & Chen, G. (2019). An extended overview of natural gas use embodied in the world economy and supply chains: Policy implications from a time series analysis. Energy Policy, 137, 111068. DOI: 10.1016/j.enpol.2019.111068

Karaca, C. (2012). The Relationship Between Economic Development and Environmental Pollution: An Empirical Analysis on Developing Countries. *Çukurova University Journal of Social Sciences Institute, 21*(3), 139–156.

Keeling, C. D., Whorf, T. P., & Revelle, R. (1976). *Atmospheric Carbon Dioxide Concentrations and Their Changes: A Historical Overview*. Carbon Dioxide and Climate: A Scientific Assessment, 7-44.

Khalid, A. A., Mahmood, F., & Rukh, G. (2016). Impact of climate changes on economic and agricultural value added share in GDP. *Asian Management Research Journal*, 1(1), 35–48.

Kjellstrom, T., Kovats, R. S., Lloyd, S. J., Holt, T., & Tol, R. S. J. (2009). The Direct Impact of Climate Change on Regional Labor Productivity. *Archives of Environmental & Occupational Health*, 64(4), 217–227. DOI: 10.1080/19338240903352776 PMID: 20007118

Klein, R. J., & Nicholls, R. J. (2004). *Coastal Vulnerability and Sea-Level Rise: The Role of Coastal Defenses*. Coastal Systems and Continental Margins, 123-134.

Kumar, A. (2019). Globalization and Environmental Impacts: Challenges and Opportunities.

Kuznets, S. (1955). Economic growth and income inequality. *The American Economic Review*, 45(1), 1–28.

Mahato, A. (2014). Climate Change and its Impact on Agriculture. *International Journal of Scientific and Research Publications*, 4(4), 1–6.

Ministry of Development. (2013). Tenth Development Plan (2013-2018).

Pall, P., Allen, M. R., & Stone, D. A. (2007). Human Contribution to the Length of the 2004 European Heatwave. *Nature*, 449(7164), 804–808. PMID: 17943116

Parmesan, C., & Yohe, G. (2003). A Globally Coherent Fingerprint of Climate Change Impacts Across Natural Systems. *Nature*, 421(6918), 37–42. DOI: 10.1038/nature01286 PMID: 12511946

Pata, U. K., Yurtkuran, S., & Kalça, A. (2016). Energy Consumption and Economic Growth in Turkey: ARDL Bounds Testing Approach. *Marmara University Journal of Economic and Administrative Sciences*, 38(2), 255–271.

Ramsey, J. D., Burford, C. L., Beshir, M. Y., & Jensen, R. C. (1983). Effects of Workplace Thermal Conditions on Safe Work Behavior. *Journal of Safety Research*, 4(3), 105–114. DOI: 10.1016/0022-4375(83)90021-X

Reti, M. J. (2007). An Assessment of the Impact of Climate Change on Agriculture and Food Security in the Pacific: A Case Study in Vanuatu. FAO SAPA, ftp://ftp.fao.org/docrep/fao/011/i0530e/i0530e02.pdf

Robine, J. M., Cheung, S. L. K., & Le Roy, S. (2008). Death Toll Exceeded 70,000 in Europe Heatwave. *Nature*, 455(7210), 43–44.

Ryan, S. J., Carlson, C. J., & Mordecai, E. A. (2019). Global Expansion and Redistribution of Aedes-borne Virus Transmission Risk. *Nature Microbiology*, 4(6), 1060–1067. PMID: 30921321

Sarkodie, S. A., Strezov, V., Weldekidan, H., Asamoah, E. F., Owusu, P. A., & Doyı, I. N. Y. (2019). Environmental sustainability assessment using dynamic autoregressive-distributed lag simulations-nexus between greenhouse gas emissions, biomass energy, food and economic growth. *The Science of the Total Environment*, 668, 318–332. DOI: 10.1016/j.scitotenv.2019.02.432 PMID: 30852209

Shahbaz, M., Mahalik, M. K., Shah, S. H., & Sato, J. R. (2016). Time-varying analysis of CO2 emissions, energy consumption, and economic growth nexus: Statistical experience in the next 11 countries. *Energy Policy*, 98, 33–48. DOI: 10.1016/j.enpol.2016.08.011

Shahzadi, A., Yaseen, M. R., & Anwar, S. (2019). Relationship between Globalization and Environmental Degradation in Low Income Countries: An Application of Kuznets Curve. *Indian Journal of Science and Technology*, 12(19), 1–13. Advance online publication. DOI: 10.17485/ijst/2019/v12i19/143994

Trenberth, K. E., Dai, A., & van der Schrier, G. (2014). *Global Warming and Changes in Drought*. In: Climate Extremes and Society, 287-316. DOI: 10.1038/nclimate2067

UNFCCC. (1992). *United Nations Framework Convention on Climate Change*. United Nations.

UNFCCC. (1997). *Kyoto Protocol to the United Nations Framework Convention on Climate Change*. United Nations.

UNFCCC. (2015). *Paris Agreement*. United Nations.

Uzmen, R. (2007). *Global Warming and Climate Change: Is Humanity Facing a Great Disaster?* (2nd ed.). Bilge Kültür Sanat.

Walker, G., & King, S. D. (2009). *Our Planet is Heating Up: How Can We Deal with Global Warming?* (Akpınar, Ö., Trans.). Boğaziçi University Press.

WHO. (2021). *Heatwaves and Health: Guidance for the Public*. World Health Organization (WHO)._https://data.worldbank.org/indicator/NY.GDP.PCAP.CD?locations=ET

Chapter 7
Economic Costs of Climate Change and How These Costs Can Be Mitigated

Shashank Mittal
O.P. JIndal Global University, India

Preet Kanwal
https://orcid.org/0009-0006-5114-8381
Lovely Professional University, India

Hewawasam P. G. D. Wijethilak
https://orcid.org/0009-0006-9611-5735
University of Colombo, Sri Lanka

ABSTRACT

This chapter explores the economic costs of climate change and strategies for mitigating these impacts. It examines the diverse and significant financial burdens associated with climate-related disasters, disruptions, and long-term environmental changes. The chapter outlines key areas of concern, including regional and sectoral variations in economic costs, and the use of economic models and methods to assess these impacts. It discusses various mitigation strategies, such as transitioning to a low-carbon economy and enhancing climate resilience and emphasizes the importance of international cooperation in addressing global climate challenges. The chapter concludes with an outlook on future developments and the implications for practical, managerial, ethical, and societal dimensions. Effective climate action requires comprehensive, collaborative approaches and a commitment to sustain-

DOI: 10.4018/979-8-3693-5792-7.ch007

ability and equity.

INTRODUCTION

Climate change - perhaps the most major challenge facing this 21st century. It is the greatest challenge not only to the environment but also to economies all over the world. The climate system of Earth has started changing, mainly instigated through human activities by means of burning fossil fuel, deforestation, and industrial processes, which release greenhouse gases into the atmosphere. These changes come in the form of a rise in global temperature, melting polar ice, rising sea-levels, and frequent along with intense weather events (Abidoye & Odusola, 2015). Graver than the environmental effects of climatic change are the economic dimensions attached to such changes. Therefore, for policymakers to formulate an effective policy and strategy in overcoming the impacts of such climate change and ensuring sustained economic growth, there is a need to understand the economic costs that emanate from climate change.

The economic costs of climate change are diversified into direct, indirect, and long-term costs with implications that differ across various sectors and regions. In general, direct costs involve immediate financial impacts usually resulting from extreme weather events such as hurricanes, floods, and wildfires that destroy infrastructure, property, and human life. For instance, hurricanes can flatten entire cities, while rebuilding afterward is extremely costly (Ceesay et al., 2020). Floods can rupture transport networks, damage agricultural lands, which diminish productivity and economic output. Besides the above-mentioned direct damages, there are considerable indirect costs due to a general disruption of economic activities. The outcome includes less agricultural yield from changed weather patterns, increased healthcare costs from heat-related illnesses and diseases, and disruption of global supply chains, which could impact international trade and economic stability.

What is even worse is that its long-term economic costs are now threatening to sabotage even the very basics of global economic growth. Ecosystems and biodiversity will be seriously threatened with increased temperature rise that could mean significant economic loss. Loss of biodiversity will disrupt key ecosystems, which are vital for food production, water purification, and disease regulation, among other services. The end result will be economic devastation to livelihoods that depend on such natural systems (Eboli et al., 2010). Sea-level rise and coastal erosion, on one hand, will pose monumental risks, especially to small island nations and a large number of coastal cities due to the extremely high cost of adaptation or relocation. Apart from that, climate change is also further complicating the fragile economic landscape, since its vulnerable population-especially in developing countries-is drastically affected. This, in turn, contributes to increased migration and social unrest, which further destabilizes economic stability (Ceesay, 2022).

Regional and sectoral variations in the impacts of climate change must be put into context in order to better comprehend its full economic cost. Yet, not all regions are the same in terms of being affected by climate change: low-lying coastal areas and arid zones are not quite as robust as other regions. This also extends to various economic sectors. Agriculture, for instance, is very susceptible to patterns of temperature and precipitation, whereas the energy sector may be equally burdened with aspects such as shifting demands and supply dynamics as the world shifts toward renewable sources of energy. It is due to this that such a difference needs to be duly recognized so that mitigation and adaptation measures can be suitably targeted and effective (Fankhauser & Tol, 2005).

Such are the economic costs of climate change that a multi-faceted response will be required, involving mitigation-a reduction in greenhouse gas emissions-to limit unavoidable impacts that can, in turn, be managed through adaptation. Policymakers will have to consider options ranging from carbon pricing and regulatory measures, through investments in resilient infrastructure, to technological innovation. Further, international cooperation is called for; climate change is a problem that knows no borders. Knowing the economic dimensions of climate change and how to act upon them shall pave the way for mitigating the effects of this scourge, achieving sustainable economic growth, and building a resilient tomorrow for all (Kellie-Smith & Cox, 2011).

Understanding the Economic Costs of Climate Change

The cost of climate change has to be regarded as an economic consequence, taking into consideration various ways in which due-to-climate changes affect different economies of the world. These may be divided into direct and indirect impacts, with each category having short- and long-term components; each of these carries unique challenges for policymakers, businesses, and communities. A more detailed look at the above-mentioned categories can help understand the wide-ranging scope of different economic consequences that climate change poses and the need for all-rounded mitigation and adaptation strategies (Kumar & Balaji, 2019).

The **direct costs** involve the immediate financial consequences of climate-related events and phenomena. These normally would emanate from more spectacular events like hurricanes, flooding, wildfires, and droughts. Hurricanes can cause massive destruction to infrastructure, homes, and businesses, leading to immense financial losses and calling for extensive rebuilding efforts. Hurricane Harvey caused over $125 billion in damage in the United States in 2017, demonstrating the incredible direct economic impact these events can cause (Warsame et al., 2023). Adding to the calamity, flooding destroys transportation networks, ruins agricultural lands, and contaminates water supplies. Currently fed by rising temperatures and sustained drought, wildfires have the potential to burn hundreds of acres of land and property, costing billions of dollars in damages with great implications for local economies dependent on forestry, tourism, and agriculture. These direct costs burden already strapped public and private finances, raise insurance premiums, and further load disaster response and recovery systems.

Indirect costs are the far-reaching consequences of the economic disruption occasioned by climate change. In as much as these costs manifest themselves in rather subtle ways, they still can have similarly significant impacts on economic stability and growth. For example, shifting weather patterns and increased temperatures can reduce agricultural output, increase food prices, and lead to food insecurity, especially in those economies that depend significantly on agriculture. This, in turn, has the potential to make world food markets more volatile, and hence will affect economies around the globe. Moreover, climate change can lead to increased health problems, which mean higher healthcare costs and reduced productivity of workers. For example, warmer temperatures might exacerbate the spread of diseases, including malaria and dengue fever, in tropical regions, thereby causing additional stress on healthcare systems and economies. Externally, the global supply chain disruptions because of the climate change-related events reduce trade and economic activities as companies attempt to handle uncertainties and adapt to shifting environmental conditions. An example is that the 2011 flood in Thailand severely upset the global supply chains of the electronic and automotive industries, leading to estimated losses of over $45 billion (Zhao, 2023).

Long-term Costs: These will be the long-lasting impacts of climate change on economies and societies. These costs are even more worrying because they change economic landscapes and act as an inhibitor to sustainable development. As the temperature of the Earth continues to rise, it pushes ecosystems and biodiversity into dangers never seen before, with huge economic consequences. Such loss in biodiversity can disrupt essential ecosystem services occurring within an economy, like pollination, water purification, and carbon sequestration (Abidoye & Odusola, 2015). Furthermore, increasing sea levels threaten coastal communities and infrastructure in ways that endanger the capacity of these communities and infrastructure for survival. Low-lying coasts and Small Island Developing States are particularly at risk due to the direct threat on native lands and livelihoods. Economic costs can hence range from conservation to the translocation of these communities, with estimates such as hundreds of billions of dollars annually required globally by 2050 for adaptation pertaining to sea-level rise. In addition, other social costs of climate change include heightened migration and displacement, leading to further destabilization of economies through increased social unrest, especially in developing countries that have limited resources for adaptation and building resilience (Ceesay et al., 2020).

These economic costs could be understood only by considering regional and sectoral variability in climate impacts. There is a regional difference in the impact of climate change; some regions like the Arctic and small island states are prone to immediate and serious consequences, while others are not. In the same vein, different economic sectors are differently affected. Agriculture, for instance, is very susceptible to change in temperatures and rainfall and other extreme weather conditions (Eboli et al., 2010). Energy industries are faced with two most prevalent challenges: the shift to renewable sources of energy and the implications of climatic disturbances on energy infrastructure. All these variations are relevant to note in view of specific and efficient mitigation and adaptation strategies related to climate change (Ceesay, 2022).

In sum, the economic costs of climate change are manifold and deeply far-reaching from all points of view, considering the aspects of society and the global economy. Once these costs and their implications are better understood, we will be better prepared for the challenges ahead by being able to put in place strategies that would foster resiliency, sustainability, and economic stability (Fankhauser & Tol, 2005).

Regional and Sectoral Variations in Economic Costs

Economic costs of climate change vary around the world but range very significantly by region and economic sector. These variations are driven by geographical, socio-economic, and environmental factors that determine the vulnerability and resilience of various areas and industries towards climate impacts. Understanding these regional and sectoral variations will be very critical in targeted policy and intervention processes that could effectively mitigate adverse effects on climate change and support economic development in a sustainable manner.

Regional Economic Costs of Climate Change

1. Comparison between Developed and Developing World: Most of the economic impacts of climate change occur in the developing world as a result of the combination of geographic exposure and restricted adaptive capacity. Most of the developing nations are situated in regions with exceedingly high vulnerability to climate-related hazards such as tropical storms, droughts, and heatwaves. Moreover, most of the developing countries are unable to cope and recover from weather variability and extreme events due to the high demand for finance, infrastructure, and technology (Kellie-Smith & Cox, 2011). This accentuates the economic cost of climate change as developing economies bear a disproportionately high share.

This, in turn, exacerbates socio-economic inequities and acts as an impediment to sustainable development. In contrast, most developed countries have stronger infrastructures, good emergency response systems, and large financial resources to invest in mitigation and adaptation strategies. However, in even developed countries, significant economic losses might also be suffered in specific areas, such as coasts and agricultural zones, due to climate change (Kumar & Balaji, 2019).

2. **Fragile Geographic Regions:** On account of their geographic physical features and exposure to risks propelled by climate, some geographic regions are more vulnerable to economic impacts of climate change. For example, SIDS are likely to face complex challenges that have to do with increasing sea level, coastal erosion, and the frequency of extreme weather. Economically, this will be catastrophic for those regions in terms of money generated through tourism, fisheries, and overall economic stability (Zhao, 2023). Low-lying coasts and deltas, such as Bangladesh and the Netherlands, face a very high risk of flooding and storm surges that may cause massive infrastructure destruction and loss in agriculture. Equally, global warming has rapidly altered the Arctic region as a result of warming; the melting of ice and permafrost is reported to threaten local ecosystems, indigenous communities, and economic activities such as fishing and shipping (Abidoye & Odusola, 2015).

3. **Climate-sensitive areas:** Some regions are relatively more climate-sensitive on account of their natural resource base and climate-dependent industries. For example, sub-Saharan Africa depends largely on agriculture, which is a sector sensitive to temperature, precipitation, and extreme weather conditions. Droughts, over an extended period with erratic rain, may drastically affect crop yields and livestock and result in food insecurity, enhanced prices of food, and loss of livelihood. Changes in the precipitation pattern would place at risk those economies that are dependent on the monsoon in Asia, considering the potential effect on agriculture, water resources, and energy production. In North America and Europe, a shift of growth seasons and crop viability on temperate zones would bring mixed economic impacts depending on the local adaptations (Casas-Cuestas, 2024).

Sectoral Variations in Economic Costs

1. **Agriculture** represents one of the most vulnerable spheres in terms of climate change, as directly or indirectly, the approach depends on weather and climatic conditions. Changes in temperature, precipitation, and frequency of extreme phenomena are expected to influence crop yields, productivity of livestock, and general trends in agricultural output. For instance, increased temperatures and heatwaves invariably lower both the quality and quantity of various crops, which are the common staple food of the greater part of the world's population, such as wheat, maize, and rice. Droughts can lead to water shortages that impact irrigation and then affect agricultural

yields (Ceesay et al., 2020). Conversely, excessive rainfall and flooding can wipe out crops and soil fertility. Some of the economic costs of these impacts include increased decrease in farmers' incomes, increased food prices, and unpredictable global food markets.

2. Energy: The energy sector is also vulnerable to the changing climate conditions in both energy production and demand. As temperatures rise, energy demand for cooling continues to increase and especially in urban centres. This may sometimes result in a rise in expenditure on electricity, load stress, and sometimes failure of power grids. The same climatic change, however, might affect energy production, particularly hydropower and thermal power plants. Low discharges from the rivers and water scarcity may affect hydropower generation, while extreme heat reduces the efficiency in thermal plants and enhances demand for cooling water. Similarly, the shift in the economy to renewable energy sources is brought about by policies on climate and reduction of greenhouse gases. This too presents economic challenges and opportunities that go with investments in new technologies and infrastructures that are constantly needed (Ceesay, 2022).

3. Insurance and Financial Services: Insurance and financial services are very vulnerable sectors considering the economic costs linked to the impacts of climate change due to their close exposure to various climate-related risks. For instance, increasing intensities of hurricanes, flooding, and wildfires might increase claims and payouts, which may be critical for the profitability and stability of insurers. This might lead to increased premiums, reduced coverage, and even withdrawal of insurers from a high-risk area-a development with consequences affecting property owners, businesses, and communities (Eboli et al., 2010). Through their investments in vulnerable industries and regions, the financial sector is also exposed to climate-related risks. Physical risks involve damage to assets due to extreme events in weather, while transition risks refer to shifts into a low-carbon economy which might pose financial stability and value.

4. Tourism: The industry is so vulnerable because it depends on nature and some cultural activities that are very susceptible to climate-related impacts. Moreover, coastal tourism destinations like the Caribbean and Mediterranean are especially vulnerable to increased sea levels, erosion of beaches, and storm and hurricane frequencies that could cripple infrastructure and deter tourists and their revenues. Decreased snowfall and shorter winters may also affect mountain tourism-skiers who go to the Alps or Rockies-on the viability of winter sports and businesses involved. Climate change can also alter biodiversity and ecosystems, making nature-based tourism less appealing, hence affecting the local economy dependent on ecotourism (Fankhauser & Tol, 2005).

5. Health: The health sector is one of the sectors that bear a big economic cost because of the impacts of climate change on public health. Increased temperatures and heat waves elevate the rate of heat-related illnesses and mortality, especially among the vulnerable populations such as children and the elderly or those with pre-existing health conditions. Climate change facilitates the spread of some vector-borne diseases, such as malaria and dengue fever, since changing weather patterns create conditions more favorable to disease vectors. The higher temperatures increase air pollution and allergens that could have an impact on respiratory health, leading to higher health costs and lost productivity. These health effects have direct economic costs attributed to healthcare provision and indirect costs through a reduction in workforce productivity and overall economic output (Gao, 2024).

Thus, regional and sectoral differences in the economic costs of climate change should also represent the complexity and multi-dimensionality that surrounds this global challenge. These differences in vulnerability and capacity across regions and sectors call for individualized approaches in response. By grasping these differences, targeted mitigation and adaptation strategies will, therefore, enable stakeholders and policymakers to manage the economic risks of climate change and support sustainable economic growth and development.

Economic Models and Methods for Assessing Climate Change Costs

It is a very difficult task to consistently assess the economic costs of climate change, owing to its manifold impacts and in view of the uncertainty characteristics of future climate scenarios. Economic models and methods therefore play a critical role in such estimations, providing frameworks for analyzing how changes in climate variables affect various economic sectors and overall economic output (Gao, 2024). These models represent critical instruments for policymakers, businesses, and researchers in that they provide the means by which potential economic damages from climate change can be estimated and mitigation and adaptation strategies can be evaluated for effectiveness. The section that follows discusses in some detail the major types of economic models currently in use for the cost assessment of climate change, along with their methodologies and respective strengths and weaknesses.

1. Integrated Assessment Models (IAMs)

Integrated Assessment Models are the most highly utilized tool in the evaluation of the economic costs of climate change. IAMs draw on a wide range of disciplines, including climate science, economics, and energy systems knowledge, to provide an integrated framework for the analysis of interactions among human activities, the climate system, and economic outcomes. These typically include a climate module, which models the consequences of greenhouse gas emissions on global temperatures and climate variables, and an economic module that models the economic consequences from these changes in sectors such as agriculture, health, and energy (Nordhaus, 1993; Weyant, 2017; van Vuuren ET AL., 2011).

IAMs are especially useful in assessing the long-term costs of climate change since they can model impacts of various paths of emissions and policy interventions for long periods. By integrating climate and economic dynamics, IAMs help policy makers understand the trade-offs between the costs of mitigation actions-such as reduced emissions-and the benefits of avoiding future climate damages (van Vuuren ET AL., 2011). Examples of widely used IAMs include the Dynamic Integrated Climate-Economy DICE model, the FUND Climate Framework for Uncertainty, Negotiation and Distribution model, and the Model for the Assessment of Greenhouse-gas Induced Climate Change MAGICC.

Strengths:

IAMs provide an integrated way of evaluating the cost of climate change through the interrelations between emissions, climate change, and economic consequences. These models provide a good amount of information on the implications of various policy choices over the long term and can be used to identify cost-effective strategies for reducing emissions and mitigating climate impacts.

Limitations:

IAMs are, by nature, complex models that depend on many assumptions which are directly related to their outputs about future socio-economic and technological changes. Thus, they may inherently bring in substantial uncertainty into their results. The accurateness in IAMs further depends on the soundness of the underlying data on climate and economy. Such data is often scanty, especially for developing countries and specific sectors.

2. Computable General Equilibrium (CGE) Models

Computable General Equilibrium models are yet another important tool to assess the economic cost of climate change. These latter models simulate how economies would respond to policy, technology, or exogenous shock changes, like climate change, by modeling sector and agent interactions in the economy, such as households, firms, and governments. Hence, the CGE models are quite useful in the analyses of distributional impacts related to climate change at various economic sectors and regional levels, apart from testing the various policy measures in carbon pricing or, for that matter, emissions trading schemes (Dixon & Jorgenson, 2013).

The CGE models represent the behavior of economic agents and markets through a system of equations that equilibrate supply and demand across all markets. The CGE model will, therefore, be able to capture the interdependencies between sectors and regions in considerable detail, hence its strong ability to depict implications of climate change on economic output, employment, and trade, besides various adaptation and mitigation strategies (Bosello et al., 2012).

Strengths:

Strengths include the fact that CGE models have a great deal of detail about the economy and can model complex interactions between sectors and regions. They are well-suited to analyse the distributional impacts of climate change and for evaluating the economic effects of particular policy measures (Zhai & Zhuang, 2012).

Limitations:

The data requirement is very extensive and sensitive to many of the assumptions made about the way in which economic agents behave and markets function. The estimates from CGE models may be very sensitive to model specification, particularly with respect to key parameters such as substitution elasticities and labor supply responses. More commonly, CGE models tend to assume perfect competition and market clearing, and this may be a poor approximation in any particular market, let alone under changing climate conditions.

3. Sector-Specific Models

Sector-specific models focus on the impact that climate change is having or will have on very specific industries, such as agriculture, energy, health, and transportation. Such models provide deeper levels of analysis that assess how given economic activities are influenced by climate change due to increased temperature, change in precipitation, or extreme weather events (Mendelsohn, & Dinar, 2009). Agricultural models would project the implication of altered temperature and precipitation patterns on crop yields, while energy models may assess the consequence of increased temperatures on electricity demand for cooling or the viability of hydroelectric generation (Auffhammer & Aroonruengsawat, 2011).

The added value of the sector-specific models, in addition to conveying focused adaptation action, is their ability to understand localized and sectoral impacts of climate change. As long as these models relate to specific sectors, more detailed data with sector-specific knowledge at higher resolution can be incorporated for better and more relevant estimates in terms of the cost of climate change (Kocornik-Mina & Rodriguez-Vega, 2018).

Strengths:

Sector-specific models detail the impacts of climate change on particular industries or sectors. This allows more precise estimates of economic costs and better-informed adaptation strategies. They can be site-specific, with modifications to suit regional contexts, hence increasing relevance for local decision-making.

Limitations:

Sector-specific models have limited capabilities in terms of capturing broader economic interactions and feedback effects between sectors. As such, they may fail to represent economy-wide impacts of climate change or indirect effects of sectoral changes on other parts of the economy. This kind of incorporation within a larger economic framework may be cumbersome to integrate and often involves additional data and assumptions.

4. Econometric Models

Econometric models apply statistical techniques to data from the past in order to estimate relationships between climate variables and economic outcomes. Commonly, such models are used to estimate the impacts of near- to medium-term climate change on the economy, like how temperature, precipitation, or extreme events in the past have affected economic indicators such as GDP, productivity, or employment (Deschênes & Greenstone, M2011).

Econometric models are of particular usefulness in assessing the empirical relationships between climate change and economic performance and, therefore, provide a clear vision of how economies have responded to past climate variability and extremes. These models allow for an additional element: an estimation of the potential economic impacts of future climate scenarios, using historical patterns (Dell et al., 2014).

Strengths:

Econometric models use data-driven methods to estimate the economic costs of climate change and exploit a historical observation set to determine empirical relationships and quantify impacts. They can be very informative about the short- to medium-term impacts resulting from climate variability and extremes on economic performance.

Limitations:

Econometric models, by their very nature, rely on relationships from the past, which may not accurately represent unprecedented future climate change or thresholds and nonlinearities. They are also sensitive to data quality and availability, and results may be sensitive to the choice of statistical methods and model specifications (Burke & Emerick, 2016).

5. Cost-Benefit Analysis (CBA) and Cost-Effectiveness Analysis (CEA)

CBA and CEA are the major methods of analyzing the economic efficiency of alternative options for climate change mitigation and adaptation strategies. In the case of CBA, the costs of a certain policy or project are weighed against the expected benefits in financial terms. This methodology will help the policymakers to identify if the benefits from the proposed action outweigh the cost it would incur, besides selecting those options which are economically efficient to address climate change (Tol, (2005)

In contrast, CEA considers the costs of different ways of reaching a given target or objective. CEA is particularly useful where there are options that have many benefits which cannot readily be monetized, for example, with many policy measures or technologies.

Strengths:

CBA and CEA offer a formalized method to compare the economic efficiency of various strategies against climate change, which can definitely help policymakers prioritize their actions to achieve maximum benefits or desired outcomes at minimum cost. These are widely applied in policy evaluation and decision-making processes (Stern, 2007).

Limitations:

Benefits and costs related to nonmarket values, including biodiversity, ecosystem services, and social equity, can be hard to measure and thus limit the current state of CBA and CEA. These methods also generally depend on discount rates, assumptions about future economic conditions, and valuation about values that are intangible, embedding uncertainty and subjectivity into the analysis (Anthoff & Tol, 2013).

In closing, economic models and methods for appraising the costs of climate change provide a useful means of understanding the complex, multi-disciplinary impacts of the effects of climate change on economies, as well as informing policy. Each model and method has strengths and weaknesses, and the appropriate approach in any given case depends on context, objectives, and data availability. These tools, when put to use, can help policymakers and other stakeholders understand the economic risks of climate change and design mitigation strategies that will help in reducing the impact of climate change, besides attaining sustainable economic development.

Strategies of Mitigation towards Reducing Economic Costs

Mitigation strategies intended to reduce the cost of climate change economically focus on methods applied toward lowering the emission of greenhouse gases and global warming, as this consequently reduces damages and long-term economic impacts. These are not only very crucial for addressing climate change but equally vital for sustained economic growth and development (Warsame et al., 2023). Effective mitigation involves technological innovation, policy interventions, market mechanisms, and international cooperation on a multifaceted track. The following sections discuss some mitigation strategies that can be put to use in the reduction of the economic costs of climate change toward a resilient and sustainable global economy (Kumar & Balaji, 2019).

1. Transition to Renewable Energy Sources

Transitioning from fossil fuel dependence to renewable sources of energy-let's say, hydroelectric, geothermal, solar, and wind-is one of the best ways to cut down emissions. First, because of the rapid advancement of technology and a economies of scale effect, the cost of renewable energies has become more competitive against that of fossil fuels. Further, investment in renewable-energy infrastructure will bring carbon emissions down and greatly improve energy security, opening up more jobs for emerging industries. For example, solar and wind have become the fastest-growing sources of power generation globally, as a clean and sustainable alternative to coal and natural gas. In addition to the reduction of emissions, economic advantages from the switch to renewable energy are also typical. By diversifying their energy mix, each country has the power to lessen dependence on imported fossil fuels, enhance energy security, and provide price stability. Besides this, operating costs for renewable projects are considerably lesser when compared to conventional power plants, hence saving consumers and businesses over the long haul. This sector will provide a considerable number of employments in manufacturing, installation, and maintenance, making it very important from the point of view of economic growth and social development (Zhao, 2023).

2. Energy Efficiency Improvements

Another key mitigation strategy entails increasing energy efficiency to reduce greenhouse gas emissions and decrease the economic cost of climate change. Energy efficiency is the use of less energy to accomplish tasks with greater productivity and with less waste of energy. Residential, commercial, industrial, and transportation sectors can also adopt this measure (Kulkarni et al., 2022).

Energy efficiency in the residential and commercial sectors can be achieved by thermally insulated buildings, energy-efficient electrical appliances, smart building technologies, and optimization of usage patterns. In the industrial sector, huge scope exists for reduction of energy consumption and resultant emissions by upgrading of machinery and industrial processes to their more energy-efficient alternatives. The transport sector, which accounts for a high share of greenhouse gas emissions, offers a number of opportunities for improvement on fuel-efficient vehicles, electric mobility, and better public transport systems.

Energy efficiency measures reduce emissions but also create significant economic returns by means of reduced energy expenditures for households and businesses, which in turn can invest their savings in other parts of the economy to drive innovation and growth. Energy efficiency enhancements can also contribute to energy security by reducing energy imports and making the state less vulnerable to energy price volatility (Klusak et al., 2021).

3. Carbon Pricing and Market Mechanisms

Carbon pricing is done on market principles, attributing a price to greenhouse gas emissions. This creates an economic incentive for businesses and individuals to cut down carbon emissions. This internalizes the environmental cost due to the emissions through prices, signaling to the market greener technologies and practices. Carbon pricing has commonly been pursued through two routes: carbon tax and cap-and-trade systems.

It puts a direct price on carbon by charging a fee on the carbon content of fuels. This provides certainty on the carbon cost and can be quite straightforward to implement. The revenues raised through carbon taxes can be invested in renewable energies, energy efficiency programs, or other programs related to climate change, or returned directly to households and businesses as part of a strategy to mitigate the new burden due to increased energy prices (Kellie-Smith & Cox, 2011).

Cap-and-trade systems, also referred to as emissions trading schemes, establish a limit, or cap, on the total quantity of greenhouse gas emissions allowed from covered sources. Companies are either allocated or must purchase emission allowances-a tradable instrument granting its holder the right to emit a defined amount of carbon. They are also allowed to sell the surplus permits to other companies if the companies emit below their permitted limit, therefore creating an economic incentive to reduce emissions. The cap-and-trade system provides certainty about total emissions but creates allowance prices that may fluctuate due to market conditions.

However, the pricing of carbon and its market-based approaches can be a very viable tool in pursuance of emission reduction and stimulating innovation for low-carbon technologies. By creating a monetary incentive to reduce emissions, these mechanisms go a step in aligning economic and environmental objectives in furthering this transition into a low-carbon economy (Hutton, 2011).

4. Reforestation and Land Use Management

Reforestation and improved land use management are key means of reducing global warming and its economic costs. Forests feature importantly in the Earth's carbon cycle, absorbing carbon dioxide from the atmosphere and storing it both in biomass and soil. When forests are destroyed, the stored carbon is released into the atmosphere, adding to the stock of greenhouse gases.

Reforestation is a process of planting trees on land that has already been deforested or degraded, thus restoring the ecosystem and enhancing carbon sequestration. Adoption of sustainable land use management practices-such as agroforestry, sustainable agriculture, and conservation tillage-further contributes to better storage of soil carbon and reduces emissions from land-use change and agriculture (Hutton, 2011).

Besides sequestering carbon, forests and sustainable land management practices provide a suite of ecosystem services underpinning economic and social development. This includes the maintenance of biodiversity, regulation of water cycles, prevention of soil erosion, and livelihoods for hundreds of millions of people worldwide. Investments in activities related to reforestation and sustainable land use mitigate climate change, protect key ecosystems, and enhance the resilience of rural communities to climate impacts (Haer et al., 2018).

5. Technological Innovation and Research

Any attempt to cut the economic costs of climate change, therefore, would have two important components: technological innovation and research. It is perceived that new technological innovations can attain a reduction in GHG emissions and adapt better to climate change with greater efficiency and at lower costs. Investment in R&D could act as a driver for innovation in a number of key low-carbon economy fields, such as renewable energies, energy storage, CCS, and climate-smart agriculture (Fankhauser & Tol, 2005).

The key to widespread renewable energy source adoption, like solar and wind, depends on further breakthroughs in battery technology and energy storage, because the source of these energies is intermittent by nature (Gao, 2024). Carbon capture and storage technologies are technologies that capture carbon dioxide emissions from industrial processes and power plants and store them underground to prevent them from entering the atmosphere. Innovations in sustainable agriculture and food systems can contribute to farming practice emissions reduction, improve soil health, and enhance food security.

Governments, businesses, and research institutions lead the process of technological innovation and support R&D processes. The public policies that could make a difference in low-carbon technology development and deployment include grants, subsidies, tax incentives, and public-private partnerships that may help the world mitigate climate change.

6. International Cooperation and Frameworks of Climate Policy

Climate change is a global challenge that requires an integrated approach from the international plane. International cooperation and climate policy frameworks are the essentials in mitigating GHG emissions while providing, at the same time, an economics-based response to the challenge of climate change. The Paris Agreement, reached in 2015, represents an unprecedented effort to bring countries together in trying to combat climate change and restrict global warming to well below 2 degrees Celsius above pre-industrial levels.

International cooperation can help share knowledge, technology, and finance for the effective undertaking of mitigation and adaptation measures, especially by developing countries. Climate finance refers to the transfer by developed countries to developing countries that may support low-carbon development and strengthen resilience against adverse climate change impacts (Eboli et al., 2010).

multilateral climate policy frameworks, such as the UNFCCC and the IPCC, have provided a platform for dialogue, negotiation, and cooperation on the issues of climate change. By offering a framework for international cooperation that inherently calls for more ambitious climate policy action, the development of these frameworks accelerates the efforts of the international community in reducing emissions to mitigate economic costs of climate change.

7. Sustainable Urban Planning and Infrastructure Development

The role of urban planning and infrastructure development is critical in mitigation of the effects of climate change and reductions in economic cost. It is found that cities are emitting a substantial percentage when considering global energy use and carbon emissions. The issue of sustainable urban development is very important because of the high growth rate of urban populations (Ceesay et al., 2020; Ceesay, 2022).

Sustainable urban planning involves designing cities and communities in ways to reduce emissions, improve resilience from climate impacts, and improve the quality of life. This might include investment in or promotion of public transportation, energy-efficient building development, green space creation, and sustainable waste management. Energy-efficient lighting, smart grids, and intelligent transportation systems are examples of smart city technologies which would further reduce emissions and add sustainability to the aforementioned strategies (Casas-Cuestas, 2024).

Climate-resilient infrastructure will protect the community from flooding, heatwaves, and the sea-level rise brought about by climate change. Through investment in green infrastructure and urban planning, cities decrease their carbon footprint and increase their resiliency while providing a healthier and more livable environment for their inhabitants.

8. Behavioral Change and Public Awareness

Climate change mitigation also involves changing people's behavior and increasing public awareness of the need to reduce emissions and pursue sustainable practices. Public education and raising awareness will grant individuals the opportunity to understand how their personal actions are affecting the environment, thus encouraging lifestyle choices that minimize energy consumption, reduce waste, and use low-carbon transport.

This could also be facilitated at the level of governments, businesses, and civil society organizations through the provision of information, incentives, and support that assure citizens of sustainable behavior. Some of the key incentives to drive individual behavior toward more sustainable behaviors include energy-efficient appliance labelling (Abidoye & Odusola, 2015), recycling programs, and incentives for electric vehicles. With increased awareness among the general public and an instilled culture of sustainability, society can help in this global undertaking of minimizing climate change and its resultant economic costs.

Finally, this calls for mitigation strategies to be broad and multi-dimensional if the economic costs of climate change are to be reduced and sustainability ensured in the future. Indeed, strategies that will favor renewable energy, energy efficiency, carbon pricing, reforestation, technological innovation, international cooperation, sustainable urban planning, and behavioral change will no doubt see policymakers and stakeholders effectively deal with the challenges of climate change and support economic growth and development (Zhao, 2023).

The Role of International Cooperation

International cooperation is key in addressing the global challenge of climate change and, correspondingly, lowering its economic costs. In light of the cross-border nature of greenhouse gas emissions and the interconnectedness of economies worldwide, no single country can tackle the issue of climate change on its own. International cooperation allows countries to take collective action to realize mutual benefits and efficiently share resources and knowledge for better synergies in mitigation and adaptation to climate change. This section looks at various dimensions of international cooperation, including the rationale, how it works, and how it contributes specifically to global resilience to climate change (Warsame et al., 2023).

1. The need for international cooperation

Climate change is a global problem, affecting both developed and developing countries irrespective of their geographical location. These multifaceted effects of climate change include increased temperatures, rising sea levels, extreme weather events, and loss of biodiversity. They respect no borders, and for the most part, bear massive implications for global security, economic stability, and public health. It is, therefore, not easy to emphasize the need for cooperation at the level of the world in helping to confront such challenges, thereby making sure there is a sustainable future for all (Kumar & Balaji, 2019).

Such expertise, resources, and technological innovations could be integrated in the development of much more effective climate mitigation and adaptation strategies if countries work together. International cooperation contributes to building trust between nations for the common concern of shared responsibility in addressing the challenges posed by climate change. This is an important aspect, especially for developing countries, which do not have the required financial resources, technical expertise, and institutional capacity to tackle climate change. This cooperation provides a platform through which such nations can gain access to mechanisms that will, subsequently help them build resilience and shift towards low carbon development pathways (Kulkarni et al., 2022).

2. Main Mechanisms of International Cooperation

Mechanisms of international cooperation on climate change include international agreements, multilateral organizations, and partnerships between countries, businesses, and civil society. These can provide pathways through which dialogue, negotiation, and collaboration platforms are created for countries to align their efforts and commitments toward common goals within the climate context (Casas-Cuestas, 2024).

a. International Agreements:

International agreements set the frameworks for the current global action on climate change. The most notable happens to be the adoption of the Paris Agreement under the United Nations Framework Convention on Climate Change in 2015. The Paris Agreement was a collective, globe-trotting landmark for every nation in its committed responsible global stewardship to cap the anticipated global warming into a well-below 2°C level above the pre-industrial era, besides making crosscutting efforts towards a 1.5°C limit. Under the Paris Agreement, countries are committed to providing NDCs by submitting targets and plans on their climate action, which are subjected to periodic review for enhancement. It is a mechanism that encourages countries to ratchet up ambitions over time, while ensuring transparency and accountability in the climate action being taken (Kellie-Smith & Cox, 2011).

b. Multilateral Organizations and Climate Funds:

International collaborative climate change policies depend gravely on the multilateral organizations that come in, such as the UNFCCC, the World Bank, and the Green Climate Fund. These organizations provide a negotiation platform that aids countries in capacity building and technical assistance to realize viable climate policies and strategies. Funding through funds like the GCF and GEF has, instead, granted financial assistance to developing countries in place of increasing their mitigation and adaptation actions. This provides a bridge in the financing gap toward climate action and allows countries to invest in renewable energy, energy efficiency, and climate-resilient infrastructure while addressing other critical concerns (Kulkarni et al., 2022).

c. Bilateral and Regional Partnerships:

Bilateral and regional partnerships represent another significant tool of international cooperation for climate change. This partnership between the countries allows for collaboration on a project or initiative, sharing best practices and technologies, and offering support for each other's efforts in emissions reduction and building resilience (Haer et al., 2018). For example, the European Union has entered into several regional partnerships, such as the EU-Africa Partnership and the EU-ASEAN Dialogue, with the aim of promoting climate action and sustainable development. While multilateral platforms represent one source of cooperation, bilateral cooperation, such as through the U.S.-China Climate Leaders Declaration, also provides a vital role in cultivating collaboration among major emitters in support of global climate objectives (Warsame et al., 2023).

d. Public-Private Partnerships and Civil Society Engagement:

Public-private partnerships and civil society engagement are important facets of international cooperation on climate change. Such partnerships bring together governments, businesses, NGOs, and other stakeholders to capitalize on their unique strengths and resources in support of climate action. Multi-sectoral collaboration, such as the Powering Past Coal Alliance, where governments, businesses, and NGOs rule out coal power; such groupings act as strong enablers for climate action. Civil society organizations also provide a significant role in raising awareness of the need for larger policy ambitions on climate issues and holding governments and businesses accountable for their pledges (Warsame et al., 2023).

3. Role of Stakeholders in Enhancing Climate Resilience

International cooperation not only plays an important role in mitigation-reducing emissions-but also in adaptation, which will be needed to enhance climate resilience, especially in vulnerable regions and communities. Climate resilience would refer to the capacity of society to anticipate, prepare for, and respond to climate-related risks and hazards. Building resilience involves investments in infrastructure, technology, and capacity-building in addition to integrating climate risk into planning and decision-making processes (Klusak et al., 2021).

a. Sharing Knowledge and Best Practices:

One of the most significant benefits resulting from international cooperation in shared knowledge, best practices, and lessons learned comes from the various countries and contexts. Countries can create more appropriate and innovative approaches toward climate adaptation, tailored to their particular needs and circumstances, based on such information exchange. These include the sharing of experiences in managing water resources, protection of coastal areas, and putting in place early warning systems against extreme weather conditions. This allows countries to draw upon the collective knowledge and know-how of the international community to build their adaptive capacity and reduce their vulnerability to climate impacts (Kulkarni et al., 2022).

b. Access to Climate Finance and Technology:

In developing countries, access to climate finance and technology is necessary for building resilience to climate change. International cooperation ensures the mobilization of financial resources and transfers technologies that allow developing countries to adapt to climate change. Such an example includes the grant and concessional loan provided by GCF to developing countries for implementing climate-resilient projects and programs. This has been made possible through international agreements and partnerships that facilitate technology transfer, enabling countries access to advanced technologies on climate monitoring, renewable energy, and climate-smart agriculture. International cooperation gives both financial and technological support to bridge gaps between developing and developed countries and, therefore, allows each nation to participate in the fight against global climate action (Haer et al., 2018).

c. Enhanced Institutional Capacity:

Indeed, institutional capacity building is one of the crucial components for enhanced climate resilience, especially in developing countries whose technical know-how, data, and governance arrangements may be limiting in their effective management of climate risk. International cooperation will further augment the process through capacity-building programs like training, technical assistance, and institution strengthening. These efforts assist developing countries in competency building, knowledge, and tool development to integrate climate considerations into national planning and policy frameworks; enhance climate risk assessment and management; and improve disaster preparedness and response. Through institutional capacity building, international cooperation makes societies resilient and adaptive (Klusak et al., 2021).

4. Challenges and Opportunities in International Cooperation

While much progress has been made on international cooperation regarding climate change, a set of challenges also faces this cooperation if it is going to be effective and actually make any difference. Among the more significant challenges are questions of equity and fairness given the different countries' historical responsibilities for greenhouse gas emissions and their differing capacities to address climate change. Developing countries are often among those that can least contribute to emissions and are highly vulnerable to the impacts of climate; therefore, their strive for ambitious goals in meeting climate policy and building resilience requires further support from developed countries (Casas-Cuestas, 2024).

The other challenges involve ramping up the political will and ambition to meet targets as set out in international agreements. Admittedly, though commitments to do so are under the Paris Agreement, global emissions keep growing, with the world currently off track to meet targets of both 1.5 and 2 degrees. This challenge equally demands action and increased ambition on the part of all countries, especially major emitters, with great alignment of national policies to global climate goals (Ceesay, 2022).

On the other hand, international cooperation also carries immense opportunities to increase action at a global level pertaining to climate change. With varied collaborations and innovations, countries can create new solutions and strategies that involve lower emissions, hence building resilience. International cooperation can drive forward the transition to the low-carbon economy, unlock new economic opportunities, and further advance sustainable development (Warsame et al., 2023).

In a nutshell, it is the cornerstone of international efforts to address climate change and minimize the costs thereof. It will enable countries to put in place effective mitigation and adaptation mechanisms, build resilience, and ensure a sustainable and equitable future for all. For full realization of the potential of this international cooperation, there is a real need for strengthening the existing mechanisms, addressing the challenges, and seizing opportunities for collaboration and innovation (Zhao, 2023).

Future Outlook

The interplay of scientific improvements, policy elaboration, technological change, and societal alteration shapes the future outlook in addressing climate change and its economic costs. While the world is confronted with ever-increasing impacts of climate change, there is a growing opportunity and an urgent need for humanity to alter its path in a sustainable direction. The next couple of decades will, accordingly, be quite crucial for the effectiveness of global efforts toward mitigating greenhouse gas emissions and adapting to a changing climate. This chapter explores the probable paths, challenges, and opportunities ahead in the search for economic cost reduction due to climate change and the making of a resistant and sustainable world (Casas-Cuestas, 2024).

1. Accelerating the Transition to a Low-Carbon Economy

This transition pathway to a low-carbon economy is very intrinsic in the response to climate change at a global level. Its construction entailed basic changes in energy systems, industry, transportation, agriculture, and use of land for minimal GHG emissions with a minimum impact on the environment. Acceleration through policy interventions, technological innovation, and consumer behavior is necessary in this transition for the future.

a. Renewable Energy Expansion:

Therefore, the future of energy production and consumption is oriented toward renewable energy sources such as solar, wind, hydro, and geothermal. As all these technologies are currently improving at an incredible rate, the costs have been successively reduced lately, which is the main reason renewable energy starts to be increasingly competitive with conventional fossil fuels (Zhao, 2023). The proliferation of renewable energy system infrastructure will be key to cutting emissions and meeting global climate goals. Countries set ambitious renewable energy targets and investment in large-scale projects, further showing commitments to the decarbonization of the energy sector.

b. Electrification/Decarbonization of Key Sectors: Besides this, electricity and decarbonization of key sectors like transport, industry, and buildings will play other crucial roles towards the path of a low-carbon future. Electrification of transport, advanced battery storage, and green hydrogen are some of the key enablers that would help bring down emissions related to transport and heavy industry. Energy-efficient building design, smart grids, use of sustainable materials, among others, contribute to minimizing the carbon footprint from the built environment.

c. Circular Economy and Sustainable Practices:

In fact, while there is much left for the future to be explored, for instance, the wide diffusion of circular economy principles promising not to produce waste but instead looks for maximum resource efficiency by keeping product and material utilization for longer periods than usual with sustainable production and consumption patterns, businesses and consumers alike will reduce environmental impact while contributing to a more sustainable economy. More circular business models, such as product-as-a-service, recycling, and remanufacturing, are emerging; these are expected to eventually have the potential to reduce GHG emissions and resource use by magnitudes.

2. Climate Resilience and Adaptation

While mitigation is important in order to keep global warming within thresholds, the very fact that climate change already exists brings into the forefront the importance of making adaptation a critical aspect of future policies. In simple terms, building resilience to climate change includes a process for preparing and adapting to the impacts of climate change that are already observable or projected over the coming decades.

a. Climate-Resilient Infrastructure

While this is happening, increased frequency and severity call for responsive investments in climate-resilient infrastructure that will be better equipped to deal with the rainy day, quite literally, caused by turbulent ocean waves and other destructive events such as storms, floods, and droughts (Kulkarni et al., 2022). Resilient infrastructure investments, including flood defenses, stormwater management, and resilient transportation networks, lower the economic costs of disasters and help safeguard communities and their economies against future shock.

b) Nature-Based Solutions:

Recognition continues to build for nature-based solutions in enhancing climate resilience, leveraging natural systems and processes as remarkably effective and efficient adaptation strategies. Examples range from wetland and mangrove restoration to protect coasts against storm surges, sustainable agriculture practices to improve soil health and water retention, to urban green spaces for mitigating heat island effects. Working with nature and not against it in this aspect means that society will be resilient due to the impacts of climate change while biodiversity is preserved together with ecosystem services (Zhao, 2023).

c. Integration of Climate Risk into Planning and Decision-Making:

Looking ahead, in all aspects of planning and decision-making-from urban development and investment in infrastructure to agriculture and water management-there needs to be great attention to risk with respect to climate change. This involves the preparation of solid data and the development of tools with which to study climate modeling and the assessment of risk to understand potential impacts and to develop adaptation strategies. By proactively incorporating climate risk into the governance and business decision-making cycle, governments and businesses alike would reduce their vulnerability and enhance resilience to future climate-related risks (Kulkarni et al., 2022).

3. Leveraging Technological Innovation and Digital Transformation

Technological innovation and digital transformation will be crucial to the shape of future climate action and how economic costs are reduced related to climate policies. New opportunities for enhancing climate monitoring, energy efficiency, and sustainable practices have opened up with advances in digital technologies like AI, big data, IoT, and blockchain (Kulkarni et al., 2022).

a. Smart Technologies for Energy Management:

Efficiency in use and emission of energy will further be optimized by deploying smart technologies that offer real-time data and analytics, hence offering better energy management: smart grids, smart meters, and IoT devices. This will enable efficient resource utilization, reduced energy waste, and easy integration of renewable energy sources into the grid. In the ongoing digital revolution, further large-scale deployment of such smart technologies is bound to remain a driving force for energy efficiency and emission reduction (Casas-Cuestas, 2024).

b. AI and Machine Learning for Climate Modeling and Prediction:

It is here that AI and ML can revolutionize climate science by radically improving the accuracy and finesse of climate models and predictions. It is in fact such abilities of these technologies to analyze large volumes of data for underlying patterns and trends, to flesh out the understanding of climate dynamics, and to inform proper mitigation and adaptation strategies that carry real potential. This thus makes it possible for researchers and policymakers to better anticipate climate risks and develop targeted responses that avoid unnecessary economic costs and protect vulnerable communities.

c. Blockchain for Supply Chain Transparency and Carbon Tracking:

The application of blockchain technology to supply chains brings about a level of transparency and accountability necessary to ensure that products are sourced and produced in a sustainable manner. Additionally, the blockchain can, with its secure and immutable transaction record, track carbon footprints, verify emissions reductions, and thereby support carbon markets and trading schemes. As businesses and consumers increasingly call for more transparency and sustainability, blockchain could be one of the tools to help achieve a reduction in emissions and ethical best practice across global supply chains.

4. Enhancing Global Climate Governance and Policy Frameworks

The future of climate action will be pegged to the gold standard that shall be Delhi's global climate governance and policy frameworks. The urgency for hard-hitting and effective international efforts on the crisis becomes a call voiced for its redress on flagrant impacts of climate change; hence, a call for minimization of economic costs (Casas-Cuestas, 2024).

a. Building on the Paris Agreement:

In fact, though the Paris Agreement sets a base on which global climate actions can be advanced, its success is hinged on the willingness of countries to increase commitments and policies. As a matter of fact, the future needs greater national engagements, more vigorous monitoring and reporting architectures, and increased support for developing countries toward the fulfillment of agreement goals. a. Strengthen the Paris Agreement and fully implement it to contribute to holding global warming to as low a level as possible and drastically cut the economic costs of climate change. (Casas-Cuestas, 2024).

b. Promoting Climate Justice and Equity

As the world moves on to take action in climate concerns, it will be relevant to make sure that equity and justice concerns are at the forefront in all actions related to the subject matter. It is an understanding that, while being unequal in its impacts, climate change ensures that the affected groups are in possession of the resources and other support to adapt and thrive. This is one kind of justice that demands accountability from the major emitters so that all countries can have the opportunity to participate in and benefit from the low-carbon economy (Ceesay, 2022).

c. Developing International Collaboration and Partnerships:

Future climate action will also rely on collaboration and the development of international cooperation among countries. That is, an informal network and coalitions of governments, businesses, civil society organizations, and other stakeholders. Working in partnerships with shared action, the global community can enhance accelerated progress toward the goals for climate and reduce climate change-related economic costs (Kulkarni et al., 2022).

5. Adopting a Proactive and Adaptive Attitude

The future prospect of halting climate change and its economic costs involves the notion of being proactive and adaptive. It simply means coming to terms with the fact that climate change is not something to be tackled in the distant future but, as a matter of fact, now. It also involves being prepared to adapt to changed circumstances, as well as a continuous process of learning from new information, experiences, and innovations (Casas-Cuestas, 2024).

a. Proactive investment in climate solutions:

Proactive investments in climate solutions, including renewable energy, efficiency in energy use, sustainable agriculture, and climate-resilient infrastructure, are the ways to go in both mitigation and building resilience. In so doing, countries stand to evade much larger economic costs of climate inaction and thus present other opportunities for growth and development (Ceesay, 2022).

b. Continuous Learning and Innovation:

The future of climate action will also require iterative learning and innovation. New technologies, data, and insights are continually emerging; it will be critical to adapt and refine strategies, tactics, and policies to maximize their impact and effectiveness. That is the cultivation of an innovative culture, the encouragement of experimenting, and openness to new ideas and approaches with the view to helping to address the climate crisis.

c) Engaging and Empowering Communities:

Finally, any action to address climate change must involve and empower the community to take action for a difference. This involves creating awareness, capacity building, and availing the necessary tools and other resources to support local climate initiatives and solutions. As the global community involves and gives more opportunities for local communities to act within decision-making processes, it also nurtures an approach that will be more inclusive, equitable, and effective in responding to climate change.

In sum, the future prospect of responding to climate change and its economic toll is fraught and inspiring. The world could effectively address the economic cost of climate change through accelerating the low-carbon transition of economies, enhancing climate resilience, leveraging technological innovation, enhancing global climate governance, and embracing proactive and adaptive mentalities toward the handling of this challenge. The road forward would require collective action, innovation, and determination, but the opportunities for positive change are within grasp if tackled decisively and in concert (Kellie-Smith & Cox, 2011).

Implications

The different dimensions that emanate on addressing climate change and its economic costs have to do with practical, managerial, ethical, and societal dimensions. It is of paramount importance to recognize these dimensions in formulating effective strategies and thereby making sure that climate action is inclusive and equitable. This section looks at the various types of implications that are important and have an impact on parties concerned (Eboli et al., 2010).

Practical Implications

In practice, this means that climate change requires immediate and sustained action to take in order to mitigate its impacts as well as adapt to its consequences. The increasing costs of disasters from climate change, through extreme weather events, floods, and heatwaves, show the urgent need for investing in prevention and resilient infrastructure. Essentially, this means a call for investments in disaster-resilient infrastructure: flood defenses, sustainable urban planning, and disaster response systems (Eboli et al., 2010). Another aspect is that businesses and governments have to develop ways of incorporating climate risk assessment into operational and strategic planning (Fankhauser & Tol, 2005). To this end, companies may have to redesign supply chains in preparation for probable climatic disruptions and make investments in technologies that guarantee better efficiency of resources with reduced emissions (Ceesay, 2022).

The practical implications even reach as far as new technologies and new ways of doing things that contribute to the attainment of a low carbon economy. The reason is quite straightforward: rapidly developing renewable energy technologies, energy-efficient building designs, and sustainable agriculture-all of them call for transformed industries and consumer choices. This transformation shall need new competencies and skills, supportive policies, and incentive systems that spur the adoption of climate-friendly technologies. Besides that, the practical considerations also involve the need for robust systems of monitoring and reporting that could track progress and compel accountability toward climate action (Eboli et al., 2010).

Managerial Implications

On the managerial level, addressing climate change means integrating environmental sustainability into core business strategies and decisions. Managers would have to effectively navigate the fragile balance between economic performance and environmental accountability when ensuring that a climate factor is included in organizational objectives and practices. This may involve setting ambitious targets on sustainability, such as reducing greenhouse gas emissions, improving energy efficiency, and adopting circular economy principles (Ceesay et al., 2020).

Active management also entails continuous stakeholder engagement to develop climate strategies that take into account the expectations and requirements of the stakeholders. The managers will have to transparently disclose the risks and opportunities related to climate, enhance a culture of sustainability within the organization, and drive the business to create innovation for sustainable competitive advantages in the low-carbon economy. For instance, companies that are proactively addressing the climate change risks and integrating sustainability into their operation tend to enjoy a competitive advantage through an improved brand reputation, customer loyalty, and green financing (Fankhauser & Tol, 2005).

Furthermore, managers need to deal with adaptation to the challenges of climate adaptation through strategy formulation that protects their assets and operations from such changes in climate (Ceesay, 2022). This will involve assessing vulnerability to climate impacts, resilience investment, and adaptation of business models to changing environmental conditions. Strategic risk management and scenario planning are potent tools to enable one to envision the prospect of disruptions and ensure the long-term continuity of one's businesses amidst these growing uncertainties within climate (Ceesay, 2022).

Ethical Considerations

Climate change has deep and complex ethical dimensions. Because it affects vulnerable and marginalized communities the most, questions arise as to justice and equity. These ethical considerations can only be appreciated if all climate action is inclusive and has equitable elements within it, with particular focus on supporting those who bear the largest impacts from climate changes. It means assuming responsibility for historic emissions by main emitters and providing financial and technical support to developing countries, which are least responsible for global emissions, yet they bear the full force of climate impacts (Gao, 2024).

Ethical issues also include the responsibility of business and governments to not contribute further to environmental degradation but take active steps to mitigate carbon emissions. This involves adopting practices for ethical sourcing, waste and pollution minimization, and making sure their climate strategies are coherent and compatible with wider social and environmental goals. Lifecycle impacts from products and services developed by companies should, therefore, be those that offer the least negative effect on ecosystems and communities.

There is also the moral requirement of intergenerational equity: one must be in a situation where one does not compromise the ability of future generations to meet their needs. It denotes the making of decisions that would bring forth long-term sustainability and resilience rather than immediate-term benefits. Ethical climate action has to be committed to principles of fairness, transparency, and accountability, with emphasis on protection of human rights and social well-being (Haer et al., 2018).

Societal Implications

The impacts of climate change range far and wide and hit at most aspects of life. Climate change may enhance social inequality; consequently, health vulnerabilities, displacement, and economic hardship are all left at the doorstep of the most vulnerable. The implications for addressing this must adopt a holistic approach which would investigate the social dimensions of climate action in view of public health, community resilience, and social cohesion.

On the social scale, climate change may influence public attitudes and behaviors in the way people and communities conceptualize and respond to environmental challenges. This could be further enhanced by raising awareness and education on climate change, which in turn will increase the level of public participation and support in climate policies and initiatives. Societal changes in lifestyle habits related to energy consumption, transportation modes, and buying habits can make a difference in lessening climate impacts and creating a resilient society (Hutton, 2011).

Other considerations in terms of impacts on society include a displaced role of climate change, which may contribute to the social and political changes. With increased vulnerability to climate impacts, there could be a rise in clamor for more definitive action from the governments and institutions through the enactment of appropriate climate policies. This could lead to the change in the governance structure, policy framework, and international cooperation mechanisms. Societal movements-such as climate activism and grassroots advocacy-push for stronger action on climate and make leaders more accountable for their promises. Concluding, the implications from addressing climate change and its economic cost are multifaceted and interwoven (Kellie-Smith & Cox, 2011). The practical implications address the pressing needs to act and invest in climate resilience and technology. Managerial implications pertain to issues of integrating sustainability into business strategy and operations. Ethical implications relate to matters of justice, equity, and responsibility. Societal implications cover the wide-ranging ramifications on community, behavior, and governance. The better these implications are understood and accounted for, the more stakeholders will be able to forge a more efficient and just response to the climate crisis-one befitting the needs of a sustainable future for all (Klusak et al., 2021).

CONCLUSION

The economic costs of climate change are gigantic and need comprehensive and multilevel concerted actions by the different strata of society. From extreme weather to rising sea levels, from disruption of agriculture to loss of biodiversity, a wide-ranging impact from climate change stresses the need for urgent action. As discussed in detail throughout this chapter, not only is the economic cost of climate change huge, but very diverse, cutting across different regions and sectors in different ways. They should, therefore, understand what costs these are, how they are varying in their region and sector, and use economic models in their assessment with a view to mitigation and adaptation.

International cooperation cannot be emphasized enough. Indeed, it is widely acknowledged that global action must be advanced through cooperative efforts inspired and anchored in international agreements, coordination organizations, and partnerships representative of the truly global nature of climate change. This way, different countries will be able to share various resources, knowledge, and technologies in the context of tackling common climate objectives. In turn, this would address issues of equity and justice, ensuring that vulnerable and developing nations are guided along the pathway of building resilience to and transitioning toward low-carbon economies.

The future of climate change mitigation is bleak and, at the same time, promising. There are various ways to avert economic costs and ensure a sustainable future: rapid transition to a low-carbon economy, enhancement of climate resilience, and technological innovation. This calls for proactive and adaptive measures that include solid investment in climate solutions, incorporation of climate risk into the decision-making process, and engaging the community.

The practical, managerial, ethical, and societal implications of climate change are indeed multifaceted. Practical responses include increased investment in resilient infrastructure and new technologies that can reduce vulnerability and enhance sustainability. Managers also have to embed the climate issue into their strategies and business operations if they are to remain viable. Ethically, justice and equity should form the core to ensure that actions taken are not only fair but also encompassing of inclusiveness. The societal implications are larger and encompass public attitudes, behaviors, and governance structures that emphasize the need for collective action or systemic change. Combating the economic costs of climate change therefore requires a banded effort and collaboration between governments, businesses, and people. The approach to solving this will require innovative solutions, international collaboration, and ethical and equitable commitment from the world community if the world is seriously going to manage the intricacies of climate change toward resilience and sustainability. While the challenge is evident, it is within the means of our determination and collective action to reduce impacts, lessen economic costs, and build a better world for current and future generations.

REFERENCES

Abidoye, B., & Odusola, A. (2015). Climate change and economic growth in Africa: An econometric analysis. *Journal of African Economies*, 24(2), 277–301. DOI: 10.1093/jae/eju033

Anthoff, D., & Tol, R. S. (2013). The uncertainty about the social cost of carbon: A decomposition analysis using FUND. *Climatic Change*, 117(3), 515–530. DOI: 10.1007/s10584-013-0706-7

Auffhammer, M., & Aroonruengsawat, A. (2011). Simulating the impacts of climate change, prices and population on California's residential electricity consumption. *Climatic Change*, 109(S1), 191–210. DOI: 10.1007/s10584-011-0299-y

Bosello, F., Eboli, F., & Pierfederici, R. (2012). Assessing the economic impacts of climate change: An updated CGE point of view. *FEEM Working Paper No. 2012.056*. https://doi.org/DOI: 10.2139/ssrn.2153483

Burke, M., & Emerick, K. (2016). Adaptation to climate change: Evidence from US agriculture. *American Economic Journal. Economic Policy*, 8(3), 106–140. DOI: 10.1257/pol.20130025

Casas-Cuestas, M. (2024). How much and for how long could the annual cost of atmospheric greenhouse gas (CO2e) abatement between 1960 and 2020 through carbon pricing be estimated? *Preprint*. https://doi.org/DOI: 10.21203/rs.3.rs-4571476/v1

Ceesay, E., & Fanneh, M. M. (2022). Economic growth, climate change, and agriculture sector: ARDL bounds testing approach for Bangladesh (1971-2020). *Economics Management and Sustainability*, 7(1), 95–106. DOI: 10.14254/jems.2022.7-1.8

Ceesay, E., Oladejo, H., Abokye, P., & Ugbor, O. (2020). Econometrics analysis of the relationship between climate change and economic growth in selected West African countries. *Energy and Environment Research*, 10(2), 39. DOI: 10.5539/eer.v10n2p39

Dell, M., Jones, B. F., & Olken, B. A. (2014). What do we learn from the weather? The new climate-economy literature. *Journal of Economic Literature*, 52(3), 740–798. DOI: 10.1257/jel.52.3.740

Deschênes, O., & Greenstone, M. (2011). The economic impacts of climate change: Evidence from agricultural output and random fluctuations in weather. *The American Economic Review*, 97(1), 354–385. DOI: 10.1257/aer.97.1.354

Dixon, P. B., & Jorgenson, D. W. (Eds.). (2013). *Handbook of Computable General Equilibrium Modeling (Vol. 1A)*. North-Holland., DOI: 10.1016/B978-0-444-59568-3.09997-0

Eboli, F., Parrado, R., & Roson, R. (2010). Climate-change feedback on economic growth: Explorations with a dynamic general equilibrium model. *Environment and Development Economics*, 15(5), 515–533. DOI: 10.1017/S1355770X10000252

Fankhauser, S., & Tol, R. (2005). On climate change and economic growth. *Resource and Energy Economics*, 27(1), 1–17. DOI: 10.1016/j.reseneeco.2004.03.003

Gao, S. (2024). An exogenous risk in fiscal-financial sustainability: Dynamic stochastic general equilibrium analysis of climate physical risk and adaptation cost. *Journal of Risk and Financial Management*, 17(6), 244. DOI: 10.3390/jrfm17060244

Haer, T., Botzen, W., Roomen, V., Connor, H., Zavala-Hidalgo, J., Eilander, D., & Ward, P. (2018). Coastal and river flood risk analyses for guiding economically optimal flood adaptation policies: A country-scale study for Mexico. *Philosophical Transactions. Series A, Mathematical, Physical, and Engineering Sciences*, 376(2121), 20170329. DOI: 10.1098/rsta.2017.0329 PMID: 29712799

Hutton, G. (2011). The economics of health and climate change: Key evidence for decision making. *Globalization and Health*, 7(1), 18. DOI: 10.1186/1744-8603-7-18 PMID: 21707990

Kellie-Smith, O., & Cox, P. (2011). Emergent dynamics of the climate–economy system in the Anthropocene. *Philosophical Transactions of the Royal Society A: Mathematical, Physical and Engineering Sciences, 369*(1938), 868-886. DOI: 10.1098/rsta.2010.0305

Klusak, P., Agarwala, M., Burke, M., Kraemer, M., & Mohaddes, K. (2021). Rising temperatures, falling ratings: The effect of climate change on sovereign creditworthiness. SSRN *Electronic Journal*. DOI: 10.2139/ssrn.3811958

Kocornik-Mina, A., & Rodriguez-Vega, L. (2018). Transport and climate change: A review. *Transport Policy*, 67, 101–114. DOI: 10.1016/j.tranpol.2017.05.009

Kulkarni, S., Hof, A., Wijst, K., & Vuuren, D. (2022). Disutility of climate change damages warrants much stricter climate targets. *Preprint*. DOI: 10.21203/rs.3.rs-1788130/v1

Kumar, P., & Balaji, N. (2019). Assessing the impact of variations in weather conditions on socio-economic status. *International Journal of Social Sciences & Economic Environment*, 4(2), 12–17. DOI: 10.53882/IJSSEE.2019.0402003

Mendelsohn, R., & Dinar, A. (2009). *Climate change and agriculture: An economic analysis of global impacts, adaptation and distributional effects*. Edward Elgar Publishing. DOI: 10.4337/9781849802239

Nordhaus, W. D. (1993). Rolling the "DICE": An optimal transition path for controlling greenhouse gases. *Resource and Energy Economics*, 15(1), 27–50. DOI: 10.1016/0928-7655(93)90017-O

Stern, N. (2007). *The Economics of Climate Change: The Stern Review*. Cambridge University Press. DOI: 10.1017/CBO9780511817434

Tol, R. S. (2005). The marginal damage costs of carbon dioxide emissions: An assessment of the uncertainties. *Energy Policy*, 33(16), 2064–2074. DOI: 10.1016/j.enpol.2004.04.002

van Vuuren, D. P., Edmonds, J., Kainuma, M., Riahi, K., & Weyant, J. (2011). A special issue on the RCPs. *Climatic Change*, 109(1-2), 1–4. DOI: 10.1007/s10584-011-0157-y

Warsame, A., Sheik-Ali, I., Hussein, H., & Barre, G. (2023). Assessing the long- and short-run effects of climate change and institutional quality on economic growth in Somalia. *Environmental Research Communications*, 5(5), 055010. DOI: 10.1088/2515-7620/accf03

Weyant, J. P. (2017). Some contributions of integrated assessment models of global climate change. *Review of Environmental Economics and Policy*, 11(1), 115–137. DOI: 10.1093/reep/rew018

Zhai, F., & Zhuang, J. (2012). Agricultural impact of climate change: A general equilibrium analysis with special reference to Southeast Asia. *Climate Change Economics (Singapore)*, 3(2). Advance online publication. DOI: 10.1142/S2010007812500088

Zhao, Y., & Liu, S. (2023). Effects of climate change on economic growth: A perspective of the heterogeneous climate regions in Africa. *Sustainability (Basel)*, 15(9), 7136. DOI: 10.3390/su15097136

Chapter 8
Impacts of Climate Change on the Tourism Sector

Fatma Fehime Aydin
https://orcid.org/0000-0002-7026-6889
Van Yuzuncu Yil University, Turkey

Cemalettin Levent
Independent Researcher, Turkey

ABSTRACT

Climate change, one of the most prominent problems all over the world in recent years, has significant impacts on many sectors as well as being an environmental problem. The tourism sector is one of the sectors most affected by climate change. The main objective of this study is to investigate the impact of climate change on tourism in selected countries of the world using panel data analysis based on the data set for the period 2004-2019. Based on the findings of this study, countries should adopt practices that will encourage tourist inflow to the country to increase tourism revenues, but at the same time, they should bring environmentally sensitive practices to the agenda so that both future generations and current generations can live in healthier environments.

INTRODUCTION

Climate change, one of the most important problems facing today's world, is an important problem that scientists are focused on solutions. Greenhouse gas emissions are thought to be among the most important causes of climate change. As a result

DOI: 10.4018/979-8-3693-5792-7.ch008

of the industrialization policies implemented, greenhouse gas emissions into the atmosphere increase, and the temperature values on earth, which are in a natural balance, deteriorate (Şanlı et al., 2017, p. 201). CO_2 emissions, which have an important share among greenhouse gas emissions since the industrial revolution, have increased by approximately 35% and increased from 280 parts per million (ppm) to approximately 385 parts per million (ppm). If this process continues in this way, it is thought that the level of CO_2 emissions will double in the 2050s (Aras, 2022, p. 3648). As a result of this process, increasing CO_2 emissions will trigger the problem of climate change, which is a global human development problem, as a result, the continuity of ecosystems will be prevented and this will cause social and economic negativities on people and countries (Somuncu, 2018, p. 757).

The tourism sector, which was included in the development plans of countries due to industrialization and urbanization after the Second World War, has become one of the fastest growing sectors in the world. The tourism sector, which is described by the World Travel and Tourism Council (WTTC) as "the world's largest sector and the largest employer creating employment and prosperity", has an important place in the development of the economy (Karadeniz et al., 2018, p. 171). The tourism sector, which has an important place in terms of economic growth and development, is also one of the sectors most affected by climate change.

The negative impact of climate change, which has become a global problem, on the tourism sector causes countries to be negatively affected economically. Since the effects of climate change on tourism are expected to increase over time, studies on the relationship between climate change and tourism are increasing in the literature. The main objective of this study is to investigate the impact of climate change on tourism in selected countries of the world using panel data analysis method based on the data set for the period 2004-2019.

THEORETICAL FRAMEWORK

Today, global warming and air pollution threaten the environment and human health. Since tourism is often assumed to be an important economic sector, it is important to consider the impact of global warming and air pollution on tourism attractiveness. On the other hand, research has shown that tourism is significantly affected by climate change, and it is predicted to have a long-lasting impact. The United Nations World Tourism Organization states that the range of hazards caused by climate change such as global warming is the biggest obstacle to sustainable tourism (Zikirya et al., 2021, p. 1).

Climate change is an important factor affecting tourism. Climate plays an important role especially in the preference of touristic destinations. The main reason for this is that factors such as temperature, humidity, sunlight, radiation, and precipitation rate have an impact on human perception. Therefore, the tourism sector is affected by climate change and climate change is affected by the activities arising from the tourism sector (Özekici & Silik, 2017, p. 59).

The tourism industry and destinations are vulnerable to climate change. Climate defines the length and quality of multi-billion-dollar tourism seasons and is important in destination choice and tourist spending. In many destinations, tourism and the environment are closely linked. Climate affects a number of environmental resources critical to tourism, such as weather, water levels and quality, wildlife productivity and biodiversity (Climate Change and Tourism Responding to Global Challenges, 2008). Therefore, tourism cannot be considered independently of climate change. The table below describes the impact of climate change on tourism destinations.

Table 1. Impacts of Climate Change on Tourism Destinations

Climate Changes	Impacts on Tourism
Higher temperatures	Changing seasonality, heat stress for tourists, cooling costs, changes in plant-wildlife-insect populations and distribution, infectious disease ranges
Declining snow cover and shrinking glaciers	Lack of snow in winter sports destinations, increased snow making costs, shortened winter sports seasons, reduced landscape aesthetics
Increasing frequency and intensity of extreme storms	Increased risk, insurance costs/loss of insurability, business interruption costs for tourism facilities
Decreased precipitation and increased evaporation in some regions	Water scarcity, competition for water between the tourism sector and other sectors, desertification, increased forest fires threatening infrastructure and affecting demand
Increased frequency of heavy rainfall in some areas	Flood damage to historical architecture and cultural assets, damage to tourism infrastructure, changing seasonality
Sea level rise	Coastal erosion, loss of beach areas, increased costs of protecting and maintaining waterfronts
Rising sea surface temperatures	Increased coral bleaching, marina resources and aesthetic degradation in diving and snorkeling destinations
Changes in terrestrial and marine biodiversity	Loss of natural attractions and species from destinations, increased risk of disease in tropical-subtropical countries
More frequent and widespread forest fires	Loss of natural beauty, increased flood risk, damage to tourism infrastructure
Changes in the soil (e.g. moisture levels, erosion and acidity)	Loss of archaeological assets and other natural resources, with impacts on destination attractions

Source: Climate Change and Tourism Responding to Global Challenges (2008) p. 61.

In Table 1, the effects of climate change on tourism destinations are explained in detail in the report titled "Climate Change and Tourism Responding to Global Challenges" published by the United Nations World Tourism Organization (UNWTO) in 2008. In this context, the table shows that higher temperatures, less snow, shrinking glaciers, more frequent and more effective storms, irregular rainfall, sea level rise and fall, forest fires and soil changes have significant impacts on tourism destinations.

LITERATURE REVIEW

Climate change, one of the most pressing global environmental issues facing the world today, has significant social, economic and environmental implications. The tourism sector is one of the most vulnerable sectors to climate change due to its reliance on natural resources (Urioste-Stone et al., 2015, p. 34). In this context, various studies have been conducted on the impact of climate change on tourism. One of these studies is the study conducted by Sevim and Ünlüönen (2010). The impact of climate change on tourism in accommodation establishments in Antalya province was investigated through a survey. According to the findings of the study, it was concluded that a significant portion of accommodation establishments are affected by climate change.

In the study conducted by Aydemir and Şenerol in 2014, the effects of climate change on the tourism sector in Turkey were investigated. According to the results of the study in which the Delphi survey method was applied; it is seen that the effects of climate change are beginning to be felt in the tourism sector and it is estimated that climate change will have positive and negative effects on the tourism sector.

In the study conducted by Jebli et al. on Tunisia in 2014, the relationship between the number of international tourists and CO_2 emissions per capita was investigated. According to the results obtained from the Johansen Cointegration and Granger Causality analysis applied in the study; it was concluded that there is a mutual causality between the number of international tourists and CO_2 emissions per capita.

Priego, Rosselló and Gallego (2015) investigated the impact of climate change on domestic tourism. According to the findings of the study, climate is an important factor in determining domestic tourism flows in Spain.

In the study conducted by Şahin on APEC countries in 2018, the relationship between tourism, economic growth and the environment was investigated. According to the results of the panel data analysis applied in the study; it was concluded that there is a unidirectional causality relationship from tourism to CO_2 emissions.

The study conducted by Akın et al. in 2018 investigated the relationship between tourism and low carbon economy in Turkey. According to the findings of the time series analysis applied in the study, it was determined that there is a causality relationship between the share of tourism activities in exports and GDP and CO_2 emissions.

Seetanah and Fauzel (2018) investigated the impact of climate change on tourism demand based on 18 developing island countries. According to the findings of the study, there is a mutual causality relationship between climate change and tourism demand.

Du and Ng (2018) investigated the impact of climate change on tourism for Turkey, Spain and Greece. According to the findings of the study, it is determined that the negative impact of climate change on Greece, Spain and Turkey is greater than its effects on other types of economies.

The study conducted by Tandoğan and Genç in 2019 investigated the relationship between tourism and carbon emissions in Turkey. The study is based on Rals-Engle and Granger Cointegration analysis. According to the findings of the study, it was concluded that there is a mutual causality relationship between the number of tourists and carbon emissions.

Dereli, Boyacıoğlu and Terzioglu (2019) investigated the relationship between climate change and the tourism sector using dynamic panel data analysis method. According to the findings of the study, there is a negative relationship between the share of tourism in GDP and per capita carbon dioxide emissions.

Atasoy and Güneysu Atasoy (2020) investigated the impact of climate change on tourism in Turkey using causality analysis. According to the findings of the study, it was concluded that climate change has a negative and significant impact on the tourism sector. On the other hand, it is determined that climate change indicators have a unidirectional and negative effect on international tourism revenues.

In the study by Koçak, Ulucak and Ulucak (2020) the impact of tourism developments on CO_2 emissions was investigated by advanced panel data analysis method. According to the findings of the study, it was determined that tourist inflow to the country has an increasing effect on CO_2 emissions, but revenues from tourism have a decreasing effect on CO_2 emissions.

In the study conducted by Ngxongo (2021) in the Drakensberg region in KwaZulu-Natal, the impact of climate change on the visitor destination was investigated by surveying 347 respondents. According to the econometric findings of the study, it was concluded that climate change is a determining factor in tourists' decision-making, behavior, and spending habits. However, the most important climate parameter for tourists was found to be mild climate conditions as well as summer season.

In the study conducted by Yurtkuran in 2021, the relationship between tourism and CO_2 emissions was investigated based on the 10 countries with the highest number of tourist arrivals. According to the findings of the Fourier Toda-Yamamoto

Granger analysis, it was found that there is unidirectional causality from tourism to CO_2 emissions in Mexico, Turkey and Spain, unidirectional causality from CO_2 emissions to tourism in China and Germany, and reciprocal causality between tourism and CO_2 emissions in the USA. However, there is no causality relationship between tourism and carbon emissions in the UK, Thailand, Italy, and France.

In the study conducted by Arı in 2021, the relationship between CO_2, tourism, renewable energy, and economic growth was analyzed for Turkey. According to the findings of the FMOLS analysis applied in the study, it was concluded that developments in the tourism sector did not have a statistically significant effect on CO_2 emissions.

When the literature review is evaluated in general, it is determined that there is a mutual interaction between climate change and the tourism sector, and as the number of tourists entering the country increases, this increases greenhouse gas emissions, negatively affects the environment and leads to climate change. However, it is seen that this negative impact on the environment can be eliminated when the revenues generated from tourism are utilized under favorable conditions. On the other hand, it is also found that climatic conditions have significant impacts on the tourism sector, and in some countries, there may be a decrease in tourism revenues due to the negative effects of climate change on the tourism sector.

DATA SET, METHODOLOGY AND ANALYSIS FINDINGS

The purpose of the study, the data used in the study, the model, the methods applied, and the findings of the analysis are explained in the following sub-headings.

Data Set

The aim of this study is to investigate the impact of climate change on tourism on selected countries of the world (USA, Germany, United Kingdom, China, France, Spain, Italy, Mexico, Thailand, and Turkey) based on the data set for the period 2004-2019 using panel data analysis method. While determining the country group in the study, the 10 countries with the highest number of tourist arrivals according to the data obtained from the WTO were selected.

Table 2. Definition and Source of Variables

Variables	Definition	Period	Method	Country Group	Source
CO_2 Emissions	CO_2 emissions (metric tons per capita)	2004-2019	Panel data analysis	Selected World Countries	World Bank
Tourism	International tourism, number of arrivals	2004-2019	Panel data analysis	Selected World Countries	World Bank

Table 2 presents the definition and source of the variables, the methodology of the study, the data range and the countries used as the basis. Detailed information about the data obtained from the World Bank for the variables used in the study is given. For example, CO_2 emissions, which show the climate change variable in the study, are formed as a result of burning fossil fuels and cement production. CO_2 emissions from humans occur due to the combustion of fossil fuels and cement production. Therefore, statistical data on CO_2 emissions, which constitute the largest share of greenhouse gases, include gases from fossil fuel combustion and cement production, but do not include emissions from land use such as deforestation (World Bank, 2020). In this context, the burning of carbon-containing fuels causes both global warming and anthropogenic climate change.

The tourism variable is calculated as the number of international inbound tourists (overnight visitors), the number of tourists traveling to a country other than their country of residence and usual environment for a period of no more than 12 months. In the absence of data on the number of tourists, we instead use the number of visitors, which includes tourists, same-day visitors, cruise passengers and crew members. Sources and data collection methods for inbound tourists in this context vary from country to country. In some circumstances, data are obtained from border statistics (police, immigration, etc.) and supplemented by border surveys. In some countries, the number of inbound tourists is limited to air arrivals, while in others it is limited to those staying in hotels. In some countries, however, it includes arrivals of non-resident citizens, while in others it does not (World Bank, 2020).

In this study, the following model is estimated to investigate the impact of climate change on tourism:

$$Y_{it} = \beta_0 + \beta_{1it} X_{1t} + \varepsilon_{it} \tag{1}$$

Y_{it}: CO_2 Emissions (Dependent Variable)
β_0: Constant Term
β_1 : Coefficient of the independent variable (X_1)
X_1: Number of Incoming Tourists
ε_i: Error Terms

In the study, CO_2 emissions were taken as the dependent variable as an indicator of the climate variable. On the other hand, as an indicator of tourism, the number of tourist arrivals was included in the study as an independent variable.

Method and Analysis Findings

In this study, the relationship between climate change and tourism is first tested in terms of cross-sectional dependence. After determining the cross-sectional relationship between the variables in the analysis of the study, other analyses are applied.

In the cross-section dependence test, the Breusch and Pagan (1980) LM test is used when the time dimension (T) is larger than the cross-sectional dimension (N). The Pesaran (2004) CD_{LM} (Cross Sectional Dependence Lagrange Multiplier) test is used when the cross-sectional dimension is larger than the time dimension, and the CD_{LM2} test is used when the cross-sectional and time dimensions are equal. Moreover, Pesaran et al. (2008) developed an LM (bias-adjusted CD Test) that can be applied in a heterogeneous panel (Şaşmaz et al., 2020, p. 6).

Table 3. Cross-Section Dependence and Homogeneity Tests

Test	Statistic Value	P Value
Cross-section dependence test		
LM	107,50	0,00
LM_{adj}	12,94	0,00
CD_{LM}	3,195	0,00
Homogeneity tests		
Δ	12,96	0,00
$Δ_{adj}$	14,38	0,00

Table 3 shows the results obtained from cross-section dependence and homogeneity tests. According to the results of the cross-section dependence test, all P values are less than the critical value of 0.05. Accordingly, it is concluded that there is cross-section dependence between the series.

According to the homogeneity test results, all P values are less than the critical value of 0.05. According to these results, the hypothesis of homogeneity of slope is rejected and it is concluded that there is country-specific heterogeneity.

Which generation of unit root tests will be applied in the study is determined according to the results of cross-section dependence. In cases where there is no cross-section dependence in the series, first generation unit root tests are used, while in the presence of cross-section dependence, second generation unit root tests are used (Çınar, 2011, p. 74). In this study, CIPS unit root test, which is the second-generation unit root test, is used since there is cross-section dependence among the series. The CIPS statistic is an extension of the t statistic developed by Im, Pesaran and Shin (IPS) (2003) to include cross-section dependence. The calculation of the CIPS test statistic is presented in the equation below:

$$CIPS = N^{-1} \sum_{i=1}^{N} CADF_i \qquad (2)$$

Pesaran (2007) calculated the critical values of the CIPS statistic (Bektaş, 2017, p. 60).

Table 4. CIPS unit root test (2004-2019)

Panel CIPS test	Constant			Constant and Trend		
Logtourism	-1,89			-1,91		
ΔLogtourism	-3,07[a]			-3,10[b]		
LogCO$_2$	-1,37			-1,57		
ΔLogCO$_2$	-3,02[a]			-3,24[a]		
Critical Values	10%	5%	1%	10%	5%	1%
	-2,21	-2,34	-2,6	-2,74	-2,88	-3,15

[a] Indicates that it is statistically significant at 1% significance level and [b] indicates statistically significant at 5% significance level.

Table 4 presents the CIPS unit root tests. If the calculated CIPS values are greater than the critical values in absolute value, the null hypothesis of a unit root in the series is rejected and the null hypothesis of no unit root in the series is accepted (Pesaran, 2007, pp. 265-312). When the results of CIPS unit root test are analyzed, it is observed that both variables are non-stationary at the level in the constant and constant trend models but become stationary when first differences are taken. Since all series become stationary when first differences are taken, the next step is to examine whether there is a cointegration relationship between the series.

The Dumitrescu-Hurlin (2012) panel causality test is a simple version of the Granger causality test for inhomogeneous panel data models with constant coefficients. The developers of this test consider heterogeneity in two ways. These are the heterogeneity of the regression model used to assess Granger causality and the heterogeneity of causal links. Dumitrescu-Hurlin (2012) panel causality test requires the series to be stationary at the same level (Ahmed et al., 2022, p. 11). In the study by Dumitrescu-Hurlin (2012), it is stated that the causality relationship obtained for the countries analyzed by panel data analysis can also be accepted for other countries. Dumitrescu-Hurlin (2012) panel causality test is presented in equation 3 (Dumitrescu and Hurlin, 2012, p. 1457): (3)

The results of Dumitrescu-Hurlin (2012) panel causality test applied to reveal the causality relationship between x and y variables calculated according to Equation 3 are given in the table below.

Table 5. Dumitrescu-Hurlin (2012) panel causality test

Null Hypothesis	Wald Statistic	P Value
Logtourism is not the cause of $LogCO_2$	1,44	0,32
$LogCO_2$ is not the cause of Logtourism	4,06	0,00

Dumitrescu-Hurlin (2012) panel causality test is presented in Table 5. According to Table 5, when the relationship between international tourism and CO_2 emissions is analyzed, it is seen that there is causality from CO_2 emissions to international tourism, but there is no causality from international tourism to CO_2 emissions.

The Westerlund test is superior to other cointegration tests because it takes into account cross-section dependence issues and uses structural dynamics instead of residual dynamics to determine the long-run relationship between variables (Ansari et al., 2020, p. 361). The alternative approaches presented by Westerlund (2007) are expressed in equation 4:

$$\Delta y_{it} = \delta'_i d_t + \alpha_i y_{i,t-j} + '_i x_{i,t-1} + \sum_{j=0}^{P_i} \gamma_{ij} \Delta x_{i,t-j} + \sum_{j=1}^{P_i} \alpha_{ij} \Delta y_{i,t-j} + e_{it} \quad (4)$$

In the equation, $d_t = (1, t)'$ is the deterministic components and $\delta_i = (\delta_{1i}, \delta_{2i})'$ is the associated vector of parameters. The error correction parameter α_i is estimated using the least squares method. When calculating the group mean test statistics G_α and G_τ, the error correction model should be estimated for each cross-section (Demir & Görür, 2020, p. 25). Group mean test statistics are shown in equation 5.

$$G_\alpha = \frac{1}{N} \sum_{i=1}^{N} \frac{T\hat{\alpha}_i}{\hat{\alpha}_i(1)} \quad G_\tau = \frac{1}{N} \sum_{i=1}^{N} \frac{\hat{\alpha}_i}{SE(\hat{\alpha}_i)} \quad (5)$$

Table 6. LogTourism-LogCO$_2$ Westerlund Cointegration Tests

Statistics	Value	Z Value	P Value
G_τ	-1,21	2,01	0,98
G_α	-1,05	3,54	1,00
P_τ	-9,94	-5,40	0,00
P_α	-7,90	-2,62	0,00

Table 6 presents the Westerlund cointegration test. According to Table 6, when the results of G_a and G_τ tests for international tourism and CO_2 emissions variables are analyzed, it is seen that H_0 hypotheses are accepted at 1% significance level since the P values of both tests are greater than 0.01. Therefore, when analyzed on a country basis, it is determined that there is no cointegration relationship between international tourism and CO_2 emissions. When the results of the P_τ and P_a tests are analyzed, it is seen that the H_0 hypothesis is rejected at the 1% significance level since the P values of both tests are less than 0.01, therefore, according to the P_τ and P_a tests, there is a cointegration relationship at the 1% significance level for the entire panel.

CONCLUSION

In recent years, the effects of climate change on tourism have been the subject of various studies and have been empirically analyzed by many researchers. In these studies, it has been determined that climate change has positive and negative effects on tourism. In the studies in the literature, CO_2 emissions as an indicator of climate change have been associated with tourism and macroeconomic indicators. In this context, the relationship between tourism and climate change has been investigated. In a significant part of the findings obtained in the literature, it has been determined that CO_2 emissions, which are the cause of climate change, cause tourism to be negatively affected. The main objective of this study is to investigate the impact of climate change on tourism on selected countries of the world (USA, Germany, UK, China, France, Spain, Italy, Mexico, Thailand and Turkey) based on the data set for the period 2004-2019 using panel data analysis method.

According to the Westerlund cointegration test, it is found that there is no cointegration relationship between international tourism and CO_2 emissions variables when analyzed on a country basis, but there is a cointegration relationship between these two variables for the panel as a whole. According to Dumitrescu and Hurlin (2012) panel causality test, when the relationship between international tourism and CO_2 emissions is analyzed, it is seen that there is a unidirectional causality relationship from CO_2 emissions to international tourism. In other words, while the change in CO_2 emissions, which is an indicator of climate change, affects the tourism sector, the change in the tourism sector does not affect CO_2 emissions. In the studies in the literature, it is generally concluded that the tourism sector negatively affects CO_2 emissions. In this study, a different conclusion was reached in this regard. This situation can be explained as follows. The number of tourists entering the country will increase CO_2 emissions in various ways, but this negative effect can be eliminated when the increase in tourism revenues is directed to clean environmental policies.

For all these reasons, countries should adopt practices that will encourage the entry of tourists to the country in order to increase tourism revenues, but at the same time, they should bring environmentally sensitive practices to the agenda both for future generations and for current generations to live in healthier environments.

REFERENCES

Ahmed, N., Sheikh, A. A., Hamid, Z., Senkus, P., Borda, R. C., Wysokińska-Senkus, A., & Glabiszewski, W. (2022). Exploring the causal relationship among green taxes, energy intensity, and energy consumption in Nordic countries: Dumitrescu and Hurlin causality approach. *Energies*, 15(14), 1–15. DOI: 10.3390/en15145199

Akin, C. S., Aytun, C., & Akin, S. (2018). *Turizm, düşük karbon ekonomisi ilişkisi: Türkiye üzerine bir uygulama.* VII. Ulusal III. Uluslararası Doğu Akdeniz Turizm Sempozyumu, İskenderun.

Ansari, M. A., Ahmad, M. R., Siddique, S., & Mansoor, K. (2020). An environment Kuznets curve for ecological footprint: Evidence from GCC countries. *Carbon Management*, 11(4), 355–368. DOI: 10.1080/17583004.2020.1790242

Aras, S. (2022). İklim değişikliğinin turizm ve açık alan rekreasyonuna etkisinin incelenmesi. *Journal of Tourism and Gastronomy Studies*, 10(4), 3645–3661.

Arı, A. (2021). Yenilenebilir enerji, turizm, CO_2 ve GSYH ilişkisinin Türkiye için analizi. *Akademik Yaklaşımlar Dergisi*, 12(2), 192–205. DOI: 10.54688/ayd.880406

Atasoy, M., & Atasoy, F. G. (2020). The impact of climate change on tourism: A causality analysis. *Turkish Journal of Agriculture-Food Science and Technology*, 8(2), 515–519. DOI: 10.24925/turjaf.v8i2.515-519.3250

Aydemir, B., & Şenerol, H. (2014). İklim değişikliği ve Türkiye turizmine etkileri: Delfi anket yöntemiyle yapılan bir uygulama çalışması. *Balıkesir Üniversitesi Sosyal Bilimler Enstitüsü Dergisi*, 17(31), 381–417. DOI: 10.31795/baunsobed.664062

Bektaş, V. (2017). Gelişmekte olan ülkelerde cari açıkların sürdürülebilirliği: Bir panel veri analizi. *Bolu Abant İzzet Baysal Üniversitesi Sosyal Bilimler Enstitüsü Dergisi*, 17(1), 51–66.

Ben Jebli, M., Ben Youssef, S., & Apergis, N. (2014). The dynamic linkage between CO_2 emissions, economic growth, renewable energy consumption, number of tourist arrivals and trade. *MPRA Paper No. 57261, Munich Personal RePEc Archive*, 1-12.

Breusch, T. S., & Pagan, A. R. (1980). The lagrange multiplier test and its applications to model specification in econometrics. *The Review of Economic Studies*, 47(1), 239–253. DOI: 10.2307/2297111

Çinar, S. (2011). Gelir ve CO_2 emisyonu ilişkisi: Panel birim kök ve eşbütünleşme testi. *Uludağ Üniversitesi İktisadi ve İdari Bilimler Fakültesi Dergisi.*, 30(2), 71–83.

De Urioste-Stone, S. M., Scaccia, M. D., & Howe-Poteet, D. (2015). Exploring visitor perceptions of the influence of climate change on tourism at Acadia National Park, Maine. *Journal of Outdoor Recreation and Tourism*, 11, 34–43. DOI: 10.1016/j.jort.2015.07.001

Demir, Y., & Görür, Ç. (2020). OECD ülkelerine ait çeşitli enerji tüketimleri ve ekonomik büyüme arasındaki ilişkinin panel eşbütünleşme analizi ile incelenmesi. *Ekoist: Journal of Econometrics and Statistic*, 32, 15–33.

Dereli, M., Boyacioğlu, E. Z., & Terzioğlu, M. K. (2019). İklim değişikliği ve turizm sektörü arasındaki ilişkinin dinamik panel veri analizi ile incelenmesi. *Türk Turizm Araştırmaları Dergisi*, 3(4), 1228–1243. DOI: 10.26677/TR1010.2019.238

Du, D., & Ng, P. (2018). The impact of climate change on tourism economies of Greece, Spain, and Turkey. *Environmental Economics and Policy Studies*, 20(2), 431–449. DOI: 10.1007/s10018-017-0200-y

Dumitrescu, E. I., & Hurlin, C. (2012). Testing for Granger noncausality in heterogeneous panels. *Economic Modelling*, 29(4), 1450–1460. DOI: 10.1016/j.econmod.2012.02.014

Im, K. S., Pesaran, M. H., & Shin, Y. (2003). Testing for unit roots in heterogeneous panels. *Journal of Econometrics*, 115(1), 53–74. DOI: 10.1016/S0304-4076(03)00092-7

Karadeniz, C.B., Sari, S., & Çağlayan, A. B. (2018). İklim değişikliğinin Doğu Karadeniz turizmine olası etkileri. *Uluslararası Bilimsel Araştırmalar Dergisi (IBAD)*, 170-179.

Koçak, E., Ulucak, R., & Ulucak, Z. Ş. (2020). The impact of tourism developments on CO_2 emissions: An advanced panel data estimation. *Tourism Management Perspectives*, 33, 100611. DOI: 10.1016/j.tmp.2019.100611

Ngxongo, N. A. (2021). The impact of climate change on visitor destination selection: A case study of the Central Drakensberg Region in Kwazulu-Natal. *Jàambá*, 13(1). Advance online publication. DOI: 10.4102/jamba.v13i1.1161 PMID: 34956552

Özekici, Y. K., & Silik, C. E. (2017). Türkiye'deki iklim değişikliği plan ve politikalarında turizm sektörünün yeri. *Gazi Üniversitesi Turizm Fakültesi Dergisi*, (2), 58–79.

Pesaran, M. H. (2004), General diagnostic tests for cross section dependence in panels. *IZA Discussion Paper Series No. 1240*.

Pesaran, M. H., Ullah, A., & Yamagata, T. (2008). A bias adjusted LM test of error cross section independence. *The Econometrics Journal*, 11(1), 105–127. DOI: 10.1111/j.1368-423X.2007.00227.x

Priego, F. J., Roselló, J., & Santana-Gallego, M. (2015). The Impact of climate change on domestic tourism: A gravity model for Spain. *Regional Environmental Change*, 15(2), 291–300. DOI: 10.1007/s10113-014-0645-5

Şahin, D. (2018). APEC ülkelerinde turizm, ekonomik büyüme ve çevresel kalite ilişkisi: panel veri analizi. *İktisadi Yenilik Dergisi*, 5(2), 32-44.

Şanlı, F. B., Bayrakdar, S., & İncekara, B. (2017). Küresel iklim değişikliğinin etkileri ve bu etkileri önlemeye yönelik uluslararası girişimler. *Süleyman Demirel Üniversitesi İktisadi ve İdari Bilimler Fakültesi Dergisi*, 22(1), 201–212.

Şaşmaz, M. Ü., Sakar, E., Yayla, Y. E., & Akküçük, U. (2020). The relationship between renewable energy and human development in OECD countries: A panel data analysis. *Sustainability (Basel)*, 12(18), 7450. DOI: 10.3390/su12187450

Seetanah, B., & Fauzel, S. (2018). Investigating the impact of climate change on the tourism sector: Evidence from a sample of Island Economies. *Tourism Review*, 74(2), 194–203. DOI: 10.1108/TR-12-2017-0204

Sevim, B., & Ünlüönen, K. (2010). İklim değişikliğinin turizme etkileri: Konaklama işletmelerinde bir uygulama. *Erciyes Üniversitesi Sosyal Bilimler Enstitüsü Dergisi*, 1(28), 43–66.

Somuncu, M. (2018). İklim değişikliği Türkiye turizmi için bir tehdit mi, bir fırsat mı? *Tücaum 30. Yıl Uluslararası Coğrafya Sempozyumu*, 748-771.

Tandoğan, D., & Genç, M. C. (2019). Türkiye'de turizm ve karbondioksit salımı arasındaki ilişki: Rals-engle ve granger eşbütünleşme yaklaşımı. *Anatolia: Turizm Araştırmaları Dergisi*, 30(3), 221–230.

The World Bank. (2023a). https://databank.worldbank.org/reports.aspx?source=2&series=EN.ATM. CO2E.PC&country

The World Bank. (2023b). https://databank.worldbank.org/reports.aspx?source=2&series=ST.INT.ARVL &country

Westerlund, J. (2007). Testing for error correction in panel data. *Oxford Bulletin of Economics and Statistics*, 69(6), 709–748. DOI: 10.1111/j.1468-0084.2007.00477.x

World Tourism Organization. (2008). *Climate Change and Tourism Responding to Global Challenges*. e-unwto.org/doi/epdf/10.18111/9789284412341

Yurtkuran, S. (2022). Gelen turist sayısının en fazla olduğu 10 ülkede turizm ile CO_2 Salımı arasındaki ilişki: Panel Fourier Toda-Yamamoto nedensellik analizi. *Erciyes Üniversitesi İktisadi ve İdari Bilimler Fakültesi Dergisi*, (61), 281–303. DOI: 10.18070/erciyesiibd.988886

Zikirya, B., Wang, J., & Zhou, C. (2021). The relationship between CO_2 emissions, air pollution, and tourism flows in China: A panel data analysis of Chinese Provinces. *Sustainability (Basel)*, 13(20), 11408. DOI: 10.3390/su132011408

Chapter 9
Economic–Financial Issues and Climate Change:
A Sustainability-Based Approach

Güven Güney
https://orcid.org/0000-0001-8324-2870
Atatürk University, Turkey

Eda Bozkurt
Atatürk University, Turkey

ABSTRACT

The situation in which the interaction of economic-financial and social structure with the environment is not evaluated as a whole causes present and future generations not to benefit equally from the opportunities brought by development. For this reason, it is very important for policy development to determine the relationship between climate change and economic-financial issues by considering the issue of climate change from a sustainable development perspective. Identifying which economic-financial events fuel climate change will help policymakers to develop measures and solutions in that area. For this purpose, in this study, considering the availability of the data set, estimations were made based on panel quantile regression for the countries in the World Bank database. Thus, with a new econometric method, the causes of climate change have been revealed from an economic and financial framework for the countries that make up each quantile.

DOI: 10.4018/979-8-3693-5792-7.ch009

INTRODUCTION

The concept of sustainable development began to be recognized towards the end of the 20th century and has been implemented all over the world with international agreements since the 1990s. As a development model, sustainable development is based on the principle of meeting the needs of present generations without compromising the ability of future generations to meet their needs. Sustainable development is addressed in three dimensions: economic, environmental and social issues. All countries need to embed the understanding of sustainability in their development goals in three dimensions. Because with the speed of globalization, especially the point reached by the economy shows that limited scarce resources are almost alarming in meeting unlimited human needs. (Meadows et al. 1972) The Limits to Growth study argues that industrialization, population, environmental pollution, food security and excessive consumption of non-renewable natural resources will cause the limits of growth to be exceeded. Exceeding the limits to growth means compromising on meeting the needs of future generations. A development strategy that addresses the relationship between socioeconomic progress and the environment needs to be well-constructed. This is because it is not clear after which limits of growth will lead to environmental disasters. In addition, the irreversibility of environmental damage poses a very important problem. At this point, the most important source guiding the sustainable development philosophy of countries is the United Nations Sustainable Development Goals. In other words, the Global Goals are one of the most important studies in the field of sustainability. In 2015, world leaders came together and agreed on 17 Global Goals to achieve the goals of ending extreme poverty, combating inequality and injustice, and fixing climate change by 2030. Climate change mitigation is a special topic among the sustainable development goals. Therefore, identifying the factors that cause climate change is the first step towards solving the problem. As it is known, the most important cause of climate change is the greenhouse gas emissions caused by human activities. The increase in greenhouse gas emissions is caused by fossil fuels used for energy production and transportation, industrial processes, electricity generation, agriculture, product use, waste sector, etc. Almost all of these factors are related to economic activities. Literature studies reveal the link between economic indicators and climate change as causal or cointegrated. In this study, the impact of economic and financial parameters on climate change is analyzed on a quantile basis. It is aimed to include all countries in the World Bank database in the analysis. However, the most important determining factor in the inclusion of countries in the panel is the availability of data. In analyses based on uninterrupted panel data, countries that violate data integrity are excluded from the analysis. A Method of Moments Quantile Regression (MMQR) analysis was conducted for 75 countries with different levels of development between 1995 and

2020. Thus, the phenomena of economy and climate change, which are the biggest global problems of recent years, have been addressed together. As it is known, good performance in terms of sustainability in these two concepts is critical for the existence of future generations. Knowing the economic and financial structure of countries and the impact of this structure on the environment reveals the necessity of multidisciplinary study of two seemingly independent research areas. In this framework, a different perspective on sustainability is provided for researchers working on both economics and environmental disciplines. In addition, studies that examine both economic and financial issues with a new econometric method by considering countries at different development levels are limited. In this respect, the study is expected to make a significant contribution to the field.

FOCUS OF THE CHAPTER

The real source of wealth for a nation is its people. The ultimate goal of development is to create an environment that enables people to live healthy, long and creative lives. What seems like a simple truth is often forgotten when economic wealth comes into play (United Nations Development Programme-UNDP, 1990). In the aftermath of the Second World War, there have been studies on the development problems of underdeveloped countries, discussing what should be understood by the terms underdeveloped, developing or other similar terms and what development means in these countries. It was generally concluded that a satisfactory definition could not be provided and the concept was left rather vague (Myrdal, 1974). In seeking an answer to the question of what development is, three basic principles should not be ignored. First, real development is much more than a matter of economics and economic growth. Second, development is a universal problem not only for third world countries but for all societies. The third is that development depends on the equitable interaction between different groups and different nations (Slim, 1995). Development is one of the most frequently used words by everyone from policy makers to economists and even international financial institutions. Although it has different meanings, development mostly refers to an improvement in people's quality of life. Initially, development was considered to be the growth of national or personal income, but it includes traditional economic elements as well as many socioeconomic and political dimensions such as poverty, inequality, unemployment, dignity, self-respect, democracy and freedom (Semasinghe, 2020).

The first remarkable attempt on the concept of sustainability was made by Meadows et al. (1972) in The Limits to Growth. Before the work of Meadows et al. (1972), Thomas Robert Malthus, in his work An Essay on the Principle of Population, had discussed the environment and resources. Malthus (1798) predicted that the world's population would eventually starve to death or at least live at a subsistence level because food production could not keep pace with population growth. Left unchecked, Malthus argued, population would increase at a geometric rate while food would increase at an arithmetic rate. Technological developments since then have shown that Malthus was wrong. Thanks to better farming techniques, the invention of new farming equipment and continuous advances in agricultural science, production has increased much faster than population and food prices are much lower today than they were two hundred years ago. The debate on Malthusian limits has continued over time and its next wave is represented by the ideas and expectations presented by the Club of Rome. Meadows et al. (1972) drew attention to the depletion of non-renewable resources and the resulting increases in commodity prices. The model assumed that population and industrial capital would continue to grow exponentially, leading to a similar increase in pollution and demand for food and non-renewable resources. The supply of both food and non-renewable resources is assumed to be constant. Given the assumptions, the model predicts a collapse in growth due to the depletion of non-renewable resources (Paul, 2008). The first official step for environmental issues at the global level was the United Nations Conference on the Human Environment held in Stockholm in 1972. It was the first world conference to make the environment a major issue. The Stockholm Declaration and Plan of Action on the Human Environment adopted a set of principles for the sound management of the environment. With 26 principles, the Stockholm Declaration placed environmental issues at the forefront of international concerns and marked the beginning of a dialogue among developing countries on the link between economic growth, air, water and ocean pollution and prosperity. The Stockholm Conference established the United Nations Environment Program (UNEP) (UN, 2024a).

A major turning point in the field of sustainable development was the publication of the Our Common Future report to the United Nations in 1987 by the World Commission on Environment and Development (WCED) under the leadership of Gro Harlem Brundtland. The Brundtland Report (1987) was a call to recalibrate institutional mechanisms at the global, national and local levels to promote economic development that would guarantee the security, well-being and survival of humankind in thinking about environment, development and governance (Sneddon, et al., 2006). The report formulated sustainable development as development that meets the needs of the current generation without compromising the ability of future generations to meet their own needs. The idea is to bequeath to future generations a world that can support their livelihoods in a way that leaves them no worse off than present generations. In other words, while in economic terms, development is equated with increasing utility, sustainable development is equated with a way of development that ensures that per capita utility does not decline over a given period of time (Pearce and Atkinson, 1998). Another important aspect of the Brundtland

Report (1987) is that it was the main theme of the UN Conference on Environment and Development (UNCED) and the Earth Summit held in Rio in 1992 (Borowy, 2013). At the UN Conference on Environment and Development in Rio de Janeiro, Agenda 21 was published. The Rio Summit emphasized how different social, economic and environmental factors are interconnected and evolve together, and that success in one sector requires sustainable action in other sectors over time. The conference also revealed that integrating and balancing economic, social and environmental dimensions requires new perceptions about the way we produce and consume, live, work and make decisions (UN, 2024b). The Rio+5 Summit in 1997, the Millennium Summit in 2000, the World Summit on Sustainable Development in 2002, the World Summit on Global Warming and the Kyoto Protocol in 2005, the high-level meeting on the Millennium Development Goals in 2008, the Millennium Development Goals Summit in 2010, the United Nations Conference on Sustainable Development in 2012, the Special Event on the Millennium Development Goals in 2013, the United Nations Summit on Sustainable Development in 2015, and the Stockholm+50 Conference in 2022 were held respectively (UN, 2024c).

The 2030 Agenda for Sustainable Development, adopted by all UN Member States in 2015, provided a common blueprint for peace and prosperity for people and the planet today and in the future. At its center are 17 Sustainable Development Goals (SDGs), an urgent call for action by all countries, developed and developing, in a global partnership. The idea that ending poverty and other deprivations should go hand in hand with strategies that improve health and education, reduce inequality and promote economic growth, while also aiming to combat climate change and protect oceans and forests (UN, 2024d).

Görüldüğü gibi sürdürülebilir kalkınma amaçlarında iklim değişikliği konusuna özel önem verilmektedir. In 2015, world leaders agreed on these 17 Global Goals to accomplish three important tasks by 2030. These three important tasks are ending poverty, combating inequality and injustice, and addressing climate change (United Nations Turkey, 2024). As can be seen, climate change is given special importance in sustainable development goals.

The United Nations Framework Convention on Climate Change (UNFCCC) defines the climate system as the totality of the atmosphere, hydrosphere, biosphere and geosphere and their interactions, and climate change as changes in climate directly or indirectly attributable to human activities that alter the composition of the global atmosphere, in addition to natural climate variability observed over comparable time periods. The UNFCCC also defines the adverse impacts of climate change as changes in the physical environment or biota resulting from climate change that have significant detrimental effects on the composition, resilience and productivity of natural and managed ecosystems, the functioning of socio-economic systems, or human health and well-being (UNFCC, 2008; United Nations, 1992). The Intergovernmental Panel on Climate Change (IPCC) defines climate change as a change in the state of the climate that can be characterized (e.g. using statistical tests) by changes in the mean and/or variability of its characteristics and that persists over a long period of time (usually decades or decades) (IPCC, 2024).

The term climate change is often used synonymously with the term global warming. Global warming refers to an average increase in the temperature of the atmosphere near the Earth's surface, which can lead to changes in global climate patterns. However, rising temperatures are only one aspect of climate change (United States Environmental Protection Agency-EPA, 2010). Figure 1 shows the change in annual surface temperature between 1901-2023 and 1994-2023. Recent warming has been much faster than the long-term average, with some places warming by 1 degree Fahrenheit or more every decade. The differences are most pronounced in the Arctic, where the rate of warming is increasing the loss of ice and snow. The most important factor increasing surface temperatures is greenhouse gases. Carbon dioxide (CO_2) from fossil fuels is the most important greenhouse gas. While CO_2 transmits short-wave rays, it traps long-wave rays and causes warming in the lower parts of the atmosphere. For this reason, it is necessary to control CO_2 emissions in sustainable development and climate change (NOAA Climate.gov, 2024).

Figure 1. Temperature change 1901-2023 (NOAA Climate.gov, 2024)

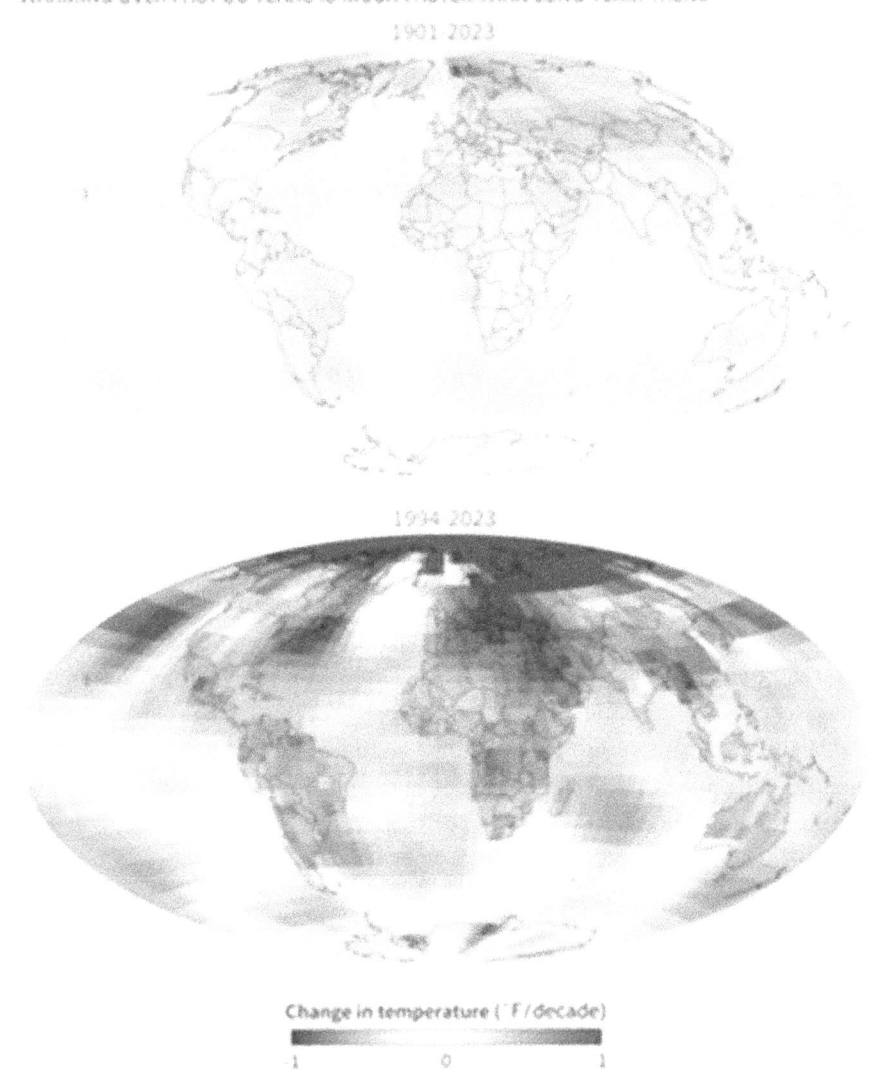

Literature Review

In the literature section, a summary of studies investigating the impact of climate change on economic and financial indicators is presented. While reviewing the literature, studies that take into account CO_2 emission, which is used as an independent variable in this study, are examined. Although the aim of the research is to determine the factors affecting climate change in economic and financial terms, studies that use social indicators such as urbanization, population, etc. as independent variables in the literature are analyzed and reported. Thus, it was determined which economic, financial and sociocultural variables were used by the researchers in the literature. The literature review was created by scanning a wide range of national and international literature. For this reason, studies are reported in tables in terms of their quantitative characteristics. Thus, it is thought to provide a convenience for the readers.

When the literature summary is analyzed, it is possible to obtain information about the results of the studies on the economic and financial determinants of climate change, by whom(s), in which year, in which cross-section and period, using which variables and methods. The impact of economic and financial factors on climate change varies across different years and the most frequently used variables are economic growth, energy consumption, population, financial development index, trade openness etc. It can be said that the most commonly used method in the analysis is Autoregressive distributed lag (ARDL) for time series and cointegration for panel. In terms of the findings of the studies, it is seen that economic and financial determinants have significant effects on climate change, albeit different in magnitude and sign.

Table 1. Literature

Author(s)/Year	Cross-Section / Period	Method	Findings
Shahbaz et al. (2013)	Indonesia /1975Q1–2011Q4	ARDL and Granger Causality	Economic growth, energy consumption increase CO_2 emissions, while for trade openness, financial development the relationship is first increasing and then decreasing.

continued on following page

Table 1. Continued

Author(s)/Year	Cross-Section / Period	Method	Findings
Yang and Zhao (2014)	India /1970-2008	Directed Acyclic Graph (DAG), Granger Causality	CO_2 emission has a unidirectional causality with energy consumption and a bidirectional causality with economic growth. Moreover, trade openness is important for CO_2 emissions.
Bento and Moutinho (2016)	Italy/1960-2011	ARDL	Renewable electricity generation per capita reduces CO_2 emissions, while international trade increases them.
Bouznit and Pablo-Romero (2016)	Algeria /1970-2010	ARDL	Economic growth, energy use, electricity consumption and imports increase CO_2 emissions.
Kais and Sami (2016)	58 countries /1990-2012	Generalized method of moments (GMM)	Energy use and income growth increase CO_2 emissions, while openness and urbanization have a negative impact.
Acheampong (2018)	116 countries/1990-2017	GMM	The impact of growth on CO_2 emissions is negative.
Mikaloy et al. (2018)	Azerbaijan /1992-2013	ARDL, Dynamic Ordinary Least Squares Method (DOLS), Fully Modified Least Squares Method (FMOLS) and Canonical Cointegration Regression (CCR)	The effect of economic growth on CO_2 is positive.
Chen et al. (2019)	China/1980-2014	ARDL and Granger Causality	Renewable energy reduces CO_2 emissions.
Farabi et al. (2019)	Malaysia and Indonesia /1971-2014	Johansen-Juselius Cointegration and Granger Causality	Increase in income and energy consumption increases CO_2 emissions
Hao et al. (2019)	29 provinces in China/ 2007-2016	GMM	Economic growth first increases and then decreases CO_2
Mahmood et al. (2019)	Tunisian/1971-2014	Nonlinear ARDL	Trade openness has an asymmetric effect on CO_2 emissions. CO_2 emissions first increase and then decrease as income increases.

continued on following page

Table 1. Continued

Author(s)/Year	Cross-Section / Period	Method	Findings
Nguyen et al. (2020)	13 G20 Countries/2000-2014	FMOLS and Panel Quantile Regression	Energy price, FDI, technology, innovation spending and trade openness inhibit CO_2 emissions while financial development increases them.
Raheem et al. (2020)	G7 Countries/1990-2014	Westerlund (2007) Cointegration, Pooled Mean Group (PMG)	Financial development is a weak determinant of CO_2 emissions.
Hailemariam et al. (2020)	17 Organisation for Economic Co-operation and Development (OECD) Countries /1945-2010	Westerlund (2007) Cointegration, DOLS, FMOLS and Common Correlated Effect Mean Group (CCEMG)	Economic growth first increases and then decreases CO_2, while income inequality increases.
Rahman and Vu (Benjamin) (2020)	Australia and Canada /1960-2015	ARDL and Granger Causality	Economic growth increases CO_2 emissions. In the short run, renewable energy and trade consumption reduce CO_2 emissions.
Dou et al. (2021)	Japan, China and South Korea /1970-2019	Panel Quantile Regression and Granger Causality	Trade openness reduces CO_2, exports reduce CO_2 emissions while imports increase them.
Khan and Ozturk (2021)	88 countries /2000-2014	GMM	Trade openness reduces CO_2; exports reduce CO_2 emissions while imports increase them. Financial development reduces CO_2 emissions.
Lv and Li (2021)	97 countries/2000-2014	Spatial Analysis	It has been determined that a country's CO_2 emissions can be affected by the financial developments of its neighbors.
Wang and Zhang (2021)	182 countries/1990-2015	Kao (1999) and Pedroni (2001) Cointegration, FMOLS	While trade openness reduces CO_2 emissions in high and upper middle-income countries, its effect is insignificant. In low-middle income countries, it increases CO_2 emissions.
Adebayo et al. (2022)	Mexico, Indonesia, Nigeria, and Turkey /1990-2018	MMQR	Economic growth and CO_2 have an inverted U-shaped relationship. Energy consumption- economic complexity increase CO_2, while there is no significant relationship with financial development..

continued on following page

Table 1. Continued

Author(s)/Year	Cross-Section / Period	Method	Findings
Kim (2022)	OECD Countires countries /1990-2018	PMG	Economic growth, ICT progress, CO_2 renewable electricity and trade openness reduce CO_2 emissions.
Lin et al. (2022)	30 provinces in China/ 2004-2015	OLS, Stochastic Impact by Regression on Population Affluence and Technology-STIRPAT	The contribution of FDI to emission reduction is heterogeneous across provinces.
Ma et al. (2022)	Developed and developing countries/1990-2020	PMG–ARDL	They suggested formulating dissimilar policies for countries with different income levels.
Guan et al. (2023)	G7 Countries	MMQR	Gross domestic product (GDP) increases greenhouse gas emissions, while energy efficiency, globalization and technology reduce them.
Jahanger et al. (2023)	France, USA, China, Japan, South Korea, Russia, Canada, Spain and United Kingdom/1990-2017	Granger Causality	Nuclear energy, internet, cell phones, economic growth squared reduce CO_2 emissions.
Luo et al. (2023)	G7 Countries	Westerlund (2007) Cointegration and MMQR	Economic growth and imports increase CO_2 emissions, while exports, energy efficiency and renewable energy production reduce emissions.
Udeagha and Breitenbach (2023)	South Africa/1960-2020	ARDL	Financial development reduces CO_2 emissions both temporarily and permanently.
Shang et al. (2024)	10 Countries /2004-2018	MMQR	Renewable energy. Increases economic growth, financial inclusion, natural resource rent, while reducing CO_2 emissions.
Sobirov et al. (2024)	14 countries in Asia/1996-2020	MMQR, FMOLS, DOLS and Pedroni Cointegration	GDP squared, renewable energy use and agriculture reduce CO_2 emissions, while urbanization and GDP increase them

ECONOMETRIC ANALYSIS

In the econometric analysis of the project, the hypothesis of which economic and financial variables will cause climate change was investigated. In the time interval covering the period 1995-2020, 75 countries were included in the analysis. Table 2 presents the countries that make up the panel.

Table 2. Countries included in the analysis

Countries	Lebanon
Algeria	Madagascar
Angola	Malaysia
Argentina	Mali
Austria	Mauritius
Belgium	Mexico
Bolivia	Morocco
Brazil	Mozambique
Bulgaria	Netherlands
Burkina Faso	New Zealand
Canada	Nicaragua
Chile	Norway
China	Oman
Colombia	Panama
Costa Rica	Paraguay
Ivory Coast	Peru
Cyprus	Philippines
Denmark	Portugal
Dominican Republic	Saudi Arabia
Ecuador	Senegal
Egypt	Singapore
Finland	South Africa
France	Spain
Gabon	Sweden
Germany	Switzerland
Ghana	Tanzania
Greece	Thailand
Guatemala	Togo

continued on following page

Table 2. Continued

Countries	Lebanon
Honduras	Tunisia
India	Türkiye
Indonesia	Uganda
Ireland	Ukraine
Italy	United Arab Emirates
Jamaica	United Kingdom
Japan	United States
Jordan	Uruguay
Kenya	Vietnam
South Korea	Zambia

The data set consisting of countries with different levels of development consists of one dependent variable and six independent variables. Countries and variables are constructed under data availability constraints. CO, GRO, FDI, FIN, GINI, ENR and ECI variables are used in the analysis. The explanations of the variables are presented in Table 3

Table 3. Information on variables

Variables	Explanations	Source
CO	CO_2 emissions (metric tons per capita)	World Bank
GRO	GDP (constant 2015 US$) in logarithm	World Bank
FDI	Foreign direct investment, net inflows (% of GDP)	World Bank
FIN	Financial development index	International Monetary Fund (IMF)
GINI	Income inequality index	World Income Inequality Database (WIID)
ENR	Renewable energy consumption (% of total final energy consumption)	World Bank
ECI	Economic complexity index	Atlas of Economic Complexity

The model established with the variables in Table 3 is as follows:

In Equation (1), i=1, 2, 3,....N denotes cross-sectional data, t=1, 2, 3, ...T denotes time and ε denotes the error term. The model is estimated based on quantile regression analysis. In quantile models, which are considered as weighted regression first introduced by Koenker and Basset Jr (1978), coefficient estimates can be made for all quantiles between the conditional quantiles of the dependent variable and the independent variables.

The MMQR method developed by Machado and Silva (2019) sets out the conditions for estimating conditional means and estimating quantiles. This method enables the use of appropriate methods for estimating conditional means. The conditional quantiles of the dependent variable are estimated by Equation (2):

$(\alpha, \beta', \delta, \gamma')$ is a differentiable vector k of known transformations of the components of Z and X with unknown parameters;

given in a panel with n units and t time periods $\{(Y_{it}, X'_{it})'\}$ according to the values for a location-scale model $Q_y(X)$ conditional quantiles $Y_{it} = \alpha_i + X'_{it}\beta + (\delta_i + Z'_{it}\gamma) U_{it}$ is estimated by the equation. $Pr\{\delta_i + Z'_{it}\gamma > 0\} = 1$ where (α_i, δ_i) parameters give individual i fixed effects X_{it} is exogenous and independently and identically distributed for any constant i. U_{it} is independent and identically distributed in time and units. X_{it} is independent of and normalized for the moment condition in the method. It is shown by equation (3).

$(\alpha_i + \delta_i q(\tau))$ scalar coefficient gives the quantile-τ fixed effect or the distributional effect at τ for individual i. That is, it shows the effect of time-invariant individual characteristics. $\int_0^1 q(\tau)d\tau$ If α_i is the average effect for individual i. According to this τth sample quantile is calculated by Equation (4) (Machado and Silva, 2019).

MMQR estimation results are presented in Table 4 and the graphical representation is presented in Figure 3.

Table 4. MMQR estimation results

Variables	Location	Scale	Quantiles								
			0.10	0.20	0.30	0.40	0.50	0.60	0.70	0.80	0.90
GRO	0.020[a] (0.000)	0.011[a] (0.000)	0.001 (0.671)	0.008[b] (0.016)	0.012[a] (0.000)	0.015[a] (0.000)	0.017[a] (0.000)	0.020[a] (0.000)	0.025[a] (0.000)	0.031[a] (0.000)	0.039[a] (0.000)
FDI	-0.005[b] (0.020)	-0.008[a] (0.000)	0.007[a] (0.002)	0.002 (0.172)	-3.790 (0.984)	-0.001 (0.324)	-0.003[c] (0.076)	-0.005[b] (0.014)	-0.009[a] (0.001)	-0.001[a] (0.000)	-0.001[a] (0.000)
FIN	0.691[a] (0.000)	0.060[c] (0.063)	0.598[a] (0.000)	0.632[a] (0.000)	0.651[a] (0.000)	0.665[a] (0.000)	0.679[a] (0.000)	0.694[a] (0.000)	0.721[a] (0.000)	0.749[a] (0.000)	0.791[a] (0.000)
GINI	-0.011[a] (0.000)	-0.003[a] (0.000)	-0.006[a] (0.000)	-0.008[a] (0.000)	-0.009[a] (0.000)	-0.010[a] (0.000)	-0.011[a] (0.000)	-0.011[a] (0.000)	-0.013[a] (0.000)	-0.014[a] (0.000)	-0.016[a] (0.000)
ENR	-0.010[a] (0.000)	-0.001[a] (0.000)	-0.012[a] (0.000)	-0.011[a] (0.000)	-0.011[a] (0.000)	-0.010[a] (0.000)	-0.105[a] (0.000)	-0.010[a] (0.000)	-0.096[a] (0.000)	-0.009[a] (0.000)	-0.008[a] (0.000)
ECI	-0.337[a] (0.005)	-0.068[a] (0.000)	-0.070[a] (0.000)	-0.032[a] (0.001)	0.010 (0.269)	-0.005 (0.613)	-0.020[c] (0.061)	-0.037[a] (0.002)	-0.067[a] (0.000)	-0.099[a] (0.000)	-0.147[a] (0.000)

Statistically significant coefficients at 1%, 5% and 10% significance levels are expressed as [a], [b], [c] respectively.

Figure 2. MMQR graph

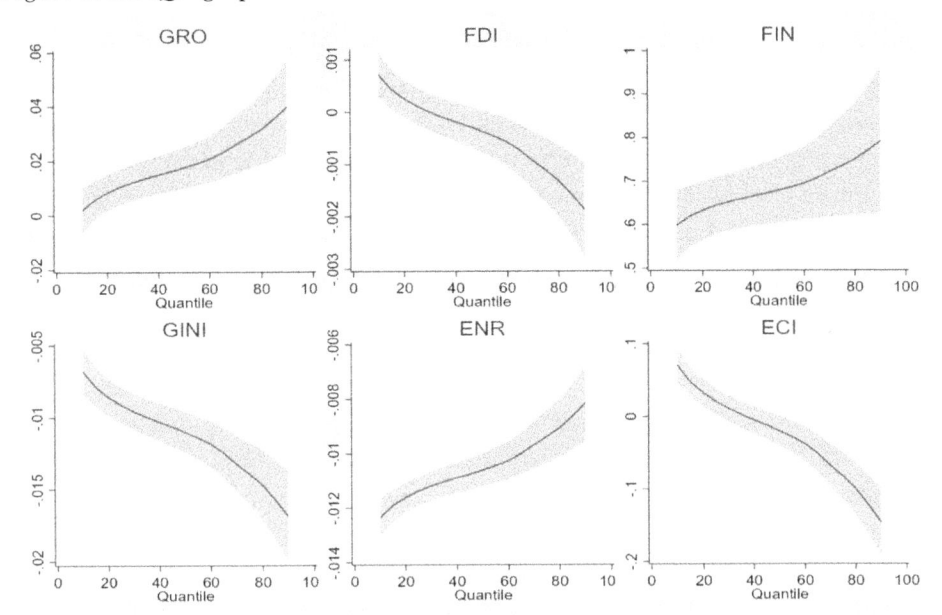

When the estimation results are analyzed, it is seen that the GRO variable is positively effective and significant for all quantiles except quantile 0.10. These results show that carbon emissions increase as the economic growth rate increases in countries. Income increases lead to diversification and expansion in the economic activities of governments, businesses and individuals. The realization of these activities using energy based on fossil fuels leads to an increase in carbon emissions and climate change. The results are similar to Adebayo et al. (2022), Jeon (2022), Guan et al. (2023), Luo et al. (2023), Beton Kalmaz and Adebayo (2024), Shang et al. (2024), Sobirov et al. (2024). The second independent variable FDI is significant except for quantiles 0.20, 0.30 and 0.40 and has a negative sign except for quantiles 0.10 and 0.20. The findings are similar to those of Nguyen et al. (2020) and Khan and Ozturk (2021). The FIN variable is significant and positively influential for all quantiles. The results are consistent with Khan and Ozturk (2021), Adebayo et al. (2022). If the financial sector has not reached a sufficient level of maturity, environmental quality is ignored as the aim of the actors in the sector is profit maximization. Thus, climate change may occur with an increase in carbon emissions. GINI is also significant and negative in all quantiles. The findings are similar to those of Hailemariam et al. (2020). As income inequality decreases, individuals will increase their consumption of energy and other carbon-intensive energy inputs, which will increase carbon dioxide emissions. As with the GINI variable, the results for the ENR variable are significant and have a negative sign in all quantiles. The findings are in line with Jeon (2022), Shang et al. (2024), Sobirov et al. (2024). In other words, as renewable energy consumption increases, carbon emissions decrease. The use of clean energy instead of fossil fuels is an important factor in preventing climate change. Finally, the ECI variable is also significant and has a negative sign except for quantiles 0.30 and 0.40. The results are supported by He et al. (2021), Caglar et al. (2022) and it can be said that a certain level of specialization in the production of goods has been reached in the countries that make up the panel and this has been transformed into environmentally friendly production.

SOLUTIONS AND RECOMMENDATIONS

It is known that various studies have been carried out in many fields such as geography, climatology, environment and meteorology on the problem of climate change and its solution. Researchers are in an effort to reveal the factors that cause climate change while focusing on the consequences of climate change. Apart from being seen as a geographical event, climate change is a concept that needs to be handled more comprehensively. This is because the factors causing the climate crisis are related to human activities that are analyzed in different disciplines. Economics is perhaps the most prominent of these fields. The impact of economic and financial policies does not only cause an increase or decrease in economic growth, inflation or deflation, exchange rate increase or decrease. For example, while the first impact of exporting or importing may seem to be on the current account balance, foreign trade-related activities have some impact on the environment. This is why the economy and climate change are two unresolved and increasingly complex problems that need different solutions. In this study, within the framework of seemingly complex relationships, solutions are sought with the MMQR model, one of the panel quantile models, for the period 1995-2020 involving 75 countries. The effects of economic growth, foreign direct investment, financial development, income inequality, renewable energy consumption and economic complexity on carbon emissions are analyzed. The findings show that economic growth and financial development increase carbon emissions while income inequality, renewable energy consumption and economic complexity indicators decrease carbon emissions. It is also found that there is no net effect for foreign direct investment.

FUTURE RESEARCH DIRECTIONS

Environmental degradation, which started with the industrial revolution, has increased unprecedentedly in recent years. The melting of glaciers and the rise in the average global temperature due to extreme temperatures have brought climate change. Climate change is recognized as the trigger of almost all environmental degradation. So, what are the factors underlying climate change? The main cause of climate change is recognized as the increase in greenhouse gas emissions as a result of human activities. However, climate change is essentially a reflection of compromising sustainable development. Climate change is seen by all societies as one of the most important threats to human and living life. The impact of this great threat on different units such as countries, regions and social classes varies. At this point, it is worth noting that the study has some limitations that could be overcome in potential future research. Obtaining older historical data on climate change or longer series that provide future projections would make the research more comprehensive. More time-extensive data would allow for broader estimates for various cross-sectional units, such as countries or regions with different levels of development. On the other hand, the research can be elaborated with forecasting experiments using different variables that measure climate change. Again, economic and financial indicators can be classified within themselves and the effects of economic indicators and financial indicators can be examined separately with two separate modeling. Countries with different levels of economic or financial development can be classified as developed, developing, underdeveloped or high-income, middle-income and low-income countries as a measure of income level and analyzed according to the characteristics of the countries. Again, countries can be analyzed individually. Thus, more specific policies and solutions can be proposed.

CONCLUSION

The most important policy recommendation that can be drawn from the findings of the study is the fact that all countries, regardless of their level of development, should first and foremost switch to environmentally friendly product production to combat environmental degradation. For this, the use of renewable energy should be increased. Of course, for many countries, abandoning the use of fossil fuels and switching to renewable energy use requires certain processes. Practices to prevent the use of fossil fuels should be implemented quickly. For example, pricing could be changed to discourage the use of fossil fuels. Technological innovations that contribute to renewable energy use and energy efficiency can be transferred. At this point, FDI investments that enable technology transfer and do not create a significant carbon emission deficit should be increased. Again, as in the case of FDI, while expanding financial markets in countries, the allocation of domestic financial resources such as loans, incentives, etc. to environmentally friendly investments should be prioritized. For a sustainable environment, exports of knowledge and skill-intensive products should be encouraged. Finally, tax exemptions and subsidies can be offered to companies that produce and export more complex and cleaner energy. Knowledge and skill-intensive product exports should be prioritized for sustainable living that mitigates climate change. Thus, countries can continue their economic growth while meeting their energy needs.

ACKNOWLEDGMENT

This study was supported by Atatürk University Coordination Unit of Scientific Research Projects with project number SCD-2023-12423.

REFERENCES

Acheampong, A. O. (2018). Economic growth, CO2 emissions and energy consumption: What causes what and where? *Energy Economics*, 74, 677–692. DOI: 10.1016/j.eneco.2018.07.022

Adebayo, T. S., Rjoub, H., Akadiri, S. S., Oladipupo, S. D., Sharif, A., & Adeshola, I. (2022). The role of economic complexity in the environmental Kuznets curve of MINT economies: Evidence from method of moments quantile regression. *Environmental Science and Pollution Research International*, 29(16), 24248–24260. DOI: 10.1007/s11356-021-17524-0 PMID: 34822076

Atlas of Economic Complexity. (2024). *Country & product complexity rankings*. https://atlas.cid.harvard.edu/rankings

Bento, J. P. C., & Moutinho, V. (2016). CO_2 emissions, non-renewable and renewable electricity production, economic growth, and international trade in Italy. *Renewable & Sustainable Energy Reviews*, 55, 142–155. DOI: 10.1016/j.rser.2015.10.151

Beton Kalmaz, D., & Adebayo, T. S. (2024). Does foreign direct investment moderate the effect of economic complexity on carbon emissions? Evidence from BRICS nations. *International Journal of Energy Sector Management*, 18(4), 834–856. DOI: 10.1108/IJESM-01-2023-0014

Borowy, I. (2013). The Brundtland Commission: Sustainable development as health issue. *Michael*, 10, 198–208.

Bouznit, M., & Pablo-Romero, M. D. P. (2016). CO_2 emission and economic growth in Algeria. *Energy Policy*, 96, 93–104. DOI: 10.1016/j.enpol.2016.05.036

Caglar, A. E., Zafar, M. W., Bekun, F. V., & Mert, M. (2022). Determinants of CO_2 emissions in the BRICS economies: The role of partnerships investment in energy and economic complexity. *Sustainable Energy Technologies and Assessments*, 51, 101907. DOI: 10.1016/j.seta.2021.101907

Chen, Y., Wang, Z., & Zhong, Z. (2019). CO_2 emissions, economic growth, renewable and non-renewable energy production and foreign trade in China. *Renewable Energy*, 131, 208–216. DOI: 10.1016/j.renene.2018.07.047

Climate, N. O. A. A. gov (2024). *Climate change: global temperature.* https://www.climate.gov/news-features/understanding-climate/climate-change-global-temperature

Dou, Y., Zhao, J., Malik, M. N., & Dong, K. (2021). Assessing the impact of trade openness on CO2 emissions: Evidence from China-Japan-ROK FTA countries. *Journal of Environmental Management*, 296, 113241. DOI: 10.1016/j.jenvman.2021.113241 PMID: 34265664

Farabi, A., Abdullah, A., & Setianto, R. H. (2019). Energy consumption, carbon emissions and economic growth in Indonesia and Malaysia. *International Journal of Energy Economics and Policy*, 9(3), 338–345. DOI: 10.32479/ijeep.6573

Guan, Z., Hossain, M. R., Sheikh, M. R., Khan, Z., & Gu, X. (2023). Unveiling the interconnectedness between energy-related GHGs and pro-environmental energy technology: Lessons from G-7 economies with MMQR approach. *Energy*, 281, 128234. DOI: 10.1016/j.energy.2023.128234

Hailemariam, A., Dzhumashev, R., & Shahbaz, M. (2020). Carbon emissions, income inequality and economic development. *Empirical Economics*, 59(3), 1139–1159. DOI: 10.1007/s00181-019-01664-x

Hao, Y., Huang, Z., & Wu, H. (2019). Do carbon emissions and economic growth decouple in China? An empirical analysis based on provincial panel data. *Energies*, 12(12), 2411. DOI: 10.3390/en12122411

He, K., Ramzan, M., Awosusi, A. A., Ahmed, Z., Ahmad, M., & Altuntaş, M. (2021). Does Globalization Moderate the Effect of Economic Complexity on CO2 Emissions? Evidence From the Top 10 Energy Transition Economies. *Frontiers in Environmental Science*, 9, 778088. DOI: 10.3389/fenvs.2021.778088

IMF. (2024). *IMF data access to macroeconomic & financial data financial development indeks database.* https://data.imf.org/?sk=f8032e80-b36c-43b1-ac26-493c5b1cd33b

IPCC. (2024). *Climate change widespread, rapid, and intensifying.* https://www.ipcc.ch/2021/08/09/ar6-wg1-20210809-pr/

Jahanger, A., Zaman, U., Hossain, M. R., & Awan, A. (2023). Articulating CO_2 emissions limiting roles of nuclear energy and ICT under the EKC hypothesis: An application of non-parametric MMQR approach. *Geoscience Frontiers*, 14(5), 101589. DOI: 10.1016/j.gsf.2023.101589

Jeon, H. (2022). CO_2 emissions, renewable energy and economic growth in the US. *The Electricity Journal*, 35(7), 107170. DOI: 10.1016/j.tej.2022.107170

Kais, S., & Sami, H. (2016). An econometric study of the impact of economic growth and energy use on carbon emissions: Panel data evidence from fifty eight countries. *Renewable & Sustainable Energy Reviews*, 59, 1101–1110. DOI: 10.1016/j.rser.2016.01.054

Khan, M., & Ozturk, I. (2021). Examining the direct and indirect effects of financial development on CO2 emissions for 88 developing countries. *Journal of Environmental Management*, 293, 112812. DOI: 10.1016/j.jenvman.2021.112812 PMID: 34058453

Kim, S. (2022). The effects of information and communication technology, economic growth, trade openness, and renewable energy on CO_2 emissions in OECD countries. *Energies*, 15(7), 2517. DOI: 10.3390/en15072517

Koenker, R., & Basset, G. S.Jr. (1978). Regression quantiles. *Econometrica*, 46(1), 33–50. DOI: 10.2307/1913643

Lin, H., Wang, X., Bao, G., & Xiao, H. (2022). Heterogeneous spatial effects of FDI on CO2 emissions in China. *Earth's Future, 10*(1), e2021EF002331. .DOI: 10.1029/2021EF002331

Luo, B., Khan, A. A., Wu, X., & Li, H. (2023). Navigating carbon emissions in G-7 economies: A quantile regression analysis of environmental-economic interplay. *Environmental Science and Pollution Research International*, 30(47), 104697–104712. DOI: 10.1007/s11356-023-29722-z PMID: 37707736

Lv, Z., & Li, S. (2021). How financial development affects CO_2 emissions: A spatial econometric analysis. *Journal of Environmental Management*, 277, 111397. DOI: 10.1016/j.jenvman.2020.111397 PMID: 33039704

Ma, W., Nasriddinov, F., Haseeb, M., Ray, S., Kamal, M., Khalid, N., & Ur Rehman, M. (2022). Revisiting the impact of energy consumption, foreign direct investment, and geopolitical risk on CO_2 emissions: Comparing developed and developing countries. *Frontiers in Environmental Science*, 10, 985384. DOI: 10.3389/fenvs.2022.985384

Machado, J. A. F., & Santos Silva, J. M. C. (2019). Quantiles via moments. *Journal of Econometrics*, 213(1), 145–173. DOI: 10.1016/j.jeconom.2019.04.009

Mahmood, H., Maalel, N., & Zarrad, O. (2019). Trade openness and CO2 emissions: Evidence from Tunisia. *Sustainability (Basel)*, 11(12), 3295. DOI: 10.3390/su11123295

Malthus, T. R. (1986). *An essay on the principle of population (1798)*. The Works of Thomas Robert Malthus, London, Pickering & Chatto Publishers, 1, 1-139.

Meadows, D. H., Randers, J., & Meadows, D. L. (2013). The limits to growth (1972). In *The future of nature* (pp. 101–116). Yale University Press. DOI: 10.2307/j.ctt5vm5bn.15

Mikayilov, J. I., Galeotti, M., & Hasanov, F. J. (2018). The impact of economic growth on CO2 emissions in Azerbaijan. *Journal of Cleaner Production*, 197, 1558–1572. DOI: 10.1016/j.jclepro.2018.06.269

Myrdal, G. (1974). What is development? *Journal of Economic Issues*, 8(4), 729–736. DOI: 10.1080/00213624.1974.11503225

Nguyen, T. T., Pham, T. A. T., & Tram, H. T. X. (2020). Role of information and communication technologies and innovation in driving carbon emissions and economic growth in selected G-20 countries. *Journal of Environmental Management*, 261, 110162. DOI: 10.1016/j.jenvman.2020.110162 PMID: 32148259

Paul, B. D. (2008). A history of the concept of sustainable development: Literature review. *The Annals of the University of Oradea. Economic Sciences Series*, 17(2), 576–580.

Pearce, D., & Atkinson, G. (1998). The concept of sustainable development: An evaluation of its usefulness ten years after Brundtland. *Environmental Economics and Policy Studies*, 134(2), 251–270. DOI: 10.1007/BF03353896

Raheem, I. D., Tiwari, A. K., & Balsalobre-Lorente, D. (2020). The role of ICT and financial development in CO2 emissions and economic growth. *Environmental Science and Pollution Research International*, 27(2), 1912–1922. DOI: 10.1007/s11356-019-06590-0 PMID: 31760620

Rahman, M. M., & Vu, X. B. (2020). The nexus between renewable energy, economic growth, trade, urbanisation and environmental quality: A comparative study for *Australia and Canada.Renewable Energy*, 155, 617–627. DOI: 10.1016/j.renene.2020.03.135

Semasinghe, W. M. (2020). Development, what does it really mean? *Acta Politica Polonica*, 49, 51–59. DOI: 10.18276/ap.2020.49-05

Shahbaz, M., Hye, Q. M. A., Tiwari, A. K., & Leitão, N. C. (2013). Economic growth, energy consumption, financial development, international trade and CO_2 emissions in Indonesia. *Renewable & Sustainable Energy Reviews*, 25, 109–121. DOI: 10.1016/j.rser.2013.04.009

Shang, T., Samour, A., Abbas, J., Ali, M., & Tursoy, T. (2024). Impact of financial inclusion, economic growth, natural resource rents, and natural energy use on carbon emissions: The MMQR approach. *Environment, Development and Sustainability*, 1–31. DOI: 10.1007/s10668-024-04513-9

Slim, H. (1995). What is development? *Development in Practice*, 5(2), 143–148. DOI: 10.1080/0961452951000157114 PMID: 12288928

Sneddon, C., Howarth, R. B., & Norgaard, R. B. (2006). Sustainable development in a post-Brundtland world. *Ecological Economics*, 57(2), 253–268. DOI: 10.1016/j.ecolecon.2005.04.013

Sobirov, Y., Makhmudov, S., Saibniyazov, M., Tukhtamurodov, A., Saidmamatov, O., & Marty, P. (2024). Investigating the Impact of Multiple Factors on CO_2 Emissions: Insights from Quantile Analysis. *Sustainability (Basel)*, 16(6), 2243. DOI: 10.3390/su16062243

Udeagha, M. C., & Breitenbach, M. C. (2023). The role of financial development in climate change mitigation: Fresh policy insights from South Africa. *Biophysical Economics and Sustainability*, 8(1), 134. DOI: 10.1007/s41247-023-00110-y

UN. (2024a). Conferences/Environment and sustainable development. https://www.un.org/en/conferences/environment/stockholm1972

UN. (2024b). Conferences/Environment and sustainable development. https://www.un.org/en/conferences/environment/rio1992

UN. (2024c). Conferences/Environment and sustainable development. https://www.un.org/en/conferences/environment/

UN. (2024d). *The 17 goals*. https://sdgs.un.org/goals

UNFCCC. (2008). *UNFCCC resource guide module 4: for preparing the national communications of non-annex i parties module 4 measures to mitigate climate change*. United Nations Framework Convention on Climate Change.

United Nation Development Programme-UNDP. (1990). *Human development report 1990*. New York Oxford Oxford University Press.

United Nations. (1992). *United Nations Framework Convention on Climate Change*. https://unfccc.int/files/essential_background/background_publications_htmlpdf/application/pdf/conveng.pdf

United Nations Türkiye. (2024). *Our work on the sustainable development goals in Türkiye*. https://turkiye.un.org/en/sdgs

United States Environmental Protection Agency-EPA. (2010). *Climate Change Indicators in the United States*. 1200 Pennsylvania Avenue, N.W. (6207J) Washington, DC 20460.

UNU WIDER. (2024). *World income inequality database (WIID)*. https://www4.wider.unu.edu/?ind=1&type=ChoroplethSeq&year=70&byCountry=false&slider=buttons

Wang, Q., & Zhang, F. (2021). The effects of trade openness on decoupling carbon emissions from economic growth–evidence from 182 countries. *Journal of Cleaner Production*, 279, 123838. DOI: 10.1016/j.jclepro.2020.123838 PMID: 32863606

World Bank. (2024). *World development indicators*. https://databank.worldbank.org/source/world-development-indicators

Yang, Z., & Zhao, Y. (2014). Energy consumption, carbon emissions, and economic growth in India: Evidence from directed acyclic graphs. *Economic Modelling*, 38, 533–540. DOI: 10.1016/j.econmod.2014.01.030

KEY TERMS AND DEFINITIONS

Climate Change: The increase in global average temperature.

Economic Development: The process by which a country's political and social welfare improves alongside its economy.

Economic Growth: An increase in the amount of production of goods and services.

Global Warming: Increase in global temperature due to greenhouse gases

Greenhouse Gas: Gaseous components such as carbon dioxide, methane, nitrous oxide and water vapor.

Method of Moments Quantile Regression (MMQR): A panel quantile regression analysis.

Sustainable Development: Development that meets the needs of the present without compromising the ability of future generations to meet their needs.

Chapter 10
Impacts of Climate Change on Employment:
An Evaluation on COVID-19 and Remote Working as an Alternative Employment Method

Berivan Özay Acar
Yüzüncü Yıl Üniversitesi, Turkey

ABSTRACT

Climate change is a growing concern around the world. This crisis causes significant impacts in various sectors such as agriculture, tourism, energy, and construction. Risks due to climate events are increasing, job losses are occurring, and workers' health is negatively affected. Especially as the frequency and severity of natural disasters increases, employment security decreases in these sectors. Recently, the COVID-19 pandemic has rapidly changed the business world and ways of doing business. Remote working has emerged as a result of this change and has been adopted as an alternative employment method. Unlike traditional office environments, the remote working model reduces greenhouse gas emissions, reduces carbon emissions caused by traffic congestion, and enables more efficient use of resources by optimizing energy consumption. Therefore, this study considers evaluating the effects of climate change on employment through the COVID-19 pandemic and the remote working model as an eco-strategy that will contribute to environmental sustainability.

DOI: 10.4018/979-8-3693-5792-7.ch010

INTRODUCTION

The issue of climate change, which is of great national and international importance, is a vitally important issue that closely concerns all living things on earth in terms of its consequences, although its causes and effects vary regionally. For this reason, it is aimed to reduce the problem of climate change through international solidarity and cooperation. Because greenhouse gas emissions from industrial processes, agricultural activities, deforestation and fossil fuel use affect and warm the whole world as a result, no matter from which region, city, country or even continent they are emitted. In the last hundred years, this warming has reached the highest levels in the history of mankind, disrupting the ecological balance. Especially in the last decade, the melting of the poles, the rise in sea levels, the decrease in precipitation regimes, the increase in forest fires due to warming, the danger of flooding in some regions and drought in others show that climate change has reached serious dimensions. July 2019 was recorded as the hottest month on record in the world, and in September 2019, the global average sea level reached a record high by 87.6 mm above the average of 1993, when satellite measurements began (Blunden and Arndt, 2020). According to research and reports, if human-induced global warming is not slowed down and prevented, the world will become uninhabitable for many species and humanity.

As the effects of climate change become more visible, environmental sensitivities are becoming increasingly important in the public opinion, and this situation creates pressure to regulate economic life within ecological criteria. In the current process, it has become inevitable that ecological standards have been institutionalized and that processes related to employment and production must be regulated in accordance with these standards. The realization that production processes based solely on consumption are not sustainable due to the fact that the environment is not an unlimited resource has led to the prominence of views that renewable resources can be used effectively to increase employment and income in society. Within the framework of these views; contrary to the negative causal relationship between environmental standards and employment growth in the literature, it is predicted that job creation strategies based on ecological concerns such as climate change can lead to employment growth in the long run. It is thought that the remote working model, which has increased in popularity all over the world with the Covid-19 pandemic, will contribute to combating climate change in various ways. It is stated that the remote working model has some positive aspects not only for employers and employees, but also for society and the environment, especially home-based remote working model can be accepted as an eco-strategy that will contribute to environmental sustainability by saving transportation costs, reducing environmental and air pollution, reducing traffic congestion and reducing traffic accidents (Saura et al., 2022). It is also emphasized that this working model can contribute to the dispersal of businesses and employees from metropolitan areas and urban centers to rural settlements by supporting balanced spatial development with lower carbon emissions (Samek Lodovici, 2021). For this reason, this study aims to evaluate the effects of climate change on employment through the recent global Covid-19 pandemic and to address the remote working model as an eco-strategy that will contribute to environmental sustainability.

Climate Change

In a general approach, climate change is defined as slow, long-term changes in climatic conditions with significant local and global impacts, regardless of the cause (Türkeş, 1997a). Although climate changes have been occurring since the formation of the Earth, they have accelerated considerably due to human activities (Kadıoğlu, 2012). According to the Intergovernmental Panel on Climate Change (IPCC) report "Climate Change 2022: Impacts, Adaptation and Vulnerability", 95 percent of climate change is caused by human activities. Since the pre-industrial period, greenhouse gas emissions from human activities (anthropogenic) such as Methase (CH_4), Carbon Dioxide (CO_2), Diazot Monoxide (N_2O) have been increasing in the atmosphere due to industrialization, fossil fuel use, urbanization, rapid population growth, deforestation, misuse of land and wastes (Türkeş, 2001). Due to the increase in greenhouse gas concentrations in the atmosphere, average global temperatures are also increasing. The late 19th century warming is referred to as the turning point of global warming, which became more pronounced after the 1980s, and historical temperature records are witnessed every year. Global warming and climate changes cause a decrease in water reserves, adversely affect natural resources and damage biodiversity (Kurnaz, 2019). It is necessary to know in detail the activities that cause greenhouse gas formation, which is the main factor of climate change, within the scope of combating climate change. According to research, 25 percent of greenhouse gas emissions such as carbon dioxide and methane gas are caused by heat and electricity use, 24 percent by improper agricultural activities and deforestation, 21 percent by the industrial sector, 14 percent by transportation activities and 6 percent by buildings (IPCC, 2014). Although the vulnerability of countries to the effects of climate change varies, the recent increase in the severity of droughts, forest fires, floods and weather events proves the deterioration in the ecosystem balance and increases the sensitivity of countries regarding the policies and measures to be taken on climate change (IPCC, 2014b:32).

Although the 7.9 billion people on Earth constitute only 0.01 percent of the living beings in the universe, they have caused the extinction of 83 percent of wild mammals, 80 percent of marine mammals, 50 percent of plants and 15 percent of fish since the existence of the first civilizations. In the century we live in, it is accepted by climate scientists that the climate system is deteriorating. It is predicted that if human beings, who have a key role in the deterioration of the ecological and natural balance, continue the activities that cause the deterioration of the climate balance in the same way and intensity and do not take the necessary measures, the vital effects of global warming and deterioration in the climate system in the future will deeply affect the lives of all living things (Sırdaş, 2003; Öztürk, 2012).

Macroeconomic Implications of Climate Change

In the context of combating climate disruptions, the climate crisis has so far been positioned as a driving force in national and global policy action planning, usually within the framework of the environmental damage it causes. However, the complex economic models and practices applied over the years have shown that the stresses and destructive effects of the climate crisis on various sectors have become one of the major threats of the era for the global economy. Scientific research predicts that trade routes may change significantly with the complete melting of the Arctic Ocean in 2030 due to global temperatures (Acar et al., 2018). In the report of the Intergovernmental Panel on Climate Change (IPCC) published in 1995, the economic dimensions of the climate crisis were discussed comprehensively for the first time. In the second assessment report of the IPCC, which explained that climate change causes cumulative net economic losses, it was emphasized that 30 percent of greenhouse gas emissions could be reduced at negative or zero cost. However, the failure to implement planned national and global climate policies over time has increased the economic damages caused by climate change. In this regard, William Nordhaus, who received the 2018 Nobel Prize in Economics, explained that an increase of 4 degrees Celsius in global temperatures would cause a loss of 4 percent in global GDP, and an increase of 6 degrees Celsius would cause a loss of around 11 percent. The effects of increases in global temperature due to climate change on global GDP are presented in Figure 1.

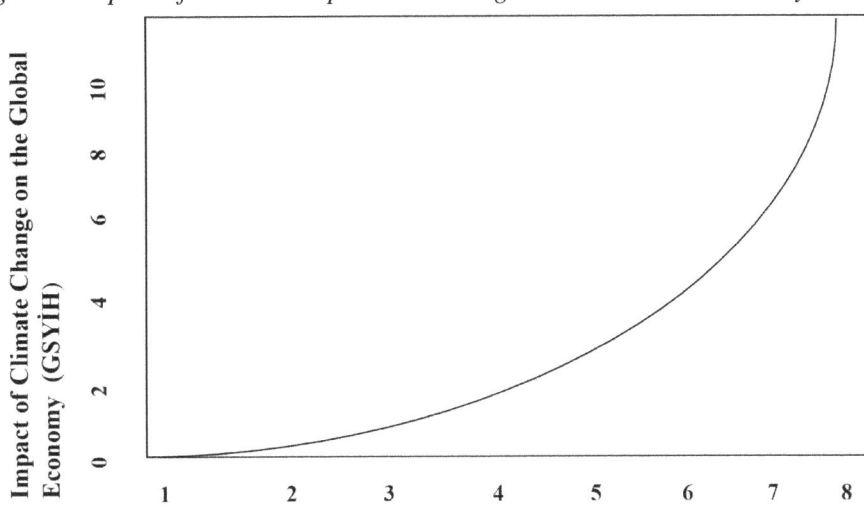

Figure 1. Impact of Global Temperature Changes on the Global Economy

Source: *Nordhaus, 2019*

Before Nordhous' statements on the climate crisis, economist Nicholas Stern's book "The Economics of Climate Change" stated that the annual cost to the global economy of preventing this situation in the scenario of inaction against changes in the climate order and global warming would correspond to 1 percent of global GDP, and explained that temperature increases and the climate crisis are a major threat to the global economy. In June 2019, Moody's Anallytics published a report titled "The Economic Impacts of Climate Change", according to which the hurricanes caused by changes in the climate pattern in 2017 cost the US economy, which is responsible for 14 percent of greenhouse gas emissions globally, approximately USD 300 billion. In the United States, the cost of extreme climate events caused by climate change since 1980 has been recorded as approximately 1.6 trillion dollars. According to a report published by Münich Re, one of the world's largest insurance companies, the cost of forest fires in California has exceeded 24 billion US dollars.

The economic risks caused by climate change can appear in different ways. Extreme increases in temperature, decreases in water resources and food supply, sea level rise due to melting glaciers, famine and possible migration waves are just some of the possible destructive effects of the change in the climate order (Roos and Hoffart, 2021: 2; IPPC, 2019: v). Efforts to transition to a low-carbon economy within the scope of combating climate change entail various economic risks. Achieving sustainable growth and slowing down the economy towards the consumption level may cause a number of economic imbalances that may feed vulnerability in the transition to a low-carbon economy. In this context, the risks and various costs caused by climate change may shake macroeconomic balances, while macroeconomic imbalances that may occur may make it difficult to implement policies.

Figure 2. Basic Relationship between Climate Change and Macroeconomic Vulnerabilities

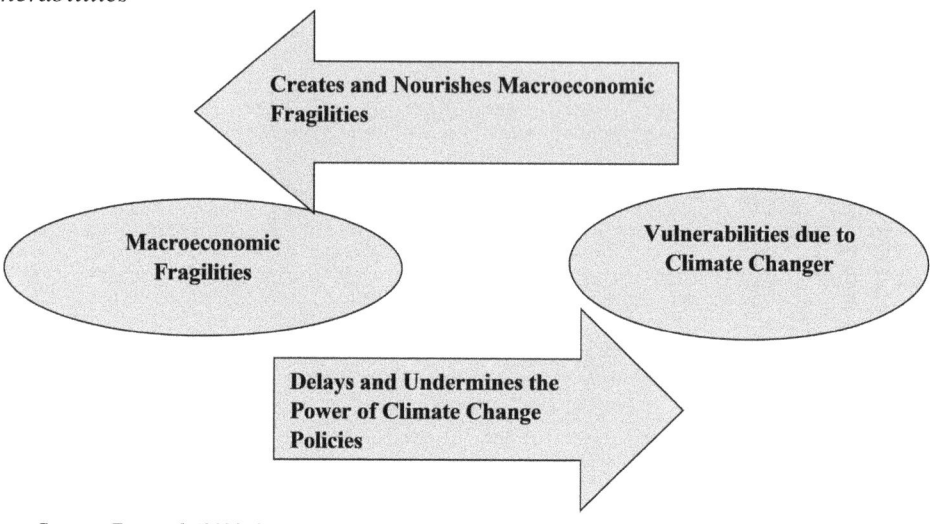

Creates and Nourishes Macroeconomic Fragilities

Macroeconomic Fragilities

Vulnerabilities due to Climate Changer

Delays and Undermines the Power of Climate Change Policies

Source: *Fayen vd, (2020: 2)*

Today's economic systems and modern economic dynamics encourage economic growth through intensive resource utilization. However, environmental damages that cannot be priced in terms of the market in the growth processes of countries damage the ecosystem and create negative externalities that harm all living things on a global scale and whose effects last for many years (Nordhaus, 2013a: 6; Hagens, 2020: 2; Nordhaus, 2019: 1991). For this reason, correcting this problem by shaping the policies necessary to prevent the risks caused by climate change within the macroeconomic system through the intervention of public power in the market is one of the current important issues in the economics literature. Macroeconomic analyses and approaches to be developed in the light of economic modeling are important in combating climate change. Therefore, it is critical to evaluate the market costs caused by the climate crisis on a sectoral basis (food and agriculture sector, tourism sector, energy sector, health sector) and to analyze the macroeconomic reflections to eliminate these costs.

Impacts of Climate Change on Employment

Jobs depend on the characteristics of the sectors they are connected to. The characteristics of the sectors are formed by the ecosystems they depend on. For this reason, jobs are directly or indirectly dependent on or affected by ecosystems (ILO, 2018). For example, jobs in the agriculture, fisheries, animal husbandry, tourism and forestry sectors are directly dependent on ecosystems, while jobs in the hydroelectricity, energy and pharmaceutical sectors interact indirectly with ecosystems. In this context, the services provided by ecological systems for free are highly valuable for humanity both economically and socially (World Bank, 2022). Since vital ecosystem health and sustainability depend on factors such as climate, air, water, plants, soil, etc., the effects of climate change pose a threat to the sustainability of ecosystem services and negatively affect jobs in sectors where ecosystem services are used directly or indirectly. Jobs are affected not only by disruptions in the climate pattern, but also by the transformations necessitated by the policies developed to combat climate change. In this context, it is expected that alternatives such as demand for renewable energy instead of environmentally damaging fossil energy and low-carbon use instead of carbon-intensive services will have an impact on the labor market. Policies to combat climate change will lead to the disappearance of some jobs and professions in some sectors and the emergence of new jobs (Deschenes and Jobs, 2013).

Approximately 1.2 billion jobs around the world are directly dependent on ecosystem services. Thus, it is stated that climate change directly or indirectly affects approximately 40 percent of the jobs in the sector with a presence in the global economy (ILO, 2018). The economic negativities caused by climate change globally affect the employment component, which is one of the main executive elements of the economy, negatively as well as affecting many sectors. In this respect, the inclusion of action plans related to climate change in employment policies will constitute an important reaction in terms of offsetting the global economic impacts caused by the climate crisis.

According to the International Labor Organization (ILO), 1.2 billion of the world's 3.2 billion jobs in 2014 were intensively or directly dependent on ecosystem conditions. These sectors include agriculture, fisheries, forestry, food, tobacco, biofuels, pharmaceutical and chemical industries, renewable energy sources and tourism. There is variation in the share of employment based on ecosystem services across G20 countries, with India, China and Indonesia having the highest rates at 52 percent, 50 percent and 40 percent, respectively, while European countries such as the UK and Germany have rates of 5 percent and 6 percent, respectively. In the European Union (EU) as a whole, the proportion of total employment directly based on ecosystem services is around 16 percent. In this context, it is stated that losses in ecosystem services due to climate change will cause disruption of activities in many sectors, which will indirectly result in more employment (job) losses (ILO, 2018). Developing employment policies in a climate-sensitive manner is important in terms of reducing the economic crises caused by climate change. In this direction, it is thought that the efforts to be made in the transportation, construction and energy sectors to achieve the goal of keeping global warming below 2°C by 2100 will positively affect employment on a global scale.

COVID-19 Pandemic and Climate Change

There is no evidence of a direct relationship between changes in climate patterns and the emergence of the Covid-19 pandemic or the transmission of the disease. However, climate change affects the environmental determinants of human and living health and may indirectly affect the pandemic by creating additional burden on health systems. Considering that pandemics throughout history have been caused by wildlife, it is obvious that damages to the natural environment and destructive human activities have a share in this situation. According to a study conducted by the Center for Disease Control and Prevention in the United States, it is stated that three-quarters of such infectious diseases that will cause pandemics are of animal origin, and the reason for this is that humans enter the habitats of animals (Figueres, 2020). The World Health Organization (WHO) has stated that extreme weather events caused by climate change will lead to the spread of vector-borne diseases such as dengue and malaria. It is stated that the abnormal warming of the planet may increase the frequency of infectious diseases and increase the spread of diseases as disease-carrying insects reach cold regions due to humidity, increased temperature and floods (WHO, 2020). It is also predicted that diseases that have not been in the air for many years will emerge and spread with the melting process of glaciers caused by global warming. According to the findings of a study published in Nature, degraded habitats harbor many viruses that can infect and sicken humans. The reason for this is thought to be that the loss of biodiversity may increase viral infections on the remaining species. In a preliminary study on Covid-19 patients, it was stated that there may be a relationship between air pollution and mortality rates because air pollution makes people more susceptible to respiratory diseases (Cui et al., 2003). In another similar study, a correlation was found between the total number of Covid-19 infections and the level of particulate matter (PM10) in the air (MC, 2020). As the Covid-19 pandemic is coming to the end of its course, the effects of climate change may manifest themselves during the relief period caused by the elimination of the measures taken for the transmission of the disease. For this reason, the issue of climate change requires improved perspectives and a continuous effort. It is thought that the common denominator of climate change and the Covid-

19 pandemic stems from the environments that have been changed and damaged by human hands, and the main difference between them is the rapid global response given by almost the whole world during the Covid-19 outbreak and the declared pandemic period and the search for a remedy with an effort above the social power, while the abstract nature of climate change effects and the helplessness and inaction felt in the face of this situation (Reynolds, 2019). The issue of climate change and its impacts is a public health problem, just like the Covid-19 outbreak, which has recently been declared a pandemic. The measures taken to protect public health during the Covid-19 process have led to a large annual reduction in carbon dioxide emissions. However, this reduction is not sufficient to achieve a global balance. This reduction needs to be systematically sustained (Reynolds, 2019).

An Eco-Strategy to Contribute to Environmental Sustainability: Teleworking

Changes and developments in the economic, social and technological fields are felt in different ways in all areas of our lives, which leads to the formation of a new working culture and organization for employees and businesses (Ellison, 2004). Especially in parallel with the developments in the field of technology, changes in the needs of businesses and employees have led to the emergence of flexible working styles in terms of space and time. In addition, businesses today have started to implement flexible working models with the aim of adapting to differentiating conditions, reducing costs and continuing their activities uninterruptedly in case of any crisis such as the Covid-19 outbreak in order to survive in increasing competitive conditions. Today, the purpose of using flexible working models is to adapt the work life to changing conditions as well as to ensure employee satisfaction (Berber, 2022:2). The concept of remote working, also called "virtual office" and "teleworking", are some of the concepts that try to explain the same phenomenon with different terms. Teleworking is a type of work that enables a job to be carried out from a different location other than the central office location (Meşhur, 2010). It is the fulfillment of work-related processes such as service and production by employees outside and away from the physical location of the enterprise by providing various infrastructural systems (technological, digital, etc.) (Öztürkoğlu, 2013:121).

Since the 1970s, the telecommuting model has started to be discussed, and it has become a working model that has attracted more attention with the technological developments, telephone, computer and information communication technologies (ICT) taking more place in our lives (Klopotek, 2017). In particular, the hybrid working model, which is a mixture of telecommuting and home- and office-based working, has recently been used as a popular working theme in both social sciences and technological studies (Hunter, 2018).

In the 1980s, the telecommuting model became a flexible working model that was frequently mentioned with ideas such as saving energy, providing cost advantages, reducing air pollution caused by intensive car use during commuting, and contributing to the solution of ecological problems in parallel (Handy and Mokhtarian, 1995:100). It is thought that the telecommuting model will contribute to combating climate change in various ways:

1. Reduced Carbon Footprint: Teleworking can reduce transportation emissions by eliminating the need for employees to go to the office. This reduces traffic congestion, which in turn reduces air pollution and greenhouse gas emissions.

2. Energy Savings: It can reduce electricity use by cutting energy expenditure in offices. Reducing energy-intensive processes such as heating, cooling and lighting in offices contributes to the conservation of natural resources and the fight against climate change.

3. Sustainability Awareness: The remote working model allows employees to adopt more sustainable lifestyles and develop eco-friendly habits in their own home environment. This can encourage practices such as recycling more, using energy more efficiently and consuming fewer resources.

4. Flexibility and Travel Restriction: Remote work can reduce greenhouse gas emissions caused by air and road transportation by reducing business travel. By doing their work remotely, employees can contribute to lower carbon emissions because they do not have to travel when they do not need to travel for work.

Since the telework model supports balanced spatial development with lower carbon emissions, it is an important current working model that can be used to combat climate change, especially by contributing to the dispersion of businesses and employees from urban centers to rural settlements (Samek Lodovici, 2021). In particular, home-based remote working is considered as an eco-strategy that will contribute to ensuring ecological balance (Saura et al., 2022).

CONCLUSION

The increase in greenhouse gas emissions into the atmosphere as a result of human activities has been manifested in the recent prominence of the phenomenon of climate change, which is the disruption of the earth's climate balance. This phenomenon has recently become more pronounced, bringing with it a range of impacts such as high temperatures, droughts, extreme weather events and sea level rise. Moreover, given that climate change is likely to lead to more frequent and intense natural disasters, there is a serious concern about its vital negative impacts on societies, economies and ecosystems. For this reason, climate change stands out among the important issues that countries need to tackle immediately through global cooperation and decisive action plans.

Although there is no direct relationship between the Covid-19 pandemic, which started in China in 2019 and rapidly spread across the world, and climate change, it is believed that there are indirect interactions and links. For example, climate change may cause habitats to change, increasing the spread (of ticks and mosquitoes), leading to increased risk of infectious diseases and human infection. Furthermore, climate change can reduce the capacity to fight any epidemics, as it can lead to vulnerability and weakening of societies. Adverse conditions in food security and water resources caused by climate change can affect people's dietary habits and health status, reducing their resilience to disease.

Employment is another area affected by climate change. Climate change also has various complex impacts on employment. It may cause job losses in some sectors. Natural disasters such as forest fires, droughts and floods associated with climate change can disrupt agriculture, tourism and construction sectors, causing damages and job losses. With the effects of climate change increasing globally every day, it is expected that the green economy, which will encourage the development of environmentally friendly and sustainable industries, and the telecommuting model, one of the flexible working models that can reduce greenhouse gas emissions, will become widespread. The telecommuting model, which allows employees to work from their homes or other remote locations, can be an effective tool in combating climate change with features such as reduced foot carbon footprint, energy savings, sustainability awareness, flexibility and travel reduction. By promoting telework policies, governments can reduce the impact of global warming and climate change by enabling working individuals to reduce their environmental impact.

REFERENCES

Acar, Z., Gönencgil, B., & Gümüşoğlu, N. K. (2018). Long-term changes in hot and cold extremes in Turkey. *Coğrafya Dergisi*, (37), 57–67. DOI: 10.26650/JGEOG2018-0002

Berber, P., & Keleş, E. (2022). *Bilgi ve Teknolojileri Aracılığıyla Uzaktan Çalışma "Tele Çalışma: Kapsam ve Doğası"*. Pegem Akademi Yayınncılık.

Blunden, J., & Arndt, D. S. (2020). A Look at 2019. *Bulletin of the American Meteorological Society*, 101(7), 612–622. DOI: 10.1175/BAMS-D-20-0203.1

Cui, Y., Zhang, Z. F., Froines, J., Zhao, J., Wang, H., Yu, S. Z., & Detels, R. (2003). Air pollution and case fatality of SARS in the People's Republic of China: An ecologic study. *Environmental Health: A Global Access Science Source, 2*, 1–5. .DOI: 10.1186/1476-069X-2-1

Ellison, N. B. (2004). *Telework and social change: How technology is reshaping the boundaries between home and work*. Bloomsbury Publishing USA. DOI: 10.5040/9798216024132

Figueres, C., & Ca, T. (2020). *Our approach to covid-19 can also help tackle climate change | New Scientist*, https://www.newscientist.com/article/mg24532763-500 - our-approach-to-covid-19-can-also-help-tackle-climate-change/

Handy, S. L., & Mokhtarian, P. L. (1995). Planning for telecommuting measurement and policy issues. *Journal of the American Planning Association*, 61(1), 99–111. DOI: 10.1080/01944369508975623

Hunter, L. (2018). *Carnivores of the world* (Vol. 117). Princeton University Press.

IPCC. (2014a). AR5 Working Group II. *Climatic Change*.

IPCC. (2014b). AR5 Working Group III. *Climatic Change*.

Kadioglu, I., & Farooq, S. (2017). Potential distribution of sterile oat (Avena sterilis L.) in Turkey under changing climate. *Turkish journal of weed science, 20*(2), 1-13.

Kłopotek, M. (2017). The advantages and disadvantages of remote working from the perspective of young employees. *Organizacja i Zarządzanie: kwartalnik naukowy*, (4), 39-49.

MC. C. (2020). *Perché l'inquinamento da Pm10 può agevolare la diffusione del virus.* https://www.ilsole24ore.com/art/l-inquinamentoparticolato-ha-agevolato-diffusionecoronavirus-ADCbb0D

Meşhur, H. F. A. (2010). A research on the attitudes of organizations towards telework. *Dokuz Eylül University Journal of Faculty of Economics and Administrative Sciences*, 25(1), 1–24.

Nordhaus, W. D. (2019). Climate change: The ultimate challenge for economics. *The American Economic Review*, 109(6), 1991–2014. DOI: 10.1257/aer.109.6.1991

Nordhaus, W. D. (2021). Climate Club Futures: On the Effectiveness of Future Climate Clubs.

Öztürkoğlu, Y. (2013). Tüm Yönleriyle Esnek Çalışma Modelleri. Beykoz Akademi Dergisi, 1(1), 109-129.

Reynolds, M. (2019). Coronavirus shows the enormous scale of the climate crisis. https://www.wired.co.uk/article/coronavirus-climate-change.

Roos, M., Hoffart, F. M., Roos, M., & Hoffart, F. M. (2021). Climate change and responsibility. *Climate economics: A call for more pluralism and responsibility*, 121-155.

Samek Lodovici, M. (2021). The impact of teleworking and digital work on workers and society. .DOI: 10.2861/72272

Saura, J. R., Ribeiro-Soriano, D., & Zegarra Saldaña, P. (2022). Exploring the challenges of remote work on Twitter users' sentiments: From digital technology development to a post-pandemic era. *Journal of Business Research*, 142, 242–254. DOI: 10.1016/j.jbusres.2021.12.052

Sırdaş, S. (2002).. . *Meteorolojik Kuraklık ve Türkiye Modellemesi, İstanbul Teknik Üniversitesi Dergisi*, 2(2), 95–103.

Türkeş, M. (1997). Hava ve iklim kavramları üzerine. *TÜBİTAK Bilim ve Teknik Dergisi*, 355, 36–37.

Türkeş, M. (2001). Global climate protection, climate change framework agreement and Turkey. *Plumbing Engineering, TMMOB Chamber of Mechanical Engineers, Periodical Technical Publication*, 61, 14–29.

WHO. (2020). *Coronavirus disease (COVID-19) Pandemic*. WHO. https://www.who.int/emergencies/diseases/novel-coronavirus-2019

WHO World Health Organization. (2020). *WHO*. https://www.who.int/reproductivehealth/en/

World Bank. (2022). Country Climate and Development Report: Türkiye. The World Bank Group.

Section 3
Fighting Climate Change

Chapter 11
Methods of Combating Against the Problem of Climate Change

Enes Yalçın
İzmir Kâtip Çelebi University, Turkey

ABSTRACT

This study addresses the critical issue of climate change, examining the contributing factors and the resulting problems. Key drivers of climate change, including greenhouse gas emissions, deforestation, and industrial pollution, are analyzed. The study also explores the multifaceted impacts of climate change, such as extreme weather events, rising sea levels, and biodiversity loss. Emphasis is placed on the role of international platforms and agreements at the nation-state level, such as the Paris Agreement, in combating climate change. Additionally, the study highlights effective sub-national initiatives and practices, showcasing local government actions and community-based projects aimed at reducing carbon footprints and enhancing climate resilience. Through comprehensive analysis, this research underscores the importance of coordinated efforts at both global and local levels to mitigate the adverse effects of climate change and promote sustainable development.

INTRODUCTION

Climate change, like all other environmental issues, has been one of the frequently discussed topics since the second half of the 20th century. Considering the close relationship between each environmental issue, this situation would be considered usual. Although climate change does not have sudden destructiveness like global natural disasters and pandemics, it has a cumulative effect that affects future gener-

DOI: 10.4018/979-8-3693-5792-7.ch011

ations and makes the world more uninhabitable day by day. This cumulative effect, beyond affecting a specific region, leads to large-scale changes all over the world. As a matter of fact, the fight against climate change requires global cooperation in which all actors, ranging from states to local governments, non-governmental organizations to scientists and even the public, participate.

The fact that global warming will exceed 1.5°C throughout the year for the first time in human history in 2023 has reminded us of the need to implement global cooperation as soon as possible. Global cooperation includes various measures to reduce the effects of climate change, such as creating policies, supporting scientific research, investing in renewable energy sources, reducing greenhouse gas emissions and developing adaptation strategies. In addition, it is important to raise awareness and train all actors who create and/or implement environmental policies towards more sustainable and environmentally friendly practices. Combating climate change will be possible not with the efforts of a single country or organization, but with the whole world coming together and finding a common solution.

Considering that interests are in conflict, it is likely that many difficulties will be encountered on the path to global cooperation. The differences between the interests and policies of countries make it difficult to achieve cooperation. In addition, economic and commercial concerns cause some countries to be reluctant to combat climate change. However, to overcome these obstacles and achieve effective global cooperation, leadership, solidarity, a fair approach, inclusive policies and compromise are required. Despite this necessity, the approach of nation states to prioritize self-security and ideological baggage are seen as reasons that hinder the cooperation opportunities of country administrations. At this point, local governments and cities come to the fore as pragmatic and practical implementers.

Local governments, which increase the applicability of relevant policies by developing policies suitable for local people, have begun to implement mitigation and adaptation targets against climate change within the framework of cooperation. At the local level, there are many instruments that can be used in the combat against climate change, such as the integration of transportation with environmentally friendly practices, the design of built environment elements, the nature-friendly use of urban energy systems, urban planning principles that consider environmental health, the establishment of green areas, and waste management. These vehicles are put into use in accordance with the structure and size, climate, geographical conditions, economic structure, and potential resources of the cities.

The roles undertaken by local governments in combating climate change essentially exceed local boundaries. At the point where nation states experience insufficient cooperation and coordination due to the reasons mentioned, local governments step in and create international platforms by taking initiative in combating climate change, as in other issues. In this study, exemplary practices carried out at the level of contractual international platforms and subnational units at the level of nation states in the fight against climate change are examined.

The Concept of Climate Change and Problems Resulting From Climate Change

Industrialization and urbanization, especially since the beginning of the 18th century, have intensified day by day, leading to the destructive use of natural resources. This destruction has caused climate change, as well as consequences such as pollution, desertification, and decrease in biodiversity. Climate change is an environmental problem caused by global warming that is transboundary in terms of its formation and consequences. Global warming occurs as a result of natural or human activities and means an increase in the average temperature of the earth. Although global warming can occur naturally, the main reason is the greenhouse gas effect caused by human activities. Greenhouse gases released due to industrial production accumulate in the atmosphere as the amount of emission increases and delays the exit time of infrared radiation from the atmosphere, increasing the destructive effect of the average temperature increase.

The United Nations Framework Convention on Climate Change defines the concept of climate change as a change that occurs directly or indirectly as a result of natural changes in the climate over a long period of time and human activities that disrupt the composition of the global atmosphere. (UNFCCC, 1992, p. 3). The consequences of climate change, whose effects occur especially in the long term, whose development carries uncertainty at every stage, and which causes significant changes in both physical and human geography, are summarized as follows by Engin (2010, p. 73):

- Temperature increases in the short and long term,
- Changing rainfall patterns,
- Formation of extreme weather conditions,
- Melting glaciers and rising sea levels,
- Changes in the number of species within the ecosystem,
- Migration of human communities and economic activities from places with climate risk.

As it can be seen, climate change carries a large number of problems in social, cultural, political, and administrative dimensions, in addition to ecological problems, leading to mass migrations in the first place. Within the scope of this study, it is useful to address some of these problems. Climate change threatens freshwater ecosystems, which are vital for people living primarily in rural areas around the world, and leaves societies faced with deprivation. This situation has the potential to cause global economic and political crises in the future. At the same time, the decrease in water resources increases the rate of pollution in the existing waters, causing epidemic diseases. It is expected that national and international trade will also be disrupted due to climate change.

Tourism is a sector that offers a direct contribution to the economies of countries and provides many subsectors with life opportunities. Aside from specific branches such as health tourism and gastronomy tourism, tourism activities are primarily shaped by factors such as historical attributes, geographical location, topography, and vegetation. However, climate conditions are uniquely significant due to their direct impact on these activities. In this context, it can be said that tourism is one of the areas directly affected by climate change. For instance, small island states, whose existence largely depends on marine tourism, will be negatively affected by sea level changes due to climate change. Climate changes will also have a direct negative impact on winter tourism and skiing. In addition, seasonal shifts, especially in tourism regions, will cause job and income losses. Therefore, it is seen that tourism types, tourism durations, and tourism costs will undergo radical changes due to climate change.

Agricultural activities are also among the areas that are primarily negatively affected by climate change. Although it is thought that global warming positively affects agricultural production by leading to an increase in harvest in some countries such as Belarus, Bulgaria, Estonia, Kazakhstan, Latvia, and Russia, where the cold climate prevails (Safonov, 2019, p. 8) climate change increases agricultural costs, changes agricultural crop periods, reduces agricultural productivity, and production. With the resulting multiplier effect, food prices are also rising globally, and inflation is increasing. As a result of global climate change affecting agriculture through current trends such as floods, droughts, and increased precipitation variability, yield losses have been reported in major crops such as maize (4.1%), soybeans (4.5%), rice (1.8%), and wheat (1.8%) between 1981 and 2010. (IPCC, 2022, p. 664).

Climate change negatively impacts global public health, particularly affecting communities with limited resources, low technology, and weak infrastructure The IPCC categorizes the impacts of climate change on human health into two main types: direct and indirect effects. Direct effects include exposure to extreme heat and disasters such as floods and storms caused by the frequency and intensity of weather events. Indirect effects are characterized by disruptions in ecological systems (IPCC, 1996, p. 11). A striking example is that in 2004, climate change was attributed to 3% of deaths worldwide due to diarrhea, 3% of deaths due to malaria, and 3.8% of deaths due to dengue fever. Furthermore, it was determined that approximately 0.2% of deaths globally in that year were climate change-related, with 85% of these deaths being child fatalities (World Health Organization, 2009, p. 24).

Climate change is also known to have significant sociological impacts. This phenomenon, which can create a snowball effect, initiates a chain reaction where losses in water resources and agricultural production lead to malnutrition, health issues, and ultimately the migration of people from the affected regions. Climate change results in the depletion of natural resources, creates conflicts over scarce resources and globally displaces many people, turning them into "climate refugees" as they abandon their homes. As can be seen, the impact of climate change, contrary to what most people think, is not only limited to the fact that the world is becoming a warmer place, but also leads to the fact that the world, as a whole of systems, is facing a large number of problems.

Factors Causing Climate Change

Anthony Giddens attempts to explain the inability to take necessary steps in addressing climate change through the "Giddens Paradox." According to Giddens, society tends to sit back and wait rather than take action because climate change is not visible, tangible, or urgent within the flow of daily life. However, when climate change becomes visible and urgent through damages and natural disasters, such as extreme weather events, agricultural losses, or related health problems that disrupt the rhythm of daily life, it becomes too late to act. Giddens argues that the issue of climate change, perceived as invisible, abstract, and distant in terms of time and geography, is relegated to a sub-real position in individuals' minds and the social agenda when compared to the political and economic agenda, which is accepted as visible, tangible, and immediate (Giddens, 2009, p. 2). In combating climate change, identifying the source of the problem immediately after defining it is crucial. Climate change is shaped by both natural causes and human-induced factors. Latitude effects, the Earth's precession motion, volcanic activities, changes in the Sun's radiation levels, the general circulation of the atmosphere, local geological and geophysical processes, storms, currents, and large-scale ocean temperature changes are some of the natural causes[1].

The 5th Assessment Report of the Intergovernmental Panel on Climate Change (IPCC), published in 2014, emphasized for the first time with 95% certainty that human activities are the cause of climate change (IPCC, 2014, p. v). The use of fossil fuels, industrial processes, population growth, mining, deforestation, and the shift in land use and consumption habits against natural living stand out as the primary causes of human-induced climate change.

The primary factor leading to the prominence of human-caused reasons is the increase in greenhouse gas emissions directly causing climate change. Greenhouse gases, which naturally constitute only 1% of the atmosphere, are essential for the continuation of life on Earth[2]. Greenhouse gases, much like cholesterol, are indispensable for the survival of the human organism, yet pose a threat to life when their levels exceed normal limits. These gases facilitate the absorption of solar radiation, ensuring that the Earth's temperature remains within a viable range. However, due to the aforementioned human activities, greenhouse gases have reached levels unseen in three million years, trapping more solar radiation and thus causing the greenhouse effect. This results in global warming and ultimately leads to climate change. Carbon dioxide is the greenhouse gas that contributes most to global warming. In 1800, the atmospheric concentration of carbon dioxide was 280 parts per million (ppm), which increased to 350 ppm by 2005 (Klein et al., 2005, p. 579). According to NASA data from May 2024, this level has now reached 427 ppm. Additionally, NASA has identified other "vital signs" apart from carbon dioxide, including "methane", "Arctic Sea ice minimum extent", "ice sheets", "sea level", "ocean warming", and as expected, "global temperature", all of which show a continuous increase (NASA, 2024). Sectoral distribution of carbon dioxide emissions is as follows: The electricity and heat production sector accounts for 44.4% of total emissions, resulting in 15.11 billion tons of CO_2 released. The transport sector contributes 20.9%, emitting 7.1 billion tons of CO_2, while the manufacturing and construction sector accounts for 18.2%, with emissions totaling 6.18 billion tons of CO_2. The buildings sector contributes 8%, leading to 2.71 billion tons of CO_2 emissions, and the industrial sector accounts for 4.8%, emitting 1.63 billion tons of CO_2. Land-use change and forestry sector contributes 3.4%, resulting in 1.17 billion tons of CO_2 emissions. Other fuel combustion activities contribute 1.7%, amounting to 566.79 million tons of CO_2 emissions. Lastly, fugitive emissions account for 0.8%, resulting in 268.34 million tons of CO_2 released (Our World in Data, 2020).

Intensive and unplanned urbanization, along with many other issues, also contributes to the problem of climate change. Globally, approximately 1.5 million people are added to the urban population each week. It is projected that by 2030, the urban population ratio will reach 30%, and by 2050, three out of every four people will live in cities. (World Economic Forum, 2022, p. 8). Throughout history, cities have been established near water sources and on fertile lands. Over time, aggressive urbanization has been destroying natural structures and leading to a decline in biodiversity. Cities are the primary source of global greenhouse gas emissions, contributing approximately 50-80% of the total emissions (Gordon, 2016, p. 530).

International Platforms in Combating Climate Change: Contracts at the Nation State Level

Since climate change is a global environmental problem, the solution to the problem must also be international. Although it is known that local governments have established international platforms (contracts/cooperation organizations), especially since the last quarter of the 20th century, this section examines international platforms at the level of nation states, which are contracts in the combat against climate change.

Since the First World Climate Conference (1979), where it was announced that carbon dioxide gas would become dangerous because of dependence on fossil fuels, many international platforms have been established on the subject. In addition, it is known that two basic documents are of primary importance in the combat against climate change: United Nations Framework Convention on Climate Change (Signature year: 1992, Entry into force year: 1994) and Kyoto Protocol (Signature year: 1997, Entry into force year: 2005). Before evaluating these documents, it is necessary to mention the "Vienna Convention for the Protection of the Ozone Layer," adopted in 1985, and the "Montreal Protocol on Substances that Deplete the Ozone Layer," adopted in 1987. The Montreal Protocol, which is the continuation of the first one, is considered the most successful multilateral agreement on the environment. The Montreal Protocol, for the first time in history, envisaged the restriction of human-derived substances that deplete the ozone layer on the basis of a multilateral agreement, in the light of scientific results that were not yet certain at that time. This model applied to the ozone layer has set a kind of precedent for the climate change regime. In this context, the Montreal Protocol constituted a turning point in the creation of the UN Framework Convention on Climate Change. (Ministry of Foreign Affairs of the Republic of Turkey, 2024). In addition, the first assessment report published in 1990 by the Intergovernmental Panel on Climate Change (IPCC), a joint initiative of the World Meteorological Organization (WMO) and the United Nations Environment Programme (UNEP), is significant as it laid the foundation for the United Nations Framework Convention on Climate Change (UNFCCC). The report provided objective scientific information on climate change (Tuğaç, 2020, p. 225).

The aim of the United Nations Framework Convention on Climate Change, to which the European Union and 196 countries are parties, is to ensure that human-induced greenhouse gas emissions in the atmosphere are kept at a level that will prevent their dangerous impact on the global climate system. The Convention emphasizes that, while achieving this goal, care must be taken to ensure that ecosystems adapt to climate change, that food production is not threatened, and that economic progress continues in a sustainable manner. Under the title of "Commitments" in Article 4 of the Convention, it is pointed out that developed countries should play a leading role in combating climate change. Additionally, the principle of "common but differentiated responsibilities" and "respective capabilities" has been taken into account. (UNFCCC, 1992, pp. 4-6).

The Kyoto Protocol, which includes two commitment periods, envisages that at the end of the first commitment period (2008-2012), the greenhouse gas emissions of developed countries will be reduced by a total of 5% compared to the 1990 level (UNFCCC, 1997). For the Kyoto Protocol, which was opened for signature in March 1998, to come into force, at least 55 countries, including industrialized countries, responsible for at least 55% of the total carbon dioxide emissions in 1990, had to ratify the Protocol. As a result of long negotiations and discussions, the Protocol came into force in February 2005 with the signature and approval of 140 countries, including the European Union. (Uysal Oğuz, 2010, p. 26). The failure of the United States (USA) to ratify the protocol has been a serious obstacle to its entry into force, as the United States alone accounts for approximately one-third of the carbon dioxide emissions of developed countries.

Although the USA signed the Kyoto Protocol, complaining that its carbon sinks and the emission trade it planned to carry out with developing countries were not taken into account in the protocol negotiation processes, it refrained from becoming a party to the protocol by not submitting it to the Senate. Essentially, the reason for this decision was that at the end of 2008-2012, which is the first obligation period of the protocol, the USA would have to reduce its carbon emissions by 35% below the level estimated for 2012. The USA claimed that the cost of efficiency in energy production would be high, and that it would be much cheaper to provide emission reductions in developing countries instead (Ulueren, 2001; Murphy, 2001, p. 647).

The Bush administration, on the other hand, addressed the issue directly and declared their intention to withdraw from the protocol, unlike the Clinton administration, which signed the protocol and made it forgotten by not submitting it to the Senate for approval. According to this approach, the Kyoto Protocol is a dysfunctional agreement that will impose unacceptable costs on the US economy and exempt developing countries from taking action. The US administration has argued that combating climate change can be achieved by supporting scientific research and long-term technological change, rather than a complex and flawed international protocol (Schreurs, 2004, p. 208).

In the second commitment period of the protocol, covering 2013-2020, it was decided that the parties in the Annex-B list would reduce their emissions by at least 18% compared to 1990 levels in 2020, unlike the first commitment period. (UNFCCC, 2024b). Australia, Canada, Japan and Russia, which assumed obligations in the first commitment period, did not undertake any obligations in the second commitment period. The second commitment period of the Kyoto Protocol (Doha Amendment), which had to be accepted by 144 party countries in order to come into force, came into force on 31 December 2020 (Ministry of Foreign Affairs of the Republic of Turkey, 2024). The Kyoto Protocol, which is an important step in the fight against climate change with its concrete and restrictive regulations, is also important in that it contains a text that respects the development rights of developing countries with the principles of "common but differentiated responsibilities" and voluntary commitments. In addition, emission upper limits were determined for some developed countries with the Kyoto Protocol. Flexibility mechanisms have been developed to facilitate these countries in fulfilling their responsibilities. In return for the greenhouse gas emissions reduced through these mechanisms, businesses have obtained carbon certificates, and mandatory markets have emerged where these certificates are traded.

The Paris Agreement, which is another agreement that should be emphasized after the two basic documents mentioned, is an internationally legally binding agreement on climate change. 195+2 (Palestine and the Vatican have observer status) party country accepted the agreement at the UN Climate Change Conference held in Paris on 12 December 2015. The agreement entered into force on November 4, 2016, and the goal of the agreement was determined as "limiting the global average temperature increase to 1.5°C and making efforts to keep it below 2°C" (UNFCCC, 2024a).

The main theme of the Paris Agreement, which consists of twenty-nine articles, is the Intended Nationally Determined Contributions (INDC) submitted by the countries included in the agreement. These documents are considered road maps containing the commitments of the parties. The agreement essentially requires all parties to undertake obligations to reduce emissions. However, while more reduction commitments are expected from developed countries and these countries are expected to achieve absolute reduction, developing countries are expected to reduce according to their current capacities. This reflects the principle of "common but differentiated responsibilities". The agreement includes commitments to support vulnerable, underdeveloped countries that will be most affected by climate change, with a strong emphasis on adaptation. Besides, the Paris Agreement is notable for being the first climate change agreement to directly address human rights. It emphasizes the necessity of considering human rights, the right to health, the rights of indigenous peoples, local communities, refugees, children, persons with disabilities, people in vulnerable situations, and principles of intergenerational equity in climate change actions, recognizing climate change as a common concern of humankind (Paris Agreement, 2015, p. 2).

It is known that the European Union (EU) has become a global green leader in climate change policies, especially after the USA refused to ratify the Kyoto Protocol. The determination of the EU, whose member states accounted for nearly a quarter of the carbon dioxide emissions in 1990, to continue the ratification process of the protocol despite the opposition of the United States, positively affected the participation of the governments of countries outside Europe in the process.

Exemplary Combat Practices at the Subnational Level Against Climate Change

In combating climate change, there are fundamentally three types of interventions: avoiding climate change, avoiding "dangerous" climate change, and responding to dangerous climate change. The first is characterized by steps to reduce greenhouse gas emissions, remove carbon dioxide from the atmosphere, and increase reflectivity to avoid global average temperature increases. The second aims to ensure that rising temperatures do not affect fundamental interests. The third involves compensating for the damages that have occurred (Heyward, 2013, p. 25). Nevertheless, two approaches are prominent in addressing climate change: mitigation and adaptation. Mitigation basically means preventing or slowing down climate change. Adaptation includes policies aimed at preparing societies and economies against the possible effects of climate change. (Doğan & Tüzer, 2011, p. 159). Climate change adaptation can occur autonomously, in a planned manner, or naturally. Autonomous actions are undertaken by specific actors triggered by climate change, while planned actions are programmed and turned into policy by public and/or private actors. Natural adaptation actions emerge spontaneously from ecosystems as the impacts of climate change unfold (Tuğaç, 2020, p. 235). Traditional policies prioritizing mitigation activities in combating climate change have increasingly given way to an approach where both mitigation and adaptation are considered equally important, especially following the Paris Agreement.

The areas where mitigation and adaptation policies can basically be implemented are urban areas where greenhouse gases are intensely emitted. Due to urbanization, the resulting concretization leads to excessive heating of cities. The roads, buildings, and sidewalks in urban centers release the energy they have absorbed throughout the day during the night, thereby creating the urban heat island effect. Consequently, scientific research and administrative practices have focused on urban areas, leading to the development of various generic concepts targeting cities that aim to combat climate change and promote sustainable development: ecological city/eco-city, compact city, slow cities/cittaslow, smart cities, green cities, low carbon cities, and liveable cities. The realization of these concepts will be possible with the construction of a governance mechanism formed by a large number of actors, especially local governments, assuming responsibility.

Local government action on climate change reflects a broader, global trend that challenges the conventional wisdom that "sub-national" government units lack the ability to effectively deal with cross-border environmental challenges (Rabe, 2002, p. 3). For instance, it is known that although the USA rejected the Kyoto Protocol, some US states and cities took the initiative to reduce carbon. An example of this is the fact that Los Angeles, through its own policies, has reduced carbon energy use in its huge port by 40 percent in just five years, while the United States does not adhere to universal standards. These measures have reduced urban CO_2 emissions by 16 percent, as emissions from the port account for two-fifths of the city's total carbon pollution (Barber, 2013, p. 319).

Climate change policies have been adopted by all federated states of the USA, led by both Republicans and Democrats. At the heart of the effectiveness and creativity of state and city administrations in combating climate change lies their greater authority in the delivery and oversight of services directly affecting greenhouse gas emissions, such as electricity, water, natural gas, transportation, air quality, and solid waste management, compared to the federal government (DiPeso, 2004, pp. 111-112).

The Clean Power Plan, published by the Environmental Protection Agency (EPA) in the USA in June 2014, set carbon reduction targets that differ at the state level. These goals include minimal reductions in coal-based states such as Kentucky, West Virginia, and North Dakota. It includes higher reduction targets in states that use cleaner energy, such as Washington, California, and New Jersey. When the plan's target reduction rates for all states are combined, EPA estimates that total emissions from the electricity-generating sector will decrease by 30% from 2012 levels in 2030. (Engel, 2015, pp. 452-458). Despite some efforts to establish a "green economy" at the state level, U.S. governments do not have an agenda supporting environmentally friendly practices. For example, the Recovery and Reinvestment Act provided $16.8 billion for green energy and $27.5 billion for road and bridge construction, which would increase emissions. Similar approaches raise doubts about what the "green economy" means in the United States and how steps will be taken towards this economy. (Vezirgiannidou, 2013, p. 604).

In the case of the United Kingdom, the city of London has emerged as a leader in combating climate change by taking initiative. The Greater London Authority (GLA) has been a driving force in establishing and maintaining numerous partnerships addressing the issue of climate change within the city. Under the authority, several mitigation and adaptation-related partnerships have been established, including the London Climate Change Partnership (LCCP), the London Energy Partnership (LEP), the London Resilience Partnership (LRP), the London Hydrogen Partnership (LHP), and the Drain London Forum (Davoudi et al., 2011, p. 11).

The efforts of local governments, or in other words subnational government units, to fight climate change are not limited to some practical applications. Relevant units are trying to contribute to the solution of the problem by determining some plans and principles. For instance, the Climate Action programs established by the Institute for Local Government in California cities since 1955 to meet their information needs have demonstrated how local leaders understand, implement, and monitor the results of these innovative practices. These initiatives and resources help California cities set an example for the state, nation, and world. (Institute for Local Government, 2024).

Local governments in Turkey have also assumed legal responsibilities in combating climate change. In accordance with the Municipality Law No. 5393, adopted in 2005, local governments are responsible for the removal of rainwater and wastewater, the establishment of the necessary facilities for these, the execution of solid waste service and the transportation, separation, storage and disposal of this service, reducing the carbon footprint resulting from transportation by reducing the traffic in the city. It is obliged to prepare teams, equipment, disaster and emergency plans to protect against fire and natural disasters and to reduce their damage (Municipality Law No. 5393, 2005). The Special Provincial Administration Law No. 5302, adopted in 2005, and the Municipality Law No. 5216, adopted in 2014, are also important in terms of defining the duties that local governments should undertake in the context of combating environmental problems and climate change.

Apart from local government activities carried out within national borders, cities also combat climate change by creating international platforms. The EU Covenant of Mayors for Climate and Energy, an initiative supported by the European Commission, brings together thousands of local governments around the world, mostly in Europe, who want to secure a better future for their citizens. By participating in this initiative, established in 2008, local governments voluntarily commit to implement EU climate and energy targets. The Covenant of Mayors is designed to provide local governments with a framework for local energy and climate action based on four principles: a) consistency and transparency, b) flexibility and adaptability, c) data evaluation, and d) experience sharing and promotion. In the final analysis, the 12,006 contracting cities are committed to adopting a holistic approach to combating climate change. Within the first two years of participation, they are expected to develop a Sustainable Energy and Climate Action Plan with the goals of reducing carbon dioxide emissions by at least 40% by 2030 and increasing resilience to climate change (European Commission, 2024a; European Commission, 2024b).

Another global alliance established by local governments, focusing on the fight against climate change, is the Global Covenant of Mayors for Climate & Energy. This platform, which has 13,374 cities as members from 146 countries around the world, is the world's largest alliance in the context of city climate leadership. Almost 1.2 billion people (one in eight people worldwide) live in a city that is a member of the relevant platform (Global Covenant of Mayors for Climate & Energy, 2023, p. 11). The Global Covenant of Mayors for Climate & Energy aims to facilitate the creation of a low-carbon society by utilizing mitigation, adaptation, and energy access policies. It seeks to enhance cities' access to relevant information, establish global partnerships, assist in accessing financial institutions, and raise awareness in this field.

Local Governments for Sustainability is an international platform comprising over 2,500 local and regional governments from more than 125 countries, aiming to support sustainable urban development. Sustainable urban development is shaped around five core patterns: a) low-emission development, b) nature-based development, c) circular development, d) resilient development, and e) equitable and people-centered development. This approach serves as a guide for local and regional governments in achieving sustainable urban development (Local Governments for Sustainability, 2024).

Additionally, Energy Cities, which encompasses thousands of cities from thirty European countries and is one of the founders and supporters of the Covenant of Mayors for Climate & Energy, aims not only to develop climate policies for a decentralized, democratic, and carbon-free energy system but also to transform with a climate-neutral approach. Through this international organization, the aim is to promote interdependence among sectors, regions, economic, social, and cultural actors, and city leaders to create resilient communities and share wealth and prosperity at the local level (Energy Cities, 2024).

The C40 Cities Climate Leadership Group, an international organization currently comprising ninety-six member cities worldwide, adopts a science-based and collaborative approach to combat climate change. It aims to halve emissions by 2030, limit global warming to 1.5°C, and build healthy, equitable, and resilient communities (C40 Cities, 2024). Participation in international networks, such as the aforementioned ones, facilitates the sharing of knowledge and experience, bypassing the limitations of individual efforts. It underscores the potential for local initiatives to access expert knowledge and financial support. These efforts and initiatives demonstrate the potential of local governments to advance local climate actions.

CONCLUSION

Climate change is a phenomenon that affects the lives of all living beings, and its impacts extend beyond being merely periodic. Combating climate change requires a multi-actor and multi-level approach. Therefore, global cooperation is essential in the combat against climate change. Addressing the issue requires the participation of not only nation-states but also local governments, cities, non-profit civil society organizations, universities, sports clubs, and even the arts community. This study highlights that the global platforms and agreements established by nation-states are insufficient in combating climate change because they lack enforcement power and binding provisions. Furthermore, the responsibilities undertaken by local governments and city coalitions worldwide in this regard have been examined through established platforms, identified plans and principles, and exemplary practices. In this context, it has been determined that cities play significant roles in combating climate change, unlike nation-states, which often carry ideological baggage and exhibit flexibility in diplomacy as a form of soft power. Cities are forming alternative global platforms for climate change governance to those of nation-states.

It is noteworthy that both nation-states and cities contribute to the global combat against climate change by establishing international platforms. In addition, this study must address some of the challenges encountered in dealing with climate change and provide recommendations for overcoming these obstacles. In this context, the developed recommendations, centered around biological, social, and economic concerns, can be primarily outlined as follows:

- There is a consensus within the scientific community regarding the existence of climate change and the methods that can be developed to address it. It is imperative that this consensus is conveyed to global policymakers.

- Due to its inclusive and transboundary nature, as emphasized throughout this study, establishing global cooperation in combating climate change should be considered a prerequisite.

- Active representation in international initiatives aimed at controlling global warming should be maintained, and preemptive solution modeling should be developed for potential disasters.

- First and foremost, emphasis should be placed on raising awareness and educational activities across all sectors of society, establishing awareness of governance concepts. In this context, the transparency and accountability of public authorities are crucial.

- Alternative and renewable energy sources that do not increase greenhouse gas emissions should be utilized to reduce the primary cause of global warming, which is greenhouse gas emissions.

- Areas that are most likely to be affected by climate change should be protected first.

- Measures should be taken to prevent flooding and inundation in coastal and river areas to protect specific geographical regions.

- Special protection plans should be developed for settlement areas where the population exceeds a certain threshold, in response to the effects of climate change.

- More effective and cleaner technologies should be used in all scales and types of design.

- Efforts to mitigate the effects of climate change should prioritize support for research and technological innovations, alongside enhancing transparency and elaborating on monitoring and reporting processes. Additionally, effective global sharing of developed technologies and acquired data is crucial for combating climate change comprehensively.

- Specific protection plans tailored to each economic sector sensitive to climate conditions, such as tourism, finance, agriculture, and industry, should be developed against climate change.

- Cities that produce the highest amounts of emissions should be encouraged to assume responsibility in the combat against climate change. In this context, the relevant units must address and complete the legal-administrative regulatory gaps in combating climate change.

- While taking precautions on the relevant issue, the phenomenon of social justice should not be ignored, and especially economically disadvantaged societies around the world should be prevented from suffering.

- As an alternative to carbon markets, an increasing carbon tax that does not exclude any country and aligns with countries' responsibilities in climate change should be implemented. This would raise the cost of fossil fuels, prompting countries to reduce greenhouse gas emissions to avoid bearing this cost.

In the combat against climate change, it is essential to acknowledge the notable failure to achieve a significant reduction in greenhouse gas emissions, especially in the reduction of emissions from greenhouse gases. The exception to this reality occurred during the COVID-19 pandemic lockdowns. Measures such as quarantine, curfews, and travel bans implemented by governments worldwide led to a global decrease in energy demand, resulting in a roughly 17% reduction in global daily CO_2 emissions by early April 2020 compared to 2019 levels (Le Quéré et al., 647). In China, the country with the highest greenhouse gas emissions, a 25% decrease in emissions related to energy use was observed during the two-month lockdown period. Similarly, in Europe, energy demand decreased by 14%, corresponding to a 39% reduction in greenhouse gas emissions (Evans, 2020). It is evident that the long-debated and elusive emission reduction trends, discussed on global platforms for decades, were accessed during the COVID-19 pandemic. This situation tragically underscores the need for national actors to review their policies and, more importantly, their sincerity in addressing climate change challenges.

REFERENCES

C40 Cities. (2024). *About C40*. Retrieved June 01, 2024, https://www.c40.org/about-c40/

Barber, B. (2013). *If mayors ruled the world – Dysfunctional nations, rising cities*. Yale University Press.

Davoudi, S., Mehmood, A., & Brooks, L. (2011). *The London climate change adaptation strategy: Gap analysis*. Global Urban Research Unit.

DiPeso, J. (2004). Climate change and the states. *Environmental Quality Management*, 13(3), 111–116. DOI: 10.1002/tqem.20010

Doğan, S., & Tüzer, M. (2011). Küresel iklim değişikliği ile mücadele: Genel yaklaşımlar ve uluslararası çabalar. *Istanbul Journal of Sociological Studies*, (44), 157–194.

Energy Cities. (2024). *Our vision*. Retrieved May 21, 2024, from https://energy-cities.eu/vision-mission/

Engel, K. H. (2015). EPA's clean power plan: An emerging new cooperative federalism? *Publius*, 3(45), 452–474. DOI: 10.1093/publius/pjv025

Engin, B. (2010). İklim değişikliği ile mücadelede uluslararası işbirliğinin önemi. *Sosyal Bilimler Dergisi*, (2), 71–82.

European Commission. (2024a). *Why a covenant of mayors?* Retrieved May 21, 2024, from https://eu-mayors.ec.europa.eu/en/about

European Commission. (2024b). *Signatories* Retrieved May 21, 2024, from https://eu-mayors.ec.europa.eu/en/signatories

Evans, S. (2020, April 9). *Analysis: Coronavirus set to cause largest ever annual fall in CO2 emissions*. Retrieved June 30, 2024, from https://www.carbonbrief.org/analysis-coronavirus-set-to-cause-largest-ever-annual-fall-in-co2-emissions/

Giddens, A. (2009). *The politics of climate change*. Polity Press.

Global Covenant of Mayors for Climate & Energy. (2023). *Urban catalysts – A local climate stocktake: The 2023 global covenant of mayors impact report*. Retrieved May 21, 2024, from https://www.globalcovenantofmayors.org/wp-content/uploads/2023/12/GCoM-2023-Global-Impact-report-2023_10.12.2023.pdf

Gordon, D. J. (2016). Lament for a network? Cities and networked climate governance in Canada. *Environment and Planning. C, Government & Policy*, 34(3), 529–545. DOI: 10.1177/0263774X15614675

Heyward, C. (2013). Situating and abandoning geoengineering: A typology of five responses to dangerous climate change. *PS, Political Science & Politics*, 46(01), 23–27. DOI: 10.1017/S1049096512001436

Institute for Local Government. (2024). *Climate action*. Retrieved May 15, 2024, from https://www.ca-ilg.org/climate-action

IPCC. (1996). *Climate change 1995 - Impacts, adaptations and mitigation of climate change: Scientific-technical analyses contribution of working group ii to the second assessment report of the intergovernmental panel on climate change*. Cambridge University Press. Retrieved July 3, 2024, from https://www.ipcc.ch/site/assets/uploads/2018/03/ipcc_sar_wg_II_full_report.pdf

IPCC. (2014). *Climate change 2014 synthesis report, contribution of working groups I, II and III to the fifth assessment report of the intergovernmental panel on climate change*. IPCC, Geneva, Switzerland. Retrieved April 9, 2024, from https://www.ipcc.ch/site/assets/uploads/2018/02/SYR_AR5_FINAL_full.pdf

IPCC. (2022). *IPCC's WG II Sixth assessment report -WGII AR6, Climate change 2022: Impacts, adaptation and vulnerability working group II contribution to the sixth assessment report of the intergovernmental panel on climate change* Cambridge University Press, Cambridge, UK and New York, NY, USA. Retrieved April 22, 2024, from https://report.ipcc.ch/ar6/wg2/IPCC_AR6_WGII_FullReport.pdfDOI: 10.1017/9781009325844

Klein, R. J., Schipper, E. L. F., & Dessai, S. (2005). Integrating mitigation and adaptation into climate and development policy: Three research questions. *Environmental Science & Policy*, 8(6), 579–588. DOI: 10.1016/j.envsci.2005.06.010

Le Quéré, C., Jackson, R. B., Jones, M. W., Smith, A. J., Abernethy, S., Andrew, R. M., De-Gol, A. J., Willis, D. R., Shan, Y., Canadell, J. G., Friedlingstein, P., Creutzig, F., & Peters, G. P. (2020). Temporary reduction in daily global CO2 emissions during the COVID-19 forced confinement. *Nature Climate Change*, 10(7), 647–653. DOI: 10.1038/s41558-020-0797-x

Letcher, T. M. (2019). *Climate Change: Observed Impacts on Planet Earth*. Elsevier.

Local Governments for Sustainability. (2024). *Our pathways, our approach*. Retrieved May 21, 2024, from https://iclei.org/our_approach/

Ministry of Foreign Affairs of the Republic of Turkey. (2024). *Kyoto protokolü*. Retrieved April 20, 2024, from https://www.mfa.gov.tr/kyoto-protokolu.tr.mfa

Ministry of Foreign Affairs of the Republic of Turkey. (2024). *Viyana sözleşmesi ve Montreal protokolü*. Retrieved April 20, 2024, from https://www.mfa.gov.tr/viyana-sozlesmesi-ve-montreal-protokolu.tr.mfa

Municipality Law No. 5393. (2005). Retrieved May 21, 2024, from https://www.mevzuat.gov.tr/mevzuatmetin/1.5.5393.pdf

Murphy, S. D. (2001). U.S. rejection of Kyoto Protocol process. *The American Journal of International Law*, 95(3), 647–650. DOI: 10.2307/2668508

NASA. (2024). *Carbon dioxide*. Retrieved June 29, 2024, from https://climate.nasa.gov/vital-signs/carbon-dioxide/?intent=121

Our World in Data. (2020). CO_2 emissions by sector. Retrieved July 5, 2024, from https://ourworldindata.org/grapher/co-emissions-by-sector?time=latest&facet=none

Paris Agreement. (2015). Retrieved June 30, 2024, from https://unfccc.int/sites/default/files/english_paris_agreement.pdf

Rabe, B. G. (2002). Greenhouse & statehouse: The evolving state government role in climate change. *Pew Center on Global Climate Change*. Retrieved May 12, 2024, from https://www.c2es.org/wp-content/uploads/2002/11/states_greenhouse.pdf

Safonov, G. (2019). *Social consequences of climate change - Building climate friendly and resilient communities via transition from planned to market economies*. Retrieved May 5, 2024, from https://library.fes.de/pdf-files/id-moe/15863.pdf

Schreurs, M. A. (2004). The climate change divide: The European Union, The United States, and the future of the Kyoto Protocol. In Vig, N. J., & Faure, M. G. (Eds.), *Green giants? Environmental policies of the United States and the European Union* (pp. 207–230). MIT Press. DOI: 10.7551/mitpress/3363.003.0014

Tuğaç, Ç. (2020). Dünyada ve Türkiye'de iklim değişikliği politikaları. In Sağır, H. (Ed.), *Ekolojik kriz ve küresel çevre politikaları* (pp. 221–264). Beta.

Ulueren, M. (2001). Küresel ısınma, BM iklim değişikliği çerçeve sözleşmesi ve Kyoto Protokolü. *T.C. Dışişleri Bakanlığı Uluslararası Ekonomik Sorunlar Dergisi, 3,* Retrieved May 12, 2024, from https://www.mfa.gov.tr/kuresel-isinma-bm-iklim-degisikligi-cerceve-sozlesmesi-ve-kyto-protokolu.tr.mfa

UNFCCC. (1992). *United nations framework convention on climate change*. UN. Retrieved April 20, 2024, from https://unfccc.int/resource/docs/convkp/conveng.pdf

UNFCCC. (1997). *Kyoto protocol to the united nations framework convention on climate change*. UN. Retrieved April 20, 2024, from https://unfccc.int/documents/2409

UNFCCC. (2024a). *The Paris agreement. What is the Paris agreement?* UN. Retrieved April 27, 2024, from https://unfccc.int/process-and-meetings/the-paris-agreement

UNFCCC. (2024b). *What is the Kyoto protocol?* UN. Retrieved April 20, 2024, from https://unfccc.int/kyoto_protocol

Uysal Oğuz, C. (2010). İklim değişikliği ile mücadelede yerel yönetimlerin rolü: Seattle örneği [The role of local governments in fighting with climate change: The example of Seattle]. *Yönetim ve Ekonomi Dergisi*, 17(2), 25–41.

Vezirgiannidou, S. E. (2013). Climate and energy policy in the United States: The battle of ideas. *Environmental Politics*, 4(22), 593–609. DOI: 10.1080/09644016.2013.806632

World Economic Forum. (2022). *BiodiverCities by 2030: Transforming cities' relationship with nature*. Retrieved April 30, 2024, from https://www3.weforum.org/docs/WEF_BiodiverCities_by_2030_2022.pdf

World Health Organization. (2009). *Global health risks - Mortality and burden of disease attributable to selected major risks* (Technical report).

ADDITIONAL READING

Gates, B. (2021). *How to avoid a climate disaster: The solutions we have and the breakthroughs we need*. Knopf.

IPCC. (2021). *Special Report on the Ocean and Cryosphere in a Changing Climate*. SROCC.

IPCC. (2023). *Sixth Assessment Report (AR6)*.

Johnson, A. E., & Wilkinson, K. K. (2020). *All we can save: Truth, courage, and solutions for the climate crisis*. One World.

Klein, N. (2019). *On fire: The burning case for a green new deal*. Simon & Schuster.

Mann, M. E. (2021). *The new climate war: The fight to take back our planet*. Public Affairs.

Xu, Y., & Ramanathan, V. (2017). Well below 2°C: Mitigation strategies for avoiding dangerous to catastrophic climate changes. *Proceedings of the National Academy of Sciences of the United States of America*, 114(39), 10315–10323. DOI: 10.1073/pnas.1618481114 PMID: 28912354

KEY TERMS AND DEFINITIONS

Adaptation: Adjustments made to natural or human systems in response to actual or expected climate change, aiming to reduce harm or exploit beneficial opportunities.

Biodiversity Conservation: The protection and management of ecosystems, species, and genetic diversity to maintain their resilience and adaptability in the face of climate change impacts.

Carbon Footprint: The total amount of greenhouse gases, primarily carbon dioxide, released directly or indirectly by human activities, usually measured in equivalent tons of carbon dioxide (CO_2).

Carbon Pricing: A policy tool that puts a price on carbon emissions to incentivize polluters to reduce greenhouse gas emissions and transition to cleaner alternatives. This can be done through carbon taxes or cap-and-trade systems.

Climate Finance: Financial resources mobilized to support projects, initiatives, and policies that aim to mitigate and adapt to climate change, often involving public and private sector investments.

Climate Resilience: The ability of a system or community to anticipate, prepare for, respond to, and recover from the impacts of climate change.

Decarbonization: The process of reducing or eliminating carbon dioxide emissions from energy sources, industrial processes, transportation, and other human activities.

Emission Reduction Targets: Specific goals set by governments or organizations to reduce greenhouse gas emissions by a certain percentage over a defined period, contributing to global climate action.

Mitigation: Actions taken to reduce or prevent the emission of greenhouse gases, thus lessening the severity of climate change impacts.

Sustainable Development: Development that meets the needs of the present without compromising the ability of future generations to meet their own needs, often integrating economic, social, and environmental considerations.

ENDNOTES

[1] To gain extensive information about the natural factors causing climate change and the impacts of climate change on natural systems, see: Letcher, T. M. (2019). *Climate Change: Observed Impacts on Planet Earth*. Elsevier.

[2] "Water vapour (H_2O), carbon dioxide (CO_2), nitrous oxide (N_2O), methane (CH_4) and ozone (O_3) are the primary GHGs in the Earth's atmosphere. Human-made GHGs include sulphur hexafluoride (SF_6), hydrofluorocarbons (HFCs), chlorofluorocarbons (CFCs) and perfluorocarbons (PFCs); several of these are also O_3-depleting" (IPCC, 2022, p. 2911).

Chapter 12
Turkey's Policies to Combat Climate Change

Hande Saraçoğlu
https://orcid.org/0000-0002-1911-2032
Van Yüzüncü Yıl Üniversitesi, Turkey

Hande Saraçoğlu
Van Yüzüncü Yıl Üniversitesi, Turkey

ABSTRACT

Climate change, the most important environmental problem that concerns the whole world, is one of the biggest common problems facing humanity today. In this chapter, Turkey's policies to combat climate change will be examined in depth, and the country's trends in greenhouse gas emissions, efforts towards energy transformation, renewable energy investments, policies on forestry, and approaches to climate justice will be discussed. In addition, the challenges Turkey faces in combating climate change and suggestions for solutions to these challenges will be discussed. Thus, Turkey's current situation and future goals in combating climate change will be better understood.

INTRODUCTION

Human beings have been living intertwined with nature since the beginning of time and benefit from the resources offered by nature. By consuming these seemingly infinite resources, human beings cause irreparable environmental problems. Climate change has become one of the most important and urgent environmental problems all over the world. With the increase in greenhouse gas emissions in the atmosphere, climate patterns in our world are changing, ecosystems are damaged and air quality

DOI: 10.4018/979-8-3693-5792-7.ch012

is negatively affected. Increasing temperatures, rising sea levels, frequent extreme weather events and many other effects threaten life on a global scale. Factors such as the use of fossil fuels, industrial processes, agricultural practices and deforestation cause high levels of carbon dioxide, methane and other greenhouse gases to be released into the atmosphere. Thus, the sun's rays are trapped in the atmosphere, causing our planet to warm up. Therefore, combating climate change has become one of the top priorities of the international community. Combating climate change is important not only to protect the environment but also to ensure the well-being of societies and the sustainability of the planet.

Climate change, which is the most important environmental problem concerning the whole world, is one of the biggest common problems facing humanity today. Turkey plays an important role in combating climate change with its geographical location, climate diversity and natural resources. Turkey's policies to combat climate change are of great importance both in terms of protecting national interests and contributing to global goals. In Turkey, various policies are being developed to combat this problem and important steps are being taken with these policies. These include increasing energy efficiency, promoting renewable energy sources, protecting forests, reducing greenhouse gas emissions and strengthening infrastructure to combat climate change. Turkey also contributes to the fight against climate change within the framework of international agreements and protocols. Turkey's policies to combat climate change are developed through research and projects conducted by environmental protection foundations, universities and other institutions. Important steps are being taken with these combat policies. The development and implementation of these policies is of great importance for environmental protection and sustainability. Turkey's policies to combat climate change have an important share not only nationally but also internationally. The country's economic structure, geopolitical position, and energy requirements have a decisive impact on shaping the policies adopted to combat climate change. Turkey is a party to international agreements such as the Kyoto Protocol, the United Nations Environmental Convention on Climate Change and the Paris Agreement. In addition to these, it develops and implements its own unique strategies.

In this section, Turkey's policies to combat climate change will be examined in depth, covering the country's trends in greenhouse gas emissions, efforts towards energy transition, renewable energy investments, forestry policies and approaches to climate justice. In addition, Turkey's challenges in combating climate change and solutions to these challenges will be discussed. In this way, Turkey's current situation and future goals in combating climate change will be better understood.

1. CLİMATE CHANGE AS A GLOBAL PROBLEM

Climate change, driven primarily by human activities, is one of the most pressing global challenges of our time. The Earth's climate system is undergoing significant alterations, manifested in rising temperatures, shifting weather patterns, and increasing frequency of extreme events. This phenomenon is largely attributed to the accumulation of greenhouse gases (GHGs) in the atmosphere, a consequence of industrialization, deforestation, and the burning of fossil fuels. The scientific consensus on climate change is overwhelming. The Intergovernmental Panel on Climate Change (IPCC) has consistently reported that human activities are the dominant cause of observed warming since the mid-20th century. The primary greenhouse gases responsible for this warming are carbon dioxide (CO2), methane (CH4), and nitrous oxide (N2O). These gases trap heat in the Earth's atmosphere, leading to the "greenhouse effect," which results in a gradual increase in global temperatures. The impacts of climate change are already being felt across the globe. Rising global temperatures have led to the melting of polar ice caps and glaciers, contributing to sea level rise. Coastal communities are particularly vulnerable, as higher sea levels increase the risk of flooding and erosion. Small island nations and low-lying regions face the prospect of losing significant portions of their landmass, displacing millions of people and creating climate refugees. In addition to rising sea levels, climate change is also responsible for more frequent and intense weather events. Hurricanes, typhoons, and cyclones are becoming more powerful due to the increased energy in the atmosphere. Heatwaves are occurring with greater frequency and severity, leading to health crises, particularly among vulnerable populations such as the elderly and those with pre-existing health conditions. Moreover, droughts and changing precipitation patterns are disrupting agricultural systems, threatening food security and livelihoods around the world. The consequences of climate change are not limited to environmental and economic impacts; they also have profound social and political implications. As resources become scarcer due to changing climatic conditions, competition for water, arable land, and other essentials is likely to intensify, potentially leading to conflicts. Migration patterns are also expected to shift as people move away from areas that are becoming uninhabitable due to rising temperatures or sea levels. This can create tensions in receiving areas, as they grapple with the challenges of integrating large numbers of newcomers. Addressing climate change requires a coordinated global response. The Paris Agreement, adopted in 2015 under the United Nations Framework Convention on Climate Change (UNFCCC), represents a significant step in this direction. The agreement aims to limit global warming to well below 2°C above pre-industrial levels, with efforts to keep it to 1.5°C. Achieving these targets necessitates a rapid reduction in greenhouse gas emissions and a transition to a low-carbon economy. This includes increasing the use of renewable energy sources, improving energy efficiency, and promoting sustainable land use practices. However, progress towards these goals has been uneven. While some countries have made significant strides in reducing their emissions and transitioning to renewable energy, others continue to rely heavily on fossil fuels. Moreover, the financial and technological resources needed to combat climate change are not equally distributed, with developing nations often lacking the means to implement necessary measures. This disparity underscores the importance of international cooperation and support for climate action in less developed regions. As a result of industrialization and urbanization that has been experienced since the 1700s, there has been a rapid increase in the demand for natural resources, and many problems such as the destruction of natural resources, pollution, desertification and the decrease in biodiversity have been encountered in order to meet the increasing demand. Although all of them are important, the most important problem that we can consider as "climate change" is again the most important outcome of this process. The use of fossil fuels in urban settlements and energy systems used in industrial production has caused and continues to cause high levels of greenhouse gas emissions into the atmosphere. When these gases (carbon dioxide (CO_2), methane (CH_4), nitrogen oxide (N_2O), ozone (O_3) and fluorinated gases) are emitted into the atmosphere, many of them remain in the atmosphere for many years, up to thousands of years, and as more of the gases are added to the atmosphere, more heat is trapped. This extra heat causes higher air temperatures near the earth's surface, changes weather patterns and increases the temperature of the oceans (Kahraman ve Şenol, 2018: 353-370). Although it was towards the end of the 20th century that countries around the world accepted that climate change was an international issue and began various efforts in this direction, it is possible to date the beginning of scientific interest in the concept back to the 19th century (Göçoğlu, etc., 2023: 626-648). It can be said that global cooperation efforts on climate change intensified especially in the 1970s, based on action and thought. The establishment of Greenpeace, the emergence of environmental parties in various countries, and the United Nations Environment Programme are seen to have emerged in these years (Kaya, 2020: 165-191). Climate change is one of the most pressing and complex challenges facing the global community today. The unprecedented rise in average global temperatures, driven by human-induced greenhouse gas emissions, is leading to a host of far-reaching environmental, social, and economic consequences. From melting glaciers and rising sea levels to more frequent and severe weather events, the impacts of climate change are being felt across the planet. Addressing this global problem will require coordinated, large-scale action from nations, businesses, and individuals alike to transition to renewable energy sources, implement sustainable practices, and build resilience in vulnerable communities. Time is of the essence, as the window to mitigate the worst effects of climate change is rapidly closing.

Innovative solutions, bold policymaking, and a shared sense of urgency and responsibility will be essential to tackling this existential threat to the future of our world.

2. PRACTICES TO COMBAT CLIMATE CHANGE IN TURKISH LEGISLATION

With environmental problems gaining an international dimension, taking steps from local to global in this field and developing a multilateral cooperation system at various levels among countries has been continuing rapidly since the 1970s (Göçoğlu, et al., 2023: 627-630). Turkey's Policies to Combat Climate Change The devastating effects of climate change, which has been caused by industrialization and the spread of human-induced greenhouse gases into the atmosphere for years, have affected not only developing countries but also developed countries. At this point, the necessity to take measures has reached an alarm level. Developed, developing or undeveloped countries in different geographies of the world have been exposed to this disaster. This negative climate change, which affects the lives of the world population of 7.5 billion people, is a global problem that leads to serious environmental and socio-economic consequences (Paksoy, 2019: 155-160). Most of the environmental problems such as the greenhouse effect, ozone depletion, desertification, pollution of the seas and oceans, and deforestation are international in nature and have led the countries of the world to produce new policies, take measures and determine various strategies in terms of sustainability (Göçoğlu, et al., 2023: 627-630). Turkey's taking measures against climate change will provide many environmental, economic and social benefits. Moreover, the steps taken to combat climate change will not only be effective in combating climate change, but will also contribute to a sustainable future. The United Nations Environment Program was established in 1972, following the Environment Conference held in Stockholm, to carry out activities on the environment within the United Nations. The first concrete outputs of the program were the Vienna Convention and the Montreal Protocol signed in 1987, which aimed to protect the ozone layer. In the following years, the First World Climate Conference held by the World Meteorological Organization in 1979 could be considered as another international effort against climate change. The second of this conference was held in 1990. In 1988, the Intergovernmental Panel on Climate Change (IPCC), established by the UN and the World Meteorological Organization, could be considered one of the largest organizations on the subject of climate change on an international scale. The Rio Summit in 1992 clarified the institutional foundations of these efforts with the United Nations Framework Convention on Climate Change (UNFCCC), which was adopted in 1992 and entered into force on 21 March 1994. The Convention is an international agreement that sets general goals and rules to combat climate change. The Kyoto Protocol, which was adopted at the UNFCCC Third Parties Conference held in Kyoto in 1997, entered into force on 16 February 2005 as one of the most important steps that made the Convention concrete. The Rio Summit in 1992 clarified the institutional foundations of these efforts with the United Nations Framework Convention on Climate Change (UNFCCC), which was adopted in 1992 and entered into force on March 21, 1994. The Convention is an international agreement that sets general goals and rules to combat climate change. The Kyoto Protocol, which was adopted at the UNFCCC Third Parties Conference held in Kyoto in 1997, entered into force on February 16, 2005 as one of the most important steps that made the Convention concrete. It should be emphasized that the report prepared by the IPCC in 1990 had a significant impact on the adoption of the UNFCCC, and the report prepared in 1995 had a significant impact on the signing of the Kyoto Protocol. Thus, it is possible to say that the IPCC has become the most important authority in the world on climate change. Turkey joined the convention as the 189th country on May 24, 2004, and became a party to the Kyoto Protocol on August 26, 2009. Therefore, Turkey has now become one of the actors in the national fight against climate change (Doğan and Tüzer, 2011: 157-194). The increase in greenhouse gases, which is one of the biggest problems that paves the way for the climate crisis worldwide, has led countries to make a new, comprehensive and effective agreement on a global scale. Developed countries that have completed their industrialization without encountering any obstacles regarding greenhouse gas emissions so far, and developing countries that are still involved in this process and are currently increasing their greenhouse gas emissions in order to advance their industrialization, have become the main parties to the Paris Agreement signed in 2015. It is known that the agreement has been accepted by 195 countries. The Paris Agreement generally stipulates the following articles. The main ones are as follows:

Underdeveloped countries that will be greatly affected by climate change should be supported by other countries,

- ✓ As a result of global warming reaching 1° degree in time after the industrial revolution, the target should be to keep global warming below 2 degrees,
- ✓ While developed countries are expected to make definite emission reductions, developing countries should make reductions in accordance with their own capacities,
- ✓ Developed countries should provide technological, financial, etc. support to developing countries so that they can act sensitively to the climate in their development processes,
- ✓ It is aimed for countries to make more emission reductions every five years compared to the previous period.

According to the agreement, each country has common but differentiated responsibilities and relative capabilities. Countries that sign the agreement submit National Contribution Declarations every 5 years, and thus the situation can be assessed. Turkey, which signed the agreement on April 22, 2016, aimed to keep greenhouse gas emissions 21% less than the projected number in 2030 in the plan it presented. One of the developments that gave the Paris Agreement an important mission was undoubtedly the fact that the United States, which did not sign a protocol for greenhouse gas reduction in the Kyoto Protocol, and China, which again did not have a greenhouse gas reduction requirement, started to reduce greenhouse gas emissions thanks to this agreement (Karakaya, 2016: 1-12). Turkey, in its Climate Change Strategy 2010-2023 report, under the title of "Our National Attitude in the Context of International Climate Change Negotiations", states that our country is in the category of "middle-income developing countries" and that Turkey envisages to fulfill its duties in combating global climate change based on the common but differentiated responsibility sharing principle, which is the fundamental principle of the Convention, and the following issues within its means (Türkiye İklim Değişikliği Stratejisi, 2010)

- ✓ Turkey plans to implement emission limitation through measures that will not negatively affect its sustainable development and poverty reduction efforts. In addition, Turkey declares that it will carry out mitigation activities in accordance with its national programs and strategies in a measurable, reportable and verifiable manner.
- ✓ It has implemented many policies and measures in the energy, agriculture, forestry, transportation, industry and waste sectors in line with combating climate change, especially through development plans and many national plans, programs and strategy documents. It also wishes to contribute more to international efforts on this issue within the framework of its opportunities and potential.
- ✓ Turkey aims to continue its emission reduction actions and efforts to adapt to climate change by also benefiting from the financing and technology transfer opportunities provided to countries with similar economic development levels.

Turkey's Development Plan for the years 2019-2023 (Development Plan, 2019) includes steps regarding our fight against climate change under the heading of "environment". The prominent issues among these steps are continuing international negotiations on climate change, combating climate change in sectors that cause harmful gas emissions to the extent that national conditions allow, making the economy and society more resilient to climate change risks with capacity increase, developing emission control in various sectors, carrying out planning implementation and development studies for national and regional adaptation strategies regarding capacity increase on the subject, determining regional and city-level needs for adaptation to climate change and taking the necessary measures, developing solution proposals, and preparing Climate Change Action Plans in seven regions. On the other hand, in the context of our steps regarding combating climate change in our country, it is an important development to make an institutional change in addition to the priorities listed above. With the Presidential Decree No. 85 published in the Official Gazette dated October 29, 2021 and numbered 31643, the name of the Ministry of Environment and Urbanization was changed to the Ministry of Environment, Urbanization and Climate Change.

2.1. Energy Related Laws

Through energy-related laws, Turkey has taken steps towards sustainability and reduction of carbon emissions in the energy sector.

- ✓ Energy Efficiency Law; energy efficiency means that each unit of energy consumed is transformed into more services and products. In addition, energy efficiency is important for increasing the standard of living and service quality in buildings and reducing energy consumption in industrial enterprises without reducing the quality and quantity of production (Koçaslan, 2014: 117-133). The energy efficiency law entered into force in 2007 in order to reduce the economic burden, to use energy effectively and efficiently in order to protect the environment and to prevent waste. This law generally includes the formation of the administrative structure for the effective execution of energy efficiency activities, authorizations to be made at the point of carrying out energy efficiency activities, training and awareness raising of individuals in the society, mechanisms for increasing the use of renewable energy resources, issues related to incentives, and fines to be imposed on those who do not act in accordance with the law (Keskin, 2007: 106-112). The law is important in terms of optimizing energy use and using resources more effectively and efficiently.
- ✓ Law on the Utilization of Renewable Energy Resources for the Purpose of Electricity Generation; With this law, which entered into force in 2005, it is aimed to provide resource diversity and thus reduce greenhouse gas emissions rather than using fossil resources in electricity generation. The law was established to encourage the use of renewable energy resources, support renewable energy projects and increase the share of renewable energy resources in the energy mix (Kaplan, 2023: 297-336). The fact that renewable energy sources

offer a cleaner method of energy production compared to fossil fuels plays an important role in combating climate change by reducing air and water pollution. In addition, since the shift towards renewable energy sources will reduce external energy dependence, this will increase energy security. The use of renewable energy sources also has a great share in leaving a healthier environment for future generations.

- ✓ Law on Geothermal Resources and Natural Mineral Waters; This law, enacted in 2007, aims to ensure the sustainable use of geothermal resources and natural mineral waters and to reduce environmental impacts. The Law includes provisions on the exploration, research, development, production and protection of natural mineral waters and geothermal resources. Geothermal resources and natural mineral waters, which essentially belong to the state, are granted to private persons in exchange for licenses (Ömercioğlu, 2023: 159-173). The Law is important in terms of ensuring the sustainable use of geothermal resources and natural mineral waters, preventing the overuse of resources and encouraging their efficient use.

2.2. Laws Related to Fossil Fuels

- ✓ Turkish Petroleum Law; With the law adopted in 2013, it is aimed to ensure the exploration, development and production of petroleum resources in a fast, continuous and effective manner in accordance with national interests. The person who will have the right to petroleum must establish facilities and equipment within the framework of the license in a way that will not cause any harm to nature and the environment and will not cause any harm to the life of the people living in the region. In addition, while carrying out petroleum exploration and extraction activities, importance should be given to the protection of the quantity and quality of surface, underground, coastal and marine waters (Talu and Kocaman, 2018: 21). It is seen that the Law contains various provisions to promote the principle of sustainability in the energy sector.
- ✓ Mining Law; This law, which was adopted in 1985, includes procedures for the exploration, exploitation and ownership of rights on mines in a way that does not contradict national interests. In mining, harmony with the environment should not be ignored. In addition, the law stipulates that the necessary permits must be obtained from forests and protected areas, activities that may affect drinking and utility water resources must not be carried out, and the environment and human health must not be harmed (Yıldız, 2012: 36-42). This law has an important role in the effective and efficient use of natural resources and minimization of environmental impacts.

2.3. Laws Related to Agriculture

✓   Law on Agriculture; this law, adopted in 2006, outlines the framework of agricultural policies in Turkey. Climate change brings along many factors that negatively affect the agricultural sector. Increasing temperatures, irregular rainfall, drought, flooding, erosion are among these factors. The Law on Agriculture includes references to the protection of the environment, biodiversity and ecosystems, the development of both land and water resources, and the fight against natural disasters. With this law, sustainability in production and development and sensitivity to human health and the environment are adopted as principles (Talu and Kocaman, 2018: 22). The agricultural sector plays a critical role in combating climate change. Sustainable agricultural practices, reduction of carbon emissions, protection of soil health and effective and efficient use of water resources can be effective in combating climate change and thus a sustainable agricultural system can be created.

✓   Law on Soil Conservation and Land Use; This law, adopted in 2005, includes provisions on the protection and development of soil and the planned and efficient use of agricultural areas in accordance with sustainable development, with priority given to the environment. Soils that are not used in a sustainable manner become a source of carbon dioxide and nitrogen emissions, thus increasing the concentration of these gases in the atmosphere. Therefore, policies to both protect and increase soils such as green areas and forests in spatial plans are of great importance in combating climate change (Mutlu and Tezer, 2023: 305-324). In other words, this law has objectives such as protecting the sustainability and productivity of agricultural lands, preventing erosion, protecting water resources and ensuring the protection of biodiversity. In this way, the agricultural sector can become more resilient in combating climate change and sustainable agricultural practices can be developed.

3. POLICIES TO FIGHT CLIMATE CHANGE ON A GLOBAL DIMENSION

The issue of combating climate change is of great importance all over the world. Türkiye has become a party to international agreements on this issue and has supported the fight through various mechanisms. Turkey's fight against climate change has progressed rapidly in recent years, and Turkey has taken various steps in the fight against climate change by participating in international agreements and making strategic plans. In the international community's fight against climate change, climate change negotiations are held and these negotiations are organized in cooperation to reduce greenhouse gas emissions and combat climate change. The Paris Climate Summit, also known as the 21st Conference of the Parties, Turkey's position within the climate change regime has always been painful since the Rio Meeting held in 1992, and after years of unsuccessful negotiations, it resulted in a historic global agreement. Türkiye approved the Paris Agreement in the Turkish Grand National Assembly on October 7, 2021, regarding the fight against climate change. This agreement aims to reduce greenhouse gas emissions together with international cooperation (İzol and Kaval, 2023: 18-32). Many countries have made commitments to targets such as reducing greenhouse gas emissions, increasing the use of renewable energy, and increasing the sustainability of forests through international agreements such as the Paris Agreement. We see that Turkey's policies to combat climate change have started since the 2000s and these policies have intensified since 2009 with the approval of the Kyoto Protocol in the parliament. For the National Climate Change Strategy and Action Plan in 2010 and 2011, Turkey became a party to the Kyoto Protocol in 2009. The Kyoto Protocol is an agreement that aims to provide an international framework on climate change. The main goal of the Kyoto Protocol is to enable countries to reduce the amount of carbon they emit into the atmosphere to the levels in 1990. This protocol has been accepted as an important step in the fight against global warming and climate change. In the Kyoto Protocol, it is stated that the amount of greenhouse gases released into the atmosphere should be reduced to 5%, legislation should be rearranged to reduce greenhouse gas emissions from sources such as industry, motor vehicles and heating, encouraging the use of vehicles and technology systems that consume less energy, solar energy, nuclear energy. Some important provisions are included in this protocol, such as encouraging the trend towards environmentally friendly energy sources such as, reorganizing waste processes in high energy-consuming processes, and encouraging the use of renewable energy sources instead of fossil fuels (Şahin, 2016: 5-10). The Kyoto Protocol offers three different mechanisms, called the Flexibility Mechanism, namely the flexibility mechanism, the clean development mechanism and the joint execution mechanism. In order to achieve their goals, this protocol allows the parties to work to reduce greenhouse gas emissions outside their countries. The flexibility mechanism of the Kyoto Protocol allows countries to use different ways to reduce greenhouse gas emissions. This mechanism; It consists of three different tools: emission trading, common implementation and clean development mechanism. Emission trading, which enables the buying and selling of emission quotas between countries in order to reduce greenhouse gas emissions, is based on the principle of purchasing the quotas of countries that emit more. The joint implementation project, which aims to promote technology transfer between countries that are parties to the Kyoto Protocol, is carried out in countries where joint efforts are made to reduce greenhouse gas emissions. The Clean Development Mechanism, which is a part of the Kyoto Protocol, aims to provide financing for developing sustainable development projects and encourages developed countries to invest in projects in developing countries to reduce greenhouse gas emissions. These projects are carried out in areas such as clean energy production, energy efficiency and prevention of deforestation. The Joint Execution Mechanism, which is one of the Flexibility Mechanisms developed to be used in reducing human-induced greenhouse gases, is similar to the Clean Development Mechanism in many respects. These projects, which are carried out to encourage technology transfers between the countries that are parties to the Kyoto Protocol, are designed to reduce the greenhouse gas emissions of developed countries. It includes activities such as technology transfer and joint investments. These mechanisms aim to promote cooperation and technology transfer between countries to achieve the goals of the Kyoto Protocol (Engin, 2019:47-67).

3.1. United Nations Framework Convention on Climate Change

The United Nations Framework Convention on Climate Change (UNFCCC) is a vital international treaty that provides the foundation for global efforts to address the threat of climate change. Adopted in 1992, the UNFCCC establishes a framework for intergovernmental cooperation and action to stabilize greenhouse gas concentrations in the atmosphere at a level that would prevent dangerous human-induced interference with the climate system. The importance of the UNFCCC lies in several key aspects. Firstly, its universality is remarkable, with 197 countries as Parties, making it one of the most widely ratified international agreements and providing a truly global platform to coordinate climate action. Secondly, the UNFCCC outlines fundamental principles, such as common but differentiated responsibilities, that guide the equitable distribution of efforts among countries, with the ultimate objective of achieving stabilization of greenhouse gas concentrations to prevent dangerous interference with the climate system. Thirdly, the UNFCCC creates a formal framework for countries to regularly meet, share information, and negotiate further agreements and protocols to strengthen the global response to climate change, such as the Kyoto Protocol and the Paris Agreement. Fourthly, the UNFCCC has been instrumental in catalyzing national climate policies, emission reduction targets, and the mobilization of climate finance to support mitigation and adaptation efforts in developing countries. Lastly, the UNFCCC recognizes the importance of science in guiding climate action and mandates the use of the best available scientific knowledge through the Intergovernmental Panel on Climate Change (IPCC). Overall, the UNFCCC's universal participation, foundational principles, and framework for ongoing cooperation and policy development make it a crucial instrument in the global fight against the existential threat of climate change. Türkiye became a party to the United Nations Framework Convention on Climate Change in May 2004. The Convention came to the agenda at the Rio Conference and entered into force in 1992. The purpose of the agreement is to prevent greenhouse gas accumulations in the atmosphere and the dangerous human-induced impact on the climate system (Dağdemir, 2015: 49-70). With this agreement, the biggest step has been taken in combating climate change, which forms the basis of ongoing work in the international arena. Thus, it became the first global agreement to explicitly address climate change. The agreement aimed to reduce both carbon dioxide and other greenhouse gas emissions to 1990 levels by the year 2000. This agreement has the nature of a framework text. In other words, the basic principles of the contract have been regulated, but its details are left to the decisions taken together with the parties (Can, 2023: 170-172). The parties come together through annual meetings starting from the date of entry into force of the contract. The aim of these meetings is to accelerate the implementation of the agreement and to mutually discuss the best way to address the problem of climate change. All countries that are parties to, accept or participate in the agreement are responsible for fulfilling the relevant obligations (Akyel, 2009: 137-142). The obligations that the party countries will undertake within the scope of the agreement are included in the annexes. The annexes with the names of the countries contain the most important details of the contract. When the contract was written, the Warsaw Pact was dissolved, so the contract included three main country groups: Annex I, Annex II, and Non-Annex I. Countries included in the Annex I list are industrial countries with historical responsibility. They are obliged to prepare both detailed and frequent inventories of greenhouse gas emissions and to develop programs for reduction and adaptation. The countries in the Annex II list are developed countries with fiscal responsibility. These countries have been held responsible for both reducing greenhouse gas emissions and providing economic and technical support to developing countries in combating and adapting to climate change. Countries included in the Non-Annex I list are countries that do not have obligations such as both reduction and financial responsibility (Can, 2023: 170-172). It is seen that this framework agreement supports countries in issues such as reducing greenhouse gas emissions, combating the effects of climate change and adapting to climate change. In addition, the convention encourages global cooperation by bringing together countries around the world on this issue. It supports countries in increasing their resilience to the effects of climate change and developing adaptation strategies. In this context, the United Nations Environmental Convention has a key importance in combating climate change and has undertaken an important task in coordinating the joint efforts of countries.

3.2. Kyoto Protocol

Obligations to reduce greenhouse gas emissions after 2000 are regulated by the Kyoto Protocol. According to the Kyoto Protocol, countries included in the Annex I list are responsible for reducing their greenhouse gas emissions to at least 5% below the level in 1990 between 2008 and 2012 (Türkeş, 2006: 99-107). The Kyoto Protocol took this name because it was negotiated in Kyoto, Japan, in December 1997. However, the Protocol entered into force in 2005. The fact that the Protocol was opened for signature in 1997 and entered into force in 2005 suggested that it failed to solve many important problems. It has been argued that the failure of the Kyoto Protocol was due to the low targets set, the fact that the USA is not a party to the Protocol, and the fact that countries with high greenhouse gas emission rates do not have reduction targets because they are on the Non-Annex I list (Can, 2023: 173-175). The main aim of this protocol is to reduce the emission values of six greenhouse gases, namely carbon dioxide, nitrogen, methane, sulfur hexafluoride, perfluorocarbons (PFC) and hydrofluorocarbons (HFC) between 2008 and 2012. The Kyoto Protocol entered into force on February 16, 2005, and by December 2006, 169 countries had become parties to the protocol (Özmen, 2009: 42-46). The Kyoto Protocol is the only international framework document to ensure a joint fight against both global warming and climate change. In order for the protocol to achieve its planned goals, three market-oriented mechanisms have been put into effect (Özcan, 2020: 169-184);

- ✓ International carbon trading; This mechanism is a system that restricts or charges greenhouse gas emissions. Carbon trading is the name of an economic system created to encourage reducing carbon emissions into the atmosphere. Essentially, in this mechanism, companies or countries can emit a certain amount of carbon. If companies or countries emit below the determined rate, they can obtain extra carbon credits and sell these existing credits to other companies or countries in need. This mechanism, also known as emissions trading, allows the transfer of permitted emissions from one side to another. In the carbon trading system, a certain carbon emission quota is created for each country or company, then their carbon emissions are regularly measured and verified, parties that use emissions below the determined quota can sell the excess credits they have obtained, and these credits are received through a market mechanism. It is then sold. The existence of five basic elements such as measurement, transparency, responsibility, transferability or tradeability and finally consistency are very important in the functioning of the mechanism (Türkeş, et al. 84-100). The inclusion of carbon trading in the Protocol has led to countries whose emissions have decreased due to their economies contracting, selling this right they have, resulting in additional greenhouse gas emissions into the atmosphere. Additionally, by purchasing the rich people's right to pay with money, it has negatively affected environmental justice and equality. This mechanism has led to the establishment of the understanding that if I pay, I can pollute (Selçuk, 2023: 9-19).

- ✓ Joint execution; A project on emission reduction can be carried out jointly by an Annex I country in another Annex I country. Thanks to these projects, the homeowner who is successful in reducing emissions earns an emission reduction credit and the investor has the opportunity to sell this credit to another Annex I country. This makes it possible to work jointly and share resources to achieve emission reduction targets. The joint enforcement mechanism aims to promote cooperation and technology transfer in order to achieve the goals of the Kyoto Protocol (Binboğa, 2014: 5732-5759). Joint implementation can make it easier for participating countries to achieve the goals they have set, while also contributing to the reduction of global greenhouse gas emissions.
- ✓ Clean development mechanism; A country with a certain emission target enters into cooperation with a country whose emission target is not certain, and by developing projects in line with the issue of reducing greenhouse gas emissions in that country, it becomes entitled to receive "Certified Emission Reductions" and is deducted from the total target (Karakaya and Özçağ, 2001: 1-7). The success of the clean development mechanism seems possible with stakeholder participation. With the participation of stakeholders, the project can be more effective and sustainable.

Türkiye has not been able to participate in the meetings since it has not been a party to the Kyoto Protocol since 2005, when it came into force. In order for Turkey to have a say at this table, the "Draft Law Concerning Our Participation in the Kyoto Protocol to the United Nations Climate Change Environmental Convention" was adopted on February 5, 2009. However, since Turkey was not a party to the United Nations Framework Convention on Climate Change in 1997, the year the Kyoto Protocol was adopted, it was not included in the Annex-B list of the Protocol. In other words, there is no commitment to limit or reduce emissions (Binboğa, 2014: 5732- 5759). Turkey's membership of the Kyoto Protocol is important in terms of encouraging international cooperation in the fight against climate change. Turkey's becoming a party to this protocol enables it to cooperate with other party countries and have the opportunity to find common solutions. Turkey's becoming a party to the Protocol is important in reducing and controlling greenhouse gas emissions. In addition, by becoming a party to the Protocol, Turkey has shown that it plays an active role in the fight against climate change in the international arena, and Turkey has had the opportunity to receive support from international sources and use new technologies.

3.3. Paris Agreement

A new agreement was needed because the obligation to reduce emission rates mentioned in the Paris Agreement and the Kyoto Protocol will expire in 2020. Therefore, it was unanimously accepted by the countries that are parties to the United Nations Framework Convention on Climate Change at the 21st conference of the parties held in Paris in 2015. Türkiye signed this agreement as of the day it was opened for signature, but did not ratify it immediately. The reason for this is that Turkey is considered a developed country and therefore cannot benefit from the financing offered for technology transfer and climate change projects. The reason why Turkey is considered a developed country is that it is on the Annex I list, and although the request to leave this list continues, the law that found it appropriate to approve the agreement was passed by the Turkish Grand National Assembly and came into force on October 7, 2021 (Selçuk, 2023: 9-19).

There are 29 articles in the Paris Agreement. The main purpose of the agreement is to keep global warming below 2 degrees and, if possible, within 1.5 degrees. Thus, it is aimed to reduce greenhouse gas emissions consistently. Article 4, consisting of 19 paragraphs, requires all countries to submit their national contributions, such as commitments to reduce greenhouse gas emissions and the timelines of these commitments. It is stated that these contributions will be determined in line with the domestic legislation of the party countries and will be updated every five years. Article 6 of the agreement stipulates that technology transfers should be encouraged with financial support in order to reduce greenhouse gas emissions. Another important article, Article 9, states that developed countries should provide financial support to developing countries to combat climate change. Article 13 of the agreement includes sharing the activities carried out to reduce greenhouse gas emissions through national reports and global inventories in accordance with the principle of transparency. Finally, Article 16 of the agreement, which is another important article, states that the parties support the efforts made to adapt to climate change, cooperate in this sense and provide financial support. These articles of the agreement contain the main provisions and objectives. With this agreement, the parties promised to reduce emission volumes by certain amounts. It is envisaged to ensure cooperation between the parties in the fields of technology transfer, implementation and capacity building to reduce emission volumes (Köse, 2018: 55-81). No progress has been made yet with the policies to combat climate change that have been going on for many years. Global greenhouse gas emissions are increasing and our planet continues to warm. It is thought that the most important solution in line with policies to combat climate change is to reduce greenhouse gas emissions. However, it seems that it is not possible to eliminate the climate change problem as long as globalization and market economy aiming at the highest level of production and consumption continue (Can, 2023: 175-178). In this context, suggestions such as accelerating the clean energy transition by reducing energy production based on fossil fuels as much as possible, supporting sustainable transportation, increasing energy efficiency, protecting and reforesting forests, carrying out awareness-raising and training activities, and strengthening international cooperation are important in the context of combating climate change.

CONCLUSION

Fighting climate change is of vital importance for humanity. Because this problem has significant effects on the sustainability of our planet, the protection of natural ecosystems, the provision of water resources, agricultural production, the welfare and health of states and societies. The problem of climate change emerged when the greenhouse gases in the atmosphere increased and people disrupted natural balances as a result of excessive production and consumption. Although Turkey stands out as a country that has taken important steps in combating climate change, it has to make more efforts and develop strategic policies. The policies and measures examined in this section show that Turkey has made significant progress in reducing greenhouse gas emissions and adapting to climate change. Türkiye has also invested in renewable energy resources such as solar energy, wind energy and hydroelectric energy. These investments have contributed to reducing dependence on fossil fuels and reducing greenhouse gas emissions. In addition, Türkiye has also made important moves in energy efficiency. It has developed policies in areas such as increasing the energy efficiency of buildings and making industrial processes more efficient. Continuing and strengthening Turkey's efforts in the fight against climate change is important in terms of protecting national interests and contributing to global climate goals. Fighting climate change is the common responsibility of not only the state but also the entire society. It is necessary to strengthen policies to combat climate change and move away from fossil fuels as much as possible for a more sustainable future, and instead turn to clean energy sources. In this context, investment in resources such as solar energy, wind energy and hydroelectric energy should be paved, encouraged in these areas and more capacity should be created for them. Preserving and expanding forest areas is also of great importance in this regard. Sustainable management of water resources should be ensured, international cooperation should be strengthened, and most importantly, public awareness should be increased to combat climate change, support should be provided for training in this field, and the public should be made aware of climate change.

Implementing these recommendations to strengthen Turkey's policies to combat climate change will contribute to building a more sustainable future. Since the process of combating climate change is not easy, a long-term effort is needed. Therefore, cooperation and commitment of all stakeholders is needed. In addition to these suggestions, moves such as supporting scientific research in this field and encouraging innovative solutions will strengthen Turkey's fight against climate change. If effective policies are implemented to combat climate change, the amount of greenhouse gases released into the atmosphere will decrease, thus air quality will improve and the environment will be protected. Additionally, economic growth can be supported in areas such as the clean energy sector, energy efficiency projects, sustainable agriculture and environmentally friendly technologies. Overcoming this global problem can be easier by finding common solutions through global cooperation and solidarity. Most importantly, with the right struggle policies, a sustainable world in which future generations can live will be created. However, we do not have much time to fight climate change. By making correct and conscious decisions, producing innovative solutions, and working in cooperation with other states, our chances of achieving positive results in the fight against climate change increase. It is of great importance that we all act together to protect our world and leave a livable world to future generations.

Glossary

A

Adaption: Adjustments in natura lor human systems to respond to the effects of climate change.
Policies: A policy is a principle, rule, or course of action adopted or proposed by a government, organization, or individual.

C

Carbon dioxide (CO_2): A colorless, odorless gas produced by the burning of fossil fuels and other carbon- containing materials, as well as through natural processes like respiration and volcanic eruptions.
Carbon Pricing: Mechanisms, such as carbon taxes or emissions trading systems, that put a price on carbon dioxide emissions to incentivize emissions reductions.
Changing climate: The changing climate refers to the long-term shift in global or regional weather patterns and conditions over an extended period of time.

Clean Energy Sector: The clean energy sector refers to the industries and companies that produce or promote the use of renewable, sustainable, and environmentally- friendly sources of energy, such as solar, wind, hydroelectric, and geotermal power.

E

Economic growth: The definition of economic growth is the increase in the production of goods and services in an economy over time. It is typically measured by the percent change in a country's gross domestic product or gross national product.

Emissions: The release of substances, particularly greenhouse gases, into the atmosphere.

G

Global warming: The gradual increase in the overall temperature of the Earth's atmosphere due to the greenhouse effect.

Greenhouse gases: Greenhouse gases refer to the gases in the Earth's atmosphere that trap heat, causing the greenhouse effect and leading to climate change. Examples of greenhouse gases include carbon dioxide, methane, nitrous oxide, and fluorinated gases.

I

Initiatives: An initiative is a new plan or process to deal with a problem or to improve a situation.

K

Kyoto Protocol: An international agreement adopted in 1997 that set binding emission reduction targets for developed countries to address climate change.

M

Methane (CH4): A hydrocarbon gas that is the main component of natural gas and a potent greenhouse gas produced by activities such as agriculture, waste management, and the extraction and use of fossil fuels.

Montreal Protocol: An international agreement adopted in 1987 to phase out the production and use of ozone- depleting substances in order to reduce their abundance in the atmosphere and help protect the ozone layer.

P

Paris Agreement: An international treaty aimed at limiting global warming to well below 2 centigrade above pre-industrial levels.

R

Renewable energy: Energy sources that are naturally replenished, such as solar, wind, hydroelectric, and geothermal power.

S

Sustainability: The ability to meet the needs of the present without compromising the ability of future generations to meet their own needs.

Sustainable development: Development that meets the needs of the present without compromising the ability of future generations to meet their own needs.

U

United Nations Framework Convention on Climate Change (UNFCCC): An international environmental treaty adopted in 1992 with the goal of stabilizing greenhouse gas concentrations in the atmosphere.

REFERENCES

Akyel, Ö. (2009). *İklim Değişikliği Çerçeve Sözleşmesi ve Türkiye'deki Uygulamaları*. Ankara Üniversitesi Sosyal Bilimler Enstitüsü.

Binboğa, G. (2014). Uluslararası Karbon Ticareti ve Türkiye. Yaşar Üniversitesi E- Dergisi, 9(34), 5732-5759.

Can, F. (2023). *Çevre ve İklim Değişikliği 101*. Say Yayınları.

Dağdemir, Ö. (2015). Birleşmiş Milletler İklim Değişikliği Çerçeve Sözleşmesi ve Ekonomik Büyüme: İklim Değişikliği Politikasının Türkiye İmalat Sanayii Üzerindeki Olası Etkileri. *Ankara Üniversitesi SBF Dergisi*, 60(2), 49–70.

Doğan, S., & Tüzer, M. (2011). *Küresel İklim Değişikliği İle Mücadele: Genel Yaklaşımlar ve Uluslararası Çabalar. Sosyoloji Konferansları*. İstanbul Üniversitesi Sosyoloji Konferansları Dergisi.

Engin, I. (2019). İklim Değişikliği ile Mücadelede Mali Politikalar. Balıkesir Üniversitesi Sosyal Bilimler Enstitüsü, 47-67.

Göçoğlu, İ. D., Negiz, N., & Göçoğlu, V. (2023). Türkiye'nin İklim Değişikliği ile Mücadele Serüveni: Akademik Yazın Üzerine Bir Araştırma. *Süleyman Demirel Üniversitesi Vizyoner Dergisi*, 14(38), 620–630.

İzol, R., & Kaval, F. (2023). Kopenhag Okulu Bağlamında Türkiye'deki Siyasi Aktörlerin İklim Değişikliğine Yönelik Söylemlerinin Analizi. *Akdeniz İİBF Dergisi*, 23(1), 18–32.

Kahraman, S., & Şenol, P. (2018). İklim Değişikliği: Küresel, Bölgesel ve Kentsel Etkileri. *Akademia Sosyal Bilimler Dergisi*, 1, 353–370.

Kaplan, O. (2023). Türk Hukukunda İdarenin Yenilenebilir Enerji Kaynaklarından Elektrik Enerjisi Üretimi Yönünden İşlevlerinin İrdelenmesi. *Yaşar Hukuk Dergisi*, 5(2), 297–336.

Karakaya, E. (2016). Paris İklim Anlaşması: İçeriği Ve Türkiye Üzerine Bir Değerlendirme. *Adnan Menderes Üniversitesi Sosyal Bilimler Enstitüsü Dergisi*, 3(1), 1–12. DOI: 10.30803/adusobed.188842

Karakaya, E., & Özçağ, M. (2001). Sürdürülebilir Kalkınma ve İklim Değişikliği: Uygulanabilecek İktisadi Araçların Analizi. First Conference in Fiscal Policy and Transition Economies, University of Manas, 1-7.

Kaya, H. E. (2020). Kyoto'dan Paris'e Küresel İklim Politikaları. *Meriç Uluslararası Sosyal ve Stratejik Araştırmalar Dergisi*, 4(10), 165–191.

Keskin, T. (2007). Enerji Verimliliği Kanunu ve Uygulama Süreci. *Mühendis ve Makina*, (569), 106–112.

Koçaslan, G. (2014). Türkiye'nin Enerji Verimliliği Mevzuatı, Avrupa Birliği'ndeki Düzenlemeler ve Uluslararası-Ulusal Öneriler. C.Ü. İktisadi ve İdari Bilimler Dergisi, 15(2), 117-133.

Köse, İ. (2018). İklim Değişikliği Müzakereleri: Türkiye'nin Paris Anlaşmasını İmza Süreci. *Ege Stratejik Araştırmalar Dergisi*, 9(1), 55–81.

Mutlu, M. Y., & Tezer, A. (2023). İklim Değişikliğine Mekânsal Uyum ve Azaltım Yaklaşımlarında Toprak Ekosistem Servislerinin Rolü. *Dirençlilik Dergisi*, 7(2), 305–324.

Ömercioğlu, A. (2023). 5686 Sayılı Jeotermal Kaynaklar ve Doğal Mineralli Sular Kanunu'na Göre Tahsil Edilen İdare Payı ve Görevli Mahkeme Sorunu. *Euroasia Journal Of Social Sciences & Humanities*, 10(32), 159–173.

Özcan, B. A. (2020). Ortak Mülkiyet Çerçevesinde İklim Değişikliği Sorununun Çözümünde Kyoto Protokolü'nün Etkisi. *Akdeniz İİBF Dergisi*, 20(2), 169–184.

Özmen, M. T. (2009). Sera Gazı- Küresel Isınma ve Kyoto Protokolü. İMO Dergisi, 453(1), 42-46.

Paksoy, S. (2019). Türkiye'nin İklim Aksiyonunun Bugünkü Durumu. Çukurova Üniversitesi Sosyal Bilimler Enstitüsü Dergisi, 28(3), 155-160.

Şahin, Ö. U. (2016). Kyoto Protokolü ve Kopenhag Mutabakatının Karşılaştırmalı Analizi. [JoA]. *Journal of Awareness*, 1(1), 5–10.

Selçuk, S. F. (2023). Uluslararası İklim Değişikliği Anlaşmaları ve Türkiye'nin Tutumu. *Ulusal Çevre Bilimleri Araştırma Dergisi*, 6(1), 9–19.

Talu, N., & Kocaman, H. (2018). Türkiye'de İklim Değişikliği ile Mücadelede Politikalar, Yasal ve Kurumsal Yapı. 21.

Türkeş, M. (2006). Küresel İklimin Geleceği ve Kyoto Protokolü. *Jeopolitik*, (29), 99–107.

Türkeş, M., Sümer, U. M., & Çetiner, G. (2000). Kyoto Protokolü Esneklik Mekanizmaları. *Tesisat Dergisi*, (52), 84–100.

Türkiye İklim Stratejisi. (2010). *Türkiye İklim Değişikliği Stratejisi 2010-2023*. https://www.gmka.gov.tr/dokumanlar/yayinlar/Turkiye-Iklim-Degisikligi-Stratejisi.pdf adresinden alındı

Yıldız, T. D. (2012). *3213 Sayılı Maden Kanunu Öncesinde ve Sonrasında Türkiye'de Maden Mevzuatında Yapılan Değişikliklerin İncelenmesi*. İstanbul Teknik Üniversitesi.

Chapter 13
Policies of European Countries to Combat the Climate Crisis

Ezgi Kovancı
https://orcid.org/0000-0003-1434-2581
Adıyaman University, Turkey

ABSTRACT

The climate crisis is profoundly impacting Europe, one of the most vulnerable regions to climate change globally. Increasing temperatures, erratic rainfall patterns, and rising sea levels are intensifying, threatening human life, economies, and ecosystems across the continent. European nations are actively implementing diverse policies and initiatives, such as the Paris Agreement and the European Green Deal, and leading in renewable energy adoption and sustainable transportation. In this study, the geography of Europe, which includes industrially advanced countries, has been examined within the context of the European Union's climate policies.

INTRODUCTION

In the past decade, the impacts of climate change have become increasingly evident, with the period from 2011 to 2020 being recorded as the warmest decade in modern history. This alarming trend is underscored by the fact that by 2019, the global average temperature had risen to 1.1 °C above pre-industrial levels. This rise in temperature is accelerating at a concerning rate of approximately 0.2 °C per decade, driven predominantly by human activities. The implications of this temperature increase are profound, posing substantial risks to both the natural environment and human health. The accelerated pace of global warming has prompted an urgent

DOI: 10.4018/979-8-3693-5792-7.ch013

international response. Recognizing the severe consequences of unchecked climate change, the global community has committed to limiting the rise in global temperatures to well below 2 °C above pre-industrial levels, with a more ambitious target of capping the increase at 1.5 °C. Despite these efforts, the path to achieving these goals remains fraught with challenges. Human activities, such as the combustion of fossil fuels, widespread deforestation, and intensive livestock farming, have been the primary drivers of this unprecedented warming, contributing to more frequent and severe climate events.

The situation has reached a critical juncture. From February 2023 to January 2024, the global average temperature surged to 1.5 °C above pre-industrial levels, marking 2023 as the warmest year in over 100,000 years. This milestone highlights the urgent need for comprehensive climate action. Europe, in particular, is experiencing these changes with alarming rapidity. The continent is warming at twice the global average rate, which exacerbates the impact of climate change across the region (United Nations [UN], 2023). As a result, Europe is now grappling with a range of climate-related hazards. The frequency and intensity of heatwaves have increased, leading to severe health impacts and strain on infrastructure. Droughts are becoming more prevalent, affecting water supplies and agriculture, while floods are causing widespread damage to communities and ecosystems. These climate extremes are not only reshaping daily life but also altering the ecological balance, threatening biodiversity, and straining the resilience of natural and human systems alike. In this context, understanding and addressing the multifaceted effects of climate change on Europe is crucial. The continent's experience serves as a stark reminder of the urgent need for effective mitigation and adaptation strategies to manage the growing risks associated with a warming world. This introduction sets the stage for a deeper exploration of the specific impacts of climate change in Europe, examining how the continent is responding to these challenges and what further actions are needed to safeguard its future.

EFFECTS OF THE CLIMATE CRISIS ON EUROPE

The period from 2011 to 2020 was documented as the warmest decade on record, with the global average temperature reaching 1.1 °C above pre-industrial levels by 2019. Human-induced global warming is currently accelerating at a rate of 0.2 °C per decade. A rise of 2 °C compared to pre-industrial temperatures is associated with significant adverse impacts on both the natural environment and human health and well-being. It significantly heightens the risk of potentially catastrophic global environmental changes. As a result, the international community has acknowledged the imperative to limit global warming to well below 2 °C and to strive towards achieving a target of 1.5 °C (European Commission [EC], 2023a). Human activities such as burning fossil fuels, cutting down forests and farming livestock have resulted in unprecedented global warming. From February 2023 to January 2024, the average global temperature was 1,5 °C above pre-industrial levels. The year 2023 marked the warmest year on record globally in over 100,000 years, with temperatures 1.48 °C above pre-industrial levels, and the world's ocean temperatures also reached new highs. Europe is the fastest-warming continent, experiencing a warming rate since the 1980s that is about twice the global average (European Environment Agency [EEA], 2024a: 5). Many longstanding climate records in Europe have been shattered in recent years. The continent is now facing more frequent and intense climate hazards, including heatwaves, prolonged droughts, heavy precipitation leading to both pluvial and fluvial floods, and sea level rise causing coastal flooding.

Climate change is already affecting the daily lives of Europeans and is projected to do so in the coming years. Climate change is already affecting the daily lives of Europeans and is projected to do so in the coming years. According to the European Environment Agency (EEA), Europe is anticipated to experience warmer temperatures, with some regions becoming drier and others experiencing increased precipitation. These shifts will not only affect human health but also the ecosystems that are crucial for our well-being. However, it is not possible to say that climate change affects every region of Europe in the same way. Heatwaves, floods, droughts, and wildfires have become increasingly frequent in Europe during the summer months. According to the Copernicus Climate Change Service (C3S), May 2024 marked the 12th consecutive month with record-high temperatures. For instance, In June 2024, devastating floods in Germany resulted in several fatalities and significant economic damage. Projections indicate that snowfall will decrease in central and southern Europe, while mixed changes are expected in northern Europe (EEA, 2024b). Sea levels are forecasted to rise across all regions except the North Baltic Sea. Additionally, sea surface temperatures are anticipated to increase in all European seas, contributing to a warming trend. Furthermore, Europe's seas are expected to undergo acidification, posing further environmental challenges (EEA, 2024c). Also, climate change is anticipated to significantly alter the fundamental processes that govern tree and forest dynamics in mountainous regions like the Alps. These areas are especially vulnerable to climate impacts, experiencing temperature increases that outpace the global average, with further acceleration noted in recent decades. As the climate continues to warm rapidly, these ecosystems are likely to undergo accelerated changes in their regeneration, growth, and mortality patterns. For instance, research conducted in the Austrian Alps has indicated an anticipated reduction in the number of large trees as temperatures rise, along with notable shifts in forest composition that could impact the protective functions of Alpine forests. Similar findings have been documented in the Swiss Alps, where projections suggest a significant decline in Norway spruce populations under future climate warming scenarios (Hillebrand et al., 2023: 02). The Figure 1 illustrates the observed and projected climate changes and their impacts on the major biogeographical regions of Europe.

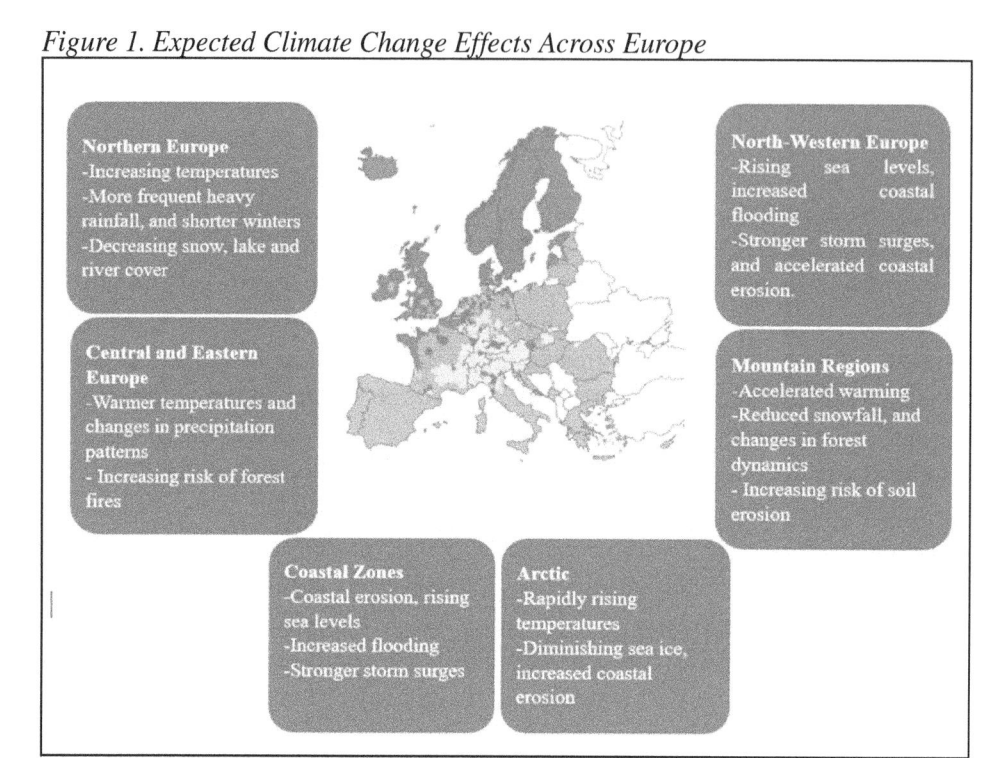

Figure 1. Expected Climate Change Effects Across Europe

Northern Europe
-Increasing temperatures
-More frequent heavy rainfall, and shorter winters
-Decreasing snow, lake and river cover

North-Western Europe
-Rising sea levels, increased coastal flooding
-Stronger storm surges, and accelerated coastal erosion.

Central and Eastern Europe
-Warmer temperatures and changes in precipitation patterns
- Increasing risk of forest fires

Mountain Regions
-Accelerated warming
-Reduced snowfall, and changes in forest dynamics
- Increasing risk of soil erosion

Coastal Zones
-Coastal erosion, rising sea levels
-Increased flooding
-Stronger storm surges

Arctic
-Rapidly rising temperatures
-Diminishing sea ice, increased coastal erosion

Source: EEA, 2012

Based on the results of the 2023 Eurobarometer survey, conducted to reveal the opinions of citizens from EU member and candidate countries, more than three quarters about 77% of EU citizens think climate change is a very serious problem at this moment. A significant number of Europeans believe that the EU (56%), national governments (56%), and business and industry (53%) are responsible for addressing climate change. Additionally, 35% of Europeans see themselves as personally responsible. More than 85% of the respondents believe it is crucial for both their national governments (86% of respondents) and the European Union (85% of respondents) to take proactive steps for energy efficiency by 2030, such as encouraging home insulation, installing solar panels, or purchasing electric cars. Furthermore, 58% of EU citizens believe that the adoption of renewable energy sources should be accelerated, energy efficiency should be improved, and the transition to a green economy should be hastened, especially in light of energy price spikes and gas supply restrictions due to Russia's actions (EC, 2023b: 7).

Europe is undergoing rapid climate changes, with recent extreme events hinting at the future under continued global warming. In July 2021, severe floods from heavy rainfall struck Belgium, Germany, Luxembourg, and the Netherlands, claiming over 230 lives and injuring more than 750 people. In 2023, flooding affected regions in southern and eastern Europe, including Italy and central Europe in May, Slovenia in August, and Bulgaria, Greece, and Türkiye[1] in September. In October 2023, Denmark experienced its worst storm surges in decades. Throughout that year, floods impacted 1.6 million people across Europe (C3S and World Meteorological Organization [WMO], 2024).

Flooding is one of the most dangerous consequences of climate change. Flooding can lead to significant loss of life and injury. Rapidly rising water levels can trap people in their homes or vehicles, making it difficult or impossible to escape. Floodwaters can cause extensive damage to homes, businesses, infrastructure, and personal property. The force of the water can destroy buildings, roads, bridges, and utilities, leading to costly repairs and long-term disruptions. Severe flooding can displace large numbers of people from their homes. Evacuations can be necessary, leading to temporary or even permanent relocation, which can be traumatic and disruptive to communities. Flooding can have a devastating impact on local and national economies. The cost of repairing damage, coupled with the loss of income for businesses that are forced to close, can lead to significant economic hardship. Floodwaters can cause soil erosion, loss of agricultural land, and contamination of water supplies with pollutants, chemicals, and waste. This can have long-term negative effects on local ecosystems and biodiversity. Flooding can lead to serious health issues, including waterborne diseases such as cholera, dysentery, and typhoid. Standing water can also become a breeding ground for mosquitoes, increasing the risk of vector-borne diseases like malaria and dengue fever. Floods can destroy crops, drown livestock, and contaminate soil, leading to food shortages and increased prices. This can affect food security and the livelihoods of farmers. The trauma and stress associated with experiencing a flood, losing possessions, or being displaced can have lasting psychological effects on individuals and communities. The recovery process after a flood can be lengthy and complex. It can take years for communities to rebuild and for individuals to regain their livelihoods, especially if they lack adequate insurance or resources. According to EEA Report (2023) currently, around 12% of the European population lives in areas potentially prone to river flooding, although many of these areas have flood defences in place. Between 2011 and 2021, the number of people living in potential riverine flood-prone areas in Europe increased by over 935,000 which is about 1.8% of the population. Additionally, 11% of healthcare facilities across Europe are located in such areas. Between 1980 and

2022, Europe recorded 5,582 flood-related deaths. Besides fatalities and injuries, those affected by flooding often suffer from mental health issues. The vulnerable population2 are the most impacted by flooding. Because they are more likely to experience significant negative impacts during events such as natural disasters, economic downturns, public health crises, and other challenging situations. Flooding also poses pollution risks: nearly 15% of industrial facilities in Europe are located in potential riverine flood-prone areas, and for urban wastewater treatment plants, this percentage is 36%. According to EEA Report an estimated 650,000 combined sewer overflows across Europe degrade water quality following heavy rainfall events (EEA, 2024d: 13). Between 1900 and 2022, Europe experienced 434 severe flooding events. France had the highest number of these incidents, with a total of 60 severe floods. Italy and Romania followed closely, each recording 53 events during this period. The number of people affected by flooding events in Europe between 1995 and 2023 is shown in Figure 2.

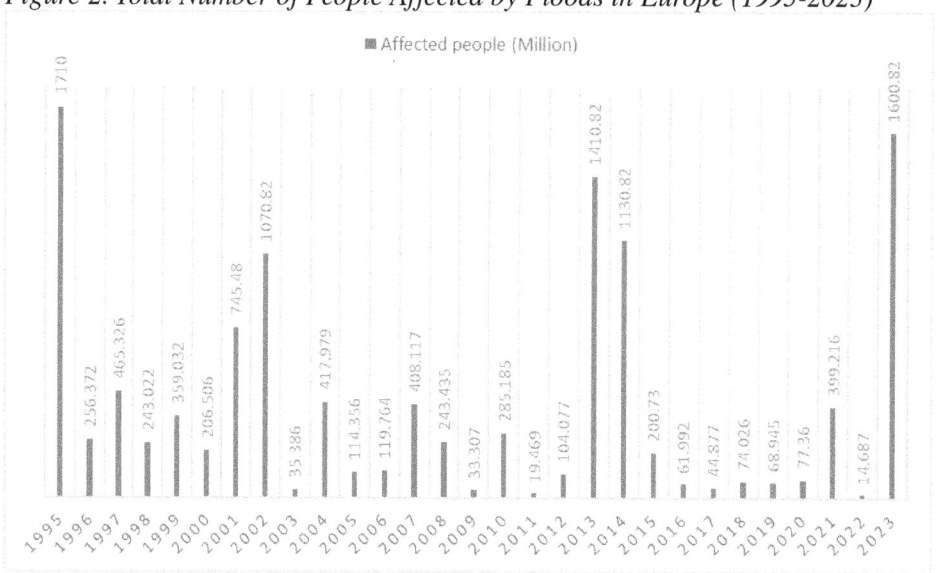

Figure 2. Total Number of People Affected by Floods in Europe (1995-2023)

Source: Our World in Data based on EM-DAT, CRED / UCLouvain, Brussels, Belgium.

Forest fires are the other severe consequence of climate change. Climate change, characterized by rising temperatures and altered precipitation patterns, is likely to increase the frequency and intensity of naturally caused forest fires. This escalation in wildfire activity is already causing severe economic, social, ecological, and environmental damage, including the destruction of homes, poor air and water quality, increased costs, and loss of life. As climate change progresses, these impacts are expected to intensify. High-severity fires may erode the resilience of forest ecosystems, leading to changes in forest structure, the loss of dominant species, and alterations in ecological functions. Additionally, increased wildfire activity poses significant risks to carbon emissions, water supply, and the wildland–urban interface (WUI). The challenges associated with current fire management practices, such as logging operations, fire suppression, and maintaining water quality in national forests, may be exacerbated by the growing severity and duration of forest fires driven by climate change (Heidari et. al., 2021: 1-2). Europe's geography has witnessed many significant forest fires. The summer of 2007 saw devastating forest fires in Greece, particularly in the regions of the Peloponnese and the island of Euboea. These fires were among the most severe in the country's history, resulting in over 67 deaths and the destruction of thousands of homes and large tracts of forest (Koutsias et. al., 2012: 41-43). Another major fire event occurred in 2009, with significant fires affecting areas in southern Greece and causing extensive damage to forests and properties. Portugal, similar to other southern European countries, is experiencing unprecedented challenges with increasingly frequent and uncontrollable fires. These fires are surpassing the capabilities of current suppression resources and raising community vulnerability. Changes in sociodemographic in rural areas, climate change, and shifts in vegetation have created exceptional conditions that necessitate adjustments in fire management strategies and planning tools. The year 2017 marked a significant milestone in the history of wildfires in Portugal, notable not only for the extensive areas burned but also for the high number of fatalities. The fires occurred at different times of the year (June and October) but were geographically close, affecting the central region of Portugal. A total of 117 deaths were reported from both incidents, with 92% of the victims located in wildland–urban interface areas (Rodrigues et. al., 2022: 1-2). In 2021, Spain experienced severe wildfires, particularly in the regions of Castilla La Mancha, Andalucía and the Canary Islands. The fires burned thousands of hectares of forest and agricultural land, prompting large-scale evacuations and extensive firefighting efforts (Fernandes and Rigolot, 2022: 233-234). While the year 2021 was less warm compared to previous years, it was marked by monthly temperature anomalies of varying signs throughout the year in the Europe region. Precipitation was also 7% below the average. Italy experienced several heatwaves in 2021, with the most intense occurring during the second week of August. During this period, temperatures in some areas of Sicily surpassed 48 °C (San-Miguel-Ayanz et. al, 2022: 49). Unfortunately, the summer of 2021 brought severe wildfires to southern Italy, including Sicily and Calabria. The fires were fuelled by extreme heat and dry conditions, causing extensive damage to forests and affecting air quality in the region. In 2022, France faced significant wildfires, particularly in the Gironde region in southwestern France. The fires burned through large areas of forest and prompted the evacuation of thousands of people.

Gironde experienced three major fires that burned 25,000 hectares, significantly exceeding the annual average for the past decade. In the summer of 2022 alone, 25,000 hectares were consumed by fire, compared to the typical 10,000 hectares (Pronto et. al., 2023: 26). In the summer of 2018, Sweden experienced its worst wildfire season in modern history. Fires affected several regions, including the areas around Ljusdal and Ockelbo. The extreme heat and drought conditions contributed to the severity of the fires. Although geographically at the intersection of Europe and Asia, Türkiye's wildfires in 2021 had significant implications for the European region. In Türkiye, the coastline extending from Hatay through the Mediterranean and Aegean regions up to Istanbul is identified as having the highest fire risk. In other words, approximately 57% (12.5 million hectares) of Türkiye's Forest area is situated in fire-sensitive regions. According to data from the General Directorate of Forestry, Department of Forest Fire Combating, the total area burned in 2021 was 139,503 hectares, with 2,793 fires reported. Forest fires predominantly occurred between March and December, with the peak months being June, July, August, and September. July recorded the highest number of fires, with 503 incidents and 104,665 hectares burned (San-Miguel-Ayanz et. al., 2022: 105). As can be seen in Figure 3, in 2023, Ukraine saw more land burned by wildfires than any other European country, with over 214,000 hectares lost to forest and wildland fires. This is significantly higher than the average of 32,000 hectares between 2006 and 2022. The rising threat of climate change has led to an increase in natural disasters in recent years, with wildfires becoming more frequent in regions that had previously never experienced them.

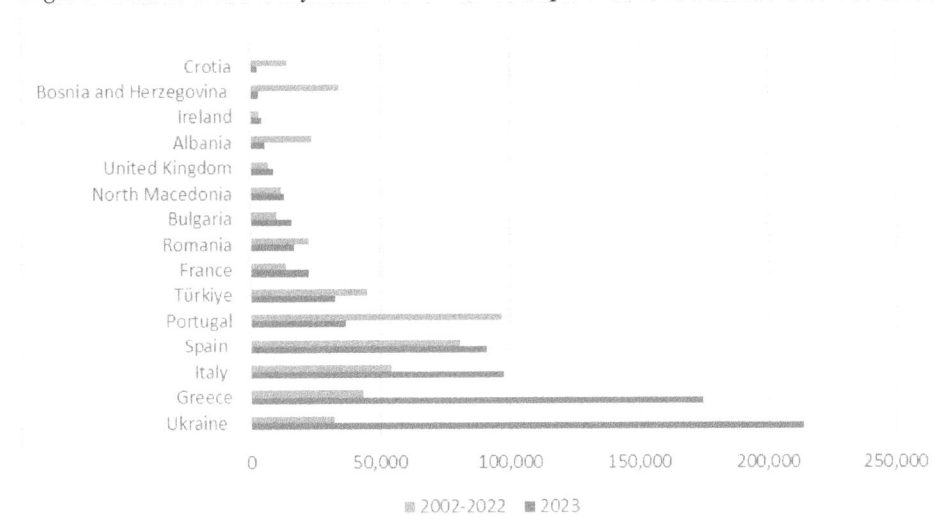

Figure 3. Land Burned by Forest Fires in Europe 2023 (hectares)

[Bar chart showing land burned by forest fires for countries including Crotia, Bosnia and Herzegovina, Ireland, Albania, United Kingdom, North Macedonia, Bulgaria, Romania, France, Türkiye, Portugal, Spain, Italy, Greece, Ukraine; comparing 2002-2022 and 2023 data; x-axis from 0 to 250,000]

Source: https://www.statista.com/.

THE HISTORY OF THE EUROPEAN UNION'S GLOBAL CLIMATE POLICIES

The EU, which ranks fourth after China, the United States (US) and India, has a significant share of global Greenhouse Gas (GHG) emissions. China, the United States, India, the EU27, Russia, and Brazil were the six largest global GHG emitters in 2022. Collectively, they account for 50.1% of the global population, 61.2% of global Gross Domestic Product (GDP), 63.4% of global fossil fuel consumption, and 61.6% of global GHG emissions. Among these top emitters, China, the United States, and India increased their emissions in 2022 compared to 2021, with India experiencing the largest relative increase at 5% (EC, 2023c: 4). The EU, as an industrialized region, has high GHG emissions largely stemming from energy production and heavy industry sectors. The use of fossil fuels holds a significant share in energy production, constituting the primary source of emissions. EEA reports frequently highlight the role of energy production in GHG emissions and emphasize the need for transitioning to renewable energy sources to meet EU climate targets (EEA, 2023). Research has shown that increasing the share of renewable energy in the EU's energy mix is crucial for reducing GHG emissions. Studies emphasize the challenges of integrating renewables into existing grids and the importance of energy storage technologies. Heavy industries such as steel, cement, and chemical production are major contributors to GHG emissions due to their energy-intensive processes (Gajdzik et. al., 2024). The literature often focuses on the decarbonization of these sectors through technological innovation and policy interventions. Several studies explore the potential of Carbon Capture and Storage (CCS) technologies in reducing emissions from heavy industries. The literature highlights the technical and economic challenges of implementing CCS at scale. Academic articles often discuss strategies for decarbonizing the heavy industry sector, including energy efficiency improvements, the use of alternative fuels, and the electrification of industrial processes (Hei Ngu, 2024: 360). The transportation sector in Europe is also a major emitter, especially through road transportation and aviation industries, which produce substantial amounts of carbon dioxide (CO_2). The transportation sector, particularly road transport and aviation, is a significant source of CO_2 emissions in Europe. The literature reviews often focus on the transition to EVs, biofuels, and other low-carbon transportation technologies. Numerous studies examine the barriers to EV adoption, such as the availability of charging infrastructure and the cost of EVs. Research indicates that widespread EV adoption could significantly reduce transportation-related GHG emissions (Liu et. al., 2023: 5). The literature highlights the challenges of reducing emissions in the aviation sector, including the development of sustainable aviation fuels and improvements in aircraft efficiency. Additionally, the agricultural sector contributes greenhouse gases such as methane (CH_4) and nitrous oxide (N_2O), primarily from livestock farming and fertilizer use. Research focuses on strategies to reduce methane emissions from livestock, such as changes in animal feed and manure management practices. The literature explores the potential of sustainable farming practices, such as precision agriculture and organic farming, in reducing N_2O emissions from fertilizers (Chataut et. al., 2023: 2-3).

In this context, the EU has established various policies and targets to reduce GHG emissions. The journey of EU to develop climate policies began in the early 1990s, marked by initial policies aimed at meeting the Kyoto Protocol targets during its first commitment period (2008-2012). Subsequently, strategies were developed to achieve goals set for the second commitment period, focusing on targets leading up to 2020. Looking forward, attention shifts to the post-2020 era, where new policies and frameworks are being introduced to further advance climate action and sustainability goals.

Shortly after the release of the first summary report of the Intergovernmental Panel on Climate Change (IPCC)[3] in 1990, climate change gained prominence at the EC in preparation for the upcoming negotiations on the United Nations Framework Convention on Climate Change (UNFCCC) later that year. EU leaders committed to stabilizing GHG emissions of the EC at 1990 levels by 2000. However, specific measures to achieve these emission reductions were not initially specified, prompting discussions on common and coordinated policies and measures (PAMs). During this early phase of climate policy development, three primary areas emerged and continue to guide policy today: reducing GHG emissions, promoting renewable energy sources (RES), and enhancing energy efficiency (Climate Policy Info Hub, 2023a).

The Process Leading to the Kyoto Protocol

The most pivotal and initial global response to address global climate change occurred with the adoption of the UNFCCC during the United Nations Conference on Environment and Development held in Rio de Janeiro in 1992. So, the UNFCCC was adopted in 1992, more than a quarter of a century ago, and today 196 countries as well as the EU are parties to it (UNFCCC, 2023). The overarching goal of the UNFCCC, described as "stabilizing greenhouse gas concentrations in the atmosphere at a level that would prevent dangerous anthropogenic interference with the climate system," mandates shared responsibilities among all Parties where a Party can be a national government or a regional economic integration organization like the EU to reduce GHG emissions and mitigate the impacts of climate change. This approach considers their respective but varied responsibilities, as well as national and regional development priorities, objectives, and unique circumstances.

In 1992, the EU adopted a strategy titled "A Community Strategy to Limit Carbon Dioxide Emissions and Improve Energy Efficiency," aimed at restricting CO2 emissions and enhancing energy efficiency. This strategy encompasses policies and measures developed as part of the EU's efforts to combat climate change. The problem has been set in the strategy that the industrialized world bears the primary responsibility for CO2 emissions. The EU alone accounts for 13% of global emissions, with per capita emissions levels approximately twice the world average (EC, 1992). While emissions from developing countries[4] have been relatively modest to date, their output is expected to increase at a faster rate compared to other regions in the coming years. Hence, it is crucial for all countries, whether developed or developing, to participate in efforts to control CO2 emissions. The objective set by the community is to stabilize CO2 emissions at 1990 levels by the year 2000. All developed countries appeared willing to align with this goal. Achieving this stabilization required reducing energy demand through increased energy efficiency and promoting fuel-switching. In the same year, efforts to address GHG emissions included discussions on a European CO2 and energy tax proposal. However, within the Community, there was disagreement over both the necessity and specifics of such a tax (EC, 1992). A group of member states, led by the United Kingdom, opposed its introduction. Despite this setback concerning CO2 taxation, softer measures focusing on energy efficiency and renewable energies were agreed upon.

In 1991, the "Specific Actions for Vigorous Energy Efficiency" (SAVE) program was launched to promote and facilitate the adoption of energy efficiency policies and programs. As part of this initiative, in 1992, common standards were established for hot water boilers, household electric refrigerators, freezers, and combinations thereof[5]. Additionally, a labelling system for household appliances was introduced to enable consumers to compare their energy consumption across different models. Starting from 1993, the SAVE Directive required member states to implement further measures aimed at limiting GHG emissions. These included energy audits for energy-intensive companies, building certification requirements, and thermal insulation standards for new buildings (Commission of the European Communities [EC], 1991). However, the directive did not specify quantified emission reduction targets, allowing member states flexibility in designing and implementing their policies.

Negotiations about climate change on a legal instrument under the UNFCCC commenced at the first Conference of Parties (COP)[6] in Berlin in 1995. In 1996, the European Community set its first long-term goal to limit the global temperature increase to below 2°C compared to pre-industrial levels. In preparation for the Kyoto summit which called COP 3 and to encourage international commitments to combat climate change by setting an example, EU Ministers agreed in early 1997 on a specific target to reduce GHG emissions by 15% by 2010, relative to 1990 levels. This target would be internally divided among the EU Member States through a "burden sharing" agreement, known as the "EU bubble", establishing specific national targets for all 15 Member States (Elzen et al., 2007: 14). However, the initial attempt at burden sharing only totalled a 9.2% reduction, with the remaining reductions planned to be achieved once an international agreement was in place.

The European Union's Kyoto Protocol Commitments and Implementations

At the climate summit in Kyoto in December 1997, the industrialised countries agreed on a set of quantitative GHG emission targets. The EC committed to an 8% reduction of a basket of six GHGs during the commitment period of 2008-2012, compared to 1990 levels. This was the highest absolute reduction target among the industrialised countries, although it was still less than the internally agreed reduction goal proposal set out in preparation for Kyoto (EC, 2018a). In response to the commitments made in Kyoto, an EU internal arrangement was agreed upon in 1998, outlining specific individual targets for each of the 15 Member States for the commitment period 2008-2012 to achieve the overall 8% reduction. In 2002, the Kyoto targets and the burden-sharing agreement were approved and became binding Community Law in 2004. Additionally, a new monitoring mechanism was established to track progress towards these targets. The EU expanded to include

Central and Eastern European countries in 2004, 2007, and 2013. Except for Malta and Cyprus, all of these countries had committed to independent GHG reduction targets under the Kyoto Protocol (EC, 2018a).

In March 2007, EU Heads of State endorsed a set of targets known as "20-20-20 by 2020" which included objectives for reducing GHG emissions (EEA, 2021), increasing renewable energy usage, and enhancing energy efficiency (EEA, 2010: 31). To achieve these goals, the EC introduced the "Climate and Energy Package" in 2008. This comprehensive package includes four main components: the Emissions Trading System (ETS Directive), the Effort-Sharing Decision (ESD), the Renewable Energy Directive (RED), and the Directive on CCS Directive (Climate Policy Info Hub, 2023b). Additionally, the Climate and Energy Package encompassed measures such as the Fuel Quality Directive and CO2 emission standards for automobiles. While specific measures for energy efficiency were not directly integrated into the package, various sector-specific strategies, including taxation, standards, and information initiatives, have been employed. Since 2012, the Energy Efficiency Directive (EED) has provided a unified framework aimed at fostering energy efficiency across the EU.

The European Union Climate Policies within the Framework of the Paris Climate Agreement

During the EC meeting on October 24, 2014 the EU endorsed new binding targets for 2030 aimed at combating climate change and promoting sustainable energy practices. These include a mandatory reduction of at least 40% in GHG emissions compared to 1990 levels within the EU. Additionally, the Council established a binding target for renewable energy, aiming for at least 27% of energy consumed in the EU to be sourced from renewables by 2030. An indicative target of at least 27% improvement in energy efficiency by 2030 was also set, with a review scheduled for 2020 to consider raising this target to 30% at the EU level. These measures are designed to build upon and strengthen the EU's existing climate and energy framework, reflecting ongoing commitments to international agreements such as the Paris Agreement and previous Kyoto Protocol obligations (UNFCCC, 2014).

The Paris Agreement, adopted in December 2015 during the 21st COP to the UNFCCC, represents a landmark global effort to combat climate change. Its primary objective is to limit the global average temperature increase to "well below" 2 °C above pre-industrial levels, with an ambitious target of 1.5 °C. Key strategies include peaking global GHG emissions as soon as possible and achieving net-zero emissions in the latter half of this century. The Agreement emphasizes both adaptation measures, such as sustainable water management and agricultural practices, and mitigation efforts, including the widespread adoption of renewable energy and shifts in societal behaviours. Furthermore, it acknowledges the critical issue of addressing "loss and damage" caused by the impacts of climate change. The EU formally ratified the Paris Agreement on October 5, 2016, thereby facilitating its entry into force on November 4, 2016 (Amanatidis and Petit, 2024). The Agreement mandates countries to set increasingly ambitious climate targets aligned with its objectives, facilitating regular assessments of global progress through a mechanism known as the Global Stocktake[7]. Acknowledging diverse national circumstances, the Agreement upholds the principle of common but differentiated responsibilities and respective capabilities (EP, 2019: 6). This principle underscores that developed nations should continue leading efforts to mitigate climate change and support actions taken by developing countries. By recognizing varying starting points and responsibilities, the Agreement aims to foster equitable participation and effective global cooperation in combating climate change.

Aligned with its commitments under the Paris Agreement, in November 2018, the EC unveiled a new long-term strategy reaffirming Europe's dedication to global climate leadership. The strategy aims to achieve net-zero GHG emissions by 2050, emphasizing a socially equitable transition that is cost-effective. Rather than introducing new policies or revising 2030 targets, the strategy sets a roadmap for EU climate and energy policies, outlining Europe's envisioned contribution towards meeting the Paris Agreement's temperature goals and aligning with the UN Sustainable Development Goals (SDG). This strategic approach initiates extensive discussions among European decision-makers on preparing for the future up to 2050, culminating in the submission of the European long-term strategy to the UNFCCC by 2020 (EC, 2018b).

EU's Commitment to the 2030 Agenda: Advancing Global Sustainable Development

The 17 Sustainable Development Goals (SDGs), adopted as a central component of the international plan for sustainable development at the United Nations in 2015, represent a global call to action to end poverty, protect the planet, and ensure prosperity for all. These goals have been embraced not only on a global scale but also by regional entities such as the EU and its member states. By adopting the SDGs, the EU and its member states have underscored their commitment to integrating these objectives into their own policies and strategies, aligning their efforts with the broader global agenda. In this context, the EU and its member states have pledged to advance the global implementation of the SDGs through a unified and coordinated approach, recognizing that collaboration and shared responsibility are essential for achieving the ambitious targets set forth by 2030. This commitment reflects the EU's recognition of the interconnected nature of sustainable development challenges, where progress in one area often supports and accelerates progress in others. The EU's approach emphasizes the importance of partnership, both within the Union and with external partners, to drive the systemic changes needed to meet the SDGs (EU, 2019).

One of the key milestones in this journey was the publication of the 2017 "European Consensus on Development" report. This document, released under the inspiring slogan "Our World, Our Dignity, Our Future," represents a renewed and collective vision for development policy within the EU. The Consensus builds upon the values and principles enshrined in the Lisbon Treaty, which provides the legal and institutional framework for the EU's external actions (EC, 2022). By taking into account this framework, the Consensus ensures that the EU's development policies are not only aligned with the SDGs but are also coherent with other EU policies, such as trade, migration, and security. The "European Consensus on Development" sets out a comprehensive strategy for achieving the SDGs by 2030, with a particular focus on fostering sustainable and inclusive growth, reducing inequalities, and promoting peace and stability. It emphasizes the EU's commitment to supporting developing countries in their efforts to implement the SDGs, recognizing that global challenges such as poverty, inequality, and environmental degradation require collective solutions. The report highlights the importance of working in partnership with developing countries, civil society, the private sector, and other international organizations to create synergies and mobilize the necessary resources for sustainable development.

Furthermore, the Consensus underscores the EU's dedication to ensuring that no one is left behind in the pursuit of the SDGs. This involves a focus on the most vulnerable and marginalized populations, including women, children, and those living in conflict-affected areas. The EU's approach to development is rooted in the principles of human rights, democracy, and the rule of law, and it seeks to promote these values as part of its global engagement. In summary, the European Union's adoption of the Sustainable Development Goals and the subsequent "European Consensus on Development" reflect a deep commitment to global sustainability and partnership (Rabinovych and Pintsch, 2023: 41). By integrating the SDGs into its policies and actions, the EU aims to contribute meaningfully to the global effort to create a more equitable, prosperous, and sustainable world by 2030. The Consensus not only outlines a vision for the future but also provides a practical roadmap for achieving these goals in collaboration with a wide range of stakeholders, ensuring that the EU remains at the forefront of global sustainable development efforts.

THE EUROPEAN GREEN DEAL

The European Green Deal, unveiled by the Commission on December 11, 2019, stands as a comprehensive framework aimed at guiding the EU towards achieving carbon neutrality by 2050. This ambitious initiative comprises a diverse range of measures supported by a detailed roadmap outlining key actions. These measures encompass significant reductions in emissions, substantial investments in cutting-edge research and innovation, and initiatives designed to protect and enhance Europe's natural environment. Through investments in green technologies, sustainable solutions, and the promotion of emerging industries, the Green Deal also functions as a strategic economic growth plan, positioning the EU as a sustainable and globally competitive entity. Central to its success lies the active engagement and commitment of the public and all stakeholders (EC, 2019). A cornerstone of the European Green Deal is the European Climate Law[8], which sets forth the objective of achieving climate neutrality by 2050. Notably, the law includes provisions to increase the EU's 2030 target for reducing greenhouse gas emissions to at least 55% below 1990 levels. Furthermore, the Commission has put forward various initiatives, including communications on the Sustainable Europe Investment Plan and the European Climate Pact. These efforts are complemented by proposals for regulations establishing the Just Transition Fund, revisions to guidelines for trans-European energy infrastructure, EU strategies for integrating energy systems and promoting hydrogen, as well as a new strategy on adapting to climate change within the EU.

A key legislative component of the Green Deal is the European Climate Law, which codifies the objective of achieving climate neutrality by 2050 into EU law. This landmark legislation sets legally binding targets for reducing GHG emissions and establishes a framework for monitoring and reporting progress. In addition to the Climate Law, the European Commission has introduced several initiatives to support the Green Deal's objectives (European Climate Law, 2021/ Document 32021R1119). This plan outlines the necessary investments to achieve the Green Deal's targets, focusing on mobilizing private and public funds for green projects and infrastructure. This initiative aims to engage citizens, businesses, and local authorities in climate action through voluntary commitments and partnerships. Proposed regulations for this fund aim to support regions and sectors that will face challenges in the transition to a low-carbon economy, ensuring a fair and inclusive transition. These revisions are designed to enhance the EU's energy infrastructure and support the integration of renewable energy sources. The strategies focus on developing a more interconnected and resilient energy system, with a particular emphasis on hydrogen as a key component of the future energy mix. Together, these efforts form a cohesive approach to tackling climate change, promoting sustainability, and driving economic growth.

The European Green Deal represents a pivotal moment in the EU's environmental and economic policy, reflecting a commitment to creating a sustainable and resilient future for Europe and the world.

On July 14, 2021, the Commission introduced a comprehensive package of legislative proposals, including new laws and amendments to existing legislation, with the goal of making the EU "Fit for 55". This package aims to facilitate the necessary transformational changes across the economy, society, and industry to achieve climate neutrality by 2050. The Fit for 55 package is a comprehensive set of proposals designed to revise and update EU legislation and introduce new initiatives to align EU policies with the climate goals agreed upon by the Council and the European Parliament. This package aims to provide a coherent and balanced framework for achieving the EU's climate objectives. It ensures a just and socially fair transition, enhances and sustains the innovation and competitiveness of EU industries, and ensures a level playing field with third-country economic operators. Additionally, it reinforces the EU's leadership in the global fight against climate change. Key components of the Fit for 55 package include (EC, 2024):

1. Revision of the EU Emissions Trading System (ETS): Extending the ETS to new sectors such as maritime transport, increasing the overall emission reduction target.
2. Effort Sharing Regulation (ESR): Setting binding annual greenhouse gas emission reduction targets for Member States from 2021 to 2030 for sectors not covered by the ETS, like buildings, road transport, and waste.
3. Renewable Energy Directive (RED): Increasing the target for renewable energy sources in the EU's energy mix to 40% by 2030.
4. Energy Efficiency Directive (EED): Raising the ambition for energy efficiency improvements to reduce overall energy consumption.
5. Carbon Border Adjustment Mechanism (CBAM): Implementing a carbon pricing mechanism on imports of certain goods to prevent carbon leakage and ensure a level playing field for EU industries.
6. ReFuelEU Aviation and FuelEU Maritime Initiatives: Promoting the use of sustainable aviation fuels and sustainable maritime fuels to reduce emissions from these sectors.
7. Revision of the Energy Taxation Directive (ETD): Adjusting the taxation of energy products to promote clean technologies and reduce fossil fuel subsidies.
8. Social Climate Fund: Establishing a fund to support vulnerable households, micro-enterprises, and transport users during the transition to a greener economy.
9. CO2 Emission Standards for Cars and Vans: Strengthening the CO2 emissions standards for cars and vans to accelerate the shift to zero-emission vehicles.

10. Land Use, Land Use Change and Forestry (LULUCF) Regulation: Enhancing the regulation on land use, land-use change, and forestry to increase carbon sequestration and reduce emissions from these sectors.

Together, these revisions are expected to enable the EU to slightly exceed its current target, achieving a 57% net reduction in emissions by 2030. In October, the EU also updated its Nationally Determined Contribution (NDC)[9]. While there were no significant changes or mention of the 57% target, the update was made to align with the Fit for 55 package and the European Climate Law. However, while the EU is on track to meet its 55% emissions reduction target by 2030, Climate Change Performance Index (CCPI) experts argue that this target is still insufficient to align with the 1.5°C goal of the Paris Agreement. A report from the newly established European Scientific Advisory Board on Climate Change (ESABCC) indicates that achieving at least 90-95% net emission cuts by 2040 is necessary for the EU to make a fair contribution to combating the climate crisis. Additionally, several systemic weaknesses persist, such as the ongoing allocation of free allowances to industry up to and beyond 2030, and the high degree of flexibility for Member States to trade, bank, and borrow their emission allowances. These issues are likely to significantly delay effective climate action (Burk et al., 2024: 20).

Overall, the European Union (EU) has made notable progress in its CCPI ranking, ascending three places to 16th position. This improvement reflects a substantial enhancement in the EU's collective climate action efforts. Among the EU member states, fourteen countries are now classified as high and medium performers in terms of their climate policies and practices. Denmark, recognized for its ambitious climate strategies and effective implementation, has achieved an impressive 4th place in the overall ranking. Estonia follows closely, securing the 5th position, demonstrating significant advancements in its climate performance. The Netherlands has shown a remarkable improvement across three out of the four CCPI categories, reflecting its high-level commitment to addressing climate change. The country has climbed five positions to reach 8th place, underscoring its effective climate policies and initiatives. Conversely, Italy has faced a significant setback, dropping 15 spots to 44th place. This decline is attributed to a notably weaker performance in the Climate Policy category compared to the previous year. The reduction in Italy's ranking highlights challenges in its climate strategy and implementation, which could have implications for its future performance. Poland remains at a lowly 55th position within the EU, reflecting a very low rating in its climate performance. The country's standing has been hindered by limited progress in climate action and renewable energy initiatives. Future improvements in Poland's ranking will heavily depend on the ambitions and effectiveness of the new Polish government. Advancements in renewable energy policies and a stronger commitment to climate action could potentially lead to a more favourable ranking in the next year's CCPI (see Figure 4). This analysis underscores the varying degrees of progress and challenges faced by EU member states in their efforts to combat climate change, with significant implications for their future performance in global climate assessments (Burk et al., 2024: 6).

Countries like Denmark and Estonia have demonstrated success by implementing ambitious and forward-thinking climate policies. These policies often include aggressive targets for reducing greenhouse gas emissions, increasing the use of renewable energy, and improving energy efficiency. It's not enough to have strong policies; effective implementation is crucial (Barker et. al., 2022). This involves translating policies into actionable programs, investing in necessary infrastructure, and ensuring compliance through robust monitoring and enforcement mechanisms. Successful countries typically have comprehensive climate strategies that address various sectors such as energy, transportation, industry, and agriculture. This holistic approach ensures that all areas contribute to climate goals. High-performing countries often benefit from strong public and political support for climate initiatives. Public awareness and political will can drive sustained action and investment in climate solutions. Investment in new technologies and innovative solutions plays a significant role in improving climate performance. For instance, advancements in renewable energy technologies or energy efficiency can lead to substantial progress. Engagement in international climate agreements and cooperation with other countries can enhance a nation's climate performance. Sharing best practices, technologies, and resources can help improve overall effectiveness. The ability to adapt and revise climate policies in response to new data or changing circumstances is also crucial. Countries that can swiftly adjust their strategies tend to perform better. In summary, the success of countries in improving their climate performance typically relies on a combination of strong policies, effective implementation, technological innovation, and broad support. Each of these factors contributes to achieving and maintaining high rankings in climate performance indices.

Table 1. Climate Change Performance Ranking of European Union Member States in 2024

4. DENMARK	24. ROMANIA	40. SLOVAK REPUBLIC
5. ESTONIA	26. FINLAND	41. SLOVENIA
8. NETHERLANDS	28. GREECE	42. CYPRUS
10. SWEDEN	29. MALTA	43. IRELAND
13. PORTUGAL	32. AUSTRIA	44. ITALY
14. GERMANY	33. LATVIA	46. BULGARIA

continued on following page

Table 1. Continued

4. DENMARK	24. ROMANIA	40. SLOVAK REPUBLIC
15. LUXEMBURG	35. CROATIA	49. HUNGARY
16. EUROPEAN UNION	37. FRANCE	52. CZECH REPUBLIC
18. SPAIN	24. ROMANIA	55. POLAND
19. LITHUANIA	39. BELGIUM	

Source: https://ccpi.org/wp-content/uploads/CCPI-2024-Results.pdf.

CONCLUSION

The EU's climate policies are widely regarded as successful, establishing it as one of the world's foremost regions in addressing climate challenges. Aligning closely with the Paris Agreement, the EU has committed to ambitious targets, notably aiming for a 55% reduction in greenhouse gas emissions by 2030. Beyond these targets, the EU pursues energy transformation through initiatives like "Fit for 55" and invests heavily in green technologies such as renewable energy, energy efficiency improvements, and electric vehicles. Advanced technologies like carbon capture and storage also feature prominently in EU strategies. Member states bolster environmental sustainability through policies promoting energy transition and investments in green infrastructure. The EU reinforces its climate agenda with effective policy tools and legal frameworks, including emissions trading systems. Globally, the EU leads efforts in climate action and collaborates extensively with other nations. Despite these achievements, challenges remain, including disparities in economic conditions and energy infrastructure among member states, posing hurdles to uniform implementation of climate goals across the EU.

The primary shortcomings in the EU's climate policies stem from disparities in implementation. There are significant economic and infrastructural differences among EU member states, complicating the uniform application of climate policies and hindering some countries' ability to meet set targets. Secondly, certain industries, particularly energy-intensive sectors and large industrial entities, exhibit resistance to stricter climate regulations, thereby slowing down policy implementation. Political processes also pose barriers to success within the EU. Internal political procedures and decision-making mechanisms sometimes impede the swift and effective formulation of climate policies. Differing national priorities and interests among member states further complicate the establishment of a cohesive approach. Additionally, some EU countries face challenges in financing climate policies, especially during economic crises or other urgent financial needs, which may obstruct or delay green investments. Globally, discrepancies or deficiencies in cooperation between the

EU's climate policies and those of other countries and regions may hinder progress towards global climate goals. These factors collectively underscore the shortcomings in the EU's climate policies and reflect significant challenges that the EU faces.

REFERENCES

Amanatidis & Petit. (2024). *Combating climate change*, Fact Sheets on the European Union European Parliament, Retrieved June 10, 2024 from https://www.europarl.europa.eu/factsheets/en/sheet/72/combating-climate-change

Barker, A., Blake, H., D'Arcangelo, F. M., & Lenain, P. (2022). Towards net zero emissions in Denmark. OECD Economics Department Working Papers No. 1705, Retrieved August 05, 2024 from https://eulacfoundation.org/system/files/digital_library/2023-07/5b40df8f-en.pdf

Burk, J., Uhlich, T., Bals, C., Höhne, N., & Nascimento, L. (2024). *Results: Monitoring Climate Mitigation Efforts of 63 Countries plus the EU – covering more than 90% of the Global Greenhouse Gas Emissions,* Climate Change Performance Index, Retrieved June 12, 2024 from https://ccpi.org/wp-content/uploads/CCPI-2024-Results.pdf

C3S & WMO. (2024). *European state of the climate 2023*, Copernicus Climate Change Service and World Meteorological Organization, Retrieved June 05, 2024 from https://climate.copernicus.eu/esotc/2023

Chataut, G., Bhatta, B., Joshi, D., Subedi, K., & Kafle, K. (2023). Greenhouse gases emission from agricultural soil: A review. *Journal of Agriculture and Food Research*, 11, 100533. DOI: 10.1016/j.jafr.2023.100533

Climate, P. I. H. (2023a). *Overview of climate targets in Europe,* Retrieved June 15, 2024 from https://climatepolicyinfohub.eu/overview-climate-targets-europe#

Climate, P. I. H. (2023b). *European Climate Policy - History and State of Play,* Retrieved June 14, 2024 from https://climatepolicyinfohub.eu/european-climate-policy-history-and-state-play.html

EC. (1991). *Energy in Europe: Energy policies and trends in the European Community,* Retrieved June 17, 2024 from https://aei.pitt.edu/79862/1/17._July_1991.pdf

EC. (1992). *A community strategy to limit carbon dioxide emissions and improve energy efficiency*, P/92/29, Retrieved June 10, 2024 from https://ec.europa.eu/commission/presscorner/detail/en/P_92_29

EC. (2018a). *Kyoto 1st commitment period (2008–12),* Retrieved June 12, 2024 from https://climate.ec.europa.eu/eu-action/international-action-climate-change/kyoto-1st-commitment-period-2008-12_en

EC. (2018b). *Communication from the Commission: to the European Parliament, the European Council, the Council, the European Economic and Social Committee, the Committee of the Regions and the European Investment Bank, A Clean Planet for all A European strategic long-term vision for a prosperous, modern, competitive and climate neutral economy,* Brussels, 28.11.2018 COM (2018) 773 final, Retrieved June 14, 2024 from https://eur-lex.europa.eu/legal-content/EN/TXT/PDF/?uri=CELEX:52018DC0773

EC. (2019). *Communication from the Commission: The European Green Deal,* Brussels, 11.12.2019 COM (2019) 640 final, Retrieved June 14, 2024 from https://eur-lex.europa.eu/legal-content/EN/TXT/?qid=1576150542719&uri=COM%3A2019%3A640%3AFIN

EC. (2022). The new European consensus on development 1our world, our dignity, our future", Retrieved August 08, 2024 from https://international-partnerships.ec.europa.eu/document/download/6134a7a4-3fcf-46c2-b43a-664459e08f51_en?filename=european-consensus-on-development-final-20170626_en.pdf

EC. (2023a). *Causes of climate change,* Retrieved June 04, 2024 from https://climate.ec.europa.eu/climate-change/causes-climate-change_en

EC. (2023b). *Special Eurobarometer 538 climate change,* Retrieved June 12, 2024 from https://europa.eu/eurobarometer/surveys/detail/2954

EC. (2023c). *GHG emissions of all world countries: JRC science for policy report,* Retrieved June 06, 2024 from https://edgar.jrc.ec.europa.eu/report_2023

EC. (2024). *Fit for 55,* Retrieved June 16, 2024 from https://www.consilium.europa.eu/en/policies/green-deal/fit-for-55/

EEA. (2010). *Tracking progress towards Kyoto and 2020 targets in Europe,* Retrieved June 15, 2024 from https://cetesb.sp.gov.br/inventario-gee-sp/wpcontent/uploads/sites/34/2014/04/eea_european.pdf

EEA. (2012). *Key observed and projected climate change and impacts for the main regions in Europe,* Retrieved June 06, 2024 from https://www.eea.europa.eu/soer/data-and-maps/figures/key-past-and-projected-impacts-and-effects-on-sectors-for-the-main-biogeographic-regions-of-europe-3

EEA. (2021). *EU achieves 20-20-20 climate targets, 55% emissions cut by 2030 reachable with more efforts and policies,* Retrieved July 02, 2024 from https://www.eea.europa.eu/highlights/eu-achieves-20-20-20

EEA. (2023). *Flexibility solutions to support a decarbonised and secure EU electricity system,* Retrieved July 08, 2024 from https://www.eea.europa.eu/publications/flexibility-solutions-to-support

EEA. (2024a). *European climate risk assessment: Executive summary,* Retrieved June 05, 2024 from https://www.eea.europa.eu/publications/european-climate-risk-assessment

EEA. (2024b). *Extreme weather: floods, droughts and heatwaves,* Retrieved July 01, 2024 from https://www.eea.europa.eu/en/topics/in-depth/extreme-weather-floods-droughts-and-heatwaves

EEA. (2024c). *Climate change impacts, risks and adaptation,* Retrieved July 03, 2024 from https://www.eea.europa.eu/en/topics/in-depth/climate-change-impacts-risks-and-adaptation

Elzen, M.G.J., Lucas, P.L., & Gujsen, A. (2007). *Exploring European countries' emission reduction targets, abatement costs and measures needed under the 2007 EU reduction objectives,* Netherlands Environmental Assessment Agency (MNP).

EP. (2019). *European policies on climate and energy towards 2020, 2030 and 2050,* Briefing, ENVI in FOCUS, Retrieved July 02, 2024 from https://www.europarl.europa.eu/RegData/etudes/BRIE/2019/631047/IPOL_BRI(2019)631047_EN.pdf

EU. (2019). *Europe's approach to implementing the Sustainable Development Goals: good practices and the way forward,* Directorate-General for External Policies Policy Department, Retrieved 08 August, 2024 from https://www.europarl.europa.eu/cmsdata/160360/DEVE%20study%20on%20EU%20SDG%20implementation%20formatted.pdf

Fernandes, P. M., & Rigolot, E. (2022). *13. Prescribed burning in the European Mediterranean Basin.* Global Application of Prescribed Fire. DOI: 10.3390/en17112476

Hei Ngu, L. (2024). Carbon capture technologies. *Encyclopedia of Sustainable Technologies (Second Edition).*

Heidari, H., Arabi, M., & Warziniack, T. (2021). Effects of climate change on natural-caused fire activity in Western U.S national forests. *Atmosphere (Basel)*, 12(8), 981. DOI: 10.3390/atmos12080981

Hillebrand, L., Marzini, S., Crespi, A., Hiltner, U., & Mina, M. (2023). Contrasting impacts of climate change on protection forests of the Italian Alps. *Frontiers in Forests and Global Change*, 6, 1240235. DOI: 10.3389/ffgc.2023.1240235

Koutsias, N., Arianoutsou, M., Kallimanis, A. S., Mallinis, G., Halley, J. M., & Dimopoulos, P. (2012). Where did the fires burn in Peloponnisos, Greece the summer of 2007? Evidence for a synergy of fuel and weather. *Agricultural and Forest Meteorology*, 156, 41–53. DOI: 10.1016/j.agrformet.2011.12.006

Liu, F., Shafique, M., & Luo, X. (2023). Literature review on life cycle assessment of transportation alternative fuels. *Environmental Technology & Innovation*, 32, 103343. DOI: 10.1016/j.eti.2023.103343

MFA. (2023). *History of Türkiye- EU relations*, Retrieved July 03, 2024 from https://www.ab.gov.tr/brief-history_111_en.html

Pronto, L., Prat-Guitart, N., & Caamaño, J. (2023). *Research for REGI Committee – Forest fires of summer 2022*, European Parliament, Policy Department for Structural and Cohesion Policies, Brussels.

Rabinovych, M., & Pintsch, A. (2023). Sustainable development: A common denominator for the EU's policy towards the Eastern partnership? *The International Spectator,* Vol. 58, No. 1, 38–57, Retrieved August 08, 2024 from https://Doi.Org/10.1080/03932729.2023.2165774

Rodrigues, A., Santiago, A., Laím, L., Viegas, D. X., & Zêzere, J. L. (2022). Rural fires—Causes of human losses in the 2017 fires in Portugal. *Applied Sciences (Basel, Switzerland)*, 12(24), 12561. DOI: 10.3390/app122412561

San-Miguel-Ayanz, J., Durrant, T., Boca, R., Maianti, P., Libertá, G., Artés-Vivancos, T., Oom, D., Branco, A., de Rigo, D., Ferrari, D., Pfeiffer, H., Grecchi, R., Onida, M., & Löffler, P. (2022). *Forest Fires in Europe, Middle East and North Africa 2022*. EUR 31269 EN, Publications Office of the European Union. DOI: 10.2760/34094

UN. (2023). *Europe warming twice as fast as other continents, warns WMO*, Retrieved August 08, 2024 from https://news.un.org/en/story/2023/06/1137867

UNFCCC. (2014). EU *Agrees 40% greenhouse gas cut by 2030*, Retrieved June 12, 2024 from https://unfccc.int/news/eu-agrees-40-greenhouse-gas-cut-by-2030

UNFCCC. (2023). *History of the convention*, Retrieved July 03, 2024 from https://unfccc.int/process/the-convention/history-of-the-convention#Essential-background

ENDNOTES

[1] Türkiye is a candidate country for the European Union (EU) and does not yet have full membership. Relations between the EU and Türkiye began with the signing of the Ankara Agreement in 1963, which established an association between Türkiye and the European Economic Community (EEC). In 1987, Türkiye applied for full EU membership and was granted candidate status in 1999. Accession negotiations began in 2005 but have progressed slowly due to various political, economic, and human rights issues. Despite these challenges, Türkiye remains an important partner for the EU in areas such as trade, security, migration, and regional stability (Ministry of Foreign Affairs [MFA], 2023).

[2] A vulnerable population refers to groups of people who are at a higher risk of experiencing adverse effects from various environmental, economic, social, or health-related factors. These groups may include elderly individuals, children, minorities and marginalized communities, homeless people, women, particularly pregnant women, migrant and refugee populations.

[3] IPCC stands for the Intergovernmental Panel on Climate Change, a scientific intergovernmental body operating under the auspices of the United Nations.

[4] In this context, "developing countries" refers to nations that are still in the process of industrializing and improving their economic status. These countries typically have lower levels of income per capita, industrial activity, and overall development compared to "developed" countries. As a result, their GHG emissions have historically been lower due to less industrialization and energy consumption. However, as these countries continue to grow economically and industrialize, their emissions are expected to rise more rapidly. This is due to increased energy demand, industrial output, urbanization, and other factors associated with economic development. Developing countries often include nations in regions such as Sub-Saharan Africa, South Asia, Southeast Asia, Latin America, and parts of the Middle East.

[5] Council Directive 92/42/EEC of 21 May 1992 on efficiency requirements for new hot-water boilers fired with liquid or gaseous fuels and Directive 96/57/EC of the European Parliament (EP) and of the Council of 3 September 1996 on energy efficiency requirements for household electric refrigerators, freezers and combinations thereof.

[6] COP stands for Conference of the Parties. It is the supreme decision-making body of the Convention. All States that are Parties to the Convention are represented at COP meetings, where they review the implementation of the Convention and any other legal instruments adopted by the COP. They also make decisions necessary to promote the effective implementation of the Convention, including institutional and administrative arrangements.

[7] Global Stocktake is a mechanism established as part of the Paris Agreement. This mechanism provides for a regular assessment process of global climate goals and progress. Global Stocktake evaluates how countries are progressing towards their climate targets and what advancements have been made globally towards achieving common objectives. This assessment process assists countries in reviewing their commitments, setting more ambitious goals, and coordinating collective efforts towards fulfilling the objectives of the Paris Agreement.

[8] Regulation (EU) 2021/1119.

[9] Under the Paris Agreement, each country's NDCs are the commitments and targets set to combat climate change. NDCs include strategies for reducing greenhouse gas emissions and adapting to climate change, tailored to each country's national circumstances and capacities.

Chapter 14
Calibrating Climate Change Curriculum Coverage in Some Modules at St. Peter's University

Sikhulile B. Msezane
University of South Africa, South Africa

Nonkanyiso Pamella Shabalala
University of South Africa, South Africa

ABSTRACT

In this qualitative research approach study, the coverage of climate change education content at St. Peters University was investigated utilising a case study research design and document analysis. This study gathered pertinent information from three modules that covered education for sustainable development using the convenience sampling method. Realist social theory was employed as an analytical and theoretical framework. The study's conclusions showed that the modules' coverage of climate change risks, hazards, mitigation, resilience, and adaptation measures is lacking. Even though climate change was discussed inequitably, the implication is that students would not be able to cascade climate change education content to citizens during teaching practice, inhibiting awareness, and acting toward a behavioural change that encourages risk identification, climate change injustices, mitigation, adaptation, and resilience.

DOI: 10.4018/979-8-3693-5792-7.ch014

INTRODUCTION AND BACKGROUND

The United Nations Framework Convention on Climate Change (UNFCCC) describes climate change as "a change in climate that is attributed directly or indirectly to human activity, alters the composition of the global atmosphere, and is in addition to the natural climate variability observed over comparable time periods" (UNFCCC, 1992). The scope of climate change content in some modules taught at one of the departments at a tertiary institution was investigated in this study in response to the definition of climate change. As a result, this study will calibrate the coverage of climate change education (CCE), which has developed parallel to, but not simultaneously with the field of scientific research on climate change (Busch, Henderson, & Stevenson, 2019). Education about climate change focuses on the issue and helps students create practical solutions. It aids students' comprehension of the causes and effects of climate change, equips them with the tools they need to cope with its effects, and gives them the authority to take the necessary steps to embrace more sustainable lifestyles (UNESCO, 2015).

Climate change and climate change education are unquestionably global issues that can be incorporated into the curriculum to promote local learning and broaden perceptions of how to lessen the effects of climate change. As demonstrated by this research, CCE is more than just understanding how to deal with climate change (UNSSC, 2023) and involves exploring issues of risks, vulnerability, adaptation, resilience, and mitigation to determine how deeply these phenomena are ingrained in the study and learning materials that teachers and students use. These communities subsequently gain knowledge about how climate change will impact them, what they can do to safeguard themselves from negative effects, and how they can lower their own carbon footprint (UNESCO, 2015). In a university, CCE will help students gain knowledge and use the right teaching methods to promote adaptation and climate change mitigation. According to the IPCC report from 2022, climate change is a long-term problem, and elements like the risk associated with environmental consequences should be understood so that people can come up with adaptation strategies and learn how to lessen the effects of climate change. The following elements will be looked at in this study to see if they are included in the study materials for students. Risk related to climate change, vulnerability, adaptation, resilience, and mitigation are only a few of these factors. The following definitions are used in this paper to gauge how much of these topics are addressed in the modules. The definition of risk in this study, which was modified from the IPCC (2022) definition, is the potential for unfavourable effects on ecological or human systems. This definition considers the variety of values and objectives connected to these systems. Risk includes vulnerability, but vulnerability also deserves significant attention on its own. The definition of vulnerability in this study is "the propensity or predisposition to be negatively impacted" and it includes a wide range of ideas and components, such as "sensitivity or susceptibility to harm" and "lack of capacity to cope and adapt. In this study, adaptation is defined as the process of adjusting to the effects of the present or anticipated climate to lessen harm or take advantage of advantageous chances. Natural systems must adapt to the influence of the actual climate; human intervention may make this process easier. In this study, resilience is defined as the ability of social, economic, and environmental systems to deal with a risky event, trend, or disturbance by reacting or reorganising in ways that preserve their core purpose, identity, and structure while also preserving their capacity for adaptation, learning, and transformation (IPCC, 2022).

Terrestrial, freshwater, coastal, and open ocean marine ecosystems have suffered significant harm from climate change, and these losses are becoming increasingly irreversible (IPCC, 2022). Impacts of climate change are greater in scope and intensity than predicted by earlier studies. For instance, according to IPCC (2022), around half of the species evaluated globally have moved poleward or, on land, also to higher altitudes. Increases in the intensity of heat extremes, mass death events on land and in the ocean, and the disappearance of kelp forests have all contributed to hundreds of local extinctions of species. According to IPCC (2022), some losses, like the first species extinctions brought on by climate change, are already irreversible. Other effects, like as the hydrological changes brought on by glacier retreat or the alterations in some mountain and Arctic ecosystems brought on by permafrost thaw, are on the verge of becoming irreversible.

The frequency and intensity of extreme weather events have increased due to climate change, which has decreased food and water security and made it more difficult to achieve the Sustainable Development Goals (IPCC, 2022). Although worldwide agricultural output has increased generally, climate change has slowed this expansion over the past 50 years. The negative effects of climate change were primarily felt in mid- and low latitude regions, but certain high latitude regions also saw benefits. The IPCC (2022) study also stated that in some maritime regions, shellfish farming and fisheries provide less food due to ocean warming and acidification. Millions of people have been subjected to severe food insecurity and decreased water security due to an increase in weather and climatic extreme events, with the greatest effects being seen in many places and/or communities in Africa, Asia, Central and South America, Small Islands, and the Arctic (IPCC, 2022).

The observed climate change in urban areas has had an influence on people's health, way of life, and essential infrastructure (IPCC, 2022). Cities, towns, and infrastructure are affected by numerous climatic and non-climate dangers, which can overlap and amplify harm. Heatwaves and other hot weather extremes have become more frequent in cities, where they have also made air pollution problems worse. The extent to which concerns of climate change injustice are covered in the teaching and learning resources will be examined during the module analysis. According to Trott et al. (2023), efforts to adopt and uphold the principles of universal human rights as well as conceptions of environmental justice are the foundations of climate justice.

Rising sea levels, droughts, and extreme weather events are already disproportionately burdening the effects of climate change caused by greenhouse gas emissions (Hickel, 2020). The youngest and oldest populations in the world as well as low-income and indigenous, black and brown and communities of colour are among the groups experiencing elevated climate-driven risks and harms (Benevolenza & DeRigne, 2019; Helldén et al., 2021). This is because climate change exacerbates age-based physiological vulnerabilities and intensifies already-existing social, economic, and health inequities. These populations frequently are excluded from decision-making and action areas related to the climate catastrophe, which adds another layer of injustice (Archer et al., 2014; Fitzgerald, 2022)

Importantly, Monroe et al. (2019) acknowledge that teachers and researchers are aware that how we approach teaching about climate change may differ from how we approach teaching about other environmental challenges. Even more than moral debates over the disposal of hazardous waste or declining biodiversity, the subject of climate change seems to powerfully resonate with held values. As a result, adults respond by defending their group identity and way of life (Monroe et al., 2019).

The research question for this paper is.

How much of the teaching on climate change is included in the education for sustainable development modules offered at St. Peter's University.

LITERATURE REVIEW

Asia and the Pacific, North America, Africa, Southern Africa, and South Africa will be highlighted in this section's discussion of the worldwide scope of climate change.

Asia and the Pacific

The infrastructure and cities in this region are also severely impacted by climate change, which is especially extreme in Pacific island and coastal zone nations. It has been observed that the Tarim Basin in Central Asia has expanded by about 0.67 percent annually in the Tian Shan, Altun, and Shan mountains because of wide distribution of glacial lake melt water over a period of 23 years (1990-2013), as well as precipitation increase and evaporation decrease due to climate change (Wang, Liu, Liu, Wei & Jiang, 2016). According to Wang et al. (2016), the growth of glacial lakes between 2000 and 2013—a result of glacial retreat and snow cover thickening since 2000—was around four times more rapid than that of the years 1990–2000. According to Li and Sheng (2012), severe meltwater from snow and glaciers feeds lakes in the high mountains of Central Asia, where water levels of these lakes remained steady despite rising temperatures. Aside from the growth of glacier-fed lakes in the Asia-Pacific region, pollution and carbon emissions in the area have limited the area's ability to continue developing (Wei, Wang & Liao, 2016).

North America

The UNEP report highlights how climate change affects human health as well as many environmental factors and, in certain circumstances, human security (UNEP, 2016). The Arctic is a location of particular concern since the effects of climate change are most pronounced in the low-temperature zones and the likelihood of future alterations is increasing (UNEP, 2016).

Africa

Climate change's effects hamper Africa's growth and development. This is since many climate-vulnerable regions of Africa are in regions where many Africans depend economically on land-based activities like agriculture and fishing. The biggest human and environmental crisis of the 21st century has been named as climate change (ISS, 2010; UNEP, 2016). Furthermore, Jeong (2006) notes that emissions from factories and automobiles are responsible for nearly two-thirds of the planet's excess carbon. More significantly, the decline in the number of trees that can absorb carbon dioxide has made it harder to reduce the high amounts of greenhouse gases in the atmosphere. Africa contributes 4% of the world's total carbon dioxide emissions, but its citizens experience the effects of global climate change more than any other country (Komolafe et al., 2014; UNEP, 2008).

Inhibited mostly by economic limitations, which make the continent vulnerable to environmental effects, the response of the continents to global climate change is rather limited. The effects of climate change are obvious in Africa, where temperatures are rising, soils are drying out, there are more pests and diseases, and ideal places for raising livestock and crops are shifting. Floods, deforestation, erosion, and rising desertification in the Sahara region are additional visible repercussions of climate change. According to the IPCC (2022), climate change will almost certainly result in yield losses in Africa. Estimated yield losses at the halfway point of the century range from 18% for southern Africa to a total yield loss of 22% for sub-Saharan Africa, with yield losses for South Africa and Zimbabwe exceeding 30%. According to the United Nations Environmental Programme (UNEP, 2008), even little changes in rainfall and water availability in many African regions might have a significant impact on agriculture, making adaptation more challenging as climate change increases and its effects worsen.

Southern Africa

Southern Africa's freshwater, marine, and terrestrial ecosystems have changed because of climate change (IPCC, 2014). The SADC (2008) notes that due to decreasing rainfall totals and rising evaporation rates, climate fluctuation in the region influences the availability of water resources and food security. One sign of the detrimental effects of climate change has been the frequency of droughts, floods, and tropical cyclones. A wide range of environmental and socioeconomic activities are negatively impacted by droughts in the region, particularly in nations like Botswana, Malawi, and Mozambique. On the other hand, tropical cyclones and floods extend the wet season in some parts of the region. Between January and March of 2012, Mozambique saw two significant cyclones that caused landfalls in southern Africa and Madagascar (Chikoore & Vermeulen, 2015). The southwest Indian Ocean islands and the coast of southeast Africa are particularly susceptible to the effects of tropical cyclones. According to Chikoore and Vermeulen (2015), Tropical Cyclone Giovanna struck Madagascar in February 2012 and killed 35 people, while Tropical Cyclone Irina's looping route claimed the lives of 65 people in Madagascar. One of the worst tropical cyclones to ever hit Southern Africa, Tropical Cyclone Idai, recently occurred because of climate change. Idai, a tropical cyclone, left more than 1300 people dead and countless others missing after wreaking havoc in Mozambique, Zimbabwe, and Malawi (Yuhas, 2019). These cyclones are unfavourable effects of climate change that have spread throughout southern Africa. Along with Madagascar, these cyclones also had an impact on nearby Swaziland and the eastern lowlands of South Africa (Chikoore & Vermeulen, 2015).

SADC (2008) found that over the previous 100 years, regional temperatures have increased by more than 0.5 degrees Celsius. Mulenga et al. (2016) go on to say that the consequences of rising temperatures and unpredictable rainfall have a detrimental effect on smallholder maize output in Zambia and more generally in the southern African region. The increase in temperatures has been attributed to several factors, including human activity (SADC, 2008). Unpredictable rainfall patterns, heat waves that interfere with agricultural development, the spread of alien weed species, and wildfires have all contributed to Swaziland's low production (Orchard et al., 2016). Methane, which is emitted during the decomposition of organic materials, has also increased in atmospheric concentration in the area (SADC, 2008).

Livestock, particularly cattle ranching in the area, is the main producer of methane. Increases in sea levels brought on by the melting of snow and ice, decreases in food yields, and the expansion of malaria to some regions of South Africa and Namibia are further effects of climate change.

South Africa

Information about the dangers, impacts, and vulnerability of climate change has been produced by the international community, but there is still a significant information vacuum in South Africa and the rest of southern Africa regarding the long-term trends of global warming (DEA, 2016). Although there is little information about long-term trends in climate change, the South African Weather Service and the National Disaster Management Centre are increasingly synthesising and communicating risks associated with observed climatic conditions, such as droughts, hailstorms, poor water quality, and high temperatures (DEA, 2016). Increased coral reef bleaching in the tropical coastal waters of Sodwana Bay in KwaZulu-Natal is the most compelling evidence of long-term climate change impacts in South Africa (Celliers & Schleyer, 2008). Climate change indicators that are now accessible in South Africa include greenhouse gas emissions, temperature, and rainfall (SAEO, 2016). In keeping with the global forecasts of a 20 percent increase in greenhouse emissions between 2000 and 2010, the nation's greenhouse emissions are rising. The fact that the energy supply sector accounts for around 50% of total emissions in South Africa shows how heavily dependent the nation's electricity supply is on coal. About 16 percent of the remaining emissions are produced by industry and construction, 16 percent by fuel emissions, 9 percent by transportation, and the remaining emissions are produced by agriculture and forestry, garbage, and other small sources. The research states that from the middle of the 1960s, the temperature has been generally warming over the past 40 years. In the western, north-eastern, and far eastern regions of the nation, the temperature increase has been more pronounced. Furthermore, according to IPCC (2022) and Nunn (2012), warm temperatures cause health problems as well as disruptions to livelihoods due to storm surges, sea level rise, and coastal flooding. The patterns of rainfall in South Africa have also undergone significant alterations because of global warming. The seasons of wet and dry weather have been more extreme due to the rainfall (SAEO, 2016).

During the El Nino Southern Oscillation circumstances, South Africa had severe droughts, especially in the Western Cape Province (DEA, 2017). According to Kemp et al. (2003), a drought is an environmental condition that denotes a protracted dry spell, typically accompanied by a lack of precipitation, during which crops wither and reservoirs are reduced in size. The Western Cape Province experienced this definition between 2017 and 2018. The effects of drought events on society and the environment are getting increasingly difficult when they coincide with rising urban water demands and social vulnerability issues, as observed in the Western Cape Province. According to the IPCC (2014) and UNESCO (2018), extreme heat also contributes to food and water shortages, ecosystem degradation, and biodiversity loss, all of which are currently being felt in Western Cape regions that are suffering from drought. The year 2015 saw above-average temperatures, making it the warmest year on record since 1951 (DEA, 2017). In comparison to average maximum temperatures in cities like Pretoria, Johannesburg, Limpopo, Mpumalanga, and the Northwest, the country's maximum temperatures were more than 5% higher. Changes from extraordinarily high amounts of rainfall and storms to drought occurrences between 2010 and 2015 are evidence of the short-term effects of climate change trends in South Africa (DEA, 2016). Wildfire incidents frequently serve as indicators of high temperature effects (DEA, 2016).

Environmental Education/Education for Sustainable Development as a Response to Environmental Impacts

According to Hill et al. (2006: 93), nature is also seen as a vulnerable natural resource that can be overexploited and damaged, endangering human existence. Both informal, public-facing methods and formal methods, such its incorporation into basic through tertiary level curriculum, have been used to promote environmental education (EE)/education for sustainable development (ESD) (Hill et al., 2006). The main responsibility of EE is to combat the epidemic of environmental impact worries by environmental knowledge, awareness, attitude modification, and the development of skills to deal with harmful environmental effects. The biophysical, human (social, economic, and political), and environmental elements are all shaped by EE/ESD. Since EE/ESD is a response to crises caused by human environmental impact, it is crucial to conduct research that focuses on the growing effects of human activity on the environment and the significance of improving knowledge, awareness, skills, attitudes, decision-making, action, and behaviour to address harmful environmental effects.

According to Kyburz-Graber (2013), EE/ESD was introduced as a new requirement for educational institutions in many nations near the end of the 20th century as a reaction to mounting evidence of environmental damage brought on by human activity. To stop environmental deterioration, numerous documents and programmes have been created. Sitarz (1994) further points out that the rise in global population and unsustainable production, consumption, and development patterns, particularly in industrialised nations, have boosted economic growth and exacerbated adverse environmental effects. Nations all around the world today acknowledge that industrialization and excessive resource use have harmed the environment and produced uncontrollable levels of waste and pollution. This study focuses on how environmental impact themes are covered in the curriculum, relevant resource materials, and examinations. South Africa has been undergoing environmental deterioration, which is like the global scenario of environmental degradation (DEA, 2017).

According to Kyburz-Graber (2013), education is crucial for resolving environmental issues. The purpose of this study was to examine how environmental impact subjects were covered in the South African curriculum. Gough (2013) emphasises that the inclusion of EE in the curriculum enables the creation of trans-cultural spaces in which researchers from other regions work to reframe and disseminate their own knowledge traditions, which is in line with the goal of this study. He adds that since EE/ESD is ever changing and evolving, there is still a lot of study to be done. This study investigates how environmental issues are covered in the curriculum for early childhood development (ESD). Gough (2013) emphasises the need for a set of values-based methodologies with a particular approach to analysing a phenomenon in EE/ESD research.

In agreement with Gough (2013), Robottom (2013) noted that the environmental education phenomena from the early 1970s to the present has been characterised by both continuity and contestation. He also emphasises the connection between "education for sustainable development" and the extraordinary persistence of environmental practise as well as the debate over the terminology used in the sector (ESD). According to John, Mei, and Guang (2013), to successfully implement a curriculum innovation like EE, the stakeholders must be meticulously and methodically prepared for the shift. They go on to say that to support, monitor, and assess the changes, the stakeholders should also initiate environmental impact study. Additionally, research and development initiatives are required to investigate topics like organisational capacity and the integration of the curriculum, as well as teacher professional development. Lotz-Sisitka (2013) sees that Post-structuralism and Critical Realism have shaped and influenced EE curriculum research in this regard. She goes on to say that this type of research should place more of an emphasis on theory creation, where new places need to be created rather than methods. Hill et al. (2006) also note that EE should not simply replicate the current realities of coexisting with nature, but also allow people to explore alternative realities, enable them to critically assess these realities, and allow them to make decisions about what the right interaction with nature should be in their local context.

Coverage of ESD Content in the Curriculum of Ten Southern Africa Countries

Teacher Education Curriculum

The ten southern African nations of Botswana, Lesotho, Malawi, Mauritius, Mozambique, Namibia, South Africa, Swaziland, Zimbabwe, and Zambia all have teacher education programmes that incorporate ESD knowledge, according to a 2017 UNESCO report on Environmental scan informing teacher education programming for ESD secondary teacher education. The majority of these nations' undergraduate (Bachelor of Education degrees) and doctoral programmes focus on integrating ESD (masters degrees, Bachelor of Education Honours degrees and Postgraduate Certificates in Education). In most countries, ESD content is covered more in pre-service programmes, however in certain countries, pre-service and in-service ESD coverage are similar (UNESCO, 2017). Through various courses in the Faculty of Education and in other faculties and fields that offer modules to teacher educators, ESD components have been included in all ten of the nations featured in this research. For instance, the Faculty of Sciences in Namibia's Biological Sciences does include ESD content, and some modules—like the Ecosystems Ecology modules—have extensive ESD understanding (UNESCO, 2017).

A course with a significant emphasis on ESD has just been designed by the University of Botswana. According to the UNESCO (2017) study, ESD is present in some institutions' teaching strategies where a focus is put on collaborative learning, problem solving, and project work. The graduate diploma student teachers' criteria for teaching practise evaluate how well students interact with ESD-focused teaching strategies in addition to the specific subject covered in science and social studies classes. ESD is included as a co-curricular activity in some colleges of teacher education (such as clubs).

South Africa

Teacher Education Curriculum

The national school curriculum has changed in South Africa, and as a result, so has the inclusion of ESD knowledge in the curriculum. Different teacher education credentials, such as bachelor's degrees, postgraduate certificates, honours degrees, master's degrees, and PhD degrees, cover ESD (UNESCO, 2017). These programmes cover a variety of environmental and sustainability subjects in relation to curricula. The framework of the teacher education curriculum helps student teachers comprehend and apply the environmental and sustainability content that is incorporated into all levels of the CAPS curriculum. According to UNESCO's environmental scan from 2017, the pre-service programmes cover ESD in teacher education to a much larger extent. Environmental education is provided as a separate course at some universities, including the University of South Africa (UNISA) and the Northwest University (UNESCO, 2017). Environmental education is an elective in the bachelor's and master's degree programmes at certain other schools, like Rhodes University. The Fundisa for Change programme, which supports teacher educators and teacher professional development, developed a teaching methodology that Rhodes University's Bachelor of Education Honours elective uses to help students develop their environmental learning and teaching context by emphasising fundamental environmental knowledge, teaching practise, and assessment (UNESCO, 2017). The method calls for subject-matter expertise, better teaching techniques, and better evaluation techniques to contextualise environmental learning within each student's unique professional and educational context.

In all South Africa's provinces, ESD is made available to in-service teachers in the form of brief courses. To educating teachers in the SADC region about ESD, the short courses are occasionally also offered to other SADC nations. A collaboration initiative called Fundisa for Change, which includes all South Africa's key environmental organisations (state, parastatal, and NGO) with a stake in teacher preparation, was formed in 2011 (Fundisa for Change, 2013). To strengthen systemic impact, the Fundisa for Change programme strives to unite sectoral activities. Its primary goal is to improve environmental education and transformative environmental learning in classrooms (Fundisa for Change, 2013). As these are "catalytic institutions" with potential long-term impact on the education and training of South African teachers, it achieves this by concentrating on building the ability of South Africa's teacher education institutions and important partners participating in teacher development.

THEORETICAL FRAMEWORK

The case study research's theoretical foundation is Realist Social Theory (RST). The theoretical and methodological framework methods are the foundation of this RST investigation. As Young (2008) and Creswell (2009) describe, people want understanding of the world in which they live and work. As a result, it is important to explore social interests and the accompanying power dynamics. The effects of the global environmental crisis, such as ozone layer loss, growing carbon dioxide levels in the atmosphere, global warming, deforestation, climate change, pollution, and incorrect toxic waste disposal, show that it is a real social issue. According to Hartas (2010), people construct their own realist interpretations of their experiences through interactions with others and their surroundings. The connections and functions of structure, culture, and agency in the inclusion of environmental impact subjects like climate change in the curriculum made up the theoretical foundation of this study. According to the critical realism paradigm, which serves as the study's foundation, social reality is stratified and emergent (Parra, Said-Hung & Montoya-Vargas, 2020; Howell, 2015).

METHODOLOGY

Research Paradigm, Approach, and Design

This study employed a constructivist research paradigm because the researchers made every effort to guarantee that the point of view of the subject being watched was understood separately from the point of view of the observer (Sobh & Perry, 2006). To better understand the occurrences in genuine natural settings, the researchers employed a qualitative study methodology (Leedy & Ormord, 2013). To provide a complete investigation of current modifications regarding a real-life occurrence, a descriptive case study design was adopted (Yin, 2014). To capture the intricacy of the dialectical interaction between ideas and practises, this study used a descriptive case study technique. Document analysis offers an efficient way for case study research, as proposed by Bowen (2009), and this study is in line with those circumstances. Document analysis provides researchers with access to a large amount of information that has already been produced and collected, such as curricular materials, course outlines, and textbooks. This can be particularly useful when studying curriculum coverage, as it allows researchers to identify patterns and gaps in coverage across multiple sources. An in-depth examination of a programme, event, activity, or process by the researchers is accomplished using a case study design, which was selected for this study (Bowen, 2009). In addition, Bassey (1999) mentions, concurring with Johnson, Christensen (2008), and Creswell (2009), that one of the purposes of using case studies is to enable inquiries into educational programmes, systems, projects, or events to verify their value, as seen in the analysis of researchers, and to convey this to interested audiences.

Documents

Three environmental education modules from the Department of Science and Technology Education were purposefully sampled for this study. The likelihood that a document will include important data for the current research is the criterion used to choose which papers to analyse. Analysing documents requires reading, skimming, and interpreting them (Bowen, 2009). In this study, the study materials utilised in the purposively sampled modules were subjected to content analysis, which involves finding meaningful and pertinent information on educating people about climate change.

Data Collection

A document analysis was utilised to gather data for this study. According to O'Leary (2014), the analysis of documents is a qualitative research technique in which the researcher analyses documents to give a study topic voice and meaning. To obtain the climate change-related content that was contained in the Education for Sustainable Development/Environmental Education courses, study guides, tutorial letters, and required books were analysed. In order to develop document analysis questions that can be answered by text analysis, a document analysis guide that uses research questions as its foundation was developed. We had to download the original documents to annotate them. We also had to assess the reliability of the documents. Finally, we investigated the context's specifics and the content of the document. On the university repository, we had access to these documents for download. According to Yin (2011), document analysis entails gathering and inspecting objects. Using information gleaned from documents, he claims, one can create a wide range of verbal, numerical, graphic, and pictorial data forms. Data from documents, according to Yin (2011), can provide essential information about things that are not immediately observable (policies, study guides, etcetera).

Document Analysis

Documents are objects that offer secondary evidence, yet they are crucial in qualitative research because the findings they produce may support the primary evidence. According to Yin (2011), gathering information from documents takes time, thus it's important to pay close attention to choosing the precise details that will be used in the analysis of the research. Yin (2011) provides examples of strategies that are required to guarantee that the right data is gathered to support the claim. These strategies are as follows: first, get a general idea of the full array of any type of object to be collected, including the range of years, size of documents, and scope of available documents; second, after some preliminary collecting, review the data and consider how the collected material is likely to fit the rest of the study. The dependability of the data gathered can be ensured by repeatedly using this strategy.

Yin (2011) emphasises that written records can supplement in-person meetings and dialogues. Before the interviews take place, these documents can be read to gather subject knowledge on the topics that will be covered in the interviews. Research documents are crucial because they allow the researcher to be aware of the availability of different research materials in advance. Documents, which cannot be changed by the researcher as is the case with qualitative interviews, can lessen the issues and obstacles of reflexivity, according to Yin (2011).

The acquired data from document analysis was interpreted in this study using a document analysis. The researchers became familiar with the data they had gathered, employed coding to organise the data into codes from the analysis, and then calibrated themselves as instruments of analysis. Then data was organized into themes and categories to interpret the findings. The analysis was validated by the other researchers to compare the documents analysed and the results of the current research.

Figure 1. Procedure of document analysis

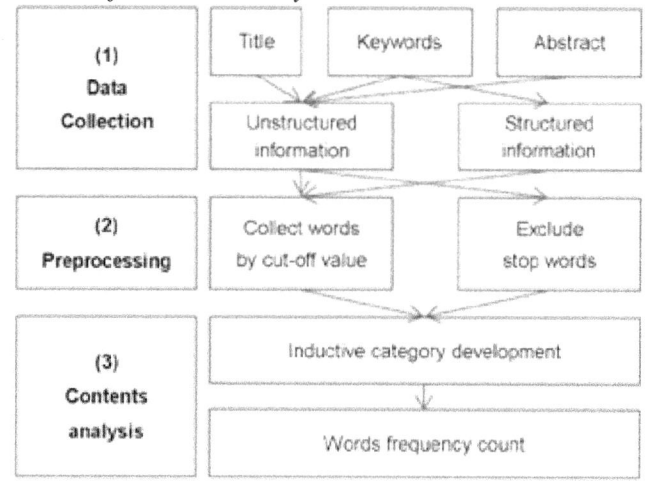

Source: Adopted from Rhie, Lim & Yun (2014)

Figure 1 illustrates how data were gathered from purposefully chosen ESD/EE modules within the department utilising the title, keywords, and abstract as a guide for this unstructured qualitative data. The objective of this study's second stage, preprocessing, was screening terms for concepts relevant to climate change, including mitigation, adaptation, and environmental impacts. For document analysis, this procedure used a deductive method. The traits of climate change inclusion in the study materials were mapped during the last stage of analysis, which also included the inductive technique.

Trustworthiness

All information was referred to preserve the veracity of the sources and to promote credibility. Some of the scholarly approved techniques the researchers used to ensure rigour and trustworthiness were credibility, transferability, dependability, confirmability, and triangulation.

Ethical Considerations

We followed the necessary ethical guidelines, as is typical of qualitative research, when carrying out this study. The research ethics committee of the institution under investigation gave authorization, thus it was our obligation as researchers to determine the extent of climate change material coverage while we were analysing this research.

FINDINGS

In this part, three modules were examined to see how thoroughly they addressed climate change education themes at the tertiary education level. The modules examined were ADD3705, ADS3705, and EED2601 written in codes.

Teaching Science, Society, and Environment in the Natural Sciences and Technology Intermediate Phase (Module Code: ADD3705)

The St. Peters University [Pseudonym] Department of Science and Technology Education offers the Teaching Science, Society, and Environment in the Natural Sciences and Technology Intermediate Phase (ADD3705) curriculum. One of the modules included in the Advanced Diploma in Natural Science Education curriculum is ADD3705.

Teachers who are already in service, not in service, or students are included in this programme. The module was created in 2021 as a year module, and for the previous three years, there have been roughly four students enrolled per year. The goal of this module, which is a component of the Intermediate Phase in Natural Sciences and Technology Education, is to acquire pedagogical content knowledge about science, technology, and society. Students who pass this module will be able to link science, society, technology, and the environment, discuss the online teaching and learning environment in an open and distance education context, and demonstrate in-depth pedagogical knowledge to teach science, society, and the environment in accordance with the curriculum and assessment policy statement for the senior phase of natural science.

The Specific Outcome of This Module Includes

To define and elucidate online learning and teaching within the context of comprehensive open and distance education. to acquire in-depth pedagogical topic knowledge to instruct in Science, Society, and the Environment in accordance with the Natural Sciences and Technology Intermediate Phase Curriculum and Assessment Policy Statement (CAPS). Through hands-on activities, increase student engagement in science, society, and the environment. enhancing students' learning and advancement in science, society, and the environment by integrating assessment with instruction.

General Environmental Content in ADD3705 Module

The module study guide's learning unit 2—which covers society and the environment—contains information on the environment. One way or another, interpersonal interactions throughout society are discussed in the environmental content of society. The emphasis, however, is on the several fundamental elements of the environment, such as the biosphere, atmosphere, lithosphere, and hydrosphere, under the idea of environment. There is also general discussion about environmental issues. The content also discusses environmental valuation, sustainability perspectives, sustainability, sustainability pillars, and sustainable development objectives. Students are also forced to consider how they may use their knowledge to address environmental problems they see in their society because of the activities.

The knowledge of environmental content is the main topic of Learning Unit 3. The following learning objectives for this subject are connected to ESD:

   Define the major terms that are important to this lesson, such as environment and sustainability.
   Describe the environmental issues in the context of your community.
   Determine what material in the Intermediate Phase of Natural Sciences and Technology relates to the environment and sustainability. CAPS
   Describe how the delivery of the Intermediate Phase Natural Sciences and Technology Curriculum can be enhanced by the utilisation of indigenous knowledge.

The main environmental challenges, environmental dangers, sustainability information, and indigenous knowledge—another method of environmental education—are the emphasis of this section.

The emphasis of Learning Unit 4 is on teaching strategies and processes for environmental and sustainability understanding. The following learning outcomes are associated with ESD.

List the educational content knowledge that is relevant to science, the environment, and society, and explain each item.

Justify the inclusion of hands-on learning opportunities in science classrooms. Finally, learning unit 5 concentrates on assessment to improve learning of science, society, and environment. As it relates to ESD, the following learning outcomes are present.

List and analyse the principles for developing alternate evaluations in the context of environmental and sustainability knowledge.

Talk about the many assessment techniques for more thorough, all-encompassing evaluations of the environmental and sustainability material in the natural sciences curriculum.

Justify the use of technology in evaluating environmental and sustainability information.

Climate Change Education Content Found in ADD3705

There is not much in this module's curriculum that addresses climate change. In general terms, the topic of climate change is discussed. The following discussion on climate change is found in chapter 3 of the module's study guide.

"Climate change is a topical issue worldwide. However, the concept of climate change is confused with the concept of climate variability. The two concepts are different. Even though people are perceptive of climate variability, it is not as noticeable as weather variability because it happens over seasons and years. Evidence includes statements such as: "The last few winters have seemed so short." or "There seem to be more heavy downpours in recent years." Climate change is slow and gradual, and unlike year-to-year variability, it is very difficult to perceive without scientific records. There will always be natural climate variability at many scales – decadal, yearly and short-term extreme events. This means that, over the long-term record, there will be ups and downs with the yearly and 30-year averages (climate change is usually determined in periods of 30 years), even if the climate is getting warmer. We cannot expect every summer to be warmer than the previous, but we can expect and plan for variability, etc"

Teaching Science, Society and Environment in the Natural Science Senior Phase (Module Code: ADS3705)

The Department of Science and Technology Education at St. Peters University offers the module Teaching Science, Society, and Environment in the Natural Science Senior Phase (ADS3705). One of the modules included in the Advanced Diploma in Natural Science Education programme is the ADS3705 module.

The module was created in 2021 as a year-long module, and for the past three years, enrollment has averaged fewer than 10 students every year. As a part of the Senior Phase in Natural Science Education, the module's goal is to develop pedagogical content knowledge related to Science, Environment, and Society. Those interested in teaching natural science in the senior phase are given access to this module. It focuses on how to teach and evaluate the school curriculum more successfully by utilising a suitable selection of tactics to improve learner engagement in a variety of circumstances.

The module explains how science, society, technology, and the environment are related. As part of the Senior Phase in Natural Science Education, the module will support teachers' growth and development of pedagogical subject knowledge relating to Science, Environment, and Society. Case studies are presented in the module; the majority of them are followed by an exercise. The activities in this module are designed to determine how well students comprehend the concepts being covered in each case study. The concept explanations contain the answers to the exercises. At least one self-assessment exercise is included in each section to gauge the students' comprehension of the subject matter. After completing this module, the student should be able to recognise that the scientific, environmental, and societal pedagogical content knowledge covered in this module is a component of the Senior Phase in Natural Science Education. Thus, it is crucial that students apply some of the information and abilities they have learned in this module to their actual teaching. Since St. Peters University offers Comprehensive Open Distance Learning, this module's students come from all over the world. For this module, the learning objectives pertaining to ESD are.

- Develop a thorough understanding of pedagogy to instruct in Science, Society, and the Environment in accordance with the Curriculum and Assessment Policy Statement (CAPS) for the Natural Science senior phase.
- Improve learner participation in science, society, and the environment through practical activities is the third specific outcome.
- To improve learning and advancement of students in science, society, and the environment, integrate assessment with instruction (specific outcome 4).

General Environmental Content in ADS3705 Module

The environmental content is introduced in learning unit 2. The main ideas of the module are intended to be covered in Learning Unit 2. Science, technology, society, and the environment are among these ideas. These are the learning objectives for this module's ESD content.

- Define the major terms that are important to this lesson, including science, technology, society, and the environment.
- Describe how science, technology, society, and the environment are related.

In this module study guide's learning unit 2, which covers society and the environment, there is information on the environment. The interpersonal ties within the society, which is a group of individuals, are covered in the environmental content on society. The emphasis, however, is on the several fundamental elements of the environment, such as the biosphere, atmosphere, lithosphere, and hydrosphere, under the idea of environment. Additionally, environmental issues like pollution, land degradation, global warming, and climate change are discussed. The content also discusses environmental valuation, sustainability perspectives, sustainability, sustainability pillars, and sustainable development objectives. The exercises force students to consider how they may use what they have learned in real-world scenarios that they see in their societies. The relationship between science, technology, and the environment is covered in this educational module.

The knowledge of environmental content is the main topic of Learning Unit 3. These are the learning objectives for this ESD-related learning unit.

- Define the major terms that are important to this learning unit, such as sustainability and the environment.
- Describe the environmental issues in the context of your community.
- Determine what material in the Intermediate Phase of Natural Sciences and Technology relates to the environment and sustainability. CAPS
- Describe how using indigenous knowledge can improve the delivery of the Senior Phase Natural Sciences Curriculum.

This section emphasises the most important environmental problems, hazards to the environment, sustainability information, and indigenous wisdom, which is another form of environmental education.

As part of the Senior Phase in Natural Sciences Education, Learning Unit 4 will explore the pedagogical subject knowledge related to Science, Environment, and Society (blended approaches to active teaching and learning). You ought to be able to after finishing this lesson:

- The instructional content knowledge connected to science, the environment, and society should be listed and explained.
- mention the cognitive and practical process abilities that students can acquire in the natural sciences.
- justify the use of hands-on activities in science instruction.

To improve learning and advancement of students studying science, society, and the environment, Learning Unit 5 will also examine how to integrate evaluation with learning. These are the ESD-related learning outcomes. The following ESD-related outcomes should be attainable for you after completing this learning unit.

- In light of understanding of the environment and sustainability, list and explain the principles for developing alternative assessments.
- Describe the many assessment techniques for more thorough, inclusive evaluations of the environmental and sustainability material in the Natural Sciences curriculum and how technology might be used to do it.
- Recognize the goals of evaluation

The Natural Sciences curriculum strives to ensure, among other things, that students are involved in advocating the change necessary for sustainable development. All citizens must have the competencies—which are thought to include knowledge, skills, beliefs, and attitudes—that support individual capability if sustainable development is to be achieved.

Climate Change Education Content in ADS3705

There is not much in this module's curriculum that addresses climate change. In general terms, the topic of climate change is discussed. The notion of climate change is examined in the study guide for this curriculum in chapter 3.

"Climate change is a topical issue worldwide. However, the concept of climate change is confused with the concept of climate variability. The two concepts are different. Even though people are perceptive of climate variability, it is not as noticeable as weather variability because it happens over seasons and years. Evidence includes statements such as: "The last few winters have seemed so short." or "There seem to be more heavy downpours in recent years." Climate change is slow and gradual, and unlike year-to-year variability, it is very difficult to perceive without scientific records. There will always be natural climate variability at many scales – decadal, yearly and short-term extreme events. This means that, over the long-term record, there will be ups and downs with the yearly and 30-year averages (climate change is usually determined in periods of 30 years), even if the climate is getting warmer. We cannot expect every summer to be warmer than the previous, but we can expect and plan for variability, etc"

Environmental Education (Module Code: EED2601)

At St. Peter's University's Department of Science and Technology Education, a module called Environmental Education (EED2601) is available. The EED2601 module is a required course for both the Bachelor of Environmental Management and Undergraduate Bachelor's degree programmes in Education (theme). The entirety of this module is taught online. This module was created in response to the growing awareness of the worldwide environmental challenges we face, including climate change, biodiversity loss, acid mine water drainage, environmental degradation, and various forms of pollution, to name a few, among citizens in all spheres of society. To encourage positive attitudes, behaviours, and actions toward the environment and to provide learners with the information and skills they need to live sustainably, this module intends to expose educators to such environmental concerns. The school environment is the ideal place to make sure this occurs, and the educators are in the best position to make this happen. The courses are designed to help teachers become knowledgeable, passionate environmental educators who will motivate students. You must master several ESD-related outcomes for this programme, including: to become aware of a variety of environmental challenges and act.

For this module, you will have to master several outcomes related to ESD:

- Understanding fundamental concepts and participating in discussions about environmental education is the first specific outcome (Education for Sustainable Development).
- To promote environmental education, specific outcome 2 calls for adopting and adapting a range of roles and tactics in response to shifting learner and learning needs and settings (Education for Sustainable Development).
- Determine various tactics for promoting environmental education (education for sustainable development) in ways that are suited for various goals and circumstances. This is the third specific outcome.
- Use creative approaches to contribute to the development of environmental education (education for sustainable development) in ways that are informed by contextual realities, the makeup of multicultural schools and classrooms,

historical legacies, social diversity, and the incorporation of indigenous knowledge.

General Environmental Content in EDD2601 Module

The main concepts and discussions surrounding concerns pertaining to environmental education are the focus of learning unit 1. An overview of the history of environmental education and education for sustainable development is provided in this lesson. It presents the fundamentals of environmental education and describes the holistic environment. Additionally, it gives a summary of the significant historical occurrences that shaped environmental education and education for sustainability. You must be able to when this learning unit is complete:

- Discuss the history and origins of environmental education as a movement
- Provide a definition of the holistic environment
- Define environmental education
- Discuss the characteristics of environmental education
- Identify and discuss the key international principles of environmental education
- Define sustainable development and explain the sustainable development goals
- Discuss the emergence of the Anthropocene and its implication for sustainability
- Discuss the key historical international events in the development of environmental education

This module covers the history of environmental education, the holistic environment, definition of environmental education, principles of environmental education, and education for sustainable development. It also covers models of sustainable development, sustainability indicators, and debates about education for sustainable development (ESD). From the Millennium Development Goals to the United Nations Decade of Education for Sustainable Development to the Sustainable Development Goals, the Anthropocene's global environmental crises and a timeline of significant historical international events that have influenced the development of environmental education.

The second learning unit focuses on adapting environmental education to changing student and learning needs and situations. It is a huge task to get students ready for work, citizenship, and life in the twenty-first century. Learners must acquire new knowledge and skills known as "21st century competencies" because of globalisation, technological development, and global environmental, political, and economic concerns. To address and support 21st century learning needs, current teaching methodologies and learning environment design are insufficient. This section makes sure that as our societies develop into knowledge societies, higher education and schools adapt to meet students' informational and competency demands so they can handle challenging societal, economic, and environmental problems. You must be able to when this learning unit is complete:

- Discuss how environmental education procedures use multiple intelligences.
- Discuss the use of various environmental education strategies in various learning environments.

Other environmental topics covered in Learning Unit 2 include diversity of learner needs and learning styles, multiple intelligences in EE, and diversity of practise contexts for EE (education about/in/for the environment).

The third learning unit's main theme is selecting the best environmental education methods for various situations. This course makes sure that students can use various environmental education techniques and approaches in a variety of educational settings. This unit's topics include wildlife conservation in southern African parks and nature reserves, the relationship between people and parks, the effects of climate change in Southern Africa, alien invasive plants in South Africa, and the effectiveness of various environmental education strategies in various contexts (theoretical, practical and experiential). Additionally, the unit covers categories of ESD methods, guiding principles for selecting ESD methods, environmental education strategies, active learning, authentic learning, problem-solving, critical thinking, and roles and positions of educators and learners regarding effective classroom practise for environmental education (theory, policy and observed practice).

The fourth learning unit focuses on creative methods for creating environmental education. This lesson emphasises the need of including diverse peoples and utilising their knowledge, customs, and practises when making decisions and taking action to ensure sustainability for both current and future generations. You ought to be able to by the end of this unit.

- Discuss the role of different cultures, knowledges, and practices in environmental education processes
- Discuss the application of indigenous knowledge in environmental education processes
- Outline and develop necessary 21st century competences among learners towards a sustainable future.

Climate Change Education Content in EED2601

Compared to the other two modules under consideration, the information on climate change covered in EED2601 is more focused. The following are discussions on climate change:

"The rising levels of greenhouse gases in the atmosphere have resulted in global warming and climate change (IPCC, 2013). Climate change has resulted in significant impacts on the global and local environment. This includes recent extreme weather events of increasing frequency and intensity which are seen through increases in average global temperature, changes in average rainfall patterns, frequent droughts and floods across the globe. The global average temperature of the Earth is currently 1o C or more above the 1880–1999 average (WEF, 2018). During 2017 record high temperatures were experienced in parts of Southern Europe, Russia, China, South America and southern Africa (WEF, 2018). This has significant impacts on agricultural production and food security in these areas. Linked to higher temperatures has been an increase in wildfire incidences across the globe. The world has also experienced frequent high impact hurricanes in the Atlantic, including Hurricane Harvey, Irma and Maria (WEF, 2018). South Africa has experience a general warming extremes over the last 40 years while cold extremes have decreased (Department of Environmental Affairs, 2012). Climate change worsens the challenges of water, energy and food security in South Africa (Von Bormann & Gulati, 2014). There is link between climate change risk and poverty, with poor, vulnerable and marginalised population groups being more vulnerable to climate change effects (United Nations Department of Economic and Social Affairs (DESA), 2016). A large proportion of South Africa's population highly vulnerable to climate change due poverty, inadequate housing and poor access to services (Department of Environmental Affairs, 2012)".

DISCUSSION

Three modules—ADD3705, ADS3705, and EED2601—that were purposefully sampled are the subject of the discussion of the findings. This section will explain the EE/ESD curriculum and how the subject of climate change education is treated in-depth in these three modules.

Learning unit 2, which examines society and environment, covers the environmental content of the ADD3705 module. The interpersonal ties within society are the main focus of the environmental material on society, whereas the various fundamental elements of the environment, such as the biosphere, atmosphere, lithosphere, and hydrosphere, are the main focus under the idea of environment. Also covered in the content are environmental issues, environmental valuation, perspectives on sustainability, sustainability, sustainability pillars, and sustainable development objectives. Students are forced to consider how they can use what they have learned in problems that they see in their societies because of the activities.

The third learning unit's emphasis is on environmental subject matter. Determining the important concepts pertinent to this learning unit is one of the unit's learning objectives. Additionally, this section emphasises indigenous knowledge, environmental and sustainability-related topics, and significant environmental risks.

The fourth learning unit focuses on the strategies and procedures for imparting knowledge about the environment and sustainability. The learning outcomes contain a list and justification of the pedagogical content knowledge connected to science, the environment, and society as well as identifying the cognitive and practical process skills that students will be able to develop in Natural Sciences and Technology. Active learning, practical work, and a Process Framework Guiding Enquiry-based Learning Practices are all used as teaching strategies.

The fifth and final learning unit emphasises assessment to further the study of science, society, and the environment. Understanding the purposes of assessment, differentiating between assessment to determine progress and assessment to enhance learning, listing and discussing guidelines for developing alternative assessments in the context of knowledge of the environment and sustainability, talking about the variety of assessment methods for more inclusive, holistic assessments of environment and sustainability content in the natural sciences curriculum, and explaining

In conclusion, the ADD3705 module gives learners in-depth pedagogical subject understanding for instructing in the areas of society, the environment, and science. As a preface to the Advanced Diploma in Education in Natural Science Education, the module also provides background information. The module's environmental material includes a range of environmental topics, including sustainability and indigenous knowledge. To improve learning and advance sustainable growth, the module also emphasises instructional strategies and assessment procedures. There is no mention of climate change education in this module.

The Advanced Diploma in Natural Science Education curriculum at St Peters University includes the module ADS3705 from the Department of Science and Technology Education. The module's learning units 2 and 3 contain the environmental content, which covers important ideas including science, technology, society, the environment, sustainability, indigenous knowledge, and environmental challenges. The module employs a variety of teaching strategies to convey information about science, the environment, and sustainability, including an active learning framework made up of encounters with inquiries, information seeking, reporting, and action taking, as well as a Process Framework Guiding Enquiry-based Learning Practices, which increases learner engagement in science, society, and the environment through practical activities. To improve learning and advancement of students in Science,

Society, and the Environment, Learning Unit 5 also explores how to combine evaluation with learning. This curriculum does not provide a clear explanation of how climate change education will be covered.

The first learning unit of the EED2601 module gives an overview of the history and progression of environmental education and education for sustainable development. It explains the holistic environment, summarises the EE/ESD concepts, and talks about the main EE/ESD worldwide principles. Additionally, it outlines the principles of sustainable development and its objectives. Learners are given an overview of the Anthropocene's worldwide environmental challenges as well as a timeline of significant international events that have shaped the field of environmental education. The second learning unit is concerned with adapting EE/ESD curriculum to changing student and learning needs and situations. The methods of environmental education involve teaching students about the diversity of learning styles, the use of various EE/ESD methodologies in various learning environments, and the many intelligences. The focus of learning unit three is on the appropriate EE/ESD strategies for various situations. It explores wildlife conservation, climate change, and alien invasive plants in South Africa, highlights the historical and contextual factors that need to be considered in environmental education processes, and describes and applies relevant methods and approaches suitable for various learning contexts. Finally, learning unit 4 focuses on creative methods for EE/ESD development. The need of including many peoples and utilising diverse knowledge, cultures, and practises when making decisions and acting toward sustainability is made clear to learners. In addition to stressing the significance of acting in climate change education, the Intergovernmental Panel on Climate Change (IPCC) also underlines that immediate action is required to stop global warming (IPCC, 2018). Students can be urged to act by their teachers by lowering their carbon footprints, getting involved in neighbourhood environmental projects, and speaking out in favour of climate-friendly legislation.

Overall, the EED2601 module seeks to inform instructors and students about environmental issues, encourage environmentally conscious attitudes and behaviours, and equip them with the knowledge and abilities needed to live sustainably. The major concepts and debates surrounding environmental education concerns should be understood by the end of the module, and students should be able to adopt and adapt a variety of roles and methods in response to evolving phenomena like climate change.

The subject of climate change is only briefly discussed as one of the foremost environmental problems facing the entire planet and the cause of it. The more complicated topics of climate change, adaptation, and mitigating measures, however, do not interest students. According to the researchers, there is a lack of theoretical or conceptual understanding of climate change in the three modules, which would lead to a lack of content knowledge of climate change education among educators and students in schools. Only conceptual knowledge of climate change is explored in the modules, and there are few activities that specifically address the subject. In support of the implicit discussion of climate change education in these courses, which contradicts the ethos of the subject as seen in Figure 1 below, where Moshou and Drinia (2023) see education about the issue as a crucial step in creating awareness and acting. Figure 1 illustrates how awareness and action result in a shift in behaviour toward the environment, which helps with climate change mitigation and adaptation.

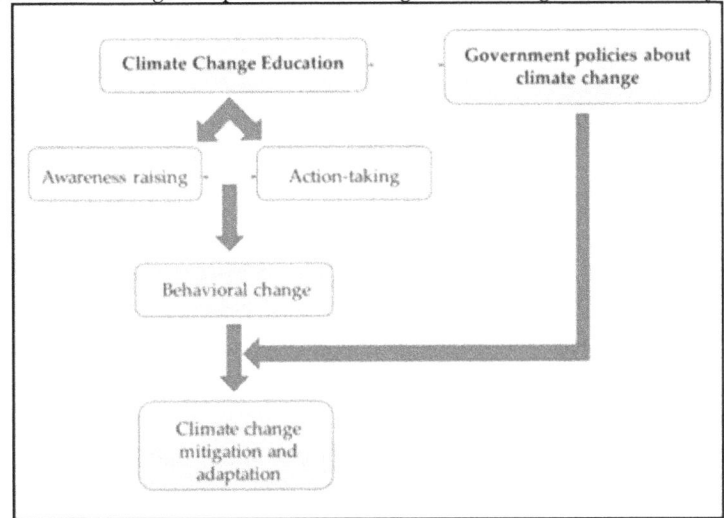

Figure 2. Climate change adaptation and mitigation strategies inclusion of education

Source: Moshou and Drinia (2023)

In higher education and in classrooms, climate change education is a critical topic that needs to be covered. In contrast to social studies instructors, who only cover climate change in 30% of their lectures, 70% of science teachers in the United States do so, according to a poll by the National Center for Science Education (National Center for Science Education, 2019). This demonstrates the necessity of closing the knowledge gap in climate change education across disciplines, as evidenced in the three modules examined in this research.

Including information on climate change in current curricula is one method to narrow the knowledge gap. Climate change education is already included in the curriculum of South Africa, nevertheless. Even teachers at higher education institutions only receive a conceptual understanding of climate change, as was already mentioned. There are numerous methods that climate change education can be investigated in more depth in institutions or rather in school lessons. For instance, climate science can be incorporated into lectures on ecology, meteorology, and geology. Teachers of social studies can concentrate on the effects of climate change on political systems, economic systems, and social systems. Readings about climate change, such the "MaddAddam" trilogy by Margaret Atwood, might be assigned by English teachers (Atwood, 2013). As previously stated, it is the duty of higher educational institutions to guarantee that the teachers they train are adequately knowledgeable about the subject matter and to develop examinations that will gauge such knowledge. According to Khatibi et al. (2021), which supports the outcomes of this study, awareness-raising about climate change is facilitated by cognitive understanding and results in behavioural change. Like this, numerous research indicates that ignorance about this pressing issue would only serve to reinforce any misconceptions that result from this ignorance (Khatibi et al., 2021; Groves & Pugh, 2002 & Nunn et al., 2014)

According to a National Wildlife Federation research, 56 percent of American teens believe that their schools are not doing enough to educate them on the topic of climate change, and 74% of teenagers in the country are concerned about it (National Wildlife Federation, 2021). Teachers can make the topic more interesting and pertinent by relating climate change to the local surroundings, cultures, and economics of their pupils. For instance, in coastal areas, educators can emphasise sea level rise and its effects on regional economies.

Finally, educators need to stress that climate change is a scientific problem rather than a political one. Only 53% of Americans, according to a survey by the Yale Program on Climate Change Communication, recognise that human activity is the primary cause of global warming, even though 70% of Americans believe that it is occurring (Yale Program on Climate Change Communication, 2021). By teaching students about the science underlying climate change, they will be better able to comprehend its origins, effects, and the need for action.

As a result, closing the knowledge gap in climate change education in schools calls for a multi-disciplinary strategy that makes the subject applicable to students' daily lives, emphasises the value of action, incorporates it into current curricula, and frames it as a scientific concern. By doing this, we can make sure that students have the information and abilities necessary to deal with one of the most important problems of our day. The study suggests including the following climate change teaching strategies into each of the three modules under consideration.

Recommended Climate Change Education Teaching Methods

i. Inquiry-Based Learning: This approach encourages students to conduct their own research, ask questions, and find out information. The National Research Council (NRC) claims that inquiry-based learning aids students in developing their capacity for critical thought and scientific reasoning (NRC, 2000).
ii. Project-Based Learning: With this strategy, students work on a climate change-related project. Project-based learning, according to research, can increase students' motivation, engagement, and information retention (Krajcik et al., 2014).
iii. Case Studies: Learning about the effects of climate change can be effectively accomplished using case studies. A study by Hmelo-Silver et al. (2007) discovered that case studies can aid students in problem-solving and engagement with complicated subjects.
iv. Collaborative teaching and learning: With this method, students collaborate in groups to address issues related to climate change. According to a study by Webb et al. (2008), collaborative learning can enhance students' critical thinking abilities and aid in the development of a better grasp of complicated subjects.
v. Multimedia and Technology-Based Learning: To make learning about climate change more dynamic and engaging, educators might leverage technology-based tools and multimedia resources. Mayer (2009) stated that multimedia can increase learners' motivation, engagement, and information retention.

Recommended Learning Strategies on Climate Change Education

i. Active Learning: This entails students actively engaging in the learning process through activities including group projects, conversations, and hands-on learning. Active learning can increase students' engagement and information retention (Freeman et al., 2014).
ii. Critical Thinking: Students can have a deeper grasp of climate change and its effects by being encouraged to think critically about it. According to a study by Ennis (2015), critical thinking can help students become better problem solvers and decision-makers.
iii. Reflection: Teachers can help students think back on what they have studied and consider how it applies to their own lives and experiences. Moon (2004) mentioned that reflection can aid students in grasping the topic and its significance more

thoroughly. Teachers can help students think back on what they have studied and consider how it applies to their own lives and experiences. Reflection can aid students in grasping the topic and its significance more thoroughly (Moon, 2004).

iv. Differentiated Instruction: This entails individualising education to match the needs of every learner, for example, by using differentiated tasks and activities. According to a study by Tomlinson (2014), differentiated education can raise student motivation and engagement.

v. Assessment for Learning: Teachers can evaluate their students' comprehension of climate change using formative assessment techniques and offer feedback to help them develop. According to Black and Wiliam (1998), formative assessment can raise students' motivation and academic performance.

CONCLUSION

The analysis of the extent of climate change content integration in a few St. Peters University modules revealed that different proponents have different levels of coverage. Analysis of the module ADD3705 and ADS3705 revealed that the education module discussed climate change in a generic and non-explicit manner. The primary climate change content in these sessions was centred on broad environmental challenges, dangers, and sustainability information. Although most environmental issues were covered in ADS3705, there was still no explicit explanation of how to educate people about climate change. Students will be prepared to adapt and take the necessary steps to embrace a more sustainable lifestyle by doing so, which will guarantee that they are aware of the causes and effects of climate change (UNESCO, 2015; IPCC, 2022). In EED2601, the climate change education content was clear and conveyed to students, in contrast to the two modules that were examined. These included significant risks related to exposure to climate risks and the susceptibility of the environment to the effects of climate change. Additionally, this module illustrated significant environmental occurrences connected to climate change, highlighting environmental difficulties brought on by climate change, and encouraging favourable attitudes and behaviours toward a sustainable environment. Although climate change was covered disproportionately, the implication is that students would not be able to cascade climate change education content that encourages awareness and acting towards behavioural change that encourages climate change risk identification, climate change injustices, mitigation, adaptation, and resilience. Even though climate change was discussed inequitably, the implication is that students would not be able to cascade climate change education content that promotes awareness and acting in the direction of a behavioural change that encourages the identification of climate change risks, the mitigation of climate change injustices, and the development of resilience. As a result, one of the objectives of Sustainable Development Goal 13 (climate change), which is to promote education, awareness-raising, and institutional and human capacity on mitigating, adapting to, and reducing the impacts of, climate change, is hindered. To incorporate climate change education into their modules, tertiary institutions can adopt some of the teaching approaches and instructional strategies that were suggested in this study. Among these techniques and approaches are case studies, collaborative teaching, multimedia and technology-based learning, active learning, critical thinking techniques, reflections, and customised instruction.

REFERENCES

Archer, D., Almansi, F., DiGregorio, M., Roberts, D., Sharma, D., & Syam, D. (2014). Moving towards inclusive urban adaptation: Approaches to integrating community-based adaptation to climate change at city and national scale. *Climate and Development*, 6(4), 345–356. DOI: 10.1080/17565529.2014.918868

Atwood, M. (2013). *MaddAddam: a novel*. Nan A. Talese/Doubleday.

Black, P., & Wiliam, D. (1998). Inside the black box: Raising standards through classroom assessment. *Phi Delta Kappan*, 80(2), 139–148.

Bowen, G. (2009). Document Analysis as a Qualitative Research Method. *Qualitative Research Journal*, 9(2), 27–40. DOI: 10.3316/QRJ0902027

Busch, K. C., Henderson, J. A., & Stevenson, K. T. (2019). Broadening epistemologies and methodologies in climate change education research. *Environmental Education Research*, 25(6), 955–971. DOI: 10.1080/13504622.2018.1514588

Celliers, L., & Schleyer, M. (2008). Coral community structure and risk assessment of high-latitude reefs at Sodwana Bay, South Africa. *Biodiversity and Conservation*, 17(13), 3097–3117. DOI: 10.1007/s10531-007-9271-6

Chikoore, H., Vermeulen, J. H., & Jury, M. R. (2015). Tropical cyclones in the Mozambique channel: January–March 2012. *Natural Hazards*, 77(3), 2081–2095. DOI: 10.1007/s11069-015-1691-0

Creswell, J. W. (2009). *Research design: Qualitative, quantitative and mixed method approaches* (3rd ed.). Sage.

Department of Environmental Affairs (DEA). (2017). *Climate change trends, risks, impacts and vulnerabilities, South Africa's 2nd annual climate change report* [Online]. Available from: https://www.environment.gov.za/sites/default/files/reports/southafrica_secondnational_ climatechnage_report2017.pdf

Ennis, R. H. (2015). Critical thinking: A streamlined conception. In *Critical thinking: A statement of expert consensus for purposes of educational assessment and instruction* (pp. 1–49). Foundation for Critical Thinking.

Fitzgerald, J. B. (2022). Working time, inequality and carbon emissions in the United States: A multi-dividend approach to climate change mitigation. *Energy Research & Social Science*, 84, 102385. DOI: 10.1016/j.erss.2021.102385

Freeman, S., Eddy, S. L., McDonough, M., Smith, M. K., Okoroafor, N., Jordt, H., & Wenderoth, M. P. (2014). Active learning increases student performance in science, engineering, and mathematics. *Proceedings of the National Academy of Sciences of the United States of America*, 111(23), 8410–8415. DOI: 10.1073/pnas.1319030111 PMID: 24821756

Fundisa for Change Programme. (2013). *Introductory core text*. Environmental Learning Research Centre, Rhodes University.

Gough, N. (2013). Thinking globally in environmental education. In Stevenson, R. B. (Eds.), *International Handbook of Research on Environmental Education* (pp. 33–44). American Educational Research Association. DOI: 10.4324/9780203813331-3

Groves, F. H., & Pugh, A. F. (2002). Cognitive Illusions as Hindrances to Learning Complex Environmental Issues. *Journal of Science Education and Technology*, 11(4), 381–390. Retrieved April 19, 2023, from. DOI: 10.1023/A:1020694319071

Hartas, D. (2010). *Educational research and inquiry: Qualitative and quantitative approaches*. Continuum International. DOI: 10.5040/9781474243834

Hickel, J. (2020). The sustainable development index: Measuring the ecological efficiency of human development in the anthropocene. *Ecological Economics*, 167, 106331. DOI: 10.1016/j.ecolecon.2019.05.011

Hill, J., Alan, T., & Woodland, W. (2006). *Sustainable development*. Ashgate.

Hmelo-Silver, C. E., Duncan, R. G., & Chinn, C. A. (2007). Scaffolding and achievement in problem-based and inquiry learning: A response to Kirschner, Sweller, and Clark. *Educational Psychologist*, 42(2), 99–107. Retrieved April 23, 2023, from. DOI: 10.1080/00461520701263368

Howell, K. E. (2015). *Empiricism, positivism and post-positivism*. Sage.

Institute for Security Studies (ISS). (2010). *Climate change and natural resources. Conflicts in Africa*. Available from: https://issafrica.org/research/monographs/climate- change- and-natural-resources- conflicts-in-africa

Intergovernmental Panel on Climate Change. (2014). *Climate change report 2014: Impacts, adaptation and vulnerability. Fifth assessment report*. New York: Cambridge University Press.

Intergovernmental Panel on Climate Change. (2018). Global warming of 1.5°C. Retrieved from https://www.ipcc.ch/sr15/

Intergovernmental Panel on Climate Change. (2022). Climate Change 2022: Impacts, Adaptation and Vulnerability. Contribution of Working Group II to the Sixth Assessment Report of the Intergovernmental Panel on Climate Change. Cambridge University Press. DOI: 10.1017/9781009325844

Jeong, H. (2006). *Globalisation and the physical environment*. CHP.

John, L. C. K., Mei, W. S., & Guang, Y. (2013). EE policies in three Chinese communities. In *International Handbook of 390 Research on Environmental Education* (pp. 178–188). American Educational Research Association. DOI: 10.4324/9780203813331-35

Johnson, B., & Christensen, L. (2008). *Educational research: Quantitative, qualitative, and mixed approaches*. Sage.

Khatibi, F. S., Dedekorkut-Howes, A., Howes, M., & Torabi, E. (2021). Can public awareness, knowledge and engagement improve climate change adaptation policies? *Discover Sustainability*, 2(1), 18. Retrieved April 22, 2023, from. DOI: 10.1007/s43621-021-00024-z

Komolafe, A. A., Adegboyega, S. A., Anifowose, A. Y. B., Akinluyi, F. O., & Awoniran, D. R. (2014). Air pollution and climate change in Lagos, Nigeria: Needs for 391 proactive approaches to risk management and adaptation. *American Journal of Environmental Sciences*, 10(4), 412–423. DOI: 10.3844/ajessp.2014.412.423

Krajcik, J., Czerniak, C., & Berger, C. (2014). *Teaching science in elementary and middle school: A project-based approach*. Routledge. DOI: 10.4324/9780203113660

Kyburz-Graber, R. (2013). Socioecological approaches to environmental education and research. In *International Handbook of Research on Environmental Education* (pp. 23–57). American Educational Research Association. DOI: 10.4324/9780203813331-2

Leedy, P. D., & Ormord, J. E. (2013). *Practical Research: Planning and Design*. Prentice Hall.

Li, J., & Sheng, Y. (2012). An automated scheme for glacial lake dynamics mapping using Landsat imagery and digital elevation models: A case study in the Himalayas. *Int J Remote Sens*, 33: 5194–5213. DOI: 10.1080/01431161.2012.657370

Lotz-Sisitka, H. (2013). Curriculum research in environmental education. In Stevenson, R. B. (Eds.), *International Handbook of Research on Environmental Education* (pp. 191–193). American Educational Research Association.

Mayer, R. E. (2009). *Multimedia learning*. Cambridge University Press. DOI: 10.1017/CBO9780511811678

Monroe, M. C., Plate, R. R., Oxarart, A., Bowers, A., & Chaves, W. A. (2019). Identifying effective climate change education strategies: A systematic review of the research. *Environmental Education Research*, 25(6), 791–812. DOI: 10.1080/13504622.2017.1360842

Moon, J. A. (2004). *Reflection in learning and professional development: Theory and practice*. Routledge Falmer.

Moshou, H., & Drinia, H. (2023). Climate Change Education and Preparedness of Future Teachers—A Review: The Case of Greece. *Sustainability (Basel)*, 15(2), 1177. DOI: 10.3390/su15021177

Mulenga, B. P., Wineman, A., & Sitko, N. J. (2017). Climate trends and farmers perceptions of climate change in Zambia. *Environmental Management*, 59(2), 291–306. DOI: 10.1007/s00267-016-0780-5 PMID: 27778064

National Center for Science Education. (2019). Climate change in the classroom: A national survey of middle and high school science teachers. Available from: https://ncse.ngo/climate-change-classroom-national-survey-middle-and-high-school-science-teachers

National Research Council. (2000). Inquiry and the national science education standards: A guide for teaching and learning. National Academies Press.

National Wildlife Federation. (2021). National wildlife federation releases results of climate survey of American teenagers. Available from: https://www.nwf.org/News-and- Magazines/Media-Center/News-by-Topic/Global-Warming/2021/05-18-21-National- Wildlife-Federation-Releases-Results-of-Climate-Survey-of-American-Teenagers

Nunn, P. D. (2012). Understanding and adapting to sea-level rise. In Harris, F. (Ed.), *Global Environmental Issues* (2nd revised ed., pp. 87–104). Wiley. DOI: 10.1002/9781119950981.ch5

Nunn, P.D., Aalbersberg, W., & Lata, S. (2014). Beyond the core: community governance for climate-change adaptation in peripheral parts of Pacific Island Countries. *Reg Environ Change,14*, 221–235. DOI: 10.1007/s10113-013-0486-7

O'leary, Z. (2014). Primary data: Surveys, interviews and observation. The essential guide to doing your research project, 201-216.

Orchard, S. E., Stringer, L. C., & Manyatsi, A. M. (2016). Farmer perceptions and responses to soil degradation in Swaziland. *Land Degradation & Development*, 28(1), 46–56. DOI: 10.1002/ldr.2595 PMID: 30393450

Parra, J. D., Said-Hung, E., & Montoya-Vargas, J. (2020). (Re) introducing critical realism as a paradigm to inform qualitative content analysis in *causal* educational research. *International Journal of Qualitative Studies in Education.* DOI: 10.1080/09518398.2020.1735555

Robottom, I. (2013). Changing discourses in EE/ESD. In *International Handbook of Research on Environmental Education* (pp. 156–162). American Educational Research Association. DOI: 10.4324/9780203813331-32

Sitarz, D. (1994). *Agenda 21. The earth strategy to save our planet.* Earth Press.

Sobh, R., & Perry, C. (2006). Research Design and Data Analysis in Realism Research. *European Journal of Marketing*, 40(11), 1194–1209. DOI: 10.1108/03090560610702777

Southern Africa Environment Outlook. (2016). 2nd South Africa Environment Outlook. A report on the state of the environment. Executive Summary. Department of Environmental Affairs, Pretoria. https://www.environment.gov.za/sites/default/files/reports/environmentoutlook_execu tivesummary.pdf

Southern African Development Community. (2008). Southern Africa environmental outlook. Available from: https://www.environment.gov.za/sites/default/files/reports/environmentoutlook _chapter14.pdf

Tomlinson, C. A. (2014). *The differentiated classroom: Responding to the needs of all learners.* ASCD.

Trott, C. D., Lam, S., Roncker, J., Gray, E. S., Courtney, R. H., & Even, T. L. (2023). Justice in climate change education: A systematic review. *Environmental Education Research*, 29(11), 1–38. DOI: 10.1080/13504622.2023.2181265

United Nations Educational, Scientific and Cultural Organization. (2015). Not just hot air: putting climate change education into practice. https://unesdoc.unesco.org/images/0023/002330/233083e.pdf

United Nations Educational, Scientific and Cultural Organization. (2017). Sustainability Starts with Teachers: Environmental scan informing teacher education programming for ESD secondary teacher education. https://www.google.com/search?q=sustainability+starts+with+teachers&rlz=1C1GCE A_en-ZA996ZA996&oq=Sustainability+starts+&aqs=chrome.1.69i57j0i512l5j0i15i22i30j0i10i22i30l3.7975j0j15&sourceid=chrome&ie=UTF-8

United Nations Environment Programme. (2008a). *Africa atlas of our changing environment*. Available from: https://wedocs.unep.org/handle/20.500.11822/7717

United Nations Environment Programme. (2016). *Summary of the sixth global environmental outlook GEO-6 regional assessments: Key findings and policy messages UNEP/EA.2/INF/17*. Available from: http://www.scpclearinghouse.org/sites/default/files/summary_of_the_sixth_glo bal_environment_outlook_geo6_regional_assessments_key_findings_and_po licy_messages_unep_ea2_inf_17-2016 geo-6_summary_en.p.pdf

United Nations Framework Convention on Climate Change. (1992). FCCC/INFORMAL/84,GE.05- 62220(E), 200705. Available online: https://unfccc.int/resource/docs/convkp/conveng.pdf

Wang, J., Xing, J., Mathur, R., Pleim, J., Wang, S., Hogrefe, C., Gan, C., Wong, D. C., & Hao, J. (2017). Historical trends in PM2.5 Related premature mortality during 1990-2010 across the Northern Hemisphere. *Environmental Health Perspectives*, 125(3), 400–408. DOI: 10.1289/EHP298 PMID: 27539607

Webb, N. M., Troper, J. D., & Fall, R. (2008). Constructive activity and learning in collaborative small groups. *Journal of Educational Psychology*, 100(1), 1–13. PMID: 19578558

Wei, Y., Wang, K., Liao, H., & Tatano, H. (2016). Economics of climate change and risk of disasters in Asia-Pacific region. *Natural Hazards*, 84(S1), S1–S5. DOI: 10.1007/s11069-016-2590-8

Yale Program on Climate Change Communication. (2021). Climate opinions by state. Available from: https://climatecommunication.yale.edu/visualizations-data/ycom-us/

Yin, R. K. (2011). *Qualitative research from start to finish*. The Guilford Press.

Young, M. (2008). *Bringing knowledge back in from social constructivism to social realism in the sociology of education*. Taylor and Francis.

Yuhas, A. (2019). Cyclone Idai may be 'one of the worst' disasters in the Southern Hemisphere. *New York Times,* 19 March 2019. Available from: https://www.nytimes.com/2019/03/19/world/africa/cyclone-idai-mozambique.html

Chapter 15
Utilizing OpenStreetMap Data for Local Climate Change Assessment and Policy Formulation

Munir Ahmad
https://orcid.org/0000-0003-4836-6151
Survey of Pakistan, Pakistan

Miguel Angel Osorio Rivera
https://orcid.org/0000-0002-8641-2721
Escuela Superior Politécnica de Chimborazo, Ecuador

William Estuardo Carrillo Barahona
https://orcid.org/0000-0002-1432-9638
Escuela Superior Politécnica de Chimborazo, Ecuador

Amara Nisar
https://orcid.org/0009-0005-6573-7577
University of the Punjab, Lahore, Pakistan

Noor Ul Safa
https://orcid.org/0009-0002-3522-199X
GC Women University, Sialkot, Pakistan

ABSTRACT

This chapter has delved into the significant role that OSM data can play in local climate change assessments and policy development. Utilizing the open, comprehensive, and continually updated geospatial information available in OSM, researchers

DOI: 10.4018/979-8-3693-5792-7.ch015

and policymakers can equip themselves with essential tools to understand and tackle climate issues at the local scale. OSM offers numerous benefits for climate research, such as open access, extensive spatial coverage, and regular updates. Nevertheless, it is important to recognize the limitations, including potential data inconsistencies and gaps in certain regions. Implementing data validation techniques is crucial to ensure accurate analysis. OSM's success heavily depends on the active engagement of the global community. Ongoing contributions and efforts to enhance data quality are essential to maintain the platform's reliability and usefulness for climate research.

INTRODUCTION

Climate change is undeniably one of the most urgent and complex challenges confronting our planet today (Bell & Masys, 2020; Kabir et al., 2023; Upadhyay, 2020; Zalasiewicz & Williams, 2021). Its impacts span across various scales, from global phenomena like rising temperatures and sea levels to more localized effects such as shifts in precipitation patterns and the frequency of extreme weather events. Addressing these challenges effectively requires a comprehensive understanding of local climate variations and their specific impacts on communities worldwide. At the heart of effective climate action lies the need for detailed and localized data that can elucidate the factors influencing regional climates (Levin et al., 2023; Morshed et al., 2024). Key among these factors are land use practices and surface characteristics, which significantly influence how heat is absorbed and radiated back into the atmosphere. For instance, urban areas characterized by extensive pavement and buildings often experience higher temperatures than surrounding rural areas—a phenomenon known as the urban heat island effect (Schneider et al., 2023). Understanding these localized variations is crucial for implementing targeted mitigation and adaptation strategies that can enhance resilience to climate change impacts.

Moreover, detailed climate data enables policymakers, urban planners, and scientists to assess vulnerabilities, prioritize interventions, and develop resilient infrastructure that can withstand future climate scenarios. By integrating scientific research with geographic information systems (GIS) and other spatial data platforms, communities can better anticipate climate-related risks such as flooding, droughts, and heat waves (Garro-Quesada et al., 2023; Opach et al., 2023; Patel & Patel, 2024; Truong et al., 2023). This holistic approach empowers decision-makers to enact informed policies and investments that promote sustainable development and safeguard the well-being of present and future generations. Traditional methods of collecting detailed geospatial data, essential for understanding local climate variations and their impacts, have historically been costly and time-intensive. These methods often involve specialized equipment, extensive field surveys, and data processing, which can pose significant barriers to accessing up-to-date and comprehensive information.

However, the emergence of open-source geospatial data platforms like OpenStreetMap (OSM) presents a promising alternative. OSM operates as a collaborative project where individuals from around the world can contribute and edit geographic data using a user-friendly online editor. This inclusive approach has democratized access to geospatial information, transforming OSM into a rich and continuously evolving resource (Huang et al., 2024; Laato & Tregel, 2023). By harnessing the collective knowledge and efforts of a global community of volunteers, OSM offers a dynamic platform for mapping and documenting local environments in unprecedented detail (Sarkar & Anderson, 2022). Contributors map diverse features such as roads, buildings, land use patterns, and natural landscapes, providing a comprehensive snapshot of urban and rural areas alike. This detailed data is crucial for understanding how land use and surface characteristics influence microclimates, such as urban heat islands, and for assessing vulnerabilities to climate change impacts (Mühlhofer et al., 2024; Queiroz et al., 2024; Slovic et al., 2023). Moreover, OSM's open-access model allows researchers, policymakers, and communities to freely access, use, and build upon the data. This openness fosters innovation and collaboration in addressing climate-related challenges by enabling the development of tools, applications, and analyses that leverage OSM's extensive geographic information (Zhang et al., 2024).

Readily available and detailed information within OSM can leverage valuable insights into the spatial distribution of factors that influence local climates. Therefore, the objective of this chapter is to explore the potential of utilizing OSM data for local climate change assessment and policy formulation. We discuss the types of data available within OSM relevant to climate studies, explore methodologies for analyzing this data, and demonstrate its application in developing effective local climate change policies. The chapter is structured as follows. First, we provide a brief overview of OSM and its capabilities for climate research. Next, we discuss the methodologies for extracting and analyzing relevant data from OSM. Subsequently, we showcase examples of how OSM data can be used to inform local climate change policy formulation. Finally, we conclude by discussing the limitations and prospects of utilizing OSM data for local climate change assessments.

OPENSTREETMAP FOR CLIMATE RESEARCH

OpenStreetMap is a revolutionary open-source platform launched in 2004 that allows anyone to contribute and edit geospatial data. This collaborative approach fosters a constantly growing and detailed record of our planet, making it a valuable resource for climate research at the local level.

Open Collaboration and Data Contribution

OSM stands in stark contrast to traditional, often costly, and labor-intensive methods of collecting geospatial data by embracing a model of open collaboration (Ma et al., 2015; Mooney & Corcoran, 2014; Romão et al., 2023). This approach empowers a diverse global community of volunteers to contribute their local knowledge and expertise, thereby creating a dynamic and continuously evolving repository of geographic data. Unlike static maps that quickly become outdated, OSM's collaborative framework ensures that information is promptly updated to reflect real-world changes such as new roads, buildings, and other features. The open-access nature of OSM data is a cornerstone of its utility and impact. Researchers and practitioners across various disciplines have unrestricted access to this wealth of information, enabling them to utilize detailed and current geographic data for their studies, analyses, and applications. This accessibility fosters innovation and facilitates the development of new tools and methodologies that leverage OSM's comprehensive dataset. By harnessing the collective efforts of volunteers worldwide, OSM not only democratizes access to geospatial information but also enriches the quality and scope of available data. This inclusive approach supports a wide range of uses, from urban planning and disaster response to environmental monitoring and socioeconomic research.

Relevant Geospatial Data for Climate Research

OpenStreetMap provides a rich and versatile resource for climate research, offering a variety of geospatial data types that are particularly valuable at the local level.

Land Use Classifications

OSM allows users to tag areas with specific land use types such as forests, wetlands, urban areas, and agricultural land (Borkowska et al., 2023; Chen et al., 2021; Liu, 2021). This detailed information is essential for understanding how land cover influences local climates. For instance, researchers can analyze the distribution of forests to assess their role in carbon sequestration and their contribution to mitigating urban heat islands. The identification of urban areas helps in understanding heat retention and runoff characteristics, which are crucial for urban climate studies and planning resilient cities.

Building Footprints

OSM data includes outlines of buildings, providing insights into building density and surface area within urban environments (Herfort et al., 2023; Moradi et al., 2023; Ullah et al., 2023). This information is invaluable for studying urban heat island effects, as densely built areas with significant impervious surfaces tend to trap heat more readily. By analyzing building footprints, researchers can quantify the built environment's impact on local temperatures and energy consumption patterns, informing strategies for sustainable urban design and climate adaptation.

Waterway Features

OSM includes comprehensive information on rivers, streams, lakes, and other water bodies (Pearson et al., 2023). This data is instrumental in mapping floodplains, analyzing drainage patterns, and assessing the potential impact of climate change on water resources. By studying waterway features in OSM, researchers can evaluate hydrological processes, water availability, and the vulnerability of communities to water-related hazards under changing climatic conditions.

In addition to these core data types, OSM allows users to tag features with a wide range of additional information. This includes details on vegetation types, soil characteristics, and specific building materials, all of which contribute valuable insights for climate research projects. The flexibility and openness of OSM enable researchers to access and utilize detailed, up-to-date geospatial data that supports informed decision-making, fosters interdisciplinary collaboration, and promotes sustainable development practices in response to climate change challenges.

Advantages of Using OSM Data

OpenStreetMap data offers several key advantages that make it highly beneficial for climate research. OSM data is openly accessible to anyone, without licensing restrictions or costs (OSM, 2021). This open-access policy enables researchers to freely download and utilize large datasets for their studies, facilitating cost-effective access to valuable geospatial information. By eliminating financial barriers, OSM promotes inclusivity in research and encourages collaboration across disciplines and regions. Moreover, the extensive network of OSM contributors worldwide ensures comprehensive spatial coverage, including in remote or developing regions where traditional data sources may be limited or unavailable. This broad coverage allows researchers to conduct detailed local climate assessments across diverse geographic areas, from urban centers to rural landscapes. The availability of detailed data on land use, building footprints, elevation, and water features supports nuanced analyses of climate impacts and vulnerabilities at various scales.

OSM operates as a dynamic and collaborative platform, where data is continually updated and improved by a global community of volunteers (Budhathoki & Haythornthwaite, 2012; McGough et al., 2024). This ongoing process ensures that the dataset remains current and reflects real-world changes such as new infrastructure developments, land use modifications, and environmental conditions. For climate researchers, access to up-to-date information is critical for monitoring trends, assessing impacts, and informing adaptive strategies in response to evolving climate conditions. Furthermore, the collaborative nature of OSM includes a system of community-driven quality assurance mechanisms. New contributions undergo peer review and validation by experienced users, enhancing data reliability and accuracy. While data quality may vary across regions, the community's collective efforts help maintain high standards and address potential errors, ensuring robust datasets for scientific analysis and decision-making.

Limitations of OSM Data

OpenStreetMap data, despite its advantages, also presents several limitations that researchers should consider when utilizing it for climate research.

OSM's open contribution model allows anyone to contribute geographic data, leading to potential inconsistencies in data quality (Biljecki et al., 2023; Moradi et al., 2023; Pinho et al., 2023; Yamashita et al., 2023). Variations in contributor expertise, mapping practices, and local knowledge can result in inaccuracies or incomplete representations of features. Researchers may need to implement rigorous data validation and verification techniques to ensure the reliability and consistency of OSM data used in their analyses. This process is essential for maintaining the integrity of research findings and making informed decisions based on accurate information.

While OSM boasts extensive global coverage, data completeness can vary significantly across regions (Herfort et al., 2023; Ullah et al., 2023). Remote or less populated areas may have fewer contributors and consequently less detailed or outdated information. This limitation can pose challenges for conducting comprehensive climate assessments in these regions, requiring researchers to supplement OSM data with alternative sources or field surveys where feasible. Addressing data gaps through integrated approaches enhances the robustness of climate research outcomes and supports equitable analysis across diverse geographic contexts. Moreover, OSM primarily reflects the current state of geographic features and infrastructure. While continuously updated by contributors, historical data availability within OSM is often limited (Novack et al., 2024). This constraint makes it difficult for researchers to analyze long-term trends or changes in land use, urbanization patterns, and environmental conditions over extended periods. Incorporating historical data from additional sources or leveraging alternative datasets becomes crucial for conducting retrospective analyses and understanding the temporal dynamics of climate impacts.

METHODOLOGIES FOR UTILIZING OSM DATA

Extracting, processing, and analyzing OSM data for climate change assessments requires specific technical steps. This section explores these methodologies.

Data Extraction Techniques

When acquiring OSM data for research purposes, researchers typically utilize two primary methods to extract relevant datasets. They can use Overpass API (https://overpass-turbo.eu/) which is a powerful tool provided by OpenStreetMap to allow users to query the OSM database and extract specific geographic data based on defined criteria. Users can define queries by specifying geographic boundaries (bounding box or polygon) that encompass their study area. Moreover, researchers can filter for specific features of interest such as land use types (e.g., forests, urban areas), building footprints, waterways, points of interest (POIs), and more. Further, the Overpass API supports complex queries using Overpass QL (Query Language), enabling researchers to tailor their data extraction to meet precise research needs. Therefore, using the Overpass API provides flexibility and control over the data extraction process, allowing researchers to retrieve detailed and customized datasets directly from the OSM database.

Alternatively, researchers can access pre-downloaded and formatted OSM datasets from various platforms. These datasets are often available for specific regions or themes and can serve as a convenient starting point for research projects. Platforms such as Geofabrik (https://www.geofabrik.de/) and HOT Export Tool (https://export.hotosm.org/v3/) provide downloadable OSM data extracts in different formats (e.g., .osm, .pbf). Researchers should verify the extent and coverage of the dataset to ensure it aligns with their study area and research objectives. Depending on specific research needs, researchers may need to refine downloaded datasets by filtering, cleaning, or processing the data further to focus on relevant features or spatial resolutions.

By leveraging these data extraction techniques, researchers can effectively harness OSM's extensive geographic information to support diverse applications in climate research, urban planning, disaster management, and environmental studies. Each method offers distinct advantages in terms of customization, accessibility, and scalability, empowering researchers to conduct informed analyses and generate actionable insights from OSM's dynamic and collaborative dataset.

Data Pre-processing Steps

Once OpenStreetMap data has been acquired for research purposes, several essential pre-processing steps are necessary to ensure data quality and suitability for analysis.

Due to the open contribution nature of OSM, the dataset may contain inconsistencies such as duplicate entries, missing information, or errors in geometry. Identifying and removing duplicate entries to streamline the dataset and prevent redundancy in analysis. Moreover, addressing geometric errors or anomalies in spatial data that could affect the accuracy of subsequent analyses. Further, verifying and supplementing missing or incomplete information where feasible, ensuring comprehensive dataset coverage. Utilizing data cleaning tools and techniques, such as spatial data validation libraries or scripting languages like Python with libraries such as GeoPandas or Shapely, can automate and expedite these processes effectively.

After cleaning, filtering allows researchers to focus on specific features or themes relevant to their research objectives. Selecting features within defined geographic boundaries or regions of interest using spatial filtering techniques. Furthermore, filtering based on specific attributes or tags associated with OSM features (e.g., land use types, building classifications) to extract pertinent data subsets. By filtering the dataset, researchers can refine their analysis scope and concentrate on extracting actionable insights from targeted OSM data segments.

OSM data is available in various formats (e.g., XML, JSON, .osm, .pbf). Formatting the data into a standardized format compatible with chosen analysis tools or Geographic Information System (GIS) software is crucial. Converting OSM data from its original format to formats compatible with GIS software such as Shapefile (.shp), GeoJSON (.geojson), or File Geodatabase (.gdb). Moreover, ensuring data attributes and schema align with analysis requirements and tool capabilities for seamless integration and manipulation. Additionally, incorporating additional datasets or layers as needed to augment OSM data for comprehensive analysis and contextual understanding. Standardizing the data format facilitates efficient data handling, visualization, and analysis workflows, enabling researchers to leverage OSM's rich spatial information effectively.

Analyzing OSM Data for Climate Change

Integration with Other Climate Datasets

Integrating OSM data with climate-related datasets such as temperature records, precipitation data, or satellite-derived environmental variables enhances the understanding of local climate patterns. Researchers can overlay OSM features (e.g., land use, vegetation cover) with climate variables to visualize spatial relationships and identify correlations. Moreover, they can use spatial analysis techniques to pinpoint areas where specific OSM features (e.g., urban areas, green spaces) coincide with climate anomalies (e.g., urban heat islands, temperature variations). Furthermore, they can incorporate climate data into predictive models that leverage OSM's detailed spatial information to forecast climate impacts on local environments and communities.

Spatial Analysis Tools

Spatial overlays involve combining multiple layers of spatial data, such as OSM features and climate datasets, to analyze their spatial relationships. This technique helps researchers understand how different geographic features intersect and influence local climate conditions. Overlaying OSM land use classifications (e.g., urban areas, forests) with climate variables (e.g., temperature, precipitation) to assess how land cover affects microclimate variations. Moreover, identifying areas where vulnerable populations coincide with climate hazards (e.g., flooding, heat islands) using demographic data overlaid with climate risk assessments derived from OSM and climate datasets.

Buffer analysis creates zones or buffers around specific OSM features to study their spatial influence on surrounding areas. Creating buffers around OSM-mapped water bodies or forests can help to analyze ecological significance and influence on local climate regulation (e.g., cooling effects, biodiversity support). Furthermore, it can assess the impact radius of urban features (e.g., roads, buildings) on microclimates by creating buffers to study heat island effects and urban heat mitigation strategies.

Network analysis is another tool that can evaluate the connectivity and accessibility of OSM-mapped infrastructure networks in relation to climate vulnerabilities and adaptation strategies. Analyzing OSM transportation networks (e.g., roads, public transit) can help to identify critical routes for evacuation during extreme weather events or assess accessibility to climate-sensitive services. Moreover, using network analysis can optimize the placement of green infrastructure (e.g., parks, green corridors) mapped in OSM to enhance climate resilience and mitigate urban heat islands.

Statistical Analysis

Correlation analysis helps researchers understand the degree of association between OSM features (e.g., land use types, building densities) and climate metrics (e.g., temperature, precipitation). The first step is to collect and preprocess OSM data and relevant climate variables for analysis, ensuring consistency and compatibility. Then compute correlation coefficients (e.g., Pearson's correlation coefficient) to quantify the strength and direction of relationships between OSM features and climate metrics. Finally, assess spatial correlations using techniques like Moran's I to identify spatial patterns and dependencies between OSM-derived variables and climate data across geographic areas.

Regression models enable researchers to predict climate variables based on OSM-derived predictors, offering insights into how spatial features influence climate patterns over time. Techniques include linear regression to model linear relationships between climate variables (dependent variable) and OSM predictors (independent variables) to estimate coefficients and assess statistical significance. It also includes, multiple regression to incorporate multiple OSM features simultaneously to explore their combined impact on climate variables, adjusting for confounding factors and improving predictive accuracy. Time-series analysis can also be used to employ time-series regression models to analyze temporal trends in climate data and assess how changes in OSM features affect climate patterns over different time periods.

Spatial regression techniques account for spatial autocorrelation in OSM and climate data, ensuring robust analysis across geographic scales. Methods include spatial autocorrelation to assess and address spatial dependencies in data using spatial autocorrelation tests (e.g., Moran's I, Geary's C) to detect clustering or spatial patterns in OSM and climate variables. Moreover, spatial lag models can be employed to model spatial relationships by incorporating neighboring effects of OSM features on climate variables, capturing spatial interactions and local-scale impacts more accurately. Furthermore, Geographically Weighted Regression (GWR) can be applied to explore spatially varying relationships between OSM predictors and climate outcomes, accommodating spatial heterogeneity and local variations.

APPLICATIONS IN LOCAL CLIMATE CHANGE POLICY

The insights gleaned from analyzing OSM data can be instrumental in formulating effective local climate change policies.

Mitigating Urban Heat Islands

Problem: Urban heat islands (UHIs) occur when urban areas experience higher temperatures than their rural surroundings due to human activities, construction materials, and lack of vegetation. This phenomenon can exacerbate heat-related health issues, increase energy consumption for cooling, and worsen air quality.

Leveraging OSM Data

OSM Solution: OSM provides valuable data on land use classifications and building footprints, which are essential for identifying areas contributing to urban heat islands. OSM data can help to categorize land use types such as residential, commercial, industrial areas, and green spaces. Analyzing this data helps pinpoint areas with high building density and limited green cover, which typically contribute to UHIs. Moreover, OSM also includes detailed outlines of buildings, allowing researchers to assess building density and surface area that contribute to heat retention and radiation.

Strategies for Mitigation

Using OSM data, identify areas with high building density and low vegetation cover that contribute significantly to UHIs. Moreover, plan and implement urban greening strategies such as planting trees in these identified hotspots or creating green roofs on buildings to reduce heat absorption and improve thermal comfort. Utilize GIS and spatial modeling techniques to simulate the impact of urban greening interventions on reducing surface temperatures and cooling effects. Also, evaluate different scenarios of tree planting, green roof installations, and other interventions to determine the most effective strategies for mitigating UHIs based on local conditions. Furthermore, use findings from spatial modeling and analysis of OSM data to inform urban planning and policy decisions aimed at reducing UHIs. Additionally, involve local communities in urban greening initiatives supported by OSM data, fostering public support and participation in creating cooler, more sustainable urban environments.

Benefits of Using OSM for UHI Mitigation

- OSM provides open-access data that is freely available for researchers, policymakers, and communities to use in planning and implementing UHI mitigation strategies.
- Detailed geographic information in OSM allows for precise identification of UHI hotspots and targeted interventions, optimizing resource allocation and effectiveness.
- OSM's global coverage and scalability enable its application in diverse urban settings worldwide, supporting tailored approaches to UHI mitigation based on local context and climate conditions.

By leveraging OSM data effectively, cities can develop proactive strategies to mitigate urban heat islands, enhance urban resilience, and improve the quality of life for residents in increasingly urbanized environments.

Flood Risk Management

Flooding poses a significant threat exacerbated by climate change, leading to more frequent and severe extreme weather events. To effectively manage flood risks, it is essential to identify and understand vulnerable areas susceptible to flooding. OpenStreetMap offers valuable geospatial data that can be utilized to enhance flood risk management strategies. OSM provides elevation data that helps in identifying low-lying areas prone to flooding. This data is crucial for mapping floodplains and understanding the terrain's susceptibility to inundation during extreme weather events. OSM also includes comprehensive data on rivers, streams, lakes, and other water bodies. This information is essential for mapping hydrological networks and predicting areas susceptible to riverine or coastal flooding.

Strategies for Flood Risk Management

- Use GIS tools to overlay elevation data, waterway features from OSM, and climate projections to create comprehensive flood risk maps. These maps highlight high-risk areas susceptible to flooding, aiding in targeted mitigation and adaptation strategies.
- Based on flood risk maps, implement land-use zoning regulations that restrict development in high-risk flood zones and encourage resilient construction practices in safer areas.
- Identify areas in need of flood protection infrastructure such as levees, flood walls, and drainage systems. Use OSM data to plan and prioritize infrastructure investments that mitigate flood impacts on communities and critical infrastructure.
- Use flood risk maps to develop evacuation plans for vulnerable communities, ensuring timely and efficient evacuation routes and shelters during flood emergencies.
- Communicate flood risk information derived from OSM data to residents and stakeholders, promoting awareness and preparedness measures for mitigating flood impacts.

Benefits of Using OSM for Flood Risk Management

- OSM provides open-access geospatial data that is freely available for use in flood risk mapping and management, facilitating collaboration among researchers, policymakers, and communities.
- Detailed mapping of elevation and water features in OSM enhances the accuracy of flood risk assessments, enabling proactive planning and response strategies.
- OSM's global coverage allows its application in diverse geographical settings, supporting tailored flood risk management approaches suited to local climate and terrain conditions.

By leveraging OSM data effectively, policymakers and stakeholders can develop informed flood risk management strategies that enhance resilience, protect communities, and mitigate the adverse impacts of flooding in a changing climate.

Equitable Climate Adaptation Strategies

Climate change disproportionately affects vulnerable populations, including low-income communities and the elderly, who often lack resources to adapt to climate impacts such as extreme heat or flooding. Effective climate adaptation strategies require targeted efforts to support these vulnerable groups. OpenStreetMap offers valuable geospatial data that can be leveraged to develop equitable climate adaptation strategies: OSM includes data on building footprints, which can be used to identify and analyze the spatial distribution of housing types, including low-income housing and residences with elderly populations.

Adaptation Strategies

- Retrofit low-income housing with energy-efficient measures to reduce energy costs and improve indoor thermal comfort, particularly during heatwaves.
- Implement programs that provide cooling assistance and resources to elderly residents during periods of extreme heat.
- Direct investments in green infrastructure projects such as parks, green roofs, and urban forests in neighborhoods with vulnerable populations to enhance climate resilience and improve air quality.
- Engage local communities and stakeholders in the planning and implementation of climate adaptation measures, ensuring that solutions address the specific needs and priorities of vulnerable residents.

Benefits of Using OSM for Equitable Climate Adaptation

- OSM facilitates the integration of building footprint data with socioeconomic indicators, enabling a comprehensive assessment of vulnerability and targeted adaptation planning.
- Detailed mapping in OSM enhances the accuracy of identifying vulnerable communities and prioritizing adaptation efforts, ensuring resources are allocated where they are most needed.
- By incorporating local knowledge and community input, OSM-supported adaptation strategies can foster community resilience and promote equitable outcomes in climate adaptation.

By leveraging OSM data effectively, cities can develop and implement climate adaptation strategies that not only enhance resilience but also promote social equity by addressing the specific needs of vulnerable populations in the face of climate change.

CONCLUSION

This chapter has explored the valuable role of OSM data in local climate change assessments and policy formulation. By leveraging the open, detailed, and constantly evolving geospatial information within OSM, researchers and policymakers gain powerful tools to understand and address climate challenges at the local level. OSM offers several advantages for climate research, including open access, detailed spatial coverage, and continual updates. However, it's crucial to acknowledge limitations like potential data inconsistencies and incompleteness in certain areas. Employing data validation techniques is essential to ensure the accuracy of analysis. The ongoing success of OSM relies heavily on the active participation of its global community. Continued contributions and data quality improvement efforts are vital to maintain the platform's reliability and value for climate research.

Emerging Trends and Future Directions

As sensor networks become more prevalent in urban environments, integrating these real-time data streams with OpenStreetMap can revolutionize climate research and adaptation strategies. Sensor data on temperature, humidity, air quality, and other parameters can complement OSM's static geographic data, providing continuous updates on local climate conditions. By overlaying sensor data onto OSM layers, researchers can create dynamic maps that visualize real-time climate variations, aiding in immediate decision-making for climate adaptation and disaster response. Machine Learning Algorithms can analyze vast amounts of OSM data to detect patterns and correlations between spatial features (e.g., land use, building density) and climate variables (e.g., temperature, precipitation). Moreover, machine learning models trained on historical OSM and climate data can forecast future climate scenarios and assess the potential impact of urban development on local climate conditions.

REFERENCES

Bell, C., & Masys, A. J. (2020). Climate Change, Extreme Weather Events and Global Health Security a Lens into Vulnerabilities. In *Advanced Sciences and Technologies for Security Applications*. DOI: 10.1007/978-3-030-23491-1_4

Biljecki, F., Chow, Y. S., & Lee, K. (2023). Quality of crowdsourced geospatial building information: A global assessment of OpenStreetMap attributes. *Building and Environment*, 237, 110295. Advance online publication. DOI: 10.1016/j.buildenv.2023.110295

Borkowska, S., Bielecka, E., & Pokonieczny, K. (2023). Comparison of Land Cover Categorical Data Stored in OSM and Authoritative Topographic Data. *Applied Sciences (Basel, Switzerland)*, 13(13), 7525. Advance online publication. DOI: 10.3390/app13137525

Budhathoki, N. R., & Haythornthwaite, C. (2012). Motivation for Open Collaboration: Crowd and Community Models and the Case of OpenStreetMap. *The American Behavioral Scientist*, 57(5), 548–575. DOI: 10.1177/0002764212469364

Chen, B., Tu, Y., Song, Y., Theobald, D. M., Zhang, T., Ren, Z., Li, X., Yang, J., Wang, J., Wang, X., Gong, P., Bai, Y., & Xu, B. (2021). Mapping essential urban land use categories with open big data: Results for five metropolitan areas in the United States of America. *ISPRS Journal of Photogrammetry and Remote Sensing*, 178, 203–218. Advance online publication. DOI: 10.1016/j.isprsjprs.2021.06.010

Garro-Quesada, M. del M., Vargas-Leiva, M., Girot, P. O., & Quesada-Román, A. (2023). Climate Risk Analysis Using a High-Resolution Spatial Model in Costa Rica. *Climate (Basel)*, 11(6), 127. Advance online publication. DOI: 10.3390/cli11060127

Herfort, B., Lautenbach, S., Porto de Albuquerque, J., Anderson, J., & Zipf, A. (2023). A spatio-temporal analysis investigating completeness and inequalities of global urban building data in OpenStreetMap. *Nature Communications*, 14(1), 3985. Advance online publication. DOI: 10.1038/s41467-023-39698-6 PMID: 37414776

Huang, X., Wang, S., Yang, D., Hu, T., Chen, M., Zhang, M., Zhang, G., Biljecki, F., Lu, T., Zou, L., Wu, C. Y. H., Park, Y. M., Li, X., Liu, Y., Fan, H., Mitchell, J., Li, Z., & Hohl, A. (2024). Crowdsourcing Geospatial Data for Earth and Human Observations: A Review. In *Journal of Remote Sensing (United States)* (Vol. 4). DOI: 10.34133/remotesensing.0105

Kabir, M., Habiba, U. E., Khan, W., Shah, A., Rahim, S., los Rios-Escalante, P. R. D., Farooqi, Z. U. R., & Ali, L. (2023). Climate change due to increasing concentration of carbon dioxide and its impacts on environment in 21st century; a mini review. In *Journal of King Saud University - Science* (Vol. 35, Issue 5). DOI: 10.1016/j.jksus.2023.102693

Laato, S., & Tregel, T. (2023). Into the Unown: Improving location-based gamified crowdsourcing solutions for geo data gathering. *Entertainment Computing*, 46, 100575. Advance online publication. DOI: 10.1016/j.entcom.2023.100575

Levin, E., Beisekenov, N., Wilson, M., Sadenova, M., Nabaweesi, R., & Nguyen, L. (2023). Empowering Climate Resilience: Leveraging Cloud Computing and Big Data for Community Climate Change Impact Service (C3IS). *Remote Sensing (Basel)*, 15(21), 5160. Advance online publication. DOI: 10.3390/rs15215160

Liu, Z. (2021). Identifying urban land use social functional units: A case study using OSM data. *International Journal of Digital Earth*, 14(12), 1798–1817. Advance online publication. DOI: 10.1080/17538947.2021.1988161

Ma, D., Sandberg, M., & Jiang, B. (2015). Characterizing the heterogeneity of the openstreetmap data and community. *ISPRS International Journal of Geo-Information*, 4(2), 535–550. Advance online publication. DOI: 10.3390/ijgi4020535

McGough, A., Kavak, H., & Mahabir, R. (2024). Is more always better? Unveiling the impact of contributor dynamics on collaborative mapping. *Computational & Mathematical Organization Theory*, 30(2), 173–186. Advance online publication. DOI: 10.1007/s10588-023-09383-6

Mooney, P., & Corcoran, P. (2014). Analysis of interaction and co-editing patterns amongst openstreetmap contributors. *Transactions in GIS*, 18(5), 633–659. Advance online publication. DOI: 10.1111/tgis.12051

Moradi, M., Roche, S., & Mostafavi, M. A. (2023). Evaluating OSM Building Footprint Data Quality in Québec Province, Canada from 2018 to 2023: A Comparative Study. *Geomatics*, 3(4), 541–562. Advance online publication. DOI: 10.3390/geomatics3040029

Morshed, S. R., Fattah, M. A., Al Kafy, A., Alsulamy, S., Almulhim, A. I., Shohan, A. A. A., & Khedher, K. M. (2024). Decoding seasonal variability of air pollutants with climate factors: A geostatistical approach using multimodal regression models for informed climate change mitigation. *Environmental Pollution*, 345, 123463. Advance online publication. DOI: 10.1016/j.envpol.2024.123463 PMID: 38325513

Mühlhofer, E., Kropf, C. M., Riedel, L., Bresch, D. N., & Koks, E. E. (2024). OpenStreetMap for multi-faceted climate risk assessments. *Environmental Research Communications*, 6(1), 015005. Advance online publication. DOI: 10.1088/2515-7620/ad15ab

Novack, T., Vorbeck, L., & Zipf, A. (2024). An investigation of the temporality of OpenStreetMap data contribution activities. *Geo-Spatial Information Science*, 27(2), 259–275. Advance online publication. DOI: 10.1080/10095020.2022.2124127

Opach, T., Navarra, C., Rød, J. K., Neset, T. S., Wilk, J., Cruz, S. S., & Joling, A. (2023). Identifying relevant volunteered geographic information about adverse weather events in Trondheim using the CitizenSensing participatory system. *Environment and Planning. B, Urban Analytics and City Science*, 50(7), 1806–1821. Advance online publication. DOI: 10.1177/23998083221136557

OSM. (2021). *OpenStreetMap Copyright and License.* https://www.openstreetmap.org/copyright

Patel, R., & Patel, A. (2024). Evaluating the impact of climate change on drought risk in semi-arid region using GIS technique. *Results in Engineering*, 21, 101957. Advance online publication. DOI: 10.1016/j.rineng.2024.101957

Pearson, R. A., Smart, G., Wilkins, M., Lane, E., Harang, A., Bosserelle, C., Cattoën, C., & Measures, R. (2023). GeoFabrics 1.0. 0: An open-source Python package for automatic hydrological conditioning of digital elevation models for flood modelling. *Environmental Modelling & Software*, 170, 105842. DOI: 10.1016/j.envsoft.2023.105842

Pinho, M. G. M., Flueckiger, B., Valentin, A., Kasdagli, M. I., Kyriakou, K., Lakerveld, J., Mackenbach, J. D., Beulens, J. W. J., & de Hoogh, K. (2023). The quality of OpenStreetMap food-related point-of-interest data for use in epidemiological research. *Health & Place*, 83, 103075. Advance online publication. DOI: 10.1016/j.healthplace.2023.103075 PMID: 37454481

Queiroz, M., Lucas, F., & Sörensen, K. (2024). Instance generation tool for on-demand transportation problems. *European Journal of Operational Research*, 317(3), 696–717. Advance online publication. DOI: 10.1016/j.ejor.2024.03.006

Romão, J., Palm, K., & Persson-Fischier, U. (2023). Open spaces for co-creation: A community-based approach to tourism product diversification. *Scandinavian Journal of Hospitality and Tourism*, 23(1), 94–113. Advance online publication. DOI: 10.1080/15022250.2023.2174183

Sarkar, D., & Anderson, J. T. (2022). Corporate editors in OpenStreetMap: Investigating co-editing patterns. *Transactions in GIS*, 26(4), 1879–1897. Advance online publication. DOI: 10.1111/tgis.12910

Schneider, F. A., Ortiz, J. C., Vanos, J. K., Sailor, D. J., & Middel, A. (2023). Evidence-based guidance on reflective pavement for urban heat mitigation in Arizona. *Nature Communications*, 14(1), 1467. Advance online publication. DOI: 10.1038/s41467-023-36972-5 PMID: 36928319

Slovic, A. D., Kanai, C., Sales, D. M., Rocha, S. C., de Souza Andrade, A. C., Martins, L. S., Coelho, D. M., Freitas, A., Moran, M., Mascolli, M. A., Caiaffa, W. T., & Gouveia, N. (2023). Spatial data collection and qualification methods for urban parks in Brazilian capitals: An innovative roadmap. *PLoS ONE, 18*(8 August). DOI: 10.1371/journal.pone.0288515

Truong, P. M., Le, N. H., Hoang, T. D. H., Nguyen, T. K. T., Nguyen, T. D., Kieu, T. K., Nguyen, T. N., Izuru, S., Le, V. H. T., Raghavan, V., Nguyen, V. L., & Tran, T. A. (2023). Climate Change Vulnerability Assessment Using GIS and Fuzzy AHP on an Indicator-Based Approach. *International Journal of Geoinformatics*, 19(2), 39–53. Advance online publication. DOI: 10.52939/ijg.v19i2.2565

Ullah, T., Lautenbach, S., Herfort, B., Reinmuth, M., & Schorlemmer, D. (2023). Assessing Completeness of OpenStreetMap Building Footprints Using MapSwipe. *ISPRS International Journal of Geo-Information*, 12(4), 143. Advance online publication. DOI: 10.3390/ijgi12040143

Upadhyay, R. K. (2020). Markers for Global Climate Change and Its Impact on Social, Biological and Ecological Systems: A Review. *American Journal of Climate Change*, 09(03), 159–203. Advance online publication. DOI: 10.4236/ajcc.2020.93012

Yamashita, J., Seto, T., Iwasaki, N., & Nishimura, Y. (2023). Quality assessment of volunteered geographic information for outdoor activities: An analysis of OpenStreetMap data for names of peaks in Japan. *Geo-Spatial Information Science*, 26(3), 333–345. Advance online publication. DOI: 10.1080/10095020.2022.2085188

Zalasiewicz, J., & Williams, M. (2021). Climate change through Earth history. In *Climate Change* (3rd ed.). Observed Impacts on Planet Earth., DOI: 10.1016/B978-0-12-821575-3.00003-7

Zhang, X., An, J., Zhou, Y., Yang, M., & Zhao, X. (2024). How sustainable is OpenStreetMap? Tracking individual trajectories of editing behavior. *International Journal of Digital Earth*, 17(1), 2311320. Advance online publication. DOI: 10.1080/17538947.2024.2311320

Chapter 16
Climate Change and Sustainable Development:
How Can Climate Change Be Addressed Within the Framework of Sustainable Development Goals?

Ajay Chandel
https://orcid.org/0000-0002-4585-6406
Lovely Professional University, India

Mohit Yadav
https://orcid.org/0000-0002-9341-2527
O.P. Jindal Global University, India

Phuong Mai Nguyen
https://orcid.org/0000-0002-2704-9707
Vietnam National University, Hanoi, Vietnam

ABSTRACT

Addressing climate change within the framework of the Sustainable Development Goals (SDGs) is crucial for achieving global sustainability. This chapter explores how integrating climate action with the SDGs can drive transformative change across sectors. It examines key areas such as mitigation and adaptation strategies, the role of finance and technology, and the need for effective policy integration. Highlighting challenges including policy coherence, financial constraints, and technological gaps, the chapter also identifies opportunities through innovation, global cooperation, and inclusive transitions. By fostering resilient cities and

DOI: 10.4018/979-8-3693-5792-7.ch016

leveraging nature-based solutions, countries can enhance their climate resilience while advancing sustainable development. This comprehensive analysis underscores the importance of a coordinated, ambitious approach to climate action to ensure a sustainable and equitable future.

INTRODUCTION

Climate change is one of the most alarming crises of the modern day, whose impacts stretch to every single aspect of life on Earth. It is basically defined as a significant change in the average global temperatures and weather patterns over time, mainly because of human activities that involve the burning of fossil fuels, deforestation, and industrial processes. These, in turn, heighten the atmospheric concentration of GHGs, which ultimately lead to global warming and a change in Earth's climate systems. The effect of climate change includes but is not limited to increased sea levels, weather events of higher frequency and intensity, loss of biodiversity, disturbance of ecosystems, and many more. These changes pose formidable risks to human health, food security, water supply, economy, and life quality in general.

On the other hand, parallel with this, the world is also confronted with the challenge of sustainable development to meet the needs of the present without compromising the ability of future generations to meet their needs. That notion of sustainable development has been encapsulated in the United Nations' 2030 Agenda, with 17 SDGs985 aimed at ending poverty, protecting the planet, and ensuring prosperity for all. These SDGs, therefore, recognize a balance among social, economic, and environmental sustainability in development. Of these goals, SDG 13 focuses on climate action in itself: "Take urgent action to combat climate change and its impacts." However, the interplay between climate change and the other SDGs goes far beyond SDG 13, affecting nearly all aspects of sustainable development.

Climate change and sustainable development are inextricably linked and interdependent. On one hand, unabated climate change could undo progress on all the SDGs. For example, more temperatures and shifting rainfall can further encourage poverty (SDG 1) through its effect on agriculture and low food security levels (SDG 2), which in turn can result in malnutrition and impinge on health outcomes (SDG 3). Climate-induced disasters can wipe out infrastructure, displace communities, and corrode livelihoods, only to further cement inequality (SDG 10) and threaten peace and justice (SDG 16). In contrast, efforts toward sustainable development may mitigate or add to climate change depending on the pathway taken. Accordingly, development strategies focused on renewable energy, sustainable agriculture, and responsible consumption offer high GHG reductions with considerable potential for resilience against climate impacts, while those focused on fossil fuel-based processes and unsustainable practices accelerate further climate change.

Climate change must be addressed in the context of sustainable development in a holistic and integrated way. It requires attention to the link between climate action and the rest of the sustainable development agenda in terms of identifying synergies that would maximize positive impacts but manage trade-offs to avoid negative ones. The promotion of clean energies, for instance, is not only vital to climate mitigation (SDG 13) but also can provide economic growth (SDG 8), innovation (SDG 9), and access to affordable energy (SDG 7). Similarly, the protection and restoration of ecosystems can facilitate climate resilience, biodiversity conservation, and livelihoods maintenance. The following chapter debates the possibility of addressing climate change in the SDGs context, both its opportunities and challenges. It stresses that development planning and policy-making should consider the element of climate at all levels, ranging from the local to the global. It calls for the culmination of collaboration among governments, the private sector, civil society, and international organizations in relation to this. Climate action, linked to sustainable development goals, will lead to a resilient, fair world in which economic prosperity will march along with social inclusion and environmental sustainability. This is not only absolutely imperative in taking up the challenges brought about by climate change, but also constitutes a vital means of reaching the long-term vision of a sustainable and prosperous future for all.

Understanding the Intersections of Climate Change and the SDGs

Climate change does not exist in a vacuum from, but actually intersects very strongly with, the Sustainable Development Goals. While SDG 13, "Climate Action," is explicitly focused on addressing climate change and its consequences, it is important to note that climate change affects virtually every dimension of sustainable development. This interrelatedness suggests that in practice, addressing climate change requires more than a decrease in the emission of greenhouse gases but includes making sure sustainable development is pursued on the social, economic, and environmental fronts.

SDG 13: Climate Action

SDG 13 calls for urgent action to combat climate change and its impacts. This goal encompasses enhancing resilience and adaptive capacity to climate-related hazards and natural disasters in all countries, completely integrating climate change measures into national policies, strategies, and planning, and improving education, awareness, and human and institutional capacity on climate change mitigation, adaptation, impact reduction, and early warning. This is all-important in achieving the SDG 13 on reducing adverse impacts of climate change, which are already evident in most parts of the world and disastrous for the most vulnerable. Indeed, unless certain actions are taken about climate change, progress made on many other areas of sustainable development is likely to be undone.

Climate Change Interconnected With Other SDGs

The impacts of climate change pervade every other sector and ecosystem, thereby influencing many SDGs. For instance, climate change threatens to pull millions of people into poverty through the destruction of homes, infrastructure, and livelihoods from destructive weather events like hurricanes, floods, and droughts. This also affects SDG 2, "Zero Hunger," since disruptions in agricultural production and food supply chains result in food insecurity and malnutrition. Changes in weather patterns can reduce crop yields and shift the availability of water resources, which in turn affects the ability to achieve food security and improve nutrition.

Further, climate change is directly linked to SDG 3, "Good Health and Well-being." As the temperature rises and the pattern of precipitation alters, it spreads vector-borne diseases like malaria and dengue fever, while extreme events cause injury and death, and also psychological stress. On the other hand, climate change exacerbates the conditions of water scarcity through altered rainfall patterns, reduction in water quality, and an increase in frequency of drought and floods. This affects SDG 6 "Clean Water and Sanitation." Not only does this supply a reduced amount of drinking water, but it further causes a wide array of sanitation and hygiene problems that most definitely increase health vulnerability.

Synergies and Trade-Offs

While climate change challenges many SDGs, it also creates opportunities for synergies-in other words, struggling against climate change will spill over positively into other development goals. For example, switching to renewable energy sources will contribute to SDG 7 "Affordable and Clean Energy" through providing clean, reliable, and sustainable energy options. Besides that, this transition can also be supportive of SDG 8, "Decent Work and Economic Growth," if jobs are created with regard to renewable energies, and to SDG 9, "Industry, Innovation, and Infrastructure," for it will, after all, drive innovation and sustainable industrialization.

Not all trade-offs are beneficial, however. Some mitigation strategies might be in direct competition with food production, such as large-scale biofuel production, which therefore would link again to food security issues, SDG 2. While hydropower projects are yet another development where communities may get displaced and ecosystems disrupted for the contribution of clean energy, this would have implications on SDGs related to social inclusion and environmental protection. Such trade-offs raise an increasing need for careful planning with inclusive decision-making processes that pay attention to all stakeholders' needs and priorities, particularly those of vulnerable and marginalized groups.

Holistic Approaches to Sustainable Development

Climate change and the SDGs are so closely interrelated that a more integrated approach is relevant in light of sustainable development. This would mean integrating all levels of governance and sectors for climate action to effectively achieve policy coherence and support one another. For example, sustainable land management contributes to climate action by reducing greenhouse gas emissions while at the same time maintaining food security, biodiversity conservation, and livelihood support: SDGs 2, 15, and 1, respectively.

Resilience to climate impacts should be generated as part of attaining the SDGs. This would be by building resilient infrastructure, reducing inequalities, inclusive and sustainable urbanization in SDG 11, and enhancing efforts of disaster risks. The above actions will reduce not only the effects of climate change but contribute to a broad range of development objectives such as reduction of inequalities under SDG 10 and attainment of sustainable economic growth under SDG 8 (Acharya, 2023).

Understanding climate change and SDG interactions underlines the need for integrated approaches to development and synergistic ones. Climate change is not a standalone problem; it is a cross-cutting challenge linked with every aspect of sustainable development. Appreciation and consideration of such interlinkages would better equip policymakers, practitioners, and stakeholders to effectively craft strategies that advance climate action and sustainable development goals together. This will be importantly instrumental in achieving a sustainable, resilient, and equitable future for all.

Mitigation Strategies within the Framework of SDG

The mitigation strategies form core building blocks for global responses to climate change and have strong linkages with the Sustainable Development Goals. Mitigation strategies are targeted at reducing GHG emissions and/or enhancing carbon sinks with the express purpose of containing the magnitude of climate change and its impacts. The SDG mitigation strategies not only contribute to attaining SDG 13, "Climate Action," but they also contribute to and are interconnected with other SDGs, such as SDG 7, "Affordable and Clean Energy," SDG 11, "Sustainable Cities and Communities," and SDG 12, "Responsible Consumption and Production." Achieving this necessitates an integrated approach, combining climate action with sustainable development objectives so that mitigation is pursued in a way that promotes social equity, economic prosperity, and environmental sustainability.

Reducing Greenhouse Gas Emissions Through Sustainable Practices

Mitigation Strategies: GHG Emission Reduction through Sustainable Practices Among the major strategies for mitigation, GHG emissions from all sectors-energy, transport, industry, agriculture, and waste management-need to be reduced. Such reduction is quintessential for achieving SDG 13 and hence has a direct link with SDG 7, aimed at ensuring access to affordable, reliable, sustainable, and modern energy for all. That includes transitioning from fossil fuels to renewable resources-solar, wind, hydro, and geothermal. Actually, renewable energy decreases emissions while creating jobs, spurring economic growth, and improving energy security-so, in fact, it contributes to SDG 8, "Decent Work and Economic Growth," and SDG 9, "Industry, Innovation, and Infrastructure."

Another critical mitigation measure is the improvement in energy efficiency. Energy efficiency in all sectors reduces energy consumption and, subsequently, emissions; hence, it contributes to SDG 12 on responsible consumption and production patterns. For example, energy-efficient appliances and industrial processes lower the demand for electricity and, therefore, reduce the need for generation based on fossil fuels. Similarly, improved insulation of buildings and the use of smart technologies significantly reduce the energy consumption of residential and commercial buildings, relating to SDG 11, which targets cities and human settlements that shall be inclusive, safe, resilient, and sustainable.

Role of Renewable Energy in Sustainable Development

The development of renewable energies is central to climate mitigation and directly connects with many SDGs. Aside from their environmental benefits, investment in renewable energy technologies supports development objectives such as poverty alleviation, improvements in health, and educational improvement. For example, improved health can come as indoor air pollution from traditional biomass cooking decreases with the supplying of clean and affordable access to energy, hence supporting SDG 3. This, in turn, enhances educational opportunities in the ability to light and power schools and homes, contributing to SDG 4.

Besides, most of the renewable energy projects boost a local economy with jobs and livelihood, thus contributing to SDG 8. They can also foster innovation and infrastructure for the advancement of SDG 9. For instance, solar panel or wind turbine manufacturing can create jobs in installation and maintenance, hence stimulating local economy growth and development.

Sustainable Urban Planning and Transportation

Besides, the cities are very highly contributors in general towards GHG emissions because of energy use, transportation, and waste generation. Thus, sustainable urban planning and transportation have to constitute part of the mitigation strategies within the SDG framework. SDG 11 emphasizes the need to make cities inclusive, safe, resilient, and sustainable, including making them carbon-neutral.

This is the art of designing cities in ways that reduce energy use to a minimum, enhance public transport, and promote cycling and walking. Well-planned compact urban areas reduce private vehicle use, and consequently emissions from the transport sector. A major investment in public transport-in buses, trains, and trams-can similarly go a long way in reducing emissions in urban areas, especially systems powered by renewable energy, while guaranteeing equal mobility for all.

It also involved the promotion of green infrastructure, such as parks and green roofs, which have been helpful in mitigating climate change through the improvement in carbon sequestration and a reduction of the heat island effect, therefore contributing to the achievement of SDG 15, "Life on Land." In addition, a potential waste management strategy involving recycling and composting can reduce methane emissions produced in landfills by contributing to the achievement of SDG 11 along with SDG 12.

Agriculture and Land Use Management

Agriculture and land use are also two critical components of climate mitigation due to their potential for GHG emission and sequestration. Sustainable agricultural practices could significantly reduce agricultural GHG emissions while enhancing soil health and productivity. These also help achieve SDG 2: "Zero Hunger," particularly for food security through sustainable agriculture.

Land use management is very important for the protection and improvement of the natural carbon sink, such as forests and wetlands, which absorb CO_2 from the atmosphere. Protection and restoration of these ecosystems are vital in terms of climate mitigation; it contributes to SDG 15: conservation of terrestrial ecosystems, combat desertification, halt and reverse land degradation, and biodiversity loss. For example, reforestation and afforestation provide for carbon sequestration, a home for fauna, biodiversity, and economic gains through sustainable forestry. SDG mitigation strategies interlink in addressing climate change problems while at the same time contributing to achieving Sustainable Development. These include reducing GHG emissions, the transition toward renewable energies, efficient energy usage, and various modes of sustainable land management that will help contribute to several SDGs in building a resilient and sustainable future. Their effectiveness is based on coordination from local to national to global levels through concerted efforts of governments, business, civil society, and individuals. It is only then, through that collaboration, that we would be able to attempt reducing the climate change situation and reach all the goals designed in the 2030 Agenda for Sustainable Development.

ADAPTATION STRATEGIES AND RESILIENCE BUILDING

Among key adaptation strategies, or rather building resilience to it, is the comprehensive response to climate change within the SDGs framework. While mitigation attempts to decrease the level of emissions to constrain future climate change, adaptation focuses on how systems and practices can be altered in order to reduce the vulnerability from current and projected climate impacts. Building resilience involves increasing the preparedness for recovery by persons, communities, and ecosystems from climatic shocks and stresses. Put them all together, and you have the essential ingredients of a package that will protect human well-being, economic stability, and environmental sustainability in the face of a changing climate.

Enhancing Climate Resilience in Communities

Perhaps the most critical objective of all the adaptation strategies is to provide resilience in the communities that are most vulnerable to climate change. It is important to note that vulnerability to climate change is not spread evenly across the globe; rather, it is concentrated within low-income countries, small island developing states, and marginalized communities that lack the capacity to respond effectively to climate-related hazards. This vulnerability lies at the heart of attaining SDG 1, "No Poverty," and SDG 10, "Reduced Inequalities," as adaptation to the changing climate could keep people out of climate-induced poverty and reduce inequality in exposure to climate risks.

It will be locally driven adaptation action, taking into consideration the peculiar context and needs of each community. This involves constructing and/or strengthening structures that could mitigate the impact of weather extremes such as floods, hurricanes, and heat waves, among other disasters, in line with SDG 9 on resilient infrastructure development. This includes enhanced early warning systems and emergency preparedness plans that enable communities to be more proactive in anticipating and responding to climate hazards. This approach of equipping communities with the knowledge and tools to be better prepared for and able to cope with climate impacts supports SDG 11, "Sustainable Cities and Communities," by promoting inclusive, safe, resilient urban and rural areas.

Climate-Smart Agriculture and Food Security

Agriculture is very vulnerable to climate change because of the fact that most of its activities depend on good climatic conditions. Changes in temperature and rainfall, among other extreme weather events, seriously disturb agricultural productivity, hence threatening food security and rural livelihoods. Climate-smart agriculture has, therefore, become one of the major adaptation strategies. CSA here refers to the adoption of agricultural practices that contribute to higher resilience of the agricultural sector due to the impacts of climate change. The main objectives are to increase productivity and reduce GHG emissions-all combined, mean alignment with SDG 2-"Zero Hunger"-focusing on ending hunger, attaining food security, promoting efficient management of productive functions in agriculture.

Examples of such climate-smart agricultural practices are cropping diversification to reduce the probability of losing an entire crop, irrigation systems that waste less water, and drought-resistant seed varieties. Such measures will allow farmers to continue adapting their practices to changing conditions while maintaining consistent yields irrespective of climatic fluctuations. Agroforestry, or integrating trees into agricultural landscapes, can offer farmers supplemental income and improved soil quality, and enhance carbon sequestration to support both SDG 15, "Life on Land," and SDG 13, "Climate Action."

Water Resources Management and Conservation

Climate change will probably lead to increased severity of water scarcity and variability in many regions, hence having significant impacts on availability and quality. This therefore poses a great challenge toward the achievement of SDG 6, "Clean Water and Sanitation," which focuses on ensuring availability and sustainable management of water and sanitation for all. Efficient water resource management and conservation are, therefore, the critical components of adaptation strategies (Sarfo et al., 2019).

Adaptation in the water sector is realized through water use efficiency, modernization of irrigation, promotion of water-saving technologies, and watersaving practices by households and industries. This apart, natural systems-that is, rivers, wetlands, and aquifers-should be protected and restored to their original condition to enhance their capabilities for absorption and regulation of water flows, reducing the risk of flood and drought accordingly. The integrated water resource management approaches that coordinate the development and management of water, land, and related resources are equally important in order to meet various competing needs for water and to make sure that this precious resource is used in a sustainable manner in the face of climate change.

Protection and Restoration of Ecosystems

Healthy ecosystems are vital components of the processes of climate adaptation and resilience building inasmuch as they provide vital services that support human well-being and protect against climate impacts. For example, mangroves, coral reefs, and wetlands along coasts prevent storm surges, which again prevents floods and erosion. Similarly, the water flow and soil erosions are regulated, and the local climates remain stable with the help of forests and grasslands. This is a critical adaptation strategy that will be in line with SDG 15, "Life on Land," which focuses on the conservation, restoration, and sustainable use of terrestrial and freshwater ecosystems.

EbA is a method that builds on or enhances the natural resilience of an ecosystem to strengthen the adaptive capacity of human society. In addition to reducing vulnerability to climate change, EbA generates many co-benefits related to biodiversity conservation, carbon sequestration, and livelihood development. For example, the restoration of degraded mangrove forests protects coastal communities from storm surges, supports fisheries, and increases biodiversity.

Finance and Technology for Adaptation

Finance and technology access will be determinant factors for the effective implementation of adaptation strategies and building resilience. Most vulnerable communities and countries usually lack the financial wherewithal necessary for investing in adaptive infrastructure, technologies, and practices (Scobie, 2019). It is a fact that must be addressed if the SDG 17, "Partnerships for the Goals," which calls for international cooperation and support to reach the goals of sustainable development, is to be attained.

Climate finance refers to the public and private funding towards climatic-related projects, which are among the important mechanisms for supporting adaptation. The developed country Parties committed themselves to jointly mobilize $100 billion per year by 2020 to address the needs of developing countries in responding to climate change, including adaptation. These may be used towards financing a wide array of activities for adaptation-from resilient infrastructure building to supporting community-based adaptation initiatives.

Equally important towards the realization of this goal are technology transfer and capacity building. The developing countries require better technologies and know-how for the realization of adaptation measures, from climate-resistant crops to warning systems and adapted water management techniques. This would be further facilitated through international cooperation and partnerships in respect of the transfer of these technologies and enhancement of local capacity for such a purpose, so that all countries are uniformly prepared to meet the challenges of climate change.

Adaptation strategies and building resilience will involve the protection of sustainable development against a changing climate. These adaptation strategies contribute to a number of the SDGs through their contributions to improved community resilience and, therefore, increased climate-resilient agriculture, sustainable management of water resources, protection of ecosystems, access to finance, and technology in building a resilient and equitable future. However, these adaptation processes can only succeed when underpinned by approaches that are inclusive and participatory in nature, whereby all stakeholders-particularly vulnerable and marginal communities-are highly involved. By integrating adaptation into planning and policy for development, we can realize a greener, far more sustainable, and resilient world-one more able to confront the challenge that climate change presents.

FINANCE AND TECHNOLOGY FOR CLIMATE ACTION

It is also important to note that finance and technology play a very critical role in furthering climate action and achieving the SDGs. SDG 13 on climate action will need massive financial investments and a host of technological innovations to reduce GHG emissions, enhance resilience to climate impacts, and achieve sustainable development. Resources and technologies that would help tackle climate change are unevenly distributed across the world. In other words, there is a gap that developing countries most prone to the vagaries of a changing climate often face inaccessibility to finance and technology for effective climate action. Such gaps must be addressed in order to allow all countries to contribute to and benefit from the global effort against climate change.

The Role of Climate Finance

Climate finance is locally, nationally, or internationally tapped funding from public, private, and alternative sources to support mitigation and adaptation activities that address climate change. It is, therefore, an important facilitator of climate action in cases where developing countries face resource scarcity for investment in low-carbon technology for measures of building resilience. Effective climate finance will reduce GHG emissions, promote adaptation to the adverse impacts of climate change, and effectively shift to more sustainable development pathways.

International climate finance has been one of the most important rallying points for global climate negotiations. In the process, through the UNFCCC, developed countries are committed to mobilizing $100 billion per year by 2020 to support climate actions in developing countries. This financing is done in the context of an undertaking to implement their NDCs, comprised of pathways and commitments toward reduction of emissions and enhanced climate resilience. While there is progress in the mobilization of climate finance, there are still a number of large gaps; funding for adaptation usually lags behind that for mitigation. In effective use of climate finance, the investments under way need to be channeled in line with the SDGs and priorities for national development. This needs an integrated path through which climate action is made part of larger development strategies, since finance must support sustainable and equitable outcomes. For example, investment in renewable energy projects contributes to the attainment of SDG 13 by reducing emissions but also contributes to SDG 7, "Affordable and Clean Energy," as it increases access to cleaner sources of energy. Likewise, financing of climate-resilient infrastructure would contribute to attaining SDG 9, "Industry, Innovation and Infrastructure" through the promotion of sustainable industrialization and innovation.

Mechanisms of Innovative Financing

Innovative financing mechanisms are essential for scaling up climate finance and leveraging private sector investment. Most of the traditional sources of funding, like government budgets or international grants, can be deficient and often inadequate to meet the huge financial needs of climate action. New mechanisms currently under development to help mobilize additional resources and leverage private sector involvement include green bonds, blended finance, and carbon pricing.

Green bonds stand among the fastest-growing instruments in the climate finance landscape. These bonds are issued to raise funds for projects that would contribute positively to the environment or climate impacts-such as renewable energy, energy efficiency, and sustainable water management, among others. Green bonds present investors with opportunities for supporting climate-friendly projects and returns on their investments, making it one of the instruments that is increasingly adopted by both public and private entities.

Another innovative approach is **Blended Finance,** which mixes public and private funding in an effort to de-risk the investments in climate projects. Blended finance can attract private capital into sectors that might otherwise be considered too risky by using public funds to absorb some of the risks in the climate investments. This is particularly useful in funding large-scale renewable energy projects or climate-resilient infrastructure in developing countries where perceived risks are higher.

Carbon pricing through carbon tax and cap-and-trade system is one of the important tools economically used to offer incentives for reduction in emissions. This mechanism puts a price on carbon emissions to incentivize businesses and people to move toward low-carbon technologies and practices. The money received from carbon pricing can be reinvested into climate action or distributed in support of vulnerable communities against impacts resulting from the transition to a low-carbon economy.

The Importance of Technology in Climate Action

An Imperative Technology can serve as a powerful facilitator of climate action through various tools and solutions that reduce emissions, enhance resilience, and support sustainable development. Technological advances in the fields of renewable energy, energy efficiency, and information and communication technologies are rapidly reducing the barriers to sharply lowering greenhouse gas emissions while improving economic performance and social equity.

While **renewable energy technologies-**solar energy, wind, hydroelectric, and geothermal power-are the keys to mitigating climate change in the long term, they will replace fossil fuels with clean and sustainable resources. This will go a long way in reducing emissions and enhancing energy security. In recent times, there has been a rapid decline in the cost of renewable energy technologies; this factor has increased their competitiveness against traditional sources of energy. Investing in renewable energy not only contributes to attaining SDG 13 but also contributes to SDG 7 through increased access to affordable and clean energy.

Other very vital aspects of climate action involve **energy efficiency technologies**. Energy efficiency in buildings, transport, and industry shows high potential for reduction in energy use and, correspondingly, reduction in emissions. The usage of energy-efficient appliances and lighting, the adoption of electric vehicles, and the optimization of industrial processes tend to reduce GHG emissions while saving energy at the same time. Such technologies would also contribute to SDG 12, "Responsible Consumption and Production," because they would see resources being utilized in a much more sustainable manner.

Information and communication technologies (ICT) are also vital for climate action based on proper data collection, analysis, and decision-making. Remote sensing, GIS, and AI technologies are ushering in better climate monitoring, improved early warning systems, and adaptive management of natural resources. These thus contribute toward attaining SDG 11, "Sustainable Cities and Communities," with a view toward smart urban planning and disaster risk reduction (Sherman et al., 2016).

Technology Transfer and Capacity Building

Whereas technology holds immense promise in furthering climate action, the access to technology is lopsided across nations, especially among developing nations. Because of this very reason, technology transfer and capacity building hold the key to ensuring that all countries enjoy the accruing benefits arising from technological innovations. The Technology Mechanism under the UNFCCC, composed of the TEC and the CTCN, is suitably positioned to facilitate access to climate technologies by developing countries.

Technology transfer does not stop at the level of providing equipment or software but also involves a greater element of knowledge, skills, and practices transfer that would actually enable countries to use such technologies effectively and to keep them going. Capacity building then becomes the cornerstone for the entire process, in the absence of which the local institutions, firms, and communities will not have the expertise to adopt and adapt new technologies. This will not only enrich climate resilience but also contribute to broader development imperatives, such as improved education, SDG 4, and promote innovation, SDG 9.

Bridging the Finance-Technology Gaps

Closing the gaps in finance and technology is indispensable for the SDGs and a just transition to a low-carbon, climate-resilient future. This must include sound international cooperation and partnership that binds well in support of SDG 17, "Partnerships for the Goals." Developed countries should be bound to provide finance and technology support to developing countries to empower them to overcome various obstacles that stand in the way of climate action and sustainable development.

Other gaps might be also covered with the help of innovative approaches, such as public-private partnership. Using advantages of both the public and the private sector in this kind of partnership enables raising resources, driving innovation, and scaling up the solutions for the climate. An example is that collaboration between governments, businesses, and research institutions can accelerate the pace of development and deployment of new technologies while also ensuring that those technologies are made available and affordable to all.

Finance and technology are key enablers of enhanced climate action and SDGs. Mobilizing adequate financial resources and harnessing technological innovation are important means through which the global community can accelerate a transition toward a low-carbon, climate-resilient future. Yet this should be an approach that is inclusive and coordinated, where finance and technology are available to all countries, especially those most vulnerable to the impacts of climate change, to realize their climate and development agendas. This can be achieved through international cooperation, mechanisms of innovative financing, and capacity building to realize a much more sustainable, equitable world better prepared to handle the challenges presented by climate change.

POLICY INTEGRATION AND GOVERNANCE

If there is to be effective action on climate, then there needs to be strong policy integration and governance frameworks that align national and international efforts toward the SDGs. Policy integration means the embedding of climate considerations into broader policy agendas so as to make climate action support and enhance other development objectives. Governance, on the other hand, involves structures, processes, and institutions through which decisions about climate action are taken and implemented. Both are integral to creating coherent and comprehensive strategies that address climate change while advancing sustainable development.

Need for Integrated Policy Approaches

Climate change represents complex, interconnected issues that touch almost all aspects of society and the economy. Consequently, the challenge of climate change cannot be dealt with in isolation. Effective climate action requires a holistic approach, as climate policies will need to be combined with other relevant areas: energy, agriculture, transport, urban planning, and social development. This makes the policies reinforce each other, maximising synergies and reducing trade-offs among the various objectives.

For example, the integration of climate action into energy policy will help transition toward renewable energy, therefore reducing greenhouse gas emissions and improving energy security and access, according to SDG 7, "Affordable and Clean Energy." On the other hand, integration of climate policies into agricultural development helps to promote climate-smart agricultural practices that enhance resilience to climate impacts, improve food security, and reduce emissions, hence fostering the attainment of SDG 2, "Zero Hunger." Embedding consideration of climate factors in these and other sectors helps policy integration create a resilient development pathway.

Second, addressing the social dimensions of climate change requires integrated policy approaches. Climate change accentuates current inequality, as its impacts are not equitably distributed within the population; rather, they are felt most strongly by poor and disadvantaged groups. Linking climate action with social policies, such as poverty alleviation, health, and education, this can protect these groups, enhance social equity, and contribute to SDG 1 "No Poverty," SDG 3 "Good Health and Well-being," and SDG 4 "Quality Education." The contribution of social safety nets or targeted support programs goes a great way toward revitalizing communities through climate-related shocks and building resilience to future impacts.

Strengthening Climate Governance

Governance is a key element in ensuring effectiveness in climate action: it prescribes how decisions are taken, who participates in the decision-making process, and how policies are implemented and followed up on. A solid governance framework ensures accountability, transparency, and inclusiveness in climate action and helps governments and stakeholders work in tandem for common goals.

The nature of climate governance at the national level entails well-defined institutional arrangements with clearly established responsibilities pertaining to climate action. For example, it requires the establishment of dedicated climate agencies or departments that are responsible for the formulation of full-fledged climate strategies and action plans and for ensuring coordination across levels of government and sectors. Secondly, good governance encompasses robust legal and regulatory frameworks that stipulate explicit mandates for climate action, entrenched ambitious targets, and mechanisms for progress monitoring and compliance enforcement (Dzebo, 2023).

Another integral part of effective action is participatory governance. Wide involvement of civil society, the private sector, indigenous communities, and marginalized groups in decision-making processes ensures that diverse perspectives are considered, and policies turn out to be more inclusive and equitable. This approach lies within SDG 16, which focuses on ensuring inclusive and participative decision-making processes at all levels. For example, local communities could be involved in the design and implementation of adaptation measures that would ensure appropriateness to cultural requirements, thereby catering to their needs and enhancing community ownership and buy-in (Pattberg & Widerberg, 2015).

International Governance and Cooperation

Climate change is a global issue that requires an internationally coordinated response. International governance frameworks, including the UNFCCC and the Paris Agreement, create critical avenues through which cooperation can take place, global goals are set, and international accountability operates. The Paris Agreement, agreed upon in 2015 under the United Nations Framework Convention on Climate Change, marked a turning point in international climate governance in that it established a framework for all countries to take action on climate change according to their national circumstances and capabilities.

It requires countries under the Paris Agreement to provide Nationally Determined Contributions, through which they detail their commitments to reduce greenhouse gas emissions and build resilience to climate impacts. Other provisions under the Agreement incorporate mechanisms for transparency and accountability; these include the Enhanced Transparency Framework through which countries are to report regularly on the implementation of their NDCs and shall also be subjected to independent review. These would be trust-building mechanisms among countries and ensure delivery on the climate commitments, thereby achieving SDG 17 "Partnerships for the Goals," in that strong global partnerships are an indispensable way to sustainable development.

Similarly, international cooperation goes a long way in guaranteeing regional resilience by responding to transboundary climate impacts. Climate change pays no attention to national boundaries; hence, major cross-boundary implications for water scarcity, food insecurity, and migration. These include the African Union's Climate Change Strategy and the European Union's Green Deal, which are regional governance frameworks through which countries can come together to find solutions to mutual problems and contribute in sharing resources in responding to climate change. These frameworks spawn knowledge sharing and capacity building wherein countries learn from other country experiences and adopt best practices.

Aligning National Policies with Global Commitments

The SDGs seek to address climate change effectively within the national frameworks that are consistent with the global commitments and targets. It ensures coherence in national actions and internationally, hence allowing better coordination and effectiveness of responses towards climate change. Alignment of national policies to the global commitments may be realized by embedding the SDGs and climate targets within the national development plans, climate strategies, and sectoral policies.

Many countries have framed National Climate Change Policies and Strategies explicitly referring to the SDGs, indicating how climate action will contribute to achieving the Goals. It mostly includes explicit targets and indicators for progress toward national priorities and international commitments, such as reduced emission, renewable energy capacity increase, and enhanced resilience to climate impacts. It involves the mainstreaming of national policies to global commitments. In so doing, countries ensure that their activities on climate action are in line with the broad-based global agenda and make a contribution to the achievement of sustainable development at all levels.

Overcoming Policy and Governance Challenges

In as much as policy integration and governance have vital roles in climate action, various challenges persist. A key challenge pertains to policy and institutional fragmentation, leading to inconsistent actions and potential conflicts (Philippidis et al., 2020).

For example, policies that support subsidies for fossil fuels might work at cross-purposes with the goals of reduction in emissions and transition toward renewable energy sources, hence the requirement for policy coherence and coordination. Overcoming these challenges requires countries to seek out appropriate institutional arrangements and mechanisms of cross-sector coordination by making sure that all relevant policies and actions support each other.

Another issue is the capacity and resource challenge to good governance. In general, developing countries need to enhance their financial, technical, and institutional capacities to formulate and implement comprehensive climate policies and programs. The capacity gaps are potential issues that can be realized through specific support from the international community through means such as financial support, technology transfer, and capacity building. The reason behind this is that strengthening national institutions and enhancing governance capacities are necessary in enabling countries to make them capable of dealing with the challenge of climate change and attaining the SDGs.

The Role of Accountability and Transparency

Good climate governance incorporates accountability and transparency. Accountability and transparency ensure that governments and other actors are responsible for their actions; engender trust among countries and communities; and lead to better-informed, more inclusive decision-making. In the SDGs, it is important that governance be transparent and accountable; only then would it be possible to ensure proper utilization of resources, proper implementation of policies, and monitoring and reporting on progress.

This would also enable countries to establish a functional MRV system that would track the progress of climate action and the SDGs. Any such system would need to be so designed as to ensure timely and reliable data on emissions, adaptation measures, and development outcomes, thereby enabling policy makers and stakeholders to assess the effectiveness of climate policies in reaching informed decisions. Mechanisms of independent review can also be significant in accountability, like the process under the Paris Agreement, by providing an objective review of countries' progress and identifying specific areas where countries are lagging behind.

In effective climate action for achieving the SDGs, policy integration and governance play a very important role. Embedding climate considerations in a broader policy agenda; promoting state-of-the-art governance frameworks; and facilitating international cooperation will, in turn, allow the countries to devise coherent and comprehensive strategies, thereby stimulating their efforts toward the issue of climate change. Such difficult leaps in moving toward such ambitious goals often concern overcoming policy fragmentation and capacity constraints and making these processes more accountable and transparent. Resilient and sustainable, this world will be able to better face the adversities that climate change brings, with focused support, inclusive governance, and strong monitoring and reporting.

CASE STUDIES

The following section examines five case studies from various geographical and socio-economic contexts to explore how countries and communities around the world address climate change within the SDGs framework. Each of the cases shows a different way of integrating climate action into sustainable development and thus reveals the variety of strategies to achieve climate resilience with low-carbon growth.

1. Costa Rica: Transition to Renewable Energy

It has been an exemplary leader in demonstrating how a small country can achieve notable advancements in climate action by integrating policy and governance. During the past two decades, Costa Rica has undertaken a firm direction to transition into renewable energy, aligning its national energy policy with SDG 7, "Affordable and Clean Energy," and SDG 13, "Climate Action." Therefore, today, nearly 99% of the country's electricity comes from renewable sources, mainly hydroelectricity and wind, supplemented by geothermal energy (Flagg, 2018).

This transition has been championed by sturdy political will, well-structured policy frameworks, and major investments in renewable energy infrastructure (Richmond-Navarro et al., 2019; John & Derakhshi, 2022). Several incentive programs have been put in place by the government to catalyze private sector investments in renewables. Such efforts reduce Costa Rica's GHG emissions while bringing in other co-benefits: providing employment opportunities, diversifying energy sources, and thereby improving the resilience of the nation's energy system against the vagaries of climate change.

2. Bangladesh: Climate Resilient Agriculture

Bangladesh is one of the most climate-vulnerable countries, with floods and cyclones prevailing over rising sea levels. Climate-resilient agriculture has especially been emphasized by the Government of Bangladesh as part of the adaptation priority and provides the basis for achieving SDG 2 on "Zero Hunger" and SDG 13. The country has already taken a number of steps toward improving the resilience of its agriculture to climate change in line with the Bangladesh Climate Change Strategy and Action Plan (Akter, 2023).

Other initiatives include the development and dissemination of climate-resilient crop varieties such as flood-tolerant rice, integrated pest management, and agroforestry. It has also invested in early warning systems and disaster risk reduction measures as a way of protecting farmers from climate-related shocks. These have helped increase agricultural productivity, improved food security, and enhanced livelihoods for millions of smallholder farmers in Bangladesh (Riyadh et al., 2021).

3. Germany: Energy Efficiency and Circular Economy

Indeed, Germany has been one of the forerunners when it comes to environmental sustainability and climate action. It is a country from which many examples can be taken up regarding integrating energy efficiency with the circular economy (Alola et al., 2023). With policies such as the "Energiewende" (Energy Transition) and

"Resource Efficiency Programme", Germany is well on track towards reducing the country's greenhouse gas emissions and using resources in a manner that is sustainable, while ensuring transition toward a low-carbon, circular economy.

The Energiewende targets increasing energy efficiency and building up renewable energies, hence being able to cover SDG 7 and SDG 13 in particular. It has resulted in substantial reductions of emissions and energy consumption within the industrial, transport, and private household sectors. Meanwhile, the Resource Efficiency Programme encourages the recycling and waste reduction of materials, with sustainable use, under SDG 12 "Responsible Consumption and Production." The aforementioned also contribute to enabling Germany to reduce its environmental footprint, provide jobs through greening of the economy, and engender innovation in green technologies (Steinfatt, 2020).

4. Rwanda: Green Growth and Climate Resilience Strategy

The Government of Rwanda has developed an ambitious national plan-the GGCRS- that embodies an integrated approach in terms of climate action with sustainable development toward attaining SDG 13 and other related goals. Launched in 2011, the GGCRS put forward a comprehensive framework for low-carbon economic growth, enhancing climate resilience, and promoting inclusive growth (Kayigema & Rugege, 2014).

The key components of the strategy involve land management, renewable energy development, and climate-resilient infrastructure construction (Oluoch et al., 2022). Various government policies and programs have also ensured promotion of green businesses, eco-tourism, and better urban planning. This helped Rwanda mark significant achievement in reduction of emissions, resilience to climate impacts, and sustainable economic development.

5. The Netherlands: Integrated Water Management

The Netherlands is one of the countries most vulnerable to flooding and sea-level rise. Hence, the country has developed an integrated approach to water management that confronts climate change while contributing to sustainable development. The Dutch Delta Programme is one comprehensive, integrated program at the national level that integrates climate adaptation, flood risk management, and spatial planning to protect the country against flooding while securing water supplies (Hartmann & Spit, 2015; Linde et al., 2011).

The Delta Programme contributes to SDG 6, "Clean Water and Sanitation," and SDG 13 by guaranteeing a sustainable management of water while raising resilience about the climate effects. Measures taken include dikes and flood-defense construction and reinforcement, wetland restoration, and innovation policies such as "Room for the River," where earmarked areas are allowed to be flooded in control, thus reducing the risk of damage elsewhere. This has been an integrated approach to make the Netherlands less vulnerable to a changing climate, protecting its citizens and infrastructure, in addition to making land and water use more sustainable (Ritzema & Loon-Steensma, 2017).

Joint Analysis

These five case studies illustrate a range of approaches that countries might consider to integrate climate action with sustainable development. Each case underlines the need for context-specific strategies that reflect priorities, challenges, and opportunities at the local level. Several common themes emerge despite differences in geography, socio-economic conditions, and governance structures (Pradhan et al., 2017).

These different cases of successful climate action depend first on **strong political commitment** and **equally firm policy frameworks**. In each case, the national government has taken up the lead with ambitious goals; developed complete strategies; identified, articulated, and implemented commensurate policies with the SDGs. Such commitment has been cardinal in mobilizing resources, engaging stakeholders, and driving progress on climate and development objectives.

It is also critical in securing sustainable outcomes, the ability to attract **cross-sectoral integration.** Indeed, these case studies show that effective climate action requires a holistic approach that integrates climate considerations across various sectors, including energy, agriculture, water management, and urban planning. It reduces negative synergies and trade-offs because policies foster the benefits of each other, allowing several SDGs to be achieved.

Third, **innovative solutions and technologies** are at the core of climate resilience and low-carbon development. Be it the transition into renewable energy in Costa Rica, integrated water management in the Netherlands, or the cases described above, the adoption and scaling up of locally relevant innovative practices and technologies remain extremely important. Innovations help not only to mitigate and adapt to climate change but also bring new economic opportunities for greater livelihoods and well-being.

Community engagement and inclusiveness are central in achieving effectiveness and sustainability in climate action and sustainable development. For one, the involvement of the local communities, businesses, and civil society ensures that decision-making processes are more inclusive, equitable, and responsive to the needs of the locals. Secondly, it also serves to engender ownership and support of the climate initiatives for greater effectiveness and sustainability.

In short, these key lessons from the case studies form a foundation for integrated, context-specific climate action approaches that are in full accord with the SDGs. Drawing inspiration from these examples, countries will be able to devise more effective strategies on how to address climate change while at the same time advancing sustainable development to ensure a resilient and equitable future for all.

CHALLENGES AND OPPORTUNITIES

Addressing climate change within the context of the Sustainable Development Goals encompasses unparalleled challenges and opportunities. As countries pursue integrating climate action with broader development imperatives, they will also have to negotiate a maze of economic, social, and political determinants. Appreciation of the challenges and utilization of opportunities are therefore critical in making any meaningful stride toward a sustainable and resilient future.

Challenges

1. Policy Coherence and Integration

Among the main issues related to climate change that arise in SDGs, the coherence and integration of this policy by various sectors and levels of governance is considered a priority. In fact, for most countries, it is a challenge to articulate climate policies with other development objectives such as economic growth, poverty reduction, and social equity. Conflicting priorities and interests at different government departments, as well as at the national and local levels, may create fragmented, piece-meal, and inconsistent policies that weaken climate actions and efforts toward sustainable development.

For example, ambitious renewable energy targets may be set within a country's SDG 7 ("Affordable and Clean Energy"); however, these can then be undercut through subsidies to fossil fuels or policies pursuing industrial development with utter disregard for the impact on the natural environment. Policy coherence requires strong political will along with coordination mechanisms that translate commitment to align policies to the SDGs at all sectors and levels of government.

2. Financial Constraints

The second big issue involves inadequate endowments to support climate action and sustainable development. There are significant funding gaps in most developing countries that impede an adequate investment in mitigation, adaptation, and resilience-building measures. While international mechanisms to address climate finance, such as the GCF and GEF, provide essential support, overall funding remains insufficient given the scale of the challenge.

Third, the process for accessing climate finance is too awkward and bureaucratic, with increasing requirements and long periods of approval that have recently emerged as discouraging elements for countries to apply for these very funds. Fundamentally, all of these fiscal challenges require rising international support; creating innovative financing mechanisms along with strengthened public-private partnerships will help mobilize resources so desperately needed for climate action and sustainable development.

3. Technological and Capacity Gaps

Technological innovation and capacity building form a vital part of accelerating climate action through the implementation of the SDGs. In reality, though-especially in the Global South-countries often do not have access to the technologies and expertise needed to address the most pressing climate issues effectively. Technological and capacity gaps can seriously hold back progress at many levels-from renewable energy deployment and sustainable agriculture to climate-resilient infrastructure and disaster risk management.

For example, a country might be so committed to emission reduction and resilience enhancement, yet, due to the lack of access to advanced technologies and competent human resources, it would be able to do very little. Such gaps can only be bridged through increased investments being made in research and development, technology transfer, and capacity building processes-particularly in developing countries and vulnerable communities.

4. Social and Cultural Barriers

Also, social and cultural factors may serve as a significant barrier to climate action and sustainable development. The impacts of climate change do not weigh on the population equally; those vulnerable and poor-the women, indigenous peoples, and low-income communities-always bear the greater share of the burden. Such inequity calls for a concern for social equity and inclusion, making sure that all voices are represented, with policies serving the interests of those who will be most affected by climate change.

These gains can sometimes be hindered by deeply entrenched social norms, power dynamics, and embedded culture. Examples include gender prejudices that hamper the effective participation of women in decision-making processes related to climate policy, and traditional land use methods that run counter to contemporary conservation practices. It is clear that such socio-cultural impediments will require conscious efforts at enforcing inclusive governance, participatory community engagement, and empowerment of marginalised groups so that they are also capable of participating in and benefiting from climate action.

5. Political and Institutional Challenges

Political and institutional factors are also important in shaping climate action and sustainable development. The important building blocks of political will, leadership, and governance structures constitute the basis for ambitious targets, fully developed policy formulation, and efficient implementation. Still, political instability, corruption, and weak institutions erode these elements and, subsequently, provide fertile ground for an environment devoid of accountability, transparency, and continuity.

For instance, frequent government or leadership changes might change priorities regarding policies at the expense of long-term strategies for climate and development. On the same breath, corruption and mismanagement reduce available resources to key climate initiatives, increasing vulnerability and slowing SDGs progress. Thus, political and institutional frameworks remain a relevant and indispensable scenario in establishing an enabling environment for climate action and sustainable development.

Opportunities

1. Synergies Between Climate Action and the SDGs

After all these challenges, significant opportunities to favorably exploit synergies between climate action and SDGs do exist. This is partly because many of the goals tend to be inherently linked; progress in one area can often undergird and support progress in other areas. For example, the promotion of renewable energies, SDG 7, can be said to contribute to economic growth, SDG 8, by creating jobs, especially in the energy sector, and helping improve health outcomes due to reduced air pollution and increased access to energy.

For example, investments in climate-resilient agriculture, vital for reaching SDG 2, will have spillover benefits for food security and reduction of poverty, hence contributing to SDG 1, while at the same time protecting ecosystems, SDG 15. By bringing these synergies to light and using them to their advantage, countries can draw up integrated strategies that achieve many goals at once; with the greatest effects, therefore, being most efficiently expedited forward toward a more sustainable future.

2. Innovative Mechanisms of Finance

To this end, it creates an enabling environment that can mobilize resources necessary for climate action and sustainable development, where innovations in financial mechanisms are increasingly recognized. Examples of such innovative financial instruments include blended finance, green bonds, and climate insurance, which can bridge funding gaps and attract investment from the private sector in climate-resilient projects.

For instance, green bonds have gained popularity to finance projects in renewable energy, energy efficiency, and other sustainable infrastructure--enabling investors to contribute to meeting environmental objectives while earning returns. Similarly, the blended finance models that combine public and private capital can de-risk investments in climate and development programs and make them more attractive to private investors. Tapping into these new innovative financing mechanisms can enable countries to unlock resources for effectively taking climate action in pursuit of the SDGs.

3. Technology and Innovation

On another level, more recent technological and innovative advances provide new opportunities to improve climate action and support sustainable development. Technological innovation-from renewable energy and energy storage technologies to climate-smart agriculture and digital solutions-is changing how countries and communities are working on climate change and with regard to achieving the SDGs.

A case in point is that the alarming reduction in prices of solar and wind energy has made renewable power viable and available. It therefore allows countries to move away from depending on fossil fuel and consequently reduce emissions. In the same vein, such climate-smart agriculture innovations as precision farming and drought-resistant crops are enabling farmers to adapt to the changing climate and improve food security. This will go a long way in harnessing innovation culture and investing in research and development for opening up new opportunities that come with climate action and sustainable development.

4. Strengthen Global Partnerships and Cooperation

Global collaboration and partnership in dealing with climate change and the SDGs are quite indispensable; at the same time, there is a great potential to be further developed over the course of the next couple of years. Both the Paris Agreement and the SDGs provide a common platform on which countries can come together to create joint responses to climate and development issues, share knowledge and best practices, and mobilize resources toward common action.

For example, international events like the Global Climate Action Summit, the United Nations Climate Change Conference (Conference of the Parties-COP), and the High-Level Political Forum on Sustainable Development provide a platform for countries to share their progress, discuss challenges, and identify new avenues of collaboration. Reinforcing this partnership and expanding international collaboration will allow countries to scale up their efforts toward accelerated progress and more sustainable and resilient outcomes.

5. Empowering Local Communities and Indigenous Knowledge

Empowerment of local communities and the use of indigenous knowledge provide a better opportunity to take up the issue of climate resilience and support sustainable development. Local communities, like indigenous peoples, often have a more complete understanding of natural conditions of their surroundings and have traditional practices that are adapted to conditions within their locality. Integration of this knowledge in policies on climate and development is bound to ensure effective solutions that culturally address the needs of building resilience and promote sustainability.

Such protection of the ecosystems, biodiversity conservation, and maintenance of livelihoods could be exemplified by community-based natural resource management and participatory land-use planning. In the same vein, valuing indigenous knowledge in climate adaptation strategies makes measures more effective and more responsive to local needs and conditions. This is where the role of local empowerment and integrating indigenous knowledge into climate action come in handy toward developing more inclusive, resilient societies that are better equipped to face the challenges brought about by climate change.

Climate change in the SDGs poses several challenges and opportunities from different perspectives. There are enormous barriers that need to be overcome with respect to coherence of policies, constraints of finance, gaps in technology, social and cultural aspects, and political and institutional dynamics. At the same time, several opportunities are available to make use of synergies, innovate, and strengthen partnerships toward better and more sustainable impacts.

Recognizing these challenges and opportunities brought about by integrated policy approaches, innovative financing mechanisms, technological advances, global cooperation, and community empowerment will increase the ability of countries to achieve considerable progress toward climate resilience and sustainable development. As in the future, building on these efforts will ensure movement toward a more inclusive, equitable, sustainable world for all.

THE FUTURE OF CLIMATE ACTION AND SUSTAINABLE DEVELOPMENT

Ahead of 2030, the challenge to address climate change within the perspective of the SDGs is continuously being outgrown. The next decade will mark an important juncture in efforts to reduce to a minimum the worst impacts of climate change and achieve sustainable development for all. Section 2 explores prevailing trends, pathways towards the future, and sets of transformational actions that will be required to build climate resilience, reduce GHG emissions, and to attain sustainability.

1. Accelerating Climate Ambition and Action

One of the urgent needs for the future is to accelerate climate ambition and action. While the commitments under the Paris Agreement form a basis in this regard, current national pledges under it are not enough to limit global warming to 1.5°C or even 2°C above pre-industrial levels. In order to avoid the worst impacts of climate change, countries need to upscale ambitions in current NDCs with more ambitious targets pertaining to the reduction of emissions.

Any climate action from now on needs to be holistic, embedding its considerations into every sphere of economic and social policy. That is, shifting from incremental adjustments towards truly transformative measures that can bring about systemic change in energy, transportation, agriculture, and industry, among other key sectors. This requires governments, businesses, and civil society to join in the real implementation of policies and measures favorable to decarbonization, the efficiency in the use of energy, and hastening the transition toward renewable sources of energy.

There is a need for increased ambition in stepping up the countries' climate policies to align with the SDGs. This calls for prioritization of the actions that contribute to multiple benefits, including reduced emissions, increased resilience, and social equity. In this respect, through more integrated and ambitious approaches to climate action, nations will have taken major strides toward the achievement of the SDGs and realization of a more sustainable future.

2. Utilizing Innovation and Technology

Innovation and technology will define the future of climate action and sustainable development in large part. Solar, wind, and battery storage-innovations that have transformed clean energy technologies-represent huge opportunities for further emission reduction. Much more investment in research and development is required to drive down costs and improve efficiency and accelerate deployment.

Besides clean energy, other emerging technologies like AI, blockchain, and the Internet of Things are opening new frontiers for climate action. For example, AI can optimize energy use, enhance climate modeling, and manage resources more effectively. Blockchain technology will enhance transparency and traceability along supply chains, hence contributing to reducing emissions and sustainability practices. IoT devices can take environmental conditions in real time and give much better and more effective natural resource monitoring and management.

As these technologies are developing further, it is also very relevant to ensure that these are accessible and affordable for all countries, especially in the Global South, by vested effort towards technology transfer, capacity building, and knowledge sharing. Innovation and technology accelerate country-level action on climate and help them achieve the SDGs effectively.

3. Enhancing Global Cooperation and Multilateralism

The global nature of challenges in climate change and sustainable development demands a strong international cooperation and multilateralism. While the SDGs and the Paris Agreement have formed the frame toward collective action, improved collaboration among countries, international organizations, businesses, and civil society would be key toward realization of their full potential.

In this respect, there is an urgent need for even closer international cooperation, on issues of common interest, with a view to rising to the challenges represented by climate change and sustainable development. This includes mobilization of finance at scale to countries in development, particularly the most vulnerable ones to climate change impacts. The international community must act in solidarity in matters related to issues of loss and damage, climate justice, and country preparedness with the right resources and support toward the full implementation of their climate pledges.

There is also the significant role that multilateral organizations can play, such as the United Nations and the World Bank and regional development banks, in helping countries make much more use of international cooperation and resource mobility for climate action. These organizations may support technical assistance, knowledge exchange, and facilitate conceiving and developing innovative financing instruments to help scale up the efforts of countries. This will, in turn, be reinforced through strengthened cooperation and multilateralism among states, thus enabling the international community to respond en bloc to climate change and hastening progress on the attainment of the SDGs.

4. Empowering Inclusive and Equitable Transitions

That transition to a low-carbon, climate-resilient future must be just and inclusive. Policies and initiatives on climate change and sustainable development should focus on giving special attention to the needs of the most vulnerable and generally marginalized societal sectors like women, indigenous peoples, and people with low incomes. A just transition would mean reduced emissions and enhanced resilience, with each person afforded an opportunity to participate in and benefit from transitioning toward a sustainable economy.

Social equity, human rights, and involvement of the community form the essence of just and inclusive transitions. Policy makers also have to pay due attention to distributional implications of policies for climate change, with measures toward mitigating potential adverse impacts on groups vulnerable to such changes. This also calls for support of workers and communities impacted by transition; investment in education and training, building skills that are needed by the green economy; and assuring that action on climate would promote social inclusion and gender equality.

This may also ensure that climate and development policies are more efficient and sustainable because the local community is empowered and indigenous knowledge is tapped. By so doing, countries would be able to build an inclusive and resilient society due to their value of the contribution of all kinds of people in dealing with the challenges of climate change and sustainable development.

5. Building Resilient and Sustainable Cities

The climate and development challenge is led by urban areas, given that cities account for more than 70 percent of global CO_2 emissions, hosting over half of the world's population. Since urbanization will be increasingly accelerated, especially in developing countries, how cities are planned, built, and managed is likely to become more and more decisive about the future of climate action and sustainable development.

Resilient and sustainable cities need holistic urban planning that considers climate risks, resource efficiency, and social equity. A great deal of investment is therefore needed in key sustainable infrastructure, such as public transport, green buildings, renewable energy systems, and green spaces, in addition to managing land use in a sustainable way. Cities should also support strategies to reduce vulnerability to climate impacts, like flooding, heatwaves, and sea-level rise, by enhancing early warning systems, resilient infrastructure investments, and community-based adaptation.

Governance and stakeholder involvement are also important in promoting resilience and sustainability within cities. It should be a collaboration between local governments and businesses with civil society in developing and implementing urban policies that make sure of sustainability and resilience. Through partnerships being enhanced and participatory processes in decision-making pursued, thereby leading to more inclusive and sustainable urban settings that help in achieving both climate action and the SDGs.

6. Integrating Nature-Based Solutions

Nature-based solutions represent one very promising pathway toward both climate change mitigation and achieving the SDGs. By harnessing ecosystems, nature-based solutions can provide a number of co-benefits through carbon storage, improved biodiversity, better water quality, and increased resilience to climate impacts. Examples of nature-based solutions include but are not limited to reforestation, wetland restoration, and sustainable agriculture that protects the mangroves along the coasts.

The integration of NBS into climate and development policies could enable many countries to concurrently achieve the following SDGs: SDG 13 ("Climate Action"), SDG 15 ("Life on Land"), and SDG 14 ("Life Below Water"). Besides, NBS can also provide cost-effective and sustainable solutions compared with traditional engineering ones for those countries that are constrained by limited financial inputs.

Realizing the full potential of nature-based solutions will require countries to step up the conservation and restoration of natural ecosystems, integrating NBS into their climate and development strategies. This calls for prioritizing biodiversity protection, attaining sustainable land and water management, and engaging the community more in overall conservation. In this way, integration of nature-based solutions will help countries build resilience against adverse climate, better ecosystem protection, and sustainable development.

We are at a very critical juncture in climate action and sustainable development. The coming decade is all that remains to make the world rise to the challenge of arresting climate change while achieving the SDGs. It will be possible for countries, with scaled-up climate ambition and action, harnessing innovation and technology, enhanced global cooperation, inclusive and just transitions, resilient and sustainable cities, and nature-based solutions, to create a future that is truly sustainable and resilient for all.

Great as the challenges are, the opportunities are even bigger. Strong political will combined with innovative solutions and collective action can indeed make the course of the world change for development and promote a more equitable and sustainable future. Moving ahead, commitment to the principles of the SDGs and the Paris Agreement will be important for shared work to create a world where people and the planet will prosper.

CONCLUSION

Confronted with the dual imperatives to fight climate change and achieve sustainable development, the obvious response is a future that sees an unprecedented rise in ambition, innovation, and collaboration. While the SDGs and the Paris Agreement provide a robust framework for such challenges, the full realization of such demands is needed beyond mere incremental changes; transformative action is necessary across all sectors of society.

The urgent necessity of faster climate action is evident. Current commitments and policies will not deliver the targets of 1.5°C or even 2°C above pre-industrial levels of global warming. In order to head off the worst effects of climate change, countries have to raise ambition in enhancing their NDCs and then implement bold measures that will drive systemic change. This is interpreted as: reducing dependence on fossil fuel, increasing energy efficiency, sustainable land use, manufacturing, and urban planning. It also entails incorporating climate action into larger development goals and can greatly advance the achievement of the SDGs while concurrently overcoming the risks from climate change.

Innovation and technology will be very important in determining the future of climate action. Clean energy technologies, especially solar and wind, are only getting better with each passing day. They will continue to create new, innovative methods to reduce greenhouse gas emissions. Newer technologies involving AI, blockchain, and IoT hold immense potential in making the climate resilient through efficient use of resources and keeping a close watch on the environmental setting. It is necessary to tap into those innovations to accelerate that progress and ensure climate solutions are truly effective and accessible. Equally important will be technology transfer and capacity building, particularly for developing countries, in support of closing gaps and ensuring every nation can make full use of technological advancement.

Global cooperation and multilateralism are indispensable in the fight against climate change for sustainable development across the globe. The Paris Agreement and the SDGs provide a framework for international cooperation, but much more is needed beyond that currently in place to catalyze action at scale, both in mobilizing resources and through effective implementation of climate strategies. More financing will be required to support developing countries-particularly the most vulnerable to the worst impacts of a changing climate-so that all nations can participate in and benefit from worldwide climate action. These multilateral institutions, among them the United Nations and regional development banks, play a very important role in providing ways of cooperation and technical assistance, and their efforts should be strengthened to assist in a coordinated response by the world community.

Sustainable development requires the core element of promoting inclusive and just transitions. A just transition toward a low-carbon economy needs to ensure fairness and inclusiveness for the benefit of all people and communities. This includes addressing social and economic inequalities, the political empowerment of subordinated groups, and ensuring that climate policies support social equity and human rights. A just transition will build social cohesion and make climate action more effective because it ensures different stakeholders are involved and committed to the process.

Resilient and sustainable cities will be at the heart of climate action and development over the next couple of decades. With over half of the population inhabiting urban centers, these centers have to grapple with diverse challenges with respect to climate change, increased vulnerability to extreme weather events, and limits on resources. The climate consideration of urban planning involves investment in green infrastructure for building resilient cities resistant to climate impacts, thus supporting sustainable growth. In fact, effective urban governance, citizen engagement, and innovative solutions lie at the very core of building cities that can contribute positively to climate action as well as the SDGs.

Lastly, nature-based solutions offer a major route to addressing climate change while making considerable contributions toward achieving sustainable development. Leveraging the power of nature in bringing about multiple benefits—such as carbon sequestration, protection of biodiversity, and resilience—will go a long way in helping countries achieve several SDGs. The integration of nature-based solutions into climate and development policies can yield an economic, feasible, and sustainable option against traditional approaches; this can provide more wholistic and resilient responses to climate change.

In all, the future of climate action and sustainable development will be challenging but promising. The way ahead requires a shared commitment to transformative action, innovative solution finding, and global cooperation. If these challenges can be overcome with ambition and imagination, along with seized opportunities brought about by rapid technological development, more inclusive forms of governance, and nature-based solutions, then a sustainable and resilient future will be achievable for everyone. It is now time for action, and by working together with a willful effort, we are able to create that future wherein people and the planet thrive in concert.

REFERENCES

Acharya, K. (2023). Navigation of climate change framework: analysis of synergy between sdgs and unfccc. DOI: 10.48001/978-81-966500-9-4_7

Akter, A., Abul Ulie, K. K., & Ali, S. S. (2023). Intercropping: Prospects and challenges in bangladesh for sustainable agriculture. *IJEAST*, 8(3), 111–120. DOI: 10.33564/IJEAST.2023.v08i03.015

Alola, A., Adebayo, T., & Olanipekun, I. (2023). Examining the energy efficiency and economic growth potential in the world energy trilemma countries. *Energies*, 16(4), 2036. DOI: 10.3390/en16042036

Dzebo, A. (2023). The paris agreement and the sustainable development goals: evolving connections. DOI: 10.51414/sei2023.036

Flagg, J. (2018). Carbon neutral by 2021: The past and present of costa rica's unusual political tradition. *Sustainability (Basel)*, 10(2), 296. DOI: 10.3390/su10020296

Hartmann, T., & Spit, T. (2015). Implementing the european flood risk management plan. *Journal of Environmental Planning and Management*, 59(2), 360–377. DOI: 10.1080/09640568.2015.1012581

John, D., & Derakhshi, E. (2022). Low carbon mobility transitions and justice: A case of costa rica. *Development*, 65(1), 71–77. DOI: 10.1057/s41301-022-00331-6 PMID: 35250210

Kayigema, V., & Rugege, D. (2014). Women's perceptions of the girinka (one cow per poor family) programme, poverty alleviation and climate resilience in rwanda. *Agenda (Durban, South Africa)*, 28(3), 53–64. DOI: 10.1080/10130950.2014.939839

Linde, A., Bubeck, P., Dekkers, J., Moel, H., & Aerts, J. (2011). Future flood risk estimates along the river rhine. *Natural Hazards and Earth System Sciences*, 11(2), 459–473. DOI: 10.5194/nhess-11-459-2011

Oluoch, S., Lal, P., Susaeta, A., Mugabo, R., Masozera, M., & Aridi, J. (2022). Public preferences for renewable energy options: A choice experiment in rwanda. *Frontiers in Climate*, 4, 874753. Advance online publication. DOI: 10.3389/fclim.2022.874753

Pattberg, P., & Widerberg, O. (2015). Transnational multistakeholder partnerships for sustainable development: Conditions for success. *Ambio*, 45(1), 42–51. DOI: 10.1007/s13280-015-0684-2 PMID: 26202088

Philippidis, G., Shutes, L., Robert, M., Ronzon, T., Tabeau, A., & Meijl, H. (2020). Snakes and ladders: World development pathways' synergies and trade-offs through the lens of the sustainable development goals. *Journal of Cleaner Production*, 267, 122147. DOI: 10.1016/j.jclepro.2020.122147 PMID: 32921933

Pradhan, P., Costa, L., Rybski, D., Lucht, W., & Kropp, J. (2017). A systematic study of sustainable development goal (sdg) interactions. *Earth's Future*, 5(11), 1169–1179. DOI: 10.1002/2017EF000632

Richmond-Navarro, G., Madriz-Vargas, R., Ureña-Sandí, N., & Barrientos-Johansson, F. (2019). Research opportunities for renewable energy electrification in remote areas of costa rica. *Perspectives on Global Development and Technology*, 18(5-6), 553–563. DOI: 10.1163/15691497-12341530

Ritzema, H., & Loon-Steensma, J. (2017). Coping with climate change in a densely populated delta: A paradigm shift in flood and water management in the netherlands. *Irrigation and Drainage*, 67(S1), 52–65. DOI: 10.1002/ird.2128

Riyadh, Z., Rahman, M., Saha, S., Ahamed, T., & Current, D. (2021). Adaptation of agroforestry as a climate smart agriculture technology in bangladesh. *International Journal of Agricultural Research, Innovation and Technology*, 11(1), 49–59. DOI: 10.3329/ijarit.v11i1.54466

Sarfo, I., Bortey, O., & Kumara, T. (2019). Effectiveness of adaptation strategies among coastal communities in ghana: the case of dansoman in the greater accra region. Current Journal of Applied Science and Technology, 1-12. DOI: 10.9734/cjast/2019/v35i630211

Scobie, M. (2019). Sustainable development and climate change adaptation: goal interlinkages and the case sids. DOI: 10.17875/gup2019-1213

Sherman, M., Berrang-Ford, L., Lwasa, S., Ford, J., Namanya, D., Llanos-Cuentas, A., Maillet, M., & Harper, S.IHACC Research Team. (2016). Drawing the line between adaptation and development: A systematic literature review of planned adaptation in developing countries. *Wiley Interdisciplinary Reviews: Climate Change*, 7(5), 707–726. DOI: 10.1002/wcc.416

Steinfatt, K. (2020). Trade policies for a circular economy. DOI: 10.30875/4445c521-en

Chapter 17
Innovation and Technology to Address the Challenges of Climate Change

Thi Minh Ngoc Luu
https://orcid.org/0000-0002-5972-7752
Vietnam National University, Hanoi, Vietnam

Mohit Yadav
https://orcid.org/0000-0002-9341-2527
O.P. Jindal Global University, India

Anugamini Srivastava
https://orcid.org/0000-0003-0617-2711
Symbiosis International University (Deemed), India

Krishan Gopal
https://orcid.org/0000-0002-0115-5659
Lovely Professional University, India

ABSTRACT

This chapter explores the pivotal role of innovation and technology in addressing the multifaceted challenges of climate change. It examines advancements in renewable energy, digital technologies, circular economy models, and climate resilience strategies, highlighting their potential to drive significant progress in climate action. The chapter also discusses the barriers to innovation adoption, including financial constraints, regulatory hurdles, and social resistance. By analyzing case studies of successful climate innovations and considering future trends, it provides a compre-

DOI: 10.4018/979-8-3693-5792-7.ch017

hensive overview of how technological and strategic innovations can mitigate and adapt to climate impacts. The chapter emphasizes the importance of collaborative efforts and inclusive approaches in fostering effective climate solutions and advancing towards a sustainable, low-carbon future.

INTRODUCTION

Climate change is the greatest challenge that humanity faces today. It has been called the most important natural threat to humankind, societies, and economies around the world. It ranges from rising temperatures to changing patterns of weather, increased frequency of extreme weather phenomena, and it would be true to assert that the impacts of climate change are being felt across the world. These represent not only environmental but also socio-economic effects that affect food security, water resources, health, and livelihoods in vulnerable communities. As the planet continues to heat up, the cry for adequate and comprehensive solutions heightens. In this regard, innovation and technology have emerged as critical tools in the fight against climate change. They now provide new avenues through which the negative impacts of those changes can be reduced and ways in which we can adapt to the changed environment, or they can be used as avenues of changing our world into a greener and resilient one (Adenle et al., 2015).

Various forms of innovations could result in giant leaps towards addressing all aspects of climate change. Technological innovation-renewable energy, energy efficiency, and carbon capture and storage, among others-is at the very core of greenhouse gas emissions reduction and at the heart of any low-carbon economy. By contrast, social and organizational innovations, including novel business models and collaborative frameworks, have a core role in facilitating systemic change and leveraging far-reaching diffusions of sustainable practices. These will not only contribute to mitigating climate impacts but also to adaptation strategies in the process of making societies resilient to unstoppable changes already occurring (Bergman et al., 2010).

The role of technology in climate action is definitely more than just mitigation; it encompasses adaptation-things that would actually enable a community or an ecosystem to adapt to the new climate and its impacts. In addition, technologies such as climate-resilient infrastructures, intelligent water management systems, and precision agriculture provide critical support for vulnerable regions to manage increasingly visible and challenged climate variability and extreme weather conditions (Dechezleprêtre et al., 2019). The present-day use of digital technologies-those using AI, machine learning, and IoT-offers new ways of monitoring changes in the environment, optimizing resources, and improving decision-making processes. These technologies provide real-time data and insights that allow proactive and informed responses to climate-related risks (Fankhaeser et al., 2008).

However, innovation and integration of technology into climate action do not come without a host of challenges. There are major financial, regulatory, and social barriers to the wide diffusion of new technologies. In most cases, innovative solutions face very high up-front costs and low funding, with related economic incentives being insufficient to favor deployment, especially in those countries where resources are particularly scarce. Sometimes the regulatory framework may set back innovation, whether through over-restriction or through failing to provide adequate support for emerging technologies. Social and cultural factors, including public perception and behavior inertia, further increase the resistance to the rate at which new technologies and practices are taken up and integrated (Gans, 2012).

Of course, despite these challenges, the potential for innovation and technology to be a driver of meaningful climate action is immense. This requires fostering an enabling environment that will support research and development, facilitate collaboration across sectors, and support scaling up successful innovations. It is about creating the policies and frameworks by governments, businesses, and civil society to support innovation and attract investment in climate-friendly technologies (Grubb, 2004). Moreover, the dissemination of awareness and education about the benefits of those technologies should be increased to have the support of the public through involvement in transition towards a sustainable future. This chapter examines how innovation and technology can be used in different ways to help deal with the challenges of climate change. We are going to discuss, in this regard, the role that different types of innovation can play in mitigating and adapting to climate impacts, ranging from technological advancement to social and organizational changes. We will discuss barriers to innovation and the adoption of technology, too, and present successful case studies of innovation-driven climate action. The potential of these innovations, along with their limitations, needs to be brought into perspective in order to appreciate the multi-faceted approach that is required in tackling climate change and what part innovation and technology can play in shaping a sustainable future (Holdren., 2006)

UNDERSTANDING CLIMATE CHANGE CHALLENGES

Climate change has to do with numerous challenges which touch on every aspect of life on Earth. Climate change, then, is primarily a result of increased levels of greenhouse gases in the atmosphere, such as carbon dioxide, methane, and nitrous oxide. These trap the sun's heat and contribute to what would otherwise be a slow rise in global temperatures, now commonly known as global warming. However, climate change is more than rising temperatures; it is a wide array of environmental disruptions, social and economic, that pose a great degree of risk to human and ecological systems (Jochem & Madlener, 2003).

Environmental Challenges

One of the most immediate and visible effects of climate change is that of the shift in weather patterns. As the temperatures around the Earth are rising, we are experiencing an increase in the instances of hurricanes, floods, heatwaves, and droughts more frequently and with much more intensity. Such extreme conditions may turn out to be disastrous for ecosystems and biodiversity (Lybbert & Sumner, 2010). For instance, temperature rise and shift in the precipitation regime might lead to habitat loss and degradation, placing a species in danger of extinction. Though coral reefs are among those ecosystems which are really sensitive to water temperature change, widespread bleaching of coral reefs has developed into a serious threat both to marine biodiversity and to people whose livelihoods depend upon these ecosystems.

Also, melting of glaciers and polar ice caps into the sea due to climatic change results in increased sea levels. The possible rise in the sea level threatens coasts around the world, as the rising seas can flood low lying lands, erode coastlines, and dislocate millions. Smaller island nations and coastal cities are particularly vulnerable because these face extinction from the increase in sea levels and storm surges. Loss of ice further affects the global climate system, since reflective surfaces of ice and snow get replaced by the darker ocean waters that absorb more heat, further accelerating warming (Matos et al., 2022).

Social Challenges

Equally important are the social aspects of climate change. Weather and extreme event changes disturb agriculture and food production, thus translating into food insecurity for those parts of the world that are already susceptible to hunger and malnutrition. Drought and erratic rainfall could destroy crops, diminish yields, and create water shortages, which affect not only rural communities dependent on agriculture but also urban populations dependent on stable supplies of food (Nordhaus, 2010).

Other health effects are due to increases in heat-related diseases, the increased dissemination of various disease-carrying insects, and poorer quality air brought about by climate change. Rising temperatures and changed precipitation can expand the ranges of vector-borne diseases such as malaria and dengue fever, while heatwaves can cause heatstroke and dehydration, especially among the elderly and people with chronic diseases. Poor air quality, driven by rising temperatures and wildfires, creates respiratory problems and other adverse health effects that most heavily burdens more vulnerable populations-including children and the elderly (Smith, 2009).

Economic Concerns

Obviously, climate change has many economic consequences and can have much larger financial losses than benefits. Some basic recovery and rebuilding costs from infrastructure, property, and business damage sustained during such extreme weather events are enormous. In addition, the agricultural sector is a big part of most economies and is very vulnerable to climate variability; droughts and floods trigger crop failure and loss of income among farmers. On the other hand, disasters, with increased frequency and severity, further deplete the already limited public resources by reallocation away from development projects toward emergency response and reconstruction initiatives (Steward, 2012).

Climate change also threatens other industries requiring good and stable weather conditions and natural attractions, such as tourism. To illustrate, higher temperatures means that ocean acidification has gradually destroyed one of the most attractive tourist coral reefs in the tropical regions. Similarly, industries depending on winter sports have been facing declining snow cover, affecting mountain resorts and local economies dependent on those industries (Su & Moaniba, 2017).

Systemic Challenges

Beyond these proximal impacts, climate change poses systemic problems that make it particularly difficult to remediate its effects. Since global systems interlink with one another, disturbance in the part of one system sends ripples throughout others. For instance, in such a way that, if climate change affects reduced agricultural productivity, it has been known to create spikes in food prices, which then go on to affect global markets and, in turn, political stability and conflict. This is becoming an increasingly scarce resource as weather patterns change precipitation and demand rises, adding to potential regional destabilization through ailing social and political tensions (Zilberman et al., 2018).

These include the poor, indigenous populations, and people from developing countries. Most of them lack proper resources and infrastructures to adapt to the changing condition; hence, they become highly vulnerable to the impacts of climate change and are not able to recover after disasters.

Need for Innovative Solutions

Clearly, this is a many-faceted problem, and impacts of this nature cannot be handled effectively with traditional approaches alone. The scaling and urgencies of the crisis require innovative solutions beyond a realm of mere incremental changes. Technological, policy, and social practices innovations will be required for the mitigation of greenhouse gas emissions, increasing resilience against climate impacts, and supporting adaptation efforts in various ways. These include all-inclusive solutions that take into consideration the needs and capacities of various communities and scalable, meaning applied on a wide basis to ensure maximum impact (Bergman et al., 2010).

The following sections explore how innovation and technology can be leveraged to help meet the challenges presented, while showcasing how new and emerging solutions will drive new approaches to climate action in building a more sustainable and resilient future.

THE ROLE OF INNOVATION IN COMBATING CLIMATE CHANGE

Innovation has, therefore, been at the heart of dealing with climate change, serving as a catalyst in developing new and improving existing solutions to find new ways of solving environmental and societal challenges (Gans, 2012). As the toll of climate change rises with greater frequency and boldness, the necessity for innovative strategies and technologies that will help mute such impacts has never been greater. In this context, innovation does not refer solely to technologies but also to a wide range of changes. Social innovations, organizational innovations, and policy innovations are no less imperative in constituting the full response to climate change (Grubb, 2004).

Technological Innovation

Technological innovation is probably the most visible, widely discussed part of ways to fight climate change. This has entailed the formulation and implementation of new technologies that abate greenhouse gas emissions, improve energy efficiency, and facilitate adaptation to climate-related changes. The development and utilization of a wide variety of renewable energy technologies, such as those generated from solar, wind, hydroelectric, and geothermal sources, have immensely reshaped the outlook of the energy sector, permeating it with much greener and more sustainable resources than fossil fuel equivalents. These are technologies that have significantly improved over the years to yield lower costs, higher efficiency, and scalability. Because of this, renewable sources are going on par with traditional sources, therefore helping the world turn to a low-carbon economy (Holdren, 2006).

Not only that, but technological innovation in energy storage, CCUS, and green transportation also holds a key role for mitigation through the reduction of emissions. The energy storage technologies, such as advanced batteries and systems for grid-scale storage, are key enablers in integrating intermittent renewable energy sources into the power grid-a precondition for the steady and reliable supply of clean energy. CCUS technologies can capture carbon dioxide from industrial processes, store it underground, or transform the gas into useful products. They represent a promising option for mitigating emissions in hard-to-abate industries, such as the cement and steel sectors (Joch.em & Madlener, 2003)

Furthermore, other sustainable transportation innovations, such as electric vehicles, hydrogen fuel cells, and improved public transit systems, also play a centric role in mitigation strategies for the transport sector-one of the biggest emitters of greenhouse gases around the world (Lybbert & Sumner, 2010). Electric vehicles, backed by ever-improving technology, enabling policies, and infrastructure development, have emerged as a fast-developing market that could potentially switch out traditional fossil fuel-based transportation with a greener, less carbon-dependent alternative.

Social and Organizational Innovation

While technological innovations are relevant, these need to be balanced with social and organizational innovations so that they can be effectively implemented and find wide diffusion. Social innovation would create new practices, behaviors, and cultural norms which reinforce sustainable lifestyles and decrease ecological footprint. It may take the form, for instance, of individual and community-led initiatives on energy conservation, waste reduction, and sustainable consumption patterns (Matos, 2022). These might be awareness creation for reducing meat to cut down on emission or the need for recycling to reduce the consumption of resources. Organizational innovation involves devising new business models, partnerships, and institutional arrangements that allow climate action. This can include coming up with models of circular economy that guarantee resource efficiency and waste reduction, the establishment of public-private partnerships in financing and scaling up renewable energy projects. Other forms of organizational innovations may entail devising novel governance structures along with decision-making procedures that integrate climate concerns into policy and planning in all levels of government and within organizations (Nordhaus, 2010).

Policy Innovation

The policy innovation again has a major role to play in all the processes to do with responding to climate change, as it will create the needed regulatory framework and incentives required toward technological and social change (Gans, 2012). Therefore, this will be in a position to accelerate transition towards a low-carbon economy by setting ambitious targets for the reduction of emissions, introduce mechanisms for carbon pricing, and give subsidies or tax incentives for clean energy technologies. Carbon pricing through carbon taxes and capandtrade systems, for instance, incentivizes economic behavior by businesses and people to reduce their carbon footprint-a key driver for investments in low-carbon technologies and practices (Smith, 2009).

Besides regulatory measures, policy innovation nowadays can also aim at designing new financial instruments and funding mechanisms supporting climate action. Green bonds, climate finance initiatives, and public funding programs each have a crucial role in mobilizing resources for research, development, and deployment of solutions. Policies that promote climate resilience and adaptation-such as zoning laws that ban construction in flood-prone areas or incentives for enhancing climate-resilient infrastructure-are rather important in enabling communities and ecosystems to adapt to the shifting climate (Steward, 2012).

Cross-Sectoral and Interdisciplinary Innovation

Combating climate change takes a holistic approach, integrating innovations across sectors and disciplines. The term "cross-sectoral" describes the coming together of various industries, such as energy, agriculture, transportation, and manufacturing, to construct wide-ranging solutions that tackle interconnected climate challenges. A classic example of this is how the integration of renewable energy technologies into agricultural industries results in reduced farm emissions and increases clean energy access to rural communities (De Stefano et al., 2016).

Interdisciplinary innovation brings expertise from various fields, such as engineering, environmental sciences, economics, social sciences, and many more, in order to develop a multifaceted solution that addresses the technical, economic, and social dimensions of climate change. This has great implications for understanding and hence managing the complex interactions between human and natural systems with regard to consequences on biodiversity, water bodies, and public health (Abdel Monem & El Ghandour, 2020).

The Importance of an Innovative Mindset and Culture

Fully applying the powers of innovation as a means of getting a cure for climate change. It is all about nurturing this innovative mindset and culture, which truly allows creativity, experimentation, and calculated risk to flourish. By so doing, an enabling environment should be established where new ideas could be tested, refined, and scaled up while failure would be positively viewed as a learning opportunity rather than a defeat. These range from investment in research and development, support for entrepreneurship, the promotion of networks and platforms that share knowledge and collaboration among Governments, Businesses, and Civil Society Organizations towards establishing a culture of innovation (Ockwell et al., 2010).

More importantly, it would be relevant to ensure that innovation is inclusive and equitable, considering the different needs and capacities of various communities, in particular those most vulnerable due to the consequence of climate change. This calls for the involvement of various stakeholders in the innovation process from the grassroots level of local communities and indigenous groups down to the women, youth, and marginalized to ensure that solutions are tailored to specific contexts and particular challenges that face them (Vushe, 2021).

Conclusion: Innovation is robust and a formidable tool in the fight against climate change-a new angle for mitigation, enhanced resilience, and adaptation to a changing world. By embracing technological, social, organizational, and policy innovations and fostering a culture of creativity and collaboration, we are better positioned to create and deploy the solutions to build a sustainable and resilient future for all (Chhetri et al., 2012).

TECHNOLOGICAL INNOVATIONS FOR CLIMATE CHANGE MITIGATION

Technological innovation can play a very important role in many aspects of dealing with climate change by providing new means and technologies to diminish GHG emissions, increase energy efficiency, and transition towards a low-carbon economy. Among such innovations are those related to renewable energy, energy storage, carbon capture technologies, and sustainable agriculture. Each one of these areas has a special role in mitigating climate change-whether it be addressing sources of emission, enhancing our energy management capability, or better use of resources (Rodima-Taylor et al., 2012).

1. Renewable Energy Technologies

Renewable energy technologies represent the leading edge of efforts to address global climate change by providing much cleaner and sustainable sources of energy compared to fossil fuels, currently the largest contributors of greenhouse gases. In the past ten years, different forms of renewable energy-solar, wind, hydroelectricity, and geothermal-have seen massive technological advancements, cost reductions, and enabling policy frameworks (Bergman et al., 2010).

Solar Energy: Photovoltaic or Solar PV is that form of technology which converts sunlight into electrical energy directly, using semiconductors. Advances in solar technology continue to drive efficiency up and production costs down, hence placing solar power among the fastest-growing renewable sources of energy in the world. Innovations such as bifacial solar panels, which are able to capture sunlight on both sides of their photovoltaic panels, and floating solar farms, utilizing the surface of bodies of water, are only a couple of examples of these two innovative ideas that have increased application and efficiency. Besides that, CSP systems, which use mirrors or lenses to concentrate sunlight on a receiver and generate thermal energy, present another method of harnessing the sun's energy, particularly for regions receiving copious sunlight (Gans, 2012).

Wind Energy: Wind is the kinetic energy created by air in motion. Wind turbines convert this energy into electricity. Advancement in technology has seen the production of larger and more efficient turbines capable of capturing more energy from the wind. Offshore wind farms build in bodies of water and obtain stronger, more consistent winds than sites on land, therefore having great potential for large amounts of clean energy generation. Floating wind turbines are a newer creation that allows the harnessing of wind energy in deeper waters, opening up more geographic opportunity for wind energy (Grubb, 2004).

Hydroelectric and Geothermal Energy: Hydroelectric power is one of the oldest methods of generating electricity from water flow and has long proved to be a reliable source of renewable energy. Innovations in small-scale and low-impact hydroelectric technologies are, therefore, permitting the harnessing of river and stream resources without the environmental impacts caused by large dams (Holdren, 2006). Geothermal energy, tapping from the Earth's internal heat, presents another valued renewable resource. Drilling technology and the technological development of enhanced geothermal systems have increased the exploitation of this resource in regions without potential natural geothermal activity.

2. Energy Storage and Smart Grids

Energy storage technologies have, with time, become very significant in providing stability and reliability for energy supply amidst the increasing relevance of renewable energy sources such as the sun and wind. Energy storing systems temporarily store excess energy that is generated at times of low demand and release this energy when demand is high or when weather conditions are not favorable for renewable energy generation.

Battery Storage: The technology of the battery, especially lithium-ion, has been fast-tracked in recent years because of demand from electric vehicles and grid storage solutions. Its development has increased energy densities, showed longer life spans, and decreased costs while large-scale battery storage technologies have become a valid solution for balancing the supply-demand of renewable sources. Power-to-X is a promising emerging technology that will provide massive-scale energy storage, alongside other emerging technologies like solid-state batteries and flow batteries offering further advances in capacity, safety, and cost-effectiveness.

Smart Grids: Smart grid technology enhances a number of aspects in electricity distribution, including efficiency and reliability, by making available advanced systems of communication, control, and automation. These grids can balance dynamic supply and demand, integrate distributed energy resources like rooftop solar panels, and utilize demand response programs, which give consumers incentives to reduce or shift energy usage during peak periods. Smart grids optimize the flow of electricity and integrate renewable sources; huge opportunities for reduction and resilience-building in energy.

3. Carbon Capture, Utilization, and Storage (CCUS)

CCUS stands for Carbon Capture, Utilization, and Storage, a suite of technologies that capture the CO_2 emitted through industrial processes and the generation of power. This prevents the CO_2 from being released into the atmosphere. The CO_2 is either stored underground in geological formations or further used to manufacture useful products, like building materials, chemicals, or fuel (Jochem & Madlener, 2003).

Carbon Capture: Technology development in carbon capture has progressed to improve efficiency and cost reduction in capturing CO_2 from flue gases and other industrial processes. Among the current technologies, post-combustion capture-involving the capture of CO_2 from exhaust gases from power plants and other industrial facilities-is the most common form. Innovations in amines and other chemical solvents have improved the efficiency of capture, reducing energy requirements of these processes. In contrast, there exist precombustion and oxy-fuel combustion-a method through which CO_2 is captured before being released to the atmosphere.

Carbon utilization is a developing area of innovation that uses captured CO2 as feedstock for producing useful products. Additionally, carbon utilization has the ability to turn CO2 into a very wide array of products, from concrete to plastics to synthetic fuels, offering an economic, market-driven incentive for capturing carbon emissions. For example, CO2 can be combined with hydrogen from renewable energy sources to produce synthetic hydrocarbons with applications as low-carbon fuels for transportation and industry (Lybbert & Sumner, 2010).

Carbon Storage: Long-term storage of captured CO2 in various geological formations is one of the established reduction technologies. Improvements in the associated monitoring and verification technologies-seismic imaging and satellite monitoring-implement greater safety and reliability for carbon storage, ensuring stored CO2 is safely prevented from leaking back into the atmosphere (Nordhaus, 2010).

4. Sustainable Agriculture and Food Production

Agriculture is one of the most significant contributors to GHG emissions and is among those sectors that are very vulnerable to the impacts of climate change. Technological innovation for sustainable agriculture contributes to mitigation through reduced emissions, increased efficiency of resources used, and enhanced resilience of food systems (Smith, 2009).

Precision agriculture is a sector in agriculture that employs a wide variety of advanced technologies like GPS, sensors, and drones to help with crop monitoring and management. It reduces waste, lowers costs, and minimizes environmental harm by optimizing inputs like water, fertilizers, and pesticides. Innovations in remote sensing and data analytics enable farmers to make data-driven decisions, improving yields and lowering emissions (Su & Moaniba, 2017).

Climate-resilient crops: Breeding and biotechnology will eventually develop crop varieties that will have more resilience due to climate change impacts on drought, heat, and pest (Steward, 2012). Resilient climate crops will help to maintain food production amid shifting weather patterns that reduce the needs of water, fertilizers, and pesticides in agriculture; hence they cut carbon emission.

Regenerative Agriculture: Regenerative agriculture focuses on a holistic approach to farming, putting emphasis on the restoration of soil health, enhancement of biodiversity, and sequestration of carbon in the soil. A number of sustainable agriculture practices, like cover cropping, no-till farming, and agroforestry, would contribute to soil fertility, water retention, and carbon capture, thus making agriculture both more viable and resilient against climate change.

5. Circular Economy Technologies

The concept of the circular economy is exactly opposite to that of the traditional linear economy, where one "takes, makes, and disposes." It thus inherently encompasses resource value and waste reduction. In this regard, the application of technology innovations within the circular economy has been supportive of extending the life of a product and reducing resources consumed, thereby allowing further material recycling and reducing emissions with less environmental impact (Pagoropoulos et al., 2017).

Waste Management and Recycling: Improving technologies in the waste management sector, such as automated sorting systems and chemical recycling, ensure that materials will be far more effectively recyclable. This cuts the demand for virgin resources and thus reduces emissions. New ideas also crop up on bioplastics and biodegradable materials to be used as alternatives to conventional plastics. This will reduce plastic waste and its environmental impact.

Product Life Extension: Product life extension technologies restore products through repair, refurbishment, and remanufacturing to extend the product life. Besides, digital technologies like IoT and blockchain facilitate transparency and traceability in supply chains, potentially managing resources more effectively and reducing the generation of waste.

The technological innovations for mitigating climate change would, in this way, be fundamental to bring about a decrease in emissions, an increase in resource efficiency, and a movement toward low-carbon sustainable development. Renewable energy, energy storage, carbon capture, sustainable agriculture, and technologies promoting the circular economy-all these forms of potential-can help meet the challenges of climate change and ensure a resilient and sustainable world (Chauhan et al., 2022)

TECHNOLOGICAL INNOVATIONS IN CLIMATE CHANGE ADAPTATION

Adaptation to climate change involves changes in systems, practices, and structures in ways that reduce adverse climate-related impacts and, where possible, takes advantage of the opportunities these changes provide. As climate continues to alter, communities throughout the world are exposed to an elevated risk of extreme weather conditions, erosion of land and sea levels, and shifting of agricultural zones (Chauhan et al., 2022). In this regard, technological innovations play an important role in making societies more resilient to such changes by means of better forecasting and improved early warning systems, while developing new materials and infrastructures suited to the changed conditions.

1. Climate-Resilient Infrastructure

Infrastructure protection through development and adaptation is pivotal for safeguarding the communities from the dire impacts of climate change. The technological innovations related to this sector encompass creation and construction of building, roadways, bridges, and other essential infrastructures that can bear natural calamities and fluctuating climatic conditions (Pagoropoulos et al., 2017).

Flood-Resilient Infrastructure: Rising Sea Levels and flood events have been exacerbated by Climate Change, making areas in Coastals and Floodplains increasingly prone to Floods. Development of flood-resilient infrastructure involves technological innovation, such as higher-order levees, sea walls, and storm surge barriers to guard against high tides and storm surges (Sen et al., 2021). It also involves integrating permeable pavements and green infrastructure into city planning, including wetlands and urban green spaces, for better handling of stormwater flow and mitigation of flooding risks. These green solutions will help absorb the excess rainwater and reduce runoff, hence lessening the potential for flooding in urban areas (Song et al., 2024).

Earthquake-Resistant Buildings: Building designs and innovations in new materials have allowed, in seismic-prone areas, the strengthening of structures to resist earthquakes. The technologies involve base isolators, which decouple the building from the ground movement, energy-dissipating devices that absorb seismic energy, among others. Advanced material applications also involve the construction of buildings with high-performance concrete and steel with capabilities to better endure seismic forces (Zhou et al., 2021).

Climate Resilient Planning for Cities: As the urban heat island effect implies, it is buildings and surfaces that absorb and retain this heat. Innovative technologies in urban planning will involve changing materials on rooftops to reflective and green roofs, increasing vegetation cover, and the structure of the buildings to allow for more ventilation and shading. Cool roofs, which are actually made from materials that reflect more sunlight and absorb less heat, and green roofs, which are covered with vegetation, can significantly reduce temperatures in urban environments (Carmin et al., 2009). Moreover, technologies inherent to a smart city-IoT-enabled sensors and climate-responsive building designs-are enabling better monitoring and management of urban heating.

2. Improved Monitoring and Early Warning Systems

The early warning systems are crucial in the reduction of impacts due to meteorological hazards and, in turn, other disasters through timely information, which enables communities to be well prepared to respond to these phenomena. Technological advances in monitoring and forecasting have guaranteed the accuracy and reliability of these systems.

Remote Sensing and Satellite Technology: Satellite technology and remote sensing supply critical data on the monitoring of the weather pattern, sea level rise, melting of glaciers among other climatic changes. Scientists monitor and predict the movement of storms, measure drought conditions, and assess ecosystem health with high-resolution satellite imagery and other data from Earth observation satellites (Guo, et al., 2015). Also, innovations in satellite technology have allowed for CubeSats-smaller satellites-and constellations of satellites that can increase the frequency and resolution of data, improving accuracy in climate models and forecasts.

Weather Forecasting: Improved computing power and machine learning algorithms have rewritten the art of weather forecasting to one that is increasingly accurate in its prediction of extreme weather events. Machine learning models can analyze vast amounts of meteorological data to find patterns in them and predict future weather conditions. This improves the accuracy of hurricane, typhoon, and other violent weather event forecasts-enabling preparedness and response.

Early Warning Systems for Natural Disasters: Early warning systems for natural disasters are based on a network of sensors, analytics, and communication that identify an imminent threat and relay that information to the community. Advanced sensor technologies-including those such as underwater pressure sensors that may detect tsunamis, or ground-based radar to monitor rainfall-enable real-time natural hazard detection and response. Integrated communication through mobile alerts, social media platforms, and others ensures that as many people as possible receive early warnings for timely evacuations or protective actions (Basher, 2006).

3. Climate-Smart Agriculture

Agriculture is very susceptible to the trend of change in climate, since both temperature changes and changes in precipitation, along with increased extreme events, may have a great influence on crop yields and productivity of livestock. These technological innovations in climate-smart agriculture seek not only to improve the resilience of agricultural systems to the changing climate backdrop but also their environmental footprint (Scherr et al., 2012).

Drought-Tolerant Varieties: Through biotechnology and improved plant breeding techniques, the emergence of drought-resistant crops that would survive under scarce water conditions takes effect. For this reason, such crops are projected to utilize water more efficiently and bear with a period of low rain and minimize or lessen crop failure due to dry spells. Genetic modification has been applied to develop heat and drought tolerance in varieties like maize and rice to ensure that farmers do not experience significant reduction in yields in the case of unfavorable weather conditions (Lipper et al., 2014).

Precision Agriculture: With precision agriculture, this modern farming practice puts together state-of-the-art technology that incorporates GPS, sensors, and data analytics for optimizations in farming practices and more judicious use of resources. Precision agriculture therefore enables a more appropriate application of water, fertilizers, and pesticides by providing real-time data on soil moisture, crop health conditions, and weather conditions, paving the way to less waste and more resilience against climate variability. Drones and remote sensing technologies also contribute to crop condition monitoring, which allows the timely detection of early signs of stress, thus intervention reduces the impact of climate-related threats (Scherr et al., 2012).

Agroforestry and Soil Management: Agroforestry is taken as the integration of trees and shrubs within the existing agricultural landscapes. Due to this fact, there are several merits that can be derived from agroforestry for climate adaptation. Trees will provide shade and wind breaks that reduce soil erosion while shielding crops from extreme weather (Lipper et al., 2014). They also contribute to maintaining the health of the soils, enhancing nutrient cycling, as well as water retention to improve agricultural system resilience to drought and flooding. Innovations in the area of soil management include conservation tillage and the use of biochar to further enhance soil structure and water retention for the improved resiliency of agricultural systems in relation to climate change.

4. Water Management Technologies

Among the impacts of climate change, particularly increased rainfall variability, longer-term droughts, and rising sea levels, water management is always one of the most important aspects of adapting to climate change. Water management will therefore have to account for technological advances that make better use of water, minimize waste, and make water systems more adaptive to the effects of climate change (Aivazidou et al., 2021).

Desalination and Water Recycling: As freshwater supplies become scarcer, desalination and water recycling emerge as technologies that can help supply clean water to arid and coastal regions. Desalination is the process of removing salt and other impurities from seawater or brackish water to make potable water. Advances in membrane technology and energy recovery systems have resulted in such a significant improvement in desalination efficiency and cost-effectiveness that it is now a feasible alternative for supplying regions with scarce water supplies (Owen, 2018). Water recycling, also known as the treatment and reutilization of wastewater, is another significant strategy in water resource conservation. Innovations in wastewater treatment, such as advanced filtration and biological treatment processes, have enabled the safe reuse of water for agriculture, industries, and even drinking purposes. Smart Irrigation Systems: Smart irrigation systems deploy sensors, weather forecasts, and data analytics to optimize water use in agriculture and landscaping. These systems can adjust irrigation scheduling and amounts to the exact plant needs by monitoring soil moisture and weather. Innovations in drip irrigation and sprinkler technology further enhance water application efficiency through the direct delivery of water to the root zone of plants, minimizing evaporative and runoff losses (Owen, 2018).

Flood Risk Management: Besides the management of water shortage, technological innovations have their role to play in mitigating flood risks. The technologies in the management of flood risks include advanced drainage systems, stormwater retention basins, and flood forecasting models that enable a community to prepare or respond to flood events. Innovations within flood modeling and mapping inspire remote sensing and GIS in order to give detailed information about flood risk areas for better planning and response strategies (Aivazidou et al., 2021).

5. Health and Climate Adaptation Technologies

Climate change is enormously affecting the sphere of public health in as much as it is capable of worsening the situation of already existing health problems and creating new health risks. Technological innovations in health and climate adaptation enable one to monitor and mitigate these risks, enhance delivery of healthcare, and improve outcomes of public health (Nhamo & Muchuru, 2019).

Vector Control Technologies: Climatic change altered the distribution of vector-borne diseases by changing the patterns of temperature and precipitation. Innovation in vector controls includes genetic modification of mosquitoes to prevent the transportation of diseases or to employ a biopesticide to reduce the spread of these diseases. Advanced monitoring systems, combining satellite data with geographic information systems, track the distribution and movement of vectors and predict the timing and location of outbreaks so interventions can be better targeted (Alcayna & O'Donnell, 2022).

Mitigation of Heat Stress: Technologies that would contribute to mitigating heat stress are becoming increasingly important in light of the increasing frequency and intensity of heatwaves for protecting the health of the populace. It is further a location where cooling technologies, such as energy-efficient air conditioners and evaporative coolers, provide comfort against extreme heat, especially among elderly and differently abled people with different health disabilities. Building design and innovative city planning reduce indoor and outdoor temperatures and, therefore, vulnerability to heat-related illnesses by using reflective surfaces, green roofs, and shading devices (Nhamo & Muchuru, 2019).

Telemedicine and Health Monitoring: Application of telemedicine or digital health technologies can, in turn, positively contribute to the delivery of the health care services, especially in the isolated and vulnerable areas that might have an increased vulnerability due to climate change. Telemedicine platforms involve health providers consulting patients remotely by reducing travel barriers and by ensuring accessibility during extreme weather conditions or natural disasters. Wearable health monitors and mobile health applications can provide continuous monitoring of health conditions, supplying early warning of potential health risks resulting from climate change such as heat stress, or respiratory problems related to poor air quality (Alcayna & O'Donnell, 2022).

6. Ecosystem-Based Adaptation

EbA involves using natural systems to reduce impacts and enhance resilience due to climate change. The technological innovations involved in EbA focus on the restoration and protection of wetlands, mangroves, and coral reefs-those systems providing nature's barriers to the effects of climate change.

Wetland Restoration: Wetlands are nature's sponges, which absorb excess water and prevent the risk of flooding. They provide vital habitats for wildlife and help in improving water quality through filtration that removes pollutants. Technological innovations in wetland restoration, such as drones for mapping and monitoring, bioengineering techniques to stabilize soil, and bioengineering techniques that help in stabilizing soil, have improved the efficacy of wetlands in providing protection against flooding and other such services (Zedler, 2000).

Mangrove and Coral Reef Restoration: Mangroves and coral reefs provide very important protection from coastline erosion and storm surges while supporting biodiversity and fisheries. Innovations in the restoration of mangroves, such as using biodegradable materials to plant seedlings, and the development of types of aquaculture non-damaging to mangroves, help restore these important ecosystems. These are identified to include coral gardening and the use of artificial structures to support coral growth as some of the powerful means of restoring coral reefs to improve their resilience to ocean acidification and warming due to climate change (Beck et al., 2022).

Green Infrastructure: Green infrastructure is that which employed natural systems, like parks or green roofs, by offering considerable environmental and social benefits. Technological innovations in the design and management of green infrastructure, such as the use of IoT sensors to monitor plant health and water consumption, enhanced the functioning of systems pertaining to cooling and flood protection, amongst others, in urban areas.

In sum, technological innovations for adaptation to climate change will no doubt be important in the process of building resilience against its vagaries through various means, such as the development of climate-resilient infrastructure, improvement of early warning systems, climate-smart agriculture, improved water management, protection of health, ecosystem-based adaptation, among others.

THE ROLE OF DIGITAL TECHNOLOGIES IN CLIMATE ACTION

Digital technologies are among the leading edges that prescribe novel solutions for most climate actions in increasing efficiency, transparency, and scalability. These range from data analytics and AI to blockchain and IoT, all of which greatly contribute to the observation of environment change, optimization of resource use, and driving sustainable practices across various fields. It is against a backdrop of the low-carbon economy, which the world has been pursuing, hand in hand with responding to climate change impacts. Digital technologies provide a strong means of acceleration and amplification through enabling more precise and data-driven decision-making and fostering greater collaboration across different stakeholders (Lawrence et al., 2017).

1. Data Analytics and Machine Learning

Data analytics and machine learning are going to revolutionize how we comprehend and act upon climate change by showing deep insights into complex environmental systems and human activities. This ranges from processing a variety of data at multiple scales-satellites, sensors, social media-to recognize patterns, forecast trends, and achieve the best responses possible.

Climate Modeling and Predictions: Advanced analytics and complex machine learning algorithms are of high importance for developing climate modeling that can simulate future climate conditions for different greenhouse gas emissions. Such models enable researchers and policy makers to see the potential impact of different factors on a global and regional scale, rising temperatures and sea levels, shifting patterns of precipitation, and increased frequency of extreme weather events. Machine learning algorithms help analyze the historical data and present trends of weather forecasting and early warning systems for improved accuracy to effectively prepare for and respond to various climate-related disasters (Hsu & Schletz, 2024).

Tracking of Carbon Footprints: Tracking carbon footprints at the level of a single individual, corporate, and nation is no more daunting a task, with machine learning and data analytics. It analyzes data from all the various activities associated with energy consumption, transportation, and waste management, highlighting opportunities for reduction in emission, thus bringing efficiency. Companies may use AI analytics to try to optimize supply chains, reduce the consumption of energy, and minimize waste produced as ways to reduce their carbon footprint. A mobile application or online platform would be able to show real-time carbon footprint tracking of the user on the consumer side and provide recommendations on how to reduce it (Lawrence et al., 2017; Hsu & Schletz, 2024).

Ecosystem Monitoring: Application of machine learning and data analytics in monitoring ecosystems and biodiversity, to gather information on how different species and their habitats are affected by climate change. Regarding remote sensing, the illustration of satellite and drone technology for collection of high-resolution imagery for vegetation, wildlife, and land use can be analyzed by machine learning algorithms to spot changes and areas at risk. This will go a long way toward developing conservation strategies and help in managing natural resources in a sustainable way (Hsu & Schletz, 2024).

2. Internet of Things (IoT)

IoT basically involves a set of devices linked for the collection, sharing, and analysis of data in real time. It thus massively opens up its scope of applications in enhanced climate action. Embedding IoT sensors and devices into different systems, such as energy grids and transportation networks, agricultural fields, and city environments, allows data to be collected that serves to optimize resource usage, reduce emissions, and build resilience due to climate change (Lawrence et al., 2017).

Smart Energy Management: IoT technologies are transforming the way we produce, distribute, and use energy by making energy management systems more efficient and flexible. For example, smart grids leverage IoT sensors and devices in real-time monitoring of energy demand and supply to allow for better integration of variable renewable sources like solar and wind energy, reducing reliance on fossil fuels. IoT-enabled smart meters and thermostats also enable consumers to track and reduce energy use, reducing waste and harmful greenhouse gas emissions. Similarly, on an industrial scale, IoT sensors can monitor equipment performance and energy use to identify areas of inefficiency and optimize processes, thus reducing energy consumption and associated emissions.

Precision Agriculture: IoT technologies play a crucial role in precision farming, letting farmers better manage their resources to increase resilience against climate change. IoT sensors could track soil moisture, temperature, and nutrient levels in real time, providing the farmer with concrete information to alter irrigation and fertilization and adjust planting schedules. Indeed, this would amount not only to higher yield and less wastage but also a reduction in stress on the environment through agricultural practice by reducing the use of water, chemical runoff, and greenhouse gas emissions. IoT-enabled weather stations and drones will monitor the weather conditions and detect any infestation of pest or disease so that intervention can be done with minimal application of harmful pesticides (Chandel et al., 2024).

Smart Cities and Urban Planning: IoT technologies are essential in Smart City development to enhance urban sustainability and climate change resilience. IoT sensors and devices will also monitor air quality, traffic patterns, waste management, and energy consumption in real time to provide invaluable data to optimize urban planning and infrastructure. For example, IoT-enabled traffic management systems reduce congestion and lower emissions by adjusting traffic signals and rerouting vehicles based on real-time data. Intelligent waste management systems make use of IoT sensors to monitor waste levels in every bin and run routes for minimum fuel consumption and, therefore, emissions. IoT technologies also contribute to urban resilience in terms of climate change through monitoring flood risks, managing water resources, and enhancing emergency response efforts.

3. Blockchain Technology

A decentralized and secure digital ledger, blockchain opens up unequalled opportunities for increased transparency, accountability, and collaboration across climate action. Indeed, through tamper-proof records of transactions and data, blockchain makes possible solutions for several of the most common pain points in climate governance, from monitoring emissions to verifying carbon credits and ensuring supply chain integrity (Chandel et al., 2023).

Carbon Credit Market: The use of blockchain technology can revolutionize carbon credit markets and make them completely transparent and secure. The traditional carbon markets are a bundle of fraud, double counting, and lack of transparency that undermine the effectiveness in emission reduction. It can do so by keeping a record of carbon credit transaction history, decentralized and immutable, in a manner that credits are real, correctly accounted for, and not double-counted. This can enhance transparency and, thus, trust among participants in carbon markets, which will lead to more businesses and individuals contributing toward greater reductions.

Supply Chain Transparency: Blockchain technology would be important in bringing more supply chain transparency and sustainability by creating a tamper-proof and verifiable record of the origin of goods, their production processes, and its effects on the environment. That is most relevant for industries that include very long and complex global supply chains, such as agriculture, textiles, or electronics, where tracing environmental and social product impacts is extremely difficult. This will include tracing and confirmation by blockchain at every stage in the value chain to make sure that business products are sourced in an environmentally sustainable manner and are ethically viable, reducing their general environmental footprint and enabling responsible consumption (Dechezleprêtre et al., 2019)

Decentralized Energy Systems: Another functionality of blockchain technology is decentralized energy systems whereby individuals and communities will be able to generate, store, and trade renewable energy in a peer-to-peer manner. This therefore means that with blockchain applied in the management of energy transactions, accurate metering, and billing, decentralized energy systems can reduce dependency on centralized utilities. This will go a long way in accelerating the movement toward a low-carbon economy, building resilience to disruptions in energy supply linked to climate change.

4. Digital Platforms and Collaboration Tools

Digital platforms and collaboration tools enable communication, knowledge sharing, and the ability to take joint action around climate change. These technologies have the potential to better connect people, organizations, and governments to enable innovation and increase action on climate.

Citizen Science and Crowdsourcing: The digital platform opens ways for citizens to contribute through science and crowdsourcing by providing data, observations, and ideas that can help in addressing climate change. For example, through mobile applications and online platforms, residents can act as environmental sensors, reporting on local conditions of air quality, wildlife sighting, or flooding that help scientists and policymakers. Crowdsourcing platforms can also be used to collect ideas and solutions from a large number of stakeholders and encourage creativity and collaboration to address climate-related issues (Gans, 2012).

Online Climate Action Networks: Through the facilitation provided by digital platforms, online networks or communities of climate action, information sharing, resources, and best practices are shared across diverse individuals and organizations. This may foster grassroots movements, campaigns, and collaboration projects which better profile local action and supplement global responses to climate change. In practice, it means that digital platforms can scale up knowledge, resources, and efforts for renewable energy cooperatives, urban farming initiatives, or zero-waste communities in entirely new ways.

Virtual Collaboration and Remote Work: Virtual collaboration and remote work, already massively accelerated by the COVID-19 pandemic, have an important role to play regarding climate action. Video conferencing, cloud computing, project management software, and other collaboration technology tools can indirectly reduce transportation-related GHG emissions by reducing the need to travel or commute. The facilitation of flexible and inclusive work through the combination of different teams and diverse stakeholders in different regions on climate action projects is also a possibility created by these tools (Su & Moaniba, 2017).

5. Artificial Intelligence and Machine Learning

AI and ML are making the space for innovations in climate action through process optimization, change prediction, and novel solution design. While vast chunks of data can be analyzed using AI and ML algorithms to identify patterns and correlations not feasible by human beings, these, in turn, enable more precise and effective climate action strategies.

Predictive Analytics for Climate Risks: Predictive analytics for climate risks involve the use of AI and ML in developing predictive tools for risks emanating from climate events and identification of vulnerabilities. These tools integrate historical climate data, satellite images, and socioeconomic data to predict the occurrence of such extreme weather events as hurricanes, floods, and droughts. Information so availed is very useful in the development of early warning systems and designing climate adaptation strategies which protect vulnerable communities and assets (Zilberman et al., 2018).

Optimization of Renewable Energy Systems: AI and ML algorithms will optimize renewable energy systems to best design for operation and maintenance. For instance, AI will predict the amount of solar and wind energy that will be produced from weather forecasts to balance supply and demand on a grid much better. ML algorithms analyze data from renewable installations to determine patterns in wear and tear to enable predictive maintenance with less downtime. These innovations contribute to the efficiency and reliability of systems producing renewable energy, therefore accelerating the transition towards a low-carbon energy future.

Carbon Capture and Storage: AI and ML play their role in the development and optimization of carbon capture and storage technologies, more commonly known as CCS. It aims at capturing carbon dioxide emission from industrial processes and storing it underground. The AI algorithms can model the behavior of CO_2 in various geological formations, which help in siting, besides developing injection strategies to optimize storage capacity and reducing associated risks. ML will also be able to analyze CCS project data for the enhancement of capture and storage processes concerning efficiency and safety (Adenle et al., 2015)

Digital technologies help drive climate action forward with novelty solutions for efficiency, transparency, and scalability in climate change mitigation and adaptation. It would ensure faster movement toward a low-carbon and sustainable economy and resilience building among all by leveraging data analytics, the Internet of Things, blockchain, Artificial Intelligence, and collaborative digital solutions.

BARRIERS TO INNOVATION AND TECHNOLOGY ADOPTION

Various types of barriers to innovation and technology adoption have been raised as arguments that will slow down the pace at which climate change could be intervened in with their huge potential. These include financial barriers, regulatory barriers, resistance to change, and other social and cultural issues. These barriers need to be addressed in order to accelerate the deployment rate for solutions to climate change and achieve meaningful progress in mitigation and adaptation. A better understanding of these challenges can help policymakers, businesses, and other stakeholders devise strategies that can help nullify their impact and thus provide a more encouraging environment for innovation and technology adoption (Su & Moaniba, 2017).

1. Financial Constraints

Probably the most serious barrier to entry in the use or adoption of new technologies is the lack of financial resources. Indeed, the innovation of technology, especially climate change development, usually requires huge investment, which can easily be a very hefty challenge for many organizations, particularly at a low resource level.

High Development Costs: High costs related to research, development, and commercialization. The startups and small enterprises may not be able to manage the required investment for the introduction of innovative solutions into the market. Large companies might also show a great deal of caution in investing in any technology due to uncertainty over returns. The increased development cost is harsh, especially for up-and-coming sectors like renewable energy, carbon capture, and advanced agricultural technologies where the technology remains evolving and markets are yet to be established (Matos et al., 2022).

Lack of access to finance: Even where promising new technologies are developed, access to finance can be an issue. Traditional sources of finance, such as bank loans or venture capital, may not be available or suitable for climate-related innovation. In most developing countries, the general state of financial markets is relatively underdeveloped and access to capital very limited. This makes access to finance one of the biggest barriers to the complete development and scaling of climate technologies, especially in those very regions of the world where the need for climate solutions is highest.

Uncertain Return on Investment: The financial risks associated with investments in new technologies could make investors wary. Many times, climate technologies involve long development timelines with uncertain returns, perceived as high-risk investments. This uncertainty is compounded by fluctuating policy environments and market conditions that make it hard for investors to predict the financial viability of climate technologies. In this respect, public funding, subsidies, and other financial incentives could bridge the financing gap and further the commercialization process of climate innovations.

2. Regulatory and Policy Challenges

The regulatory and policy frameworks can create major hurdles to the adoption of innovative technologies. Incoherent or restrictive regulations, absence of supportive policies, and bureaucratic obstacles together hamper climate solution deployment.

Fragmented Policies: Inconsistent or fragmented policies at the regional or country levels may sow uncertainty and even dampen the adoption of climate technologies. Various regulations on emissions reduction, renewable energy targets, and carbon pricing, for instance, will eventually affect market conditions of climate solutions and can challenge companies operating in different jurisdictions. Harmonization of policies can, therefore, be complemented with the establishment of clear, long-term regulatory frameworks providing a stable environment for innovation and technology adoption.

Red Tape: The complicated processes of getting required approvals through regulatory mechanisms involve a lot of time and money. In most countries, the regulatory environment is not well developed for any new technology, which means that the approval processes are extremely long and the introduction of that innovation in the market gets considerably delayed. The streamlining of regulatory procedures and clarity on guidelines with regard to approval procedures for climate technologies could help reduce these barriers and accelerate the deployment of new solutions (Abdel Monem & El Ghandour, 2020).

Lack of Standards and Certifications: There are no standard mechanisms for testing, certification, and quality standards that can make every new technology serve easily. Because no standards are in place, the performance and dependability of a solution that any innovation may offer cannot be easily judged. This fact makes many stakeholders highly skeptical and hence slow to adopt it. Building standards and applying their certifications to climate technologies can build credibility and hence help speed up their market acceptance.

3. Social and Cultural Resistance

The factors of social and cultural elements may further influence the adoption of innovative technology, especially those relating to set practices or ways of life. Resistance to change, lack of awareness, and cultural attitudes all make their marks on how new technologies are accepted and taken up.

Resistance to Change: Humans and organizations prefer to stick with what they are used to, or fear the unknown, when adopting new technologies. This kind of resistance is extremely high in fields where there are long-standing practices or where the benefits of the new technologies are not clearly visible. The way around these obstacles includes constructing appropriate communication, education, and engagement strategies that show how benefits associated with new technologies will be derived and decrease concerns related to them.

Lack of Awareness and Education: Often, the major reasons behind this trend are either ignorance or a lack of understanding regarding new technologies. Such stakeholders may not have an idea about the value proposition, applications, or even how these innovative solutions function. Education and creating awareness through outreach programs and knowledge-sharing become greater supporters of new technologies in hopes of more easily accepting those technologies.

Cultural and Social Norms: Most times, especially in cases where it is against old practices or existing values, cultural and social norms can make communities resist new technologies. There might be strong attachments in some communities to traditional sources of energy or agriculture, posing a challenge to new technologies. The engagement of communities through respect for their cultural values will help address such challenges by showing how new technologies can complement or further improve existing practices to build greater acceptance.

4. Technological and Infrastructure Barriers

Limitations of technology and infrastructure could also be a barrier to innovative solutions scaling up. Poor infrastructure, lack of technical know-how, and supportive technological capability can hinder the implementation and efficiency of climate technologies.

Gaps in Infrastructure: Due to a lack of infrastructural support in general, most regions, particularly developing ones, cannot deploy advanced technologies. As an example, renewable energy technologies including solar panels and wind turbines require reliable energy grids and storage systems-which do not exist everywhere. Investment and development in such areas could result in the creation of conditions for technology adoption.

Inadequate Technical Capacity: Full deployment of new technologies requires specific technical knowledge and expertise. In many instances, the population is not well-trained; there is often an acute shortage of people with the required expertise to deploy, operate, and maintain new innovative solutions. Training and capacity building programs can help bridge this gap by providing necessary skills to all stakeholders to effectively utilize the new technologies (Abdel Monem & El Ghandour, 2020).

Technology Transfer and Localization: For innovative technologies to be adopted effectively in varied regions, the process of technology transfer and localization may involve their adaptation or localization to suit the varying local conditions and needs. Technologies can be adapted to local climate, culture, and infrastructures. A better facilitation of such technology transfer and localization may come through collaboration between the technology developers, local stakeholders, and policy makers in ensuring that new solutions developed are appropriate and effective in varied contexts.

5. Market and Economic Factors

For innovative technologies, market and economic factors can also be an enabler or obstacle, particularly in those fields where market conditions are complex or uncertain.

Market Demand and Acceptance: The demand and acceptance of the market are the bases for the success of any new technology. Sometimes, the demand could be low, especially when there is a perception that it is an expensive or perhaps unnecessary solution. Subsidies, tax credits, public procurement policies, among others, may be created as market incentives to spur demand in this technology and, therefore, encourage its diffusion.

Economic Competitiveness: Often, the economic competitiveness of new technologies against existing solutions affects their adoptive functionality. Traditional technologies may be cheaper or better established in their use, and therefore new innovations have a hard time trying to compete with them. Showing the long-term economic benefits that new technologies can offer-such as cost savings, efficiency improvements, and environmental benefits-can help present a sound business case towards their adoption.

Resource Availability and Supply Chain: When it comes to new technologies, resource availability plays a crucial role. For example, renewable energy manufacturing currently depends on a few raw materials that either may be in limited supply or have prices that are susceptible to speculative fluctuation. This necessitates the development of a stable, sustainable supply chain for all critical materials and components needed to support the growth of climate technologies.

It is now of essence to break down barriers for innovation and technology adoption, as these are hugely needed to advance the climate action and achieve substantive progress in the mitigation and adaptation of climate change. This way, breaking financial bottlenecks, regulatory and policy complexities, social and cultural barriers, technological and infrastructure barriers, and market and economic considerations one by one, it lays a ground that would be supportive for the climate solutions to emerge and be deployed. In fact, these barriers can be overcome with collaborative efforts, targeted policies, and strategic investments in the rapid diffusion of such innovation crucial for sustainability and resilience (Abdel Monem & El Ghandour, 2020).

CASE STUDIES: SUCCESSFUL INNOVATIONS IN CLIMATE ACTION

Real-life examples of successful innovations in climate action bring out the different ways in which technology and innovation can be put to effective use in helping solve the climate challenges. These case studies present a few of the several approaches and solutions that have been put in place in different regions, testifying that innovation can indeed bring tangible results in mitigation and adaptation to climate change.

1. Solar Sister Initiative Sub-Saharan Africa

Location: Multi-country, Sub-Saharan Africa, including Uganda, Tanzania, Nigeria.

Overview: Solar Sister is one sole unique strategy to bring clean and affordable energy solutions to the rural communities of Sub-Saharan Africa through an approach that ingeniously marries solar technology with an innovative distribution model that empowers local women entrepreneurs. As a social enterprise, it sells solar lamps and clean cookstoves through a network of female entrepreneurs, which is bound to tackle two of the biggest issues-energy access and gender inequality-in one go (Samantarai & Dutta, 2023).

Innovation: Solar Sister trains women in solar technology and business skills so that they can be the solar distributors in their local areas. This model provides a sustainable solution toward a secured environment through developing economic capabilities and entrepreneurship among women. Solar lamps replace kerosene, harmful to health and environmentally hazardous, giving the community a clean and reliable source of light.

Impact: Solar Sister has effectively reached thousands of households across Sub-Saharan Africa to date, improving access to clean energy and quality of life for women and their families. With the strategy of reaching the local networks and imparting agency to women, Solar Sister had finally achieved a scalable and replicable model of sustainable development.

2. The Green Roofs Program (Chicago, USA)

Location: Chicago, Illinois, USA

Overview: The Green Roofs Program in Chicago represents an innovative urban sustainability initiative, focused on the reduction of the urban heat island effect, improving air quality, and improving the management of storm water. The program encourages the installation of green roofs, that is, roofs covered with vegetation, on various building facilities within its jurisdiction.

Innovation: The program includes some incentives for building owners in terms of installing green roofs, with grants and tax credits. Green roofs absorb rainwater, insulate buildings to reduce energy consumption, and lower the ambient temperature. They also provide additional green space in urban areas, promoting biodiversity and improving the well-being of residents Liu & Baskaran, 2003).

Impact: From the beginning of the Green Roofs Program, numerous green roofs have been installed around Chicago, greatly reducing the urban heat island effect and stormwater runoff. The benefits range from improved air quality and energy savings to increased resilience in cities to climate change. More still, the green roofs in Chicago would be considered a model for other cities that would wish to do similar initiatives in sustainability.

3. Masdar City Project, Abu Dhabi, UAE

Location: Abu Dhabi, United Arab Emirates

Overview: The Masdar City is among the most ambitious urban development projects in Abu Dhabi, aimed at being among the most sustainable cities on earth. It aims to achieve a no-carbon-emitted, no-waste urban setting with the use of cutting-edge technologies and sustainable living practices.

Innovation: Masdar City hosts a broad range of innovative technologies, from solar power and electric transportation to energy-efficient buildings. Masdar City is designed with wide renewable energy infrastructure, including large-scale solar farms, a network of electric vehicles, and a host of other clean energy technologies. The city is designed on the use of energy efficiency, water conservation, and waste reduction; its buildings are constructed to meet high sustainability standards.

Impact: Masdar City has been exemplary in setting the global benchmark of sustainable urban development. It showed the world how integrated technology and design can result in a low-carbon, resource-efficient environment. The project attracted international interest and investment. At the same time, the city is still used for testing new technologies and approaches to urban sustainability. In any case, this place expresses a proud vision for the future of eco-friendly living.

4. The TerraCycle Global Initiative (New Jersey, USA)

Location: Headquartered in Trenton, New Jersey, USA-with global operations

Overview: TerraCycle is an innovative waste management company focused on recycling hard-to-recycle materials and reducing waste. Operating on a global scale, the company provides solutions for several product and packaging waste streams that are not commonly accepted by traditional recycling programs.

Innovation: TerraCycle does an out-of-the-box business model of collecting and recycling non-recyclable waste with the involvement of companies, municipalities, and individuals. Manufacturing materials into new products through processing and upcycling activities within the company, where the trash will be turned into a useful resource (Margery et al., 2017). TerraCycle practices include programs like the "Loop" platform, which encourages the reuse of packaging items through a circular economy model.

Impact: To date, TerraCycle has been very successful in diverting several million pounds of waste from landfills and incinerators. Through the development of an array of recycling and upcycling platforms, TerraCycle has helped minimize environmental impact while "closing the loop" in moving toward a circular economy. Its global reach, combined with partnerships with major brands and organizations, has magnified its impact in driving positive change in waste management practices around the world.

5. The Barka Water Project (Barka, Oman)

Location: Barka, Oman

Overview: The Barka Water Project will be one of the most revolutionary projects in Oman, contributing to solving the problem of water shortage using the latest advances in the field of water desalination technologies. This project will diversify water resources and enhance their security and reliability in a region expected to face serious challenges with regard to the availability of this vital resource.

Innovation: The project desalinates seawater into fresh water with the help of a combination of the reverse osmosis desalination method along with renewable energy generation. Integrating solar energy in the Barka Water Project's desalination technology reduces its reliance on fossil fuels and lessens the carbon footprint associated with the production of water supplies. Energy-efficient methods of desalination and renewable sources of energy make water supplies more sustainable (Zilberman et al., 2018).

Impact: The Barka Water Project has succeeded in giving the region an assured and clean water supply, providing the very fundamentals for poor water supply into the region, thus gaining sustainability in its water management. A project that combined advanced technology with renewable energy gave an example to other regions, confronted with similar challenges with water supply, showing the potential that could be offered in the treatment of critical environmental problems (Al-Barwani & Purnama, 2011).

6. The Ocean Cleanup Project (Global)

Location: Global; ocean gyres, in particular.

Description: The Ocean Cleanup Project is an innovative initiative in which plastic waste is removed from the world's oceans. The goal of the project is to rid the world's oceans of the growing problem of plastic pollution by conducting active research on and deploying advanced technologies that capture and remove the plastic debris found floating in marine environments.

Innovation: The Ocean Cleanup Project deploys floating barriers and collection systems to accumulate plastic wastes in ocean gyres-high concentration areas. Using the current from the ocean to drive plastic debris into the central collection unit for processing and removal is facilitated by a passive system using this technology. Taking care of plastic waste on land to prevent plastic waste from entering the oceans is also part of the project.

Impact: The Ocean Cleanup Project has succeeded in picking up plastic debris from oceans and raising awareness about ocean plastic pollution. Its innovative approach proved that it is possible to clean plastic debris on a large scale and also initiated the involvement of other efforts in trying to resolve marine pollution. By incorporating technology with advocacy and education, the Ocean Cleanup Project can spearhead the path to be taken toward cleaner oceans and a healthy marine environment (Zilberman et al., 2018).

These case studies will no doubt permit one to overview successful innovations in climate action, illustrating how different technologies and approaches can stand true in their effectiveness over climate challenges. From renewable energy and sustainable urban development to waste management and water conservation, these examples truly show that innovation really does have significant potential for driving meaningful progress in both mitigating and adapting to climate change. In return, stakeholders will be able to draw out best practices, replicate successful models, and help drive the economies of the world toward a rapid pace of climate solution adoptions (Matos et al., 2022).

INNOVATION IN CLIMATE ACTION: THE FUTURE

The future of innovation in climate action is bright and full of potential as the call to climate action becomes louder. Emerging technologies, strategies, and approaches are therefore important at this point in time in advancing our efforts on mitigation and adaptation to climate change. This section looks at some key trends and future directions in climate innovation and shows how these developments might mean for the future of climate action-and progress toward a sustainable, low-carbon future (Zilberman et al., 2018).

1. Advanced Renewable Energy Technologies

Extension of Solar and Wind Energy: The future of renewable energy is poised to be captured by the enhancement of solar and wind energies. Innovations such as perovskite solar cells, which have been advanced for efficiency and lesser production costs, along with floating wind turbines that generate energy in water much deeper than in-depth capability of conventional technology, are expected to greatly enhance the deployment and efficiency of renewable energies. These technologies will continue to scale and become more reliable, making them core factors in the worldwide transition to clean energy.

Energy Storage Solution Integration: Increased use of variable renewable sources will create a demand for proper energy storage solutions. In fact, new battery technologies such as solid-state batteries and flow batteries hold immense promise in advancing the efficiency and capacity of energy storage. Moreover, this technology in grid-scale storage solutions, like pumped hydro storage and compressed air energy storage, will eventually provide an integrated solution to the intermittency in renewable energy sources for a stable and reliable supply of power (Matos et al., 2022).

Decentralized Energy Systems: Energy of the future shall be similarly decentralized to the point at which production and consumption take place as closely as possible. Innovation around microgrids and distributed energy resources will continue to evolve, enabling a community to generate and manage its own energy independently of, or in complement with, centralized power plants, making them more resilient against energy disruptions. This will facilitate various integrations of renewable energy sources and enhance the role of local communities in making decisions that concern their needs for energy (Zilberman et al., 2018).

2. Digital Technologies and Data-Driven Solutions

Artificial Intelligence and Machine Learning: These hold great potential in supporting further climate action. These technologies can analyze vast volumes of data to attain efficiency in energy use, predict environmental impacts, and understand patterns in climate variability. AI will enable the creation of more efficiency in renewable energy generation, enhance climate modeling and forecasting, and support the development of smart cities with all sustainable infrastructure.

Transparency and Accountability through Blockchain: Blockchain technology has the potential to act as a catalyzer in the fields of transparency and accountability on climate action. In this perspective, with the use of its distributed, immutable ledger, blockchain can provide enhancement in traceability of supply chains, verification of carbon credits, or emissions reductions, and full transparent reporting of climate-related data (Matos et al., 2022). This would also potentially help in the development of climate finance approaches that are decentralized, given all the systems utilize blockchain. Thus, it also accommodates much more efficient and secure transaction processing, related to investments in climate action.

Monitoring and Management-IoT of Things: IoT devices and sensors are gaining significance for environmental condition monitoring, management of energy systems, and optimal use of resources. The deployment of IoT technologies presents real-time data about air quality, water consumption, and energy use those better controls and responds to all forms of climate-related challenges. IoT-enabled smart infrastructure can support the development of resilient and adaptive urban environments.

3. Circular Economy and Sustainable Materials

Paradigm Shift: Circular Economy Models The paradigm of circular economy has been sweeping into popularity, with the increasing need for waste reduction, resource conservation, and lessening environmental impact. Circular economy models, on their end, are an innovation to the development of a new business model, including product-as-a-service and take-back schemes, which encourage reuse, recycling, and repurposing of products and materials. Such models pursue closed-loop systems that foster resource efficiency and reduce the generation of wastes (Matos et al., 2022).

Sustainable Materials and Green Chemistry: Development in sustainable materials and green chemistry will also underpin innovation for the future of climate action. Research into alternative materials, such as biodegradable plastics and low-impact building materials, will help in reducing the environmental footprint of many industries. Green chemistry practices-which seek to make safer and more sustainable chemical processes-will further enable the creation of environmentally-friendly products and technologies (Zilberman et al., 2018).

4. Climate Resilience and Adaptation Strategies

Climate Resilience Infrastructure Innovations: Building infrastructure in a manner to be resistant to the effects of climate change is a means of achieving increased resilience while reducing vulnerability. Innovations that have been created about climate-resilient infrastructure include adaptive design considerations and advanced materials integrated with nature-based solutions to make buildings, transportation systems, and urban areas more durable and adaptive. These strategies enable infrastructure effectively to respond to variations in climate conditions and events of extreme weather.

Nature-Based Solutions and Ecosystem Restoration: NbS is increasingly recognized as having the potential for addressing climate challenges while providing co-benefits related to biodiversity and ecosystem services. Innovations in reforestation and coastal protection for ecosystem restoration may sequester carbon, enhance natural resilience, and support livelihoods. The integration of nature-based solutions into climate adaptation strategies might provide cost-effective and sustainable ways of managing climate impacts (Abdel Monem & El Ghandour, 2020).

5. Collaborative and Inclusive Approaches

Knowledge Sharing for Global Collaboration: The fight against climate change requires transboundary and intersectoral collaboration. Climate action innovation in the future will be globally more collaborative, from knowledge sharing between governments, businesses, researchers, and civil society (Fankhaeser et al., 2008). International collaborative platforms and partnerships can facilitate sharing best practices, research findings, and technological solutions, which help speed up the pace toward solving global challenges and foster global innovation.

Equitable Inclusive Innovation: The climate action should be inclusive and guarantee equity in the innovations it perceives. Future innovation must address the diverse needs and perspectives of a wide array of communities that bear significant impacts of climate change. Innovative solutions through stakeholder involvement at the grassroots level, along with active responses to disparate economic and social conditions, serve to ensure that benefits from solutions aimed at addressing climate change accrue to all people (Zilberman et al., 2018).

6. Education and Capacity Building

Promote Climate Education and Trainings: Education and capacity building are very important for innovation and grooming the new generation of climate leaders. Future efforts are envisioned to include supporting the inclusion of climate education in curricula, training and development of skills, as well as research and innovation in the field of climate science and technology. More significantly, it could be further enhanced by investment in education and capacity building, enabling individuals and organizations to take on the challenge of climate action and drive innovative solution development (Matos et al., 2022).

Climate Entrepreneurship: However we cut it, the encouragement of climate entrepreneurship will ensure innovation and scaling up of solutions. That is, resources, mentorship, and funding will be channeled to entrepreneurs and startups working on climate-related technologies and solutions in the coming year. By promoting a favorable ecosystem for climate entrepreneurship, we encourage the rapid development and deployment of new technologies and business models addressing climate challenges (Fankhaeser et al., 2008).

The future of innovation in climate action holds many promises that could take our efforts on mitigation and adaptation of climate change to the next level. In renewable energy technologies, digital solutions, circular economy models, climate resilience strategies, and collaborative approaches lie the bases for developing a sustainable and resilient future. Embracing such innovations with the added emerging challenge will maximize progress toward meaningful climate action (Abdel Monem & El Ghandour, 2020).

CONCLUSION

While the impacts of climate change continue to scale upward, innovation and technology will play an increasingly vital part in taming these growing challenges. The crossing of advanced technologies, clever solutions, and joined-up efforts enables mitigation and adaptation to climate change effectively, but at every step, this journey requires collective commitment with sustained action. This ranges from renewable energy technologies and digital technologies to circular economy business models. Innovation could really be a game-changer in how we are going to tackle the climate.

By the turn of future developments, much progress will be underway in respect of energy production, environmental monitoring, and resource management, among other features. Next-generation renewable energy technologies, such as advanced solar cells and floating wind turbines, would afford greater scalability and efficiency of clean energy sources. These will be strong enablers from the digital technologies of AI, IoT, and blockchain, which enable better optimization of energy use, increased sophistication in climate modeling, and bring about transparency in environmental data. Apart from that, circular economy practices and sustainable materials contribute to less waste and encourage resource conservation in support of climate objectives.

There are, however, a set of barriers to realizing their full potential. Financial, regulatory, social, and technological problems must be resolved in order to create the enabling conditions that will see the large-scale deployment of climate solutions. This requires collaboration between governments, businesses, and civil society in establishing enabling policies, inclusive innovation, and capacity and infrastructure building.

In the future, the contribution of these innovations will be determined by how effectively we harness them and address barriers to their role. It will be important to think of a forward-looking approach that nurtures a culture of collaboration and innovation, hence speeding up the transition towards a sustainable and low-carbon future. The road to climate change is indeed complex and multi-faceted, but with continuous investment in technology and dedication toward collective action, we will be able to thread our way through and make a difference toward a resiliently sustainable world for future generations.

REFERENCES

Abdel Monem, M. A., & El Ghandour, I. A. (2020). Role of science, technology and innovation in addressing climate change challenges in Egypt. *Climate Change Impacts on Agriculture and Food Security in Egypt: Land and Water Resources—Smart Farming—Livestock, Fishery, and Aquaculture*, 59-79.

Adenle, A. A., Azadi, H., & Arbiol, J. (2015). Global assessment of technological innovation for climate change adaptation and mitigation in developing world. *Journal of Environmental Management*, 161, 261–275. DOI: 10.1016/j.jenvman.2015.05.040 PMID: 26189184

Aivazidou, E., Banias, G., Lampridi, M., Vasileiadis, G., Anagnostis, A., Papageorgiou, E., & Bochtis, D. (2021). Smart technologies for sustainable water management: An urban analysis. *Sustainability (Basel)*, 13(24), 13940. DOI: 10.3390/su132413940

Al-Barwani, H. H., & Purnama, A. (2011). A Computational Model Study of Brine Discharges from Seawater Desalination Plants at Barka, Oman. In *Expanding issues in desalination*. IntechOpen. DOI: 10.5772/21509

Alcayna, T., & O'Donnell, D. (2022). How much global climate adaptation finance is targeting the health sector? *European Journal of Public Health*, 32(Supplement_3), ckac129–146. DOI: 10.1093/eurpub/ckac129.146

Basher, R. (2006). Global early warning systems for natural hazards: systematic and people-centred. *Philosophical transactions of the royal society a: mathematical, physical and engineering sciences, 364*(1845), 2167-2182.

Beck, M. W., Heck, N., Narayan, S., Menéndez, P., Reguero, B. G., Bitterwolf, S., Torres-Ortega, S., Lange, G.-M., Pfliegner, K., Pietsch McNulty, V., & Losada, I. J. (2022). Return on investment for mangrove and reef flood protection. *Ecosystem Services*, 56, 101440. DOI: 10.1016/j.ecoser.2022.101440

Bergman, N., Markusson, N., Connor, P., Middlemiss, L., & Ricci, M. (2010). Bottom-up, social innovation for addressing climate change. *Energy transitions in an interdependent world: what and where are the future social science research agendas, Sussex*, 25-26.

Carmin, J., Roberts, D., & Anguelovski, I. (2009, June). Planning climate resilient cities: early lessons from early adapters. In *Fifth Urban Research Symposium, Cities and Climate Change: Responding to an Urgent Agenda* (pp. 28-30).

Chandel, A., Bhanot, N., & Sharma, R. (2023). A bibliometric and content analysis discourse on business application of blockchain technology. *International Journal of Quality and Reliability Management. Scopus.* Advance online publication. DOI: 10.1108/IJQRM-02-2023-0025

Chandel, A., Verma, R., Sood, K., & Grima, S. (2024). A bibliometric analysis of drones-mediated precision farming. *International Journal of Sustainable Agricultural Management and Informatics*, 10(4), 347–377. DOI: 10.1504/IJSAMI.2024.141841

Chauhan, C., Parida, V., & Dhir, A. (2022). Linking circular economy and digitalisation technologies: A systematic literature review of past achievements and future promises. *Technological Forecasting and Social Change*, 177, 121508. DOI: 10.1016/j.techfore.2022.121508

Chhetri, N., Chaudhary, P., Tiwari, P. R., & Yadaw, R. B. (2012). Institutional and technological innovation: Understanding agricultural adaptation to climate change in Nepal. *Applied Geography (Sevenoaks, England)*, 33, 142–150. DOI: 10.1016/j.apgeog.2011.10.006

De Stefano, M. C., Montes-Sancho, M. J., & Busch, T. (2016). A natural resource-based view of climate change: Innovation challenges in the automobile industry. *Journal of Cleaner Production*, 139, 1436–1448. DOI: 10.1016/j.jclepro.2016.08.023

Dechezleprêtre, A., Martin, R., & Bassi, S. (2019). Climate change policy, innovation and growth. In *Handbook on green growth* (pp. 217–239). Edward Elgar Publishing. DOI: 10.4337/9781788110686.00018

Falcke, L., Zobel, A. K., & Comello, S. D. (2024). How firms realign to tackle the grand challenge of climate change: An innovation ecosystems perspective. *Journal of Product Innovation Management*, 41(2), 403–427. DOI: 10.1111/jpim.12687

Fankhaeser, S., Sehlleier, F., & Stern, N. (2008). Climate change, innovation and jobs. *Climate Policy*, 8(4), 421–429. DOI: 10.3763/cpol.2008.0513

Gans, J. S. (2012). Innovation and climate change policy. *American Economic Journal. Economic Policy*, 4(4), 125–145. DOI: 10.1257/pol.4.4.125

Grubb, M. (2004). Technology innovation and climate change policy: An overview of issues and options. *Keio Economic Studies*, 41(2), 103–132.

Guo, H. D., Zhang, L., & Zhu, L. W. (2015). Earth observation big data for climate change research. *Advances in Climate Change Research*, 6(2), 108–117. DOI: 10.1016/j.accre.2015.09.007

Holdren, J. P. (2006). The energy innovation imperative: Addressing oil dependence, climate change, and other 21st century energy challenges. *Innovations: Technology, Governance, Globalization*, 1(2), 3–23. DOI: 10.1162/itgg.2006.1.2.3

Hsu, A., & Schletz, M. (2024). Digital technologies–the missing link between climate action transparency and accountability? *Climate Policy*, 24(2), 193–210. DOI: 10.1080/14693062.2023.2237937

Jochem, E., & Madlener, R. (2003, July). The forgotten benefits of climate change mitigation: Innovation, technological leapfrogging, employment, and sustainable development. In *Workshop on the benefits of climate policy: Improving information for policy makers*. OECD Paris.

Lawrence, S., López Ventura, J., Doody, L., & Peracio, P. (2017). Polisdigitocracy: Citizen engagement for climate action through digital technologies. *Field Actions Science Reports. The journal of field actions*, (Special Issue 16), 58-65.

Lipper, L., Thornton, P., Campbell, B. M., Baedeker, T., Braimoh, A., Bwalya, M., Caron, P., Cattaneo, A., Garrity, D., Henry, K., Hottle, R., Jackson, L., Jarvis, A., Kossam, F., Mann, W., McCarthy, N., Meybeck, A., Neufeldt, H., Remington, T., & Torquebiau, E. F. (2014). Climate-smart agriculture for food security. *Nature Climate Change*, 4(12), 1068–1072. DOI: 10.1038/nclimate2437

Liu, K., & Baskaran, B. (2003). Thermal performance of green roofs through field evaluation Proceedings of 1st North American Green Roof Conference: Greening Rooftops for Sustainable Communities, Chicago, The Cardinal Group Toronto.

Lybbert, T., & Sumner, D. (2010). Agricultural technologies for climate change mitigation and adaptation in developing countries: policy options for innovation and technology diffusion.

Margery, P., Read, S., & Lepoutre, J. (2017). TerraCycle: Outsmarting waste. In *Case Studies in Social Entrepreneurship* (pp. 114-133). Routledge.

Matos, S., Viardot, E., Sovacool, B. K., Geels, F. W., & Xiong, Y. (2022). Innovation and climate change: A review and introduction to the special issue. *Technovation*, 117, 102612. DOI: 10.1016/j.technovation.2022.102612

Matos, S., Viardot, E., Sovacool, B. K., Geels, F. W., & Xiong, Y. (2022). Innovation and climate change: A review and introduction to the special issue. *Technovation*, 117, 102612. DOI: 10.1016/j.technovation.2022.102612

Nhamo, G., & Muchuru, S. (2019). Climate adaptation in the public health sector in Africa: Evidence from United Nations Framework Convention on Climate Change National Communications. *Jàambá*, 11(1), 1–10. DOI: 10.4102/jamba.v11i1.644 PMID: 31049162

Nordhaus, W. D. (2010). Modeling induced innovation in climate-change policy. In *Technological change and the environment* (pp. 182–209). Routledge.

Ockwell, D., Watson, J., Mallett, A., Haum, R., MacKerron, G., & Verbeken, A. M. (2010). Enhancing developing country access to eco-innovation: The case of technology transfer and climate change in a post-2012 policy framework.

Owen, D. A. L. (2018). *Smart water technologies and techniques: Data capture and analysis for sustainable water management*. John Wiley & Sons. DOI: 10.1002/9781119078678

Pagoropoulos, A., Pigosso, D. C., & McAloone, T. C. (2017). The emergent role of digital technologies in the Circular Economy: A review. *Procedia CIRP*, 64, 19–24. DOI: 10.1016/j.procir.2017.02.047

Reiche, D. (2010). Renewable energy policies in the Gulf countries: A case study of the carbon-neutral "Masdar City" in Abu Dhabi. *Energy Policy*, 38(1), 378–382. DOI: 10.1016/j.enpol.2009.09.028

Rodima-Taylor, D., Olwig, M. F., & Chhetri, N. (2012). Adaptation as innovation, innovation as adaptation: An institutional approach to climate change. *Applied Geography (Sevenoaks, England)*, 33(0), 107–111. DOI: 10.1016/j.apgeog.2011.10.011

Samantarai, M., & Dutta, S. (2023). *Katherine Lucey and Solar Sister: empowering women in sub-Saharan Africa to create clean energy businesses*. The CASE Journal.

Scherr, S. J., Shames, S., & Friedman, R. (2012). From climate-smart agriculture to climate-smart landscapes. *Agriculture & Food Security*, 1(1), 1–15. DOI: 10.1186/2048-7010-1-12

Sen, M. K., Dutta, S., Kabir, G., Pujari, N. N., & Laskar, S. A. (2021). An integrated approach for modelling and quantifying housing infrastructure resilience against flood hazard. *Journal of Cleaner Production*, 288, 125526. DOI: 10.1016/j.jclepro.2020.125526

Smith, K. (2009). Climate change and radical energy innovation: the policy issues.

Song, Y., Medda, F., & Wang, M. (2024). The value of resilience bond in financing flood resilient infrastructures: A case study of Towyn. *Journal of Sustainable Finance & Investment*, 14(4), 1–24. DOI: 10.1080/20430795.2024.2366200

Steward, F. (2012). Transformative innovation policy to meet the challenge of climate change: Sociotechnical networks aligned with consumption and end-use as new transition arenas for a low-carbon society or green economy. *Technology Analysis and Strategic Management*, 24(4), 331–343. DOI: 10.1080/09537325.2012.663959

Su, H. N., & Moaniba, I. M. (2017). Does innovation respond to climate change? Empirical evidence from patents and greenhouse gas emissions. *Technological Forecasting and Social Change*, 122, 49–62. DOI: 10.1016/j.techfore.2017.04.017

Vushe, A. (2021). Proposed research, science, technology, and innovation to address current and future challenges of climate change and water resource management in Africa. *Climate Change and Water Resources in Africa: Perspectives and Solutions Towards an Imminent Water Crisis*, 489-518.

Zedler, J. B. (2000). Progress in wetland restoration ecology. *Trends in Ecology & Evolution*, 15(10), 402–407. DOI: 10.1016/S0169-5347(00)01959-5 PMID: 10998517

Zhou, Y., Shao, H., Cao, Y., & Lui, E. M. (2021). Application of buckling-restrained braces to earthquake-resistant design of buildings: A review. *Engineering Structures*, 246, 112991. DOI: 10.1016/j.engstruct.2021.112991

Zilberman, D., Lipper, L., McCarthy, N., & Gordon, B. (2018). Innovation in response to climate change. *Climate smart agriculture: building resilience to climate change*, 49-74.

Compilation of References

Abdel Monem, M. A., & El Ghandour, I. A. (2020). Role of science, technology and innovation in addressing climate change challenges in Egypt. *Climate Change Impacts on Agriculture and Food Security in Egypt: Land and Water Resources—Smart Farming—Livestock, Fishery, and Aquaculture*, 59-79.

Abid, N., Cecí, F., & Ikram, M. (2022). Green Growth and Sustainable Development: Dynamic Linkage Between Technological Innovation, ISO 14001, and Environmental Challenges. *Environmental Science and Pollution Research International*, 29(17), 1–20. DOI: 10.1007/s11356-021-17518-y PMID: 34843051

Abidoye, B., & Odusola, A. (2015). Climate change and economic growth in Africa: An econometric analysis. *Journal of African Economies*, 24(2), 277–301. DOI: 10.1093/jae/eju033

Acar, Z., Gönencgil, B., & Gümüşoğlu, N. K. (2018). Long-term changes in hot and cold extremes in Turkey. *Coğrafya Dergisi*, (37), 57–67. DOI: 10.26650/JGEOG2018-0002

Acharya, K. (2023). Navigation of climate change framework: analysis of synergy between sdgs and unfccc. DOI: 10.48001/978-81-966500-9-4_7

Acheampong, A. O. (2018). Economic growth, CO2 emissions and energy consumption: What causes what and where? *Energy Economics*, 74, 677–692. DOI: 10.1016/j.eneco.2018.07.022

Adebayo, T. S., Rjoub, H., Akadiri, S. S., Oladipupo, S. D., Sharif, A., & Adeshola, I. (2022). The role of economic complexity in the environmental Kuznets curve of MINT economies: Evidence from method of moments quantile regression. *Environmental Science and Pollution Research International*, 29(16), 24248–24260. DOI: 10.1007/s11356-021-17524-0 PMID: 34822076

Adenle, A. A., Azadi, H., & Arbiol, J. (2015). Global assessment of technological innovation for climate change adaptation and mitigation in developing world. *Journal of Environmental Management*, 161, 261–275. DOI: 10.1016/j.jenvman.2015.05.040 PMID: 26189184

Aghion, P., Caroli, E., & García-Peñalosa, C. (1999). Inequality and Economic Growth: The Perspective of The New Growth Theories. *Journal of Economic Literature*, 37(4), 1615–1660. DOI: 10.1257/jel.37.4.1615

Ahmed, N., Sheikh, A. A., Hamid, Z., Senkus, P., Borda, R. C., Wysokińska-Senkus, A., & Glabiszewski, W. (2022). Exploring the causal relationship among green taxes, energy intensity, and energy consumption in Nordic countries: Dumitrescu and Hurlin causality approach. *Energies*, 15(14), 1–15. DOI: 10.3390/en15145199

Ahsan, R., Kellett, J., & Karuppannan, S. (2014). *Climate induced migration: Lessons from Bangladesh* (Doctoral dissertation, Common Ground Publishing).

Ahsan, R. (2019). Climate-induced migration: Impacts on social structures and justice in Bangladesh. *South Asia Research*, 39(2), 184–201. DOI: 10.1177/0262728019842968

Aivazidou, E., Banias, G., Lampridi, M., Vasileiadis, G., Anagnostis, A., Papageorgiou, E., & Bochtis, D. (2021). Smart technologies for sustainable water management: An urban analysis. *Sustainability (Basel)*, 13(24), 13940. DOI: 10.3390/su132413940

Akin, C. S., Aytun, C., & Akin, S. (2018). *Turizm, düşük karbon ekonomisi ilişkisi: Türkiye üzerine bir uygulama*. VII. Ulusal III. Uluslararası Doğu Akdeniz Turizm Sempozyumu, İskenderun.

Akter, A., Abul Ulie, K. K., & Ali, S. S. (2023). Intercropping: Prospects and challenges in bangladesh for sustainable agriculture. *IJEAST*, 8(3), 111–120. DOI: 10.33564/IJEAST.2023.v08i03.015

Akyel, Ö. (2009). *İklim Değişikliği Çerçeve Sözleşmesi ve Türkiye'deki Uygulamaları*. Ankara Üniversitesi Sosyal Bilimler Enstitüsü.

Al-Barwani, H. H., & Purnama, A. (2011). A Computational Model Study of Brine Discharges from Seawater Desalination Plants at Barka, Oman. In *Expanding issues in desalination*. IntechOpen. DOI: 10.5772/21509

Alcayna, T., & O'Donnell, D. (2022). How much global climate adaptation finance is targeting the health sector? *European Journal of Public Health*, 32(Supplement_3), ckac129–146. DOI: 10.1093/eurpub/ckac129.146

Aleinikoff, T. A. (2024). Climate-Induced Displacement and the International Protection of Forced Migrants. *Social Research*, 91(2), 421–444. DOI: 10.1353/sor.2024.a930749

Ali, H., Dumbuya, B., Hynie, M., Idahosa, P., Keil, R., & Perkins, P. (2016). The social and political dimensions of the Ebola response: Global inequality, climate change, and infectious disease. *Climate change and health: improving resilience and reducing Risks*, 151-169.

Almulhim, A. I., Alverio, G. N., Sharifi, A., Shaw, R., Huq, S., Mahmud, M. J., ... Abubakar, I. R. (2024). Climate-induced migration in the Global South: an in depth analysis. *NPJ Climate Action, 3*(1), 47.

Alola, A., Adebayo, T., & Olanipekun, I. (2023). Examining the energy efficiency and economic growth potential in the world energy trilemma countries. *Energies*, 16(4), 2036. DOI: 10.3390/en16042036

Amanatidis & Petit. (2024). *Combating climate change*, Fact Sheets on the European Union European Parliament, Retrieved June 10, 2024 from https://www.europarl.europa.eu/factsheets/en/sheet/72/combating-climate-change

Anand, S., & Sen, A. (2000). Human Development and Economic Sustainability. *World Development*, 28(12), 2029–2049. DOI: 10.1016/S0305-750X(00)00071-1

Aneja, R., & Arjun, G. (2021). Estimating components of productivity growth of Indian high and medium-high technology industries: A non-parametric approach. *Social Sciences & Humanities Open*, 4(1), 1–10. DOI: 10.1016/j.ssaho.2021.100180

Ansari, M. A., Ahmad, M. R., Siddique, S., & Mansoor, K. (2020). An environment Kuznets curve for ecological footprint: Evidence from GCC countries. *Carbon Management*, 11(4), 355–368. DOI: 10.1080/17583004.2020.1790242

Anthoff, D., & Tol, R. S. (2013). The uncertainty about the social cost of carbon: A decomposition analysis using FUND. *Climatic Change*, 117(3), 515–530. DOI: 10.1007/s10584-013-0706-7

Aras, S. (2022). İklim değişikliğinin turizm ve açık alan rekreasyonuna etkisinin incelenmesi. *Journal of Tourism and Gastronomy Studies*, 10(4), 3645–3661.

Archer, D., Almansi, F., DiGregorio, M., Roberts, D., Sharma, D., & Syam, D. (2014). Moving towards inclusive urban adaptation: Approaches to integrating community-based adaptation to climate change at city and national scale. *Climate and Development*, 6(4), 345–356. DOI: 10.1080/17565529.2014.918868

Arı, A. (2021). Yenilenebilir enerji, turizm, CO_2 ve GSYH ilişkisinin Türkiye için analizi. *Akademik Yaklaşımlar Dergisi*, 12(2), 192–205. DOI: 10.54688/ayd.880406

Arrhenius, S. (1896). On the Influence of Carbonic Acid in the Air upon the Temperature of the Ground. *The London, Edinburgh and Dublin Philosophical Magazine and Journal of Science*, 41(251), 237–276. DOI: 10.1080/14786449608620846

Arrow, K., Bolin, B., Costanza, R., Folke, C., Holling, C. S., Janson, B., Levin, S., Maler, K., Perrings, C., & Pimental, D. (1995). Economic growth, carrying capacity, and the environment. *Science*, 15, 91–95. PMID: 17756719

Arslan, H. M., Khan, I., Latif, M. I., Komal, B., & Chen, S. (2022). Understanding The Dynamics of Natural Resources Rents, Environmental Sustainability, and Sustainable Economic Growth: New Insights From China. *Environmental Science and Pollution Research International*, 29(39), 58746–58761. DOI: 10.1007/s11356-022-19952-y PMID: 35368236

Ashraf, S. N., & Singh, A. K. (2022). Implications of appropriate technology and farm inputs in the agricultural sector of Gujarat: Empirical analysis based on primary data. *Agricultural Economics Research Review*, 35(2), 59–77. DOI: 10.5958/0974-0279.2022.00031.3

Askland, H. H., Shannon, B., Chiong, R., Lockart, N., Maguire, A., Rich, J., & Groizard, J. (2022). Beyond migration: A critical review of climate change induced displacement. *Environmental Sociology*, 8(3), 267–278. DOI: 10.1080/23251042.2022.2042888

Atasoy, M., & Atasoy, F. G. (2020). The impact of climate change on tourism: A causality analysis. *Turkish Journal of Agriculture-Food Science and Technology*, 8(2), 515–519. DOI: 10.24925/turjaf.v8i2.515-519.3250

Atlas of Economic Complexity. (2024). *Country & product complexity rankings.* https://atlas.cid.harvard.edu/rankings

Atta, M., Zoromba, M. A., El-Gazar, H. E., Loutfy, A., Elsheikh, M. A., El-ayari, O. S. M., Sehsah, I., & Elzohairy, N. W. (2024). Climate anxiety, environmental attitude, and job engagement among nursing university colleagues: A multicenter descriptive study. *BMC Nursing*, 23(1), 133. Advance online publication. DOI: 10.1186/s12912-024-01788-1 PMID: 38378543

Atwood, M. (2013). *MaddAddam: a novel.* Nan A. Talese/Doubleday.

Auffhammer, M., & Aroonruengsawat, A. (2011). Simulating the impacts of climate change, prices and population on California's residential electricity consumption. *Climatic Change*, 109(S1), 191–210. DOI: 10.1007/s10584-011-0299-y

Aydemir, B., & Şenerol, H. (2014). İklim değişikliği ve Türkiye turizmine etkileri: Delfi anket yöntemiyle yapılan bir uygulama çalışması. *Balıkesir Üniversitesi Sosyal Bilimler Enstitüsü Dergisi*, 17(31), 381–417. DOI: 10.31795/baunsobed.664062

Aydin, C., & Cetíntas, Y. (2022). Does The Level of Renewable Energy Matter in The Effect of Economic Growth on Environmental Pollution? New Evidence From PSTR Analysis. *Environmental Science and Pollution Research International*, 29(54), 81624–81635. DOI: 10.1007/s11356-022-21516-z PMID: 35739444

Barber, B. (2013). *If mayors ruled the world – Dysfunctional nations, rising cities*. Yale University Press.

Barbier, E. (2011, August). The Policy Challenges for Green Economy and Sustainable Economic Development. *Natural Resources Forum*, 35(3), 233–245. DOI: 10.1111/j.1477-8947.2011.01397.x

Barker, A., Blake, H., D'Arcangelo, F. M., & Lenain, P. (2022). Towards net zero emissions in Denmark. OECD Economics Department Working Papers No. 1705, Retrieved August 05, 2024 from https://eulacfoundation.org/system/files/digital_library/2023-07/5b40df8f-en.pdf

Barnett, H. J., & Morse, C. (2013). *Scarcity and Growth: The Economics of Natural Resource Availability*. Routledge. DOI: 10.4324/9781315064185

Barrier, E. B. (2017). The Concept of Sustainable Economic Development. In *The Economics of Sustainability* (pp. 87–96). Routledge. DOI: 10.4324/9781315240084-7

Barro, R. J. (2003). Determinants of Economic Growth in A Panel of Countries. *Annals of Economics and Finance*, 4, 231–274.

Bartelmus, P. (2013). The Future We Want: Green Growth or Sustainable Development? *Environmental Development*, 7, 165–170. DOI: 10.1016/j.envdev.2013.04.001

Basher, R. (2006). Global early warning systems for natural hazards: systematic and people-centred. *Philosophical transactions of the royal society a: mathematical, physical and engineering sciences*, 364(1845), 2167-2182.

Başoğlu, A. (2014). An Attempted Model and Econometric Analysis on the Economic Effects of Global Climate Change. *Karadeniz Technical University, Institute of Social Sciences*, Department of Economics, Doctoral Program.

Basu, P., Ghosh, I., & Das, P. K. (2018). Using structural equation modelling to integrate human resources with internal practices for lean manufacturing implementation. *Management Science Letters*, 8(1), 51–68. DOI: 10.5267/j.msl.2017.10.001

Baudon, P., & Jachens, L. (2021). A scoping review of interventions for the treatment of eco-anxiety. *International Journal of Environmental Research and Public Health*, 18(18), 9636. DOI: 10.3390/ijerph18189636 PMID: 34574564

Beck, M. W., Heck, N., Narayan, S., Menéndez, P., Reguero, B. G., Bitterwolf, S., Torres-Ortega, S., Lange, G.-M., Pfliegner, K., Pietsch McNulty, V., & Losada, I. J. (2022). Return on investment for mangrove and reef flood protection. *Ecosystem Services*, 56, 101440. DOI: 10.1016/j.ecoser.2022.101440

Bektaş, V. (2017). Gelişmekte olan ülkelerde cari açıkların sürdürülebilirliği: Bir panel veri analizi. *Bolu Abant İzzet Baysal Üniversitesi Sosyal Bilimler Enstitüsü Dergisi*, 17(1), 51–66.

Bell, C., & Masys, A. J. (2020). Climate Change, Extreme Weather Events and Global Health Security a Lens into Vulnerabilities. In *Advanced Sciences and Technologies for Security Applications*. DOI: 10.1007/978-3-030-23491-1_4

Ben Jebli, M., Ben Youssef, S., & Apergis, N. (2014). The dynamic linkage between CO_2 emissions, economic growth, renewable energy consumption, number of tourist arrivals and trade. *MPRA Paper No. 57261, Munich Personal RePEc Archive*, 1-12.

Bento, J. P. C., & Moutinho, V. (2016). CO_2 emissions, non-renewable and renewable electricity production, economic growth, and international trade in Italy. *Renewable & Sustainable Energy Reviews*, 55, 142–155. DOI: 10.1016/j.rser.2015.10.151

Berber, P., & Keleş, E. (2022). *Bilgi ve Teknolojileri Aracılığıyla Uzaktan Çalışma "Tele Çalışma: Kapsam ve Doğası"*. Pegem Akademi Yayıncılık.

Bergman, N., Markusson, N., Connor, P., Middlemiss, L., & Ricci, M. (2010). Bottom-up, social innovation for addressing climate change. *Energy transitions in an interdependent world: what and where are the future social science research agendas, Sussex*, 25-26.

Beton Kalmaz, D., & Adebayo, T. S. (2024). Does foreign direct investment moderate the effect of economic complexity on carbon emissions? Evidence from BRICS nations. *International Journal of Energy Sector Management*, 18(4), 834–856. DOI: 10.1108/IJESM-01-2023-0014

Bhat, N. A., & Kaur, S. (2024). Technical Efficiency Analysis of Indian IT Industry: A Panel Data Stochastic Frontier Approach. *Millennial Asia*, 15(2), 327–348. DOI: 10.1177/09763996221082199

Bilbao-Osorio, B., & Rodríguez-Pose, A. (2004). From R&D to Innovation and Economic Growth in The EU. *Growth and Change*, 35(4), 434–455. DOI: 10.1111/j.1468-2257.2004.00256.x

Biljecki, F., Chow, Y. S., & Lee, K. (2023). Quality of crowdsourced geospatial building information: A global assessment of OpenStreetMap attributes. *Building and Environment*, 237, 110295. Advance online publication. DOI: 10.1016/j.buildenv.2023.110295

Binboğa, G. (2014). Uluslararası Karbon Ticareti ve Türkiye. Yaşar Üniversitesi E- Dergisi, 9(34), 5732-5759.

Black, P., & Wiliam, D. (1998). Inside the black box: Raising standards through classroom assessment. *Phi Delta Kappan*, 80(2), 139–148.

Blunden, J., & Arndt, D. S. (2020). A Look at 2019. *Bulletin of the American Meteorological Society*, 101(7), 612–622. DOI: 10.1175/BAMS-D-20-0203.1

Boaden, P., Morrison, G., & Smith, L. (2020). Economic Impacts of Hurricane Maria on Puerto Rico. *Weather, Climate, and Society*, 12(3), 563–578.

Boldeanu, F. T., & Constantinescu, L. (2015). The Main Determinants Affecting Economic Growth. *Bulletin of the Transilvania University of Brasov. Series V, Economic Sciences*, 329–338.

Borkowska, S., Bielecka, E., & Pokonieczny, K. (2023). Comparison of Land Cover Categorical Data Stored in OSM and Authoritative Topographic Data. *Applied Sciences (Basel, Switzerland)*, 13(13), 7525. Advance online publication. DOI: 10.3390/app13137525

Borowy, I. (2013). The Brundtland Commission: Sustainable development as health issue. *Michael*, 10, 198–208.

Bosello, F., Eboli, F., & Pierfederici, R. (2012). Assessing the economic impacts of climate change: An updated CGE point of view. *FEEM Working Paper No. 2012.056*. https://doi.org/DOI: 10.2139/ssrn.2153483

Bose, P., & Lunstrum, E. (2012). Environmentally induced displacement and forced migration. *Refuge: Canada's Periodical on Refugees*, 29(2), 5–10. DOI: 10.25071/1920-7336.38163

Boulanger, P. M. (2007). Political Uses of Social Indicators: Overview and Application to Sustainable Development Indicators. *International Journal of Sustainable Development*, 10(1-2), 14–32. DOI: 10.1504/IJSD.2007.014411

Bouznit, M., & Pablo-Romero, M. D. P. (2016). CO_2 emission and economic growth in Algeria. *Energy Policy*, 96, 93–104. DOI: 10.1016/j.enpol.2016.05.036

Bowen, G. (2009). Document Analysis as a Qualitative Research Method. *Qualitative Research Journal*, 9(2), 27–40. DOI: 10.3316/QRJ0902027

Breusch, T. S., & Pagan, A. R. (1980). The lagrange multiplier test and its applications to model specification in econometrics. *The Review of Economic Studies*, 47(1), 239–253. DOI: 10.2307/2297111

Browne, S. (2012). *United Nations Development Programme and System (UNDP)*. Routledge. DOI: 10.4324/9780203806852

Budhathoki, N. R., & Haythornthwaite, C. (2012). Motivation for Open Collaboration: Crowd and Community Models and the Case of OpenStreetMap. *The American Behavioral Scientist*, 57(5), 548–575. DOI: 10.1177/0002764212469364

Burk, J., Uhlich, T., Bals, C., Höhne, N., & Nascimento, L. (2024). *Results: Monitoring Climate Mitigation Efforts of 63 Countries plus the EU – covering more than 90% of the Global Greenhouse Gas Emissions*, Climate Change Performance Index, Retrieved June 12, 2024 from https://ccpi.org/wp-content/uploads/CCPI-2024-Results.pdf

Burke, M., & Emerick, K. (2016). Adaptation to climate change: Evidence from US agriculture. *American Economic Journal. Economic Policy*, 8(3), 106–140. DOI: 10.1257/pol.20130025

Busch, K. C., Henderson, J. A., & Stevenson, K. T. (2019). Broadening epistemologies and methodologies in climate change education research. *Environmental Education Research*, 25(6), 955–971. DOI: 10.1080/13504622.2018.1514588

C3S & WMO. (2024). *European state of the climate 2023*, Copernicus Climate Change Service and World Meteorological Organization, Retrieved June 05, 2024 from https://climate.copernicus.eu/esotc/2023

C40 Cities. (2024). *About C40*. Retrieved June 01, 2024, https://www.c40.org/about-c40/

Caglar, A. E., Zafar, M. W., Bekun, F. V., & Mert, M. (2022). Determinants of CO_2 emissions in the BRICS economies: The role of partnerships investment in energy and economic complexity. *Sustainable Energy Technologies and Assessments*, 51, 101907. DOI: 10.1016/j.seta.2021.101907

Can, F. (2023). *Çevre ve İklim Değişikliği 101*. Say Yayınları.

Carmin, J., Roberts, D., & Anguelovski, I. (2009, June). Planning climate resilient cities: early lessons from early adapters. In *Fifth Urban Research Symposium, Cities and Climate Change: Responding to an Urgent Agenda* (pp. 28-30).

Casas-Cuestas, M. (2024). How much and for how long could the annual cost of atmospheric greenhouse gas (CO2e) abatement between 1960 and 2020 through carbon pricing be estimated? *Preprint.* https://doi.org/DOI: 10.21203/rs.3.rs-4571476/v1

Ceesay, E., & Fanneh, M. M. (2022). Economic growth, climate change, and agriculture sector: ARDL bounds testing approach for Bangladesh (1971-2020). *Economics Management and Sustainability*, 7(1), 95–106. DOI: 10.14254/jems.2022.7-1.8

Ceesay, E., Oladejo, H., Abokye, P., & Ugbor, O. (2020). Econometrics analysis of the relationship between climate change and economic growth in selected West African countries. *Energy and Environment Research*, 10(2), 39. DOI: 10.5539/eer.v10n2p39

Celliers, L., & Schleyer, M. (2008). Coral community structure and risk assessment of high-latitude reefs at Sodwana Bay, South Africa. *Biodiversity and Conservation*, 17(13), 3097–3117. DOI: 10.1007/s10531-007-9271-6

Cevik, M. S., & Jalles, J. T. (2022). *For whom the bell tolls: climate change and inequality.* International Monetary Fund.

Chandel, A., Bhanot, N., & Sharma, R. (2023). A bibliometric and content analysis discourse on business application of blockchain technology. International Journal of Quality and Reliability Management. *Scopus.* Advance online publication. DOI: 10.1108/IJQRM-02-2023-0025

Chandel, A., Verma, R., Sood, K., & Grima, S. (2024). A bibliometric analysis of drones-mediated precision farming. *International Journal of Sustainable Agricultural Management and Informatics*, 10(4), 347–377. DOI: 10.1504/IJSAMI.2024.141841

Charlson, F., Ali, S., Benmarhnia, T., Pearl, M., Massazza, A., Augustinavicius, J., & Scott, J. (2021). Climate change and mental health: A scoping review. *International Journal of Environmental Research and Public Health*, 18(9), 4486. DOI: 10.3390/ijerph18094486 PMID: 33922573

Chataut, G., Bhatta, B., Joshi, D., Subedi, K., & Kafle, K. (2023). Greenhouse gases emission from agricultural soil: A review. *Journal of Agriculture and Food Research*, 11, 100533. DOI: 10.1016/j.jafr.2023.100533

Chaudhuri, D. D. (2016). Impact of economic liberalization on technical efficiency of firms: Evidence from India's electronics industry. *Theoretical Economics Letters*, 6(1), 549–560. DOI: 10.4236/tel.2016.63061

Chauhan, C., Parida, V., & Dhir, A. (2022). Linking circular economy and digitalisation technologies: A systematic literature review of past achievements and future promises. *Technological Forecasting and Social Change*, 177, 121508. DOI: 10.1016/j.techfore.2022.121508

Chen, B., Tu, Y., Song, Y., Theobald, D. M., Zhang, T., Ren, Z., Li, X., Yang, J., Wang, J., Wang, X., Gong, P., Bai, Y., & Xu, B. (2021). Mapping essential urban land use categories with open big data: Results for five metropolitan areas in the United States of America. *ISPRS Journal of Photogrammetry and Remote Sensing*, 178, 203–218. Advance online publication. DOI: 10.1016/j.isprsjprs.2021.06.010

Chen, Q. (2024). The Intersection of Global Inequality and Climate Change. *International Journal of Education and Humanities*, 15(2), 378–381. DOI: 10.54097/3s08zx60

Chen, Y., Wang, Z., & Zhong, Z. (2019). CO_2 emissions, economic growth, renewable and non-renewable energy production and foreign trade in China. *Renewable Energy*, 131, 208–216. DOI: 10.1016/j.renene.2018.07.047

Chhetri, N., Chaudhary, P., Tiwari, P. R., & Yadaw, R. B. (2012). Institutional and technological innovation: Understanding agricultural adaptation to climate change in Nepal. *Applied Geography (Sevenoaks, England)*, 33, 142–150. DOI: 10.1016/j.apgeog.2011.10.006

Chikoore, H., Vermeulen, J. H., & Jury, M. R. (2015). Tropical cyclones in the Mozambique channel: January–March 2012. *Natural Hazards*, 77(3), 2081–2095. DOI: 10.1007/s11069-015-1691-0

Chisadza, C., Clance, M., Sheng, X., & Gupta, R. (2023). Climate change and inequality: Evidence from the United States. *Sustainability (Basel)*, 15(6), 5322. DOI: 10.3390/su15065322

Choudhury, A., & Shahi, S. K. (2024, June). Climate-Induced Displacement and Sustainable Development. In *NDIEAS-2024 International Symposium on New Dimensions and Ideas in Environmental Anthropology-2024 (NDIEAS 2024)* (pp. 51-64). Atlantis Press. DOI: 10.2991/978-2-38476-255-2_5

Chouinard, Y., Ellison, J., & Ridgeway, R. (2011). The Sustainable Economy. *Harvard Business Review*, 89(10), 52–62.

Cianconi, P., Betrò, S., & Janiri, D. (2020). The Impact of Climate Change on Mental Health: A Systematic Review. *Frontiers in Psychiatry*, 4(1), 1–12. DOI: 10.3389/fpsyt.2020.00074 PMID: 32210846

Çinar, S. (2011). Gelir ve CO$_2$ emisyonu ilişkisi: Panel birim kök ve eşbütünleşme testi. *Uludağ Üniversitesi İktisadi ve İdari Bilimler Fakültesi Dergisi.*, 30(2), 71–83.

Ciplet, D., Roberts, J. T., & Khan, M. R. (2015). *Power in a warming world: The new global politics of climate change and the remaking of environmental inequality.* Mit Press. DOI: 10.7551/mitpress/9780262029612.001.0001

Clemens, V., Hirschhausen, E., & Fegert, J. (2020). Report of the intergovernmental panel on climate change: Implications for the mental health policy of children and adolescents in europe—a scoping review. *European Child & Adolescent Psychiatry*, 31(5), 701–713. DOI: 10.1007/s00787-020-01615-3 PMID: 32845381

Climate Change and Power. (2002). *Economıc Instruments For European Elecetrıcıty* (Vrolijk, C., Ed.). The Royal Institute of International Affairs.

Climate, N. O. A. A. gov (2024). *Climate change: global temperature.* https://www.climate.gov/news-features/understanding-climate/climate-change-global-temperature

Climate, P. I. H. (2023a). *Overview of climate targets in Europe,* Retrieved June 15, 2024 from https://climatepolicyinfohub.eu/overview-climate-targets-europe#

Climate, P. I. H. (2023b). *European Climate Policy - History and State of Play,* Retrieved June 14, 2024 from https://climatepolicyinfohub.eu/european-climate-policy-history-and-state-play.html

Creswell, J. W. (2009). *Research design: Qualitative, quantitative and mixed method approaches* (3rd ed.). Sage.

Cui, Y., Zhang, Z. F., Froines, J., Zhao, J., Wang, H., Yu, S. Z., & Detels, R. (2003). Air pollution and case fatality of SARS in the People's Republic of China: An ecologic study. *Environmental Health: A Global Access Science Source, 2,* 1–5. .DOI: 10.1186/1476-069X-2-1

Dağdemir, Ö. (2015). Birleşmiş Milletler İklim Değişikliği Çerçeve Sözleşmesi ve Ekonomik Büyüme: İklim Değişikliği Politikasının Türkiye İmalat Sanayii Üzerindeki Olası Etkileri. *Ankara Üniversitesi SBF Dergisi*, 60(2), 49–70.

Dang, G., & Sui Pheng, L. (2015). Theories of Economic Development. *Infrastructure Investments in Developing Economies: The Case of Vietnam,* 11-26.

Darwent, D. F. (1969). Growth Poles and Growth Centers in Regional Planning-A Review. *Environment & Planning A*, 1(1), 5–31. DOI: 10.1068/a010005

Davoudi, S., Mehmood, A., & Brooks, L. (2011). *The London climate change adaptation strategy: Gap analysis.* Global Urban Research Unit.

De Stefano, M. C., Montes-Sancho, M. J., & Busch, T. (2016). A natural resource-based view of climate change: Innovation challenges in the automobile industry. *Journal of Cleaner Production*, 139, 1436–1448. DOI: 10.1016/j.jclepro.2016.08.023

De Urioste-Stone, S. M., Scaccia, M. D., & Howe-Poteet, D. (2015). Exploring visitor perceptions of the influence of climate change on tourism at Acadia National Park, Maine. *Journal of Outdoor Recreation and Tourism*, 11, 34–43. DOI: 10.1016/j.jort.2015.07.001

Debnath, R. M., & Sabastian, V. J. (2014). Efficiency in the Indian iron and steel industry – an application of data envelopment analysis. *Journal of Advances in Management Research*, 11(1), 4–19. DOI: 10.1108/JAMR-01-2013-0005

Dechezleprêtre, A., Martin, R., & Bassi, S. (2019). Climate change policy, innovation and growth. In *Handbook on green growth* (pp. 217–239). Edward Elgar Publishing. DOI: 10.4337/9781788110686.00018

Dell, M., Jones, B. F., & Olken, B. A. (2014). What do we learn from the weather? The new climate-economy literature. *Journal of Economic Literature*, 52(3), 740–798. DOI: 10.1257/jel.52.3.740

Demir, Y., & Görür, Ç. (2020). OECD ülkelerine ait çeşitli enerji tüketimleri ve ekonomik büyüme arasındaki ilişkinin panel eşbütünleşme analizi ile incelenmesi. *Ekoist: Journal of Econometrics and Statistic*, 32, 15–33.

Dempsey, N., Bramley, G., Power, S., & Brown, C. (2011). The Social Dimension of Sustainable Development: Defining Urban Social Sustainability. *Sustainable Development (Bradford)*, 19(5), 289–300. DOI: 10.1002/sd.417

Department of Environmental Affairs (DEA). (2017). *Climate change trends, risks, impacts and vulnerabilities, South Africa's 2nd annual climate change report* [Online]. Available from: https://www.environment.gov.za/sites/default/files/reports/southafrica_secondnational_ climatechnage_report2017.pdf

Dereli, M., Boyacioğlu, E. Z., & Terzioğlu, M. K. (2019). İklim değişikliği ve turizm sektörü arasındaki ilişkinin dinamik panel veri analizi ile incelenmesi. *Türk Turizm Araştırmaları Dergisi*, 3(4), 1228–1243. DOI: 10.26677/TR1010.2019.238

Deschênes, O., & Greenstone, M. (2011). The economic impacts of climate change: Evidence from agricultural output and random fluctuations in weather. *The American Economic Review*, 97(1), 354–385. DOI: 10.1257/aer.97.1.354

Dinda, S. (2004). Environmental Kuznets Curve Hypothesis: A Survey. *Ecological Economics*, 49(4), 431–455. DOI: 10.1016/j.ecolecon.2004.02.011

DiPeso, J. (2004). Climate change and the states. *Environmental Quality Management*, 13(3), 111–116. DOI: 10.1002/tqem.20010

Dixon, P. B., & Jorgenson, D. W. (Eds.). (2013). *Handbook of Computable General Equilibrium Modeling (Vol. 1A)*. North-Holland., DOI: 10.1016/B978-0-444-59568-3.09997-0

Doğan, S., & Tüzer, M. (2011). Küresel iklim değişikliği ile mücadele: Genel yaklaşımlar ve uluslararası çabalar. *Istanbul Journal of Sociological Studies*, (44), 157–194.

Doğan, S., & Tüzer, M. (2011). *Küresel İklim Değişikliği İle Mücadele: Genel Yaklaşımlar ve Uluslararası Çabalar. Sosyoloji Konferansları*. İstanbul Üniversitesi Sosyoloji Konferansları Dergisi.

Dou, Y., Zhao, J., Malik, M. N., & Dong, K. (2021). Assessing the impact of trade openness on CO2 emissions: Evidence from China-Japan-ROK FTA countries. *Journal of Environmental Management*, 296, 113241. DOI: 10.1016/j.jenvman.2021.113241 PMID: 34265664

Draper, J. (2020). *Justice in Climate-Induced Migration and Displacement* (Doctoral dissertation, University of Reading).

Draper, J., & McKinnon, C. (2018). The ethics of climate-induced community displacement and resettlement. *Wiley Interdisciplinary Reviews: Climate Change*, 9(3), e519. DOI: 10.1002/wcc.519

Du, D., & Ng, P. (2018). The impact of climate change on tourism economies of Greece, Spain, and Turkey. *Environmental Economics and Policy Studies*, 20(2), 431–449. DOI: 10.1007/s10018-017-0200-y

Dumitrescu, E. I., & Hurlin, C. (2012). Testing for Granger noncausality in heterogeneous panels. *Economic Modelling*, 29(4), 1450–1460. DOI: 10.1016/j.econmod.2012.02.014

Dzebo, A. (2023). The paris agreement and the sustainable development goals: evolving connections. DOI: 10.51414/sei2023.036

Easterly, W. (2005), National Policies and Economic Growth: A Reappraisal. *Handbook of Economic Growth, 1*, 1015-1059.

Eboli, F., Parrado, R., & Roson, R. (2010). Climate-change feedback on economic growth: Explorations with a dynamic general equilibrium model. *Environment and Development Economics*, 15(5), 515–533. DOI: 10.1017/S1355770X10000252

EC. (1991). *Energy in Europe: Energy policies and trends in the European Community,* Retrieved June 17, 2024 from https://aei.pitt.edu/79862/1/17._July_1991.pdf

EC. (1992). *A community strategy to limit carbon dioxide emissions and improve energy efficiency*, P/92/29, Retrieved June 10, 2024 from https://ec.europa.eu/commission/presscorner/detail/en/P_92_29

EC. (2018a). *Kyoto 1st commitment period (2008–12),* Retrieved June 12, 2024 from https://climate.ec.europa.eu/eu-action/international-action-climate-change/kyoto-1st-commitment-period-2008-12_en

EC. (2018b). *Communication from the Commission: to the European Parliament, the European Council, the Council, the European Economic and Social Committee, the Committee of the Regions and the European Investment Bank, A Clean Planet for all A European strategic long-term vision for a prosperous, modern, competitive and climate neutral economy,* Brussels, 28.11.2018 COM (2018) 773 final, Retrieved June 14, 2024 from https://eur-lex.europa.eu/legal-content/EN/TXT/PDF/?uri=CELEX:52018DC0773

EC. (2019). *Communication from the Commission: The European Green Deal,* Brussels, 11.12.2019 COM (2019) 640 final, Retrieved June 14, 2024 from https://eur-lex.europa.eu/legal-content/EN/TXT/?qid=1576150542719&uri=COM%3A2019%3A640%3AFIN

EC. (2022). The new European consensus on development 1our world, our dignity, our future", Retrieved August 08, 2024 from https://international-partnerships.ec.europa.eu/document/download/6134a7a4-3fcf-46c2-b43a-664459e08f51_en?filename=european-consensus-on-development-final-20170626_en.pdf

EC. (2023a). *Causes of climate change,* Retrieved June 04, 2024 from https://climate.ec.europa.eu/climate-change/causes-climate-change_en

EC. (2023b). *Special Eurobarometer 538 climate change,* Retrieved June 12, 2024 from https://europa.eu/eurobarometer/surveys/detail/2954

EC. (2023c). *GHG emissions of all world countries: JRC science for policy report,* Retrieved June 06, 2024 from https://edgar.jrc.ec.europa.eu/report_2023

EC. (2024). *Fit for 55,* Retrieved June 16, 2024 from https://www.consilium.europa.eu/en/policies/green-deal/fit-for-55/

Ediz, Ç., & Yanik, D. (2023). The effects of climate change awareness on mental health: Comparison of climate anxiety and hopelessness levels in turkish youth. *The International Journal of Social Psychiatry*, 69(8), 2157–2166. DOI: 10.1177/00207640231206060 PMID: 37874036

EEA. (2010). *Tracking progress towards Kyoto and 2020 targets in Europe,* Retrieved June 15, 2024 from https://cetesb.sp.gov.br/inventario-gee-sp/wpcontent/uploads/sites/34/2014/04/eea_european.pdf

EEA. (2012). *Key observed and projected climate change and impacts for the main regions in Europe,* Retrieved June 06, 2024 from https://www.eea.europa.eu/soer/data-and-maps/figures/key-past-and-projected-impacts-and-effects-on-sectors-for-the-main-biogeographic-regions-of-europe-3

EEA. (2021). *EU achieves 20-20-20 climate targets, 55% emissions cut by 2030 reachable with more efforts and policies,* Retrieved July 02, 2024 from https://www.eea.europa.eu/highlights/eu-achieves-20-20-20

EEA. (2023). *Flexibility solutions to support a decarbonised and secure EU electricity system,* Retrieved July 08, 2024 from https://www.eea.europa.eu/publications/flexibility-solutions-to-support

EEA. (2024a). *European climate risk assessment: Executive summary,* Retrieved June 05, 2024 from https://www.eea.europa.eu/publications/european-climate-risk-assessment

EEA. (2024b). *Extreme weather: floods, droughts and heatwaves,* Retrieved July 01, 2024 from https://www.eea.europa.eu/en/topics/in-depth/extreme-weather-floods-droughts-and-heatwaves

EEA. (2024c). *Climate change impacts, risks and adaptation,* Retrieved July 03, 2024 from https://www.eea.europa.eu/en/topics/in-depth/climate-change-impacts-risks-and-adaptation

Eghdami, S., Scheld, A. M., & Louis, G. (2023). Socioeconomic vulnerability and climate risk in coastal Virginia. *Climate Risk Management*, 39, 100475. DOI: 10.1016/j.crm.2023.100475

Ellison, N. B. (2004). *Telework and social change: How technology is reshaping the boundaries between home and work.* Bloomsbury Publishing USA. DOI: 10.5040/9798216024132

Elzen, M.G.J., Lucas, P.L., & Gujsen, A. (2007). *Exploring European countries' emission reduction targets, abatement costs and measures needed under the 2007 EU reduction objectives,* Netherlands Environmental Assessment Agency (MNP).

Emanuel, K. (2005). Increasing Destructiveness of Tropical Cyclones Over the Past 30 Years. *Nature*, 436(7051), 686–688. DOI: 10.1038/nature03906 PMID: 16056221

Emmerling, J., Andreoni, P., & Tavoni, M. (2024). Global inequality consequences of climate policies when accounting for avoided climate impacts. *Cell Reports Sustainability, 1*(1).

Energy Cities. (2024). *Our vision*. Retrieved May 21, 2024, from https://energy-cities.eu/vision-mission/

Engel, K. H. (2015). EPA's clean power plan: An emerging new cooperative federalism? *Publius*, 3(45), 452–474. DOI: 10.1093/publius/pjv025

Engin, I. (2019). İklim Değişikliği ile Mücadelede Mali Politikalar. Balıkesir Üniversitesi Sosyal Bilimler Enstitüsü, 47-67.

Engin, B. (2010). İklim değişikliği ile mücadelede uluslararası işbirliğinin önemi. *Sosyal Bilimler Dergisi*, (2), 71–82.

England, R. W. (1998). Measurement of Social Well-Being: Alternatives to Gross Domestic Product. *Ecological Economics*, 25(1), 89–103. DOI: 10.1016/S0921-8009(97)00098-0

Ennis, R. H. (2015). Critical thinking: A streamlined conception. In *Critical thinking: A statement of expert consensus for purposes of educational assessment and instruction* (pp. 1–49). Foundation for Critical Thinking.

EP. (2019). *European policies on climate and energy towards 2020, 2030 and 2050*, Briefing, ENVI in FOCUS, Retrieved July 02, 2024 from https://www.europarl.europa.eu/RegData/etudes/BRIE/2019/631047/IPOL_BRI(2019)631047_EN.pdf

EU. (2019). *Europe's approach to implementing the Sustainable Development Goals: good practices and the way forward*, Directorate-General for External Policies Policy Department, Retrieved 08 August, 2024 from https://www.europarl.europa.eu/cmsdata/160360/DEVE%20study%20on%20EU%20SDG%20implementation%20formatted.pdf

European Commission. (2024a). *Why a covenant of mayors?* Retrieved May 21, 2024, from https://eu-mayors.ec.europa.eu/en/about

European Commission. (2024b). *Signatories* Retrieved May 21, 2024, from https://eu-mayors.ec.europa.eu/en/signatories

European Enviroment Agency. (n.d.). Carbon Emission Data. https://www.eea.europa.eu/data-and-maps/daviz/atmospheric-concentration-of-carbon-dioxide-5#tabchart_5_filters=%7B%22rowFilters%22%3A%7B%7D%3B%22column-Filters%22%3A%7B%22pre_config_polutant%22%3A%5B%22CO2%20(ppm)%22%5D%7D%7D

Evans, S. (2020, April 9). *Analysis: Coronavirus set to cause largest ever annual fall in CO2 emissions*. Retrieved June 30, 2024, from https://www.carbonbrief.org/analysis-coronavirus-set-to-cause-largest-ever-annual-fall-in-co2-emissions/

Falcke, L., Zobel, A. K., & Comello, S. D. (2024). How firms realign to tackle the grand challenge of climate change: An innovation ecosystems perspective. *Journal of Product Innovation Management*, 41(2), 403–427. DOI: 10.1111/jpim.12687

Fankhaeser, S., Sehlleier, F., & Stern, N. (2008). Climate change, innovation and jobs. *Climate Policy*, 8(4), 421–429. DOI: 10.3763/cpol.2008.0513

Fankhauser, S., & Tol, R. (2005). On climate change and economic growth. *Resource and Energy Economics*, 27(1), 1–17. DOI: 10.1016/j.reseneeco.2004.03.003

Farabi, A., Abdullah, A., & Setianto, R. H. (2019). Energy consumption, carbon emissions and economic growth in Indonesia and Malaysia. *International Journal of Energy Economics and Policy*, 9(3), 338–345. DOI: 10.32479/ijeep.6573

Fernandes, P. M., & Rigolot, E. (2022). *13. Prescribed burning in the European Mediterranean Basin*. Global Application of Prescribed Fire. DOI: 10.3390/en17112476

Ferreira, V. (2017). Climate induced migration: Legal challenges. *Intergenerational responsibility in the 21st century*, 107-121.

Figueres, C., & Ca, T. (2020). *Our approach to covid-19 can also help tackle climate change | New Scientist*, https://www.newscientist.com/article/mg24532763-500-our-approach-to-covid-19-can-also-help-tackle-climate-change/

Fischer, E. M., & Schär, C. (2010). Consistent Evidence of Enhanced Summer Heat in Europe. *Geophysical Research Letters*, 37(2), L20704.

Fitzgerald, J. B. (2022). Working time, inequality and carbon emissions in the United States: A multi-dividend approach to climate change mitigation. *Energy Research & Social Science*, 84, 102385. DOI: 10.1016/j.erss.2021.102385

Flagg, J. (2018). Carbon neutral by 2021: The past and present of costa rica's unusual political tradition. *Sustainability (Basel)*, 10(2), 296. DOI: 10.3390/su10020296

Francis, A. (2019). Climate-induced migration & free movement agreements. *Journal of International Affairs*, 73(1), 123–134.

Freeman, S., Eddy, S. L., McDonough, M., Smith, M. K., Okoroafor, N., Jordt, H., & Wenderoth, M. P. (2014). Active learning increases student performance in science, engineering, and mathematics. *Proceedings of the National Academy of Sciences of the United States of America*, 111(23), 8410–8415. DOI: 10.1073/pnas.1319030111 PMID: 24821756

Fundisa for Change Programme. (2013). *Introductory core text*. Environmental Learning Research Centre, Rhodes University.

Galbraith, J. K. (1964). *Economic Development*. Harvard University Press. DOI: 10.4159/harvard.9780674333062

Galushko, V., & Gamtessa, S. (2022). Impact of climate change on productivity and technical efficiency in Canadian crop production. *Sustainability (Basel)*, 14(1), 1–21. DOI: 10.3390/su14074241

Gans, J. S. (2012). Innovation and climate change policy. *American Economic Journal. Economic Policy*, 4(4), 125–145. DOI: 10.1257/pol.4.4.125

Gao, S. (2024). An exogenous risk in fiscal-financial sustainability: Dynamic stochastic general equilibrium analysis of climate physical risk and adaptation cost. *Journal of Risk and Financial Management*, 17(6), 244. DOI: 10.3390/jrfm17060244

Garro-Quesada, M. del M., Vargas-Leiva, M., Girot, P. O., & Quesada-Román, A. (2023). Climate Risk Analysis Using a High-Resolution Spatial Model in Costa Rica. *Climate (Basel)*, 11(6), 127. Advance online publication. DOI: 10.3390/cli11060127

Gemenne, F. (2011). Climate-induced population displacements in a 4 C+ world. *Philosophical Transactions of the Royal Society A: Mathematical, Physical and Engineering Sciences, 369*(1934), 182-195.

Giddens, A. (2009). *The politics of climate change*. Polity Press.

Global Covenant of Mayors for Climate & Energy. (2023). *Urban catalysts – A local climate stocktake: The 2023 global covenant of mayors impact report*. Retrieved May 21, 2024, from https://www.globalcovenantofmayors.org/wp-content/uploads/2023/12/GCoM-2023-Global-Impact-report-2023_10.12.2023.pdf

Göçoğlu, İ. D., Negiz, N., & Göçoğlu, V. (2023). Türkiye'nin İklim Değişikliği ile Mücadele Serüveni: Akademik Yazın Üzerine Bir Araştırma. *Süleyman Demirel Üniversitesi Vizyoner Dergisi*, 14(38), 620–630.

Goff, L., Zarin, H., & Goodman, S. (2012). Climate-induced migration from Northern Africa to Europe: Security challenges and opportunities. *The Brown Journal of World Affairs*, 18(2), 195–213.

Gogoi, M., Buragohain, P. P., & Gogoi, P. (2022). Technical efficiency of organic tea growers of assam, India: A study in Dibrugarh district. *Estudios de Economía Aplicada*, 40(2), 1–10. DOI: 10.25115/eea.v40i2.6438

Goldar, B., & Sharma, A. K. (2015). Foreign investment in Indian industrial firms and impact on firm performance. *The Journal of Industrial Statistics*, 4(1), 1–18.

Gordon, D. J. (2016). Lament for a network? Cities and networked climate governance in Canada. *Environment and Planning. C, Government & Policy*, 34(3), 529–545. DOI: 10.1177/0263774X15614675

Gordon, R. (2007). Climate change and the poorest nations: Further reflections on global inequality. *U. Colo. L. Rev.*, 78, 1559.

Goudet, J., Binte Arif, F., Owais, H., Uddin Ahmed, H., & Ridde, V. (2024). Climate change and women's mental health in two vulnerable communities of bangladesh: An ethnographic study. *PLOS Global Public Health*, 4(6), e0002080. DOI: 10.1371/journal.pgph.0002080 PMID: 38935627

Gough, N. (2013). Thinking globally in environmental education. In Stevenson, R. B. (Eds.), *International Handbook of Research on Environmental Education* (pp. 33–44). American Educational Research Association. DOI: 10.4324/9780203813331-3

Green, F., & Healy, N. (2022). How inequality fuels climate change: The climate case for a Green New Deal. *One Earth*, 5(6), 635–649. DOI: 10.1016/j.oneear.2022.05.005

Greiving, S. (2013). ESPON CLIMATE-Climate Change and Territorial Effects on Regions and Local Economies. Applied Research Project, 1(4). https://www.espon.eu/export/sites/default/Documents/Projects/AppliedResearch/CLIMATE/ESPON_Climate_Final_Report-Part_C-ScientificReport.pdf

Groves, F. H., & Pugh, A. F. (2002). Cognitive Illusions as Hindrances to Learning Complex Environmental Issues. *Journal of Science Education and Technology*, 11(4), 381–390. Retrieved April 19, 2023, from. DOI: 10.1023/A:1020694319071

Grubb, M. (2004). Technology innovation and climate change policy: An overview of issues and options. *Keio Economic Studies*, 41(2), 103–132.

Guan, Z., Hossain, M. R., Sheikh, M. R., Khan, Z., & Gu, X. (2023). Unveiling the interconnectedness between energy-related GHGs and pro-environmental energy technology: Lessons from G-7 economies with MMQR approach. *Energy*, 281, 128234. DOI: 10.1016/j.energy.2023.128234

Gunasiri, H., Wang, Y., Watkins, E., Capetola, T., Henderson-Wilson, C., & Patrick, R. (2022). Hope, coping and eco-anxiety: Young people's mental health in a climate-impacted australia. *International Journal of Environmental Research and Public Health*, 19(9), 5528. DOI: 10.3390/ijerph19095528 PMID: 35564923

Guo, H. D., Zhang, L., & Zhu, L. W. (2015). Earth observation big data for climate change research. *Advances in Climate Change Research*, 6(2), 108–117. DOI: 10.1016/j.accre.2015.09.007

Haer, T., Botzen, W., Roomen, V., Connor, H., Zavala-Hidalgo, J., Eilander, D., & Ward, P. (2018). Coastal and river flood risk analyses for guiding economically optimal flood adaptation policies: A country-scale study for Mexico. *Philosophical Transactions. Series A, Mathematical, Physical, and Engineering Sciences*, 376(2121), 20170329. DOI: 10.1098/rsta.2017.0329 PMID: 29712799

Hailemariam, A., Dzhumashev, R., & Shahbaz, M. (2020). Carbon emissions, income inequality and economic development. *Empirical Economics*, 59(3), 1139–1159. DOI: 10.1007/s00181-019-01664-x

Handy, S. L., & Mokhtarian, P. L. (1995). Planning for telecommuting measurement and policy issues. *Journal of the American Planning Association*, 61(1), 99–111. DOI: 10.1080/01944369508975623

Hansen, J. (2012). Scientific Case for Avoiding Dangerous Climate Change to Protect Young People and Nature. Proceedings of the National Academy of Sciences (Submitted paper), http://arxiv.org/ftp/arxiv/papers/1110/1110.1365.pdf

Hansen, J., Sato, M., Kharecha, P., Beerling, D., Berner, R., Masson-Delmotte, V., Pagani, M., Raymo, M., Royer, D. L., & Zachos, J. C. (2008). Target Atmospheric CO2: Where Should Humanity Aim? *The Open Atmospheric Science Journal*, 2(1), 217–231. DOI: 10.2174/1874282300802010217

Hansen, N. M. (1965). Unbalanced Growth and Regional Development. *Economic Inquiry*, 4(1), 3–14. DOI: 10.1111/j.1465-7295.1965.tb00931.x

Hao, Y., Huang, Z., & Wu, H. (2019). Do carbon emissions and economic growth decouple in China? An empirical analysis based on provincial panel data. *Energies*, 12(12), 2411. DOI: 10.3390/en12122411

Harlan, S. L., Pellow, D. N., Roberts, J. T., Bell, S. E., Holt, W. G., Nagel, J., & Brulle, R. J. (2015). Climate justice and inequality. *Climate change and society. Sociological Perspectives*, 2015, 127–163.

Harris, J. (Ed.). (2001). *A survey of Sustainable Development: Social and Economic Dimensions* (Vol. 6). Island Press.

Hartas, D. (2010). *Educational research and inquiry: Qualitative and quantitative approaches*. Continuum International. DOI: 10.5040/9781474243834

Hartmann, T., & Spit, T. (2015). Implementing the european flood risk management plan. *Journal of Environmental Planning and Management*, 59(2), 360–377. DOI: 10.1080/09640568.2015.1012581

Hayes, K., & Poland, B. (2018). Addressing mental health in a changing climate: Incorporating mental health indicators into climate change and health vulnerability and adaptation assessments. *International Journal of Environmental Research and Public Health*, 15(9), 1806. DOI: 10.3390/ijerph15091806 PMID: 30131478

Heeren, A., Mouguiama-Daouda, C., & Contreras, A. (2021). On climate anxiety and the threat it may pose to daily life functioning and adaptation: a study among european and african french-speaking participants. DOI: 10.31234/osf.io/a69wp

Hei Ngu, L. (2024). Carbon capture technologies. *Encyclopedia of Sustainable Technologies (Second Edition)*.

Heidari, H., Arabi, M., & Warziniack, T. (2021). Effects of climate change on natural-caused fire activity in Western U.S national forests. *Atmosphere (Basel)*, 12(8), 981. DOI: 10.3390/atmos12080981

He, K., Ramzan, M., Awosusi, A. A., Ahmed, Z., Ahmad, M., & Altuntaş, M. (2021). Does Globalization Moderate the Effect of Economic Complexity on CO2 Emissions? Evidence From the Top 10 Energy Transition Economies. *Frontiers in Environmental Science*, 9, 778088. DOI: 10.3389/fenvs.2021.778088

Henderson, H. (1994). Paths to Sustainable Development: The Role of Social Indicators. *Futures*, 26(2), 125–137. DOI: 10.1016/0016-3287(94)90102-3

Herfort, B., Lautenbach, S., Porto de Albuquerque, J., Anderson, J., & Zipf, A. (2023). A spatio-temporal analysis investigating completeness and inequalities of global urban building data in OpenStreetMap. *Nature Communications*, 14(1), 3985. Advance online publication. DOI: 10.1038/s41467-023-39698-6 PMID: 37414776

Heyward, C. (2013). Situating and abandoning geoengineering: A typology of five responses to dangerous climate change. *PS, Political Science & Politics*, 46(01), 23–27. DOI: 10.1017/S1049096512001436

Hickel, J. (2020). The sustainable development index: Measuring the ecological efficiency of human development in the anthropocene. *Ecological Economics*, 167, 106331. DOI: 10.1016/j.ecolecon.2019.05.011

Hickman, C., Marks, E., Pihkala, P., Clayton, S., Lewandowski, R., Mayall, E., & Susteren, L. (2021). Climate anxiety in children and young people and their beliefs about government responses to climate change: A global survey. *The Lancet. Planetary Health*, 5(12), e863–e873. DOI: 10.1016/S2542-5196(21)00278-3 PMID: 34895496

Hillebrand, L., Marzini, S., Crespi, A., Hiltner, U., & Mina, M. (2023). Contrasting impacts of climate change on protection forests of the Italian Alps. *Frontiers in Forests and Global Change*, 6, 1240235. DOI: 10.3389/ffgc.2023.1240235

Hill, J., Alan, T., & Woodland, W. (2006). *Sustainable development*. Ashgate.

Hmelo-Silver, C. E., Duncan, R. G., & Chinn, C. A. (2007). Scaffolding and achievement in problem-based and inquiry learning: A response to Kirschner, Sweller, and Clark. *Educational Psychologist*, 42(2), 99–107. Retrieved April 23, 2023, from. DOI: 10.1080/00461520701263368

Holdren, J. P. (2006). The energy innovation imperative: Addressing oil dependence, climate change, and other 21st century energy challenges. *Innovations: Technology, Governance, Globalization*, 1(2), 3–23. DOI: 10.1162/itgg.2006.1.2.3

Houghton, J. (2004). *Global Warming* (3rd ed.). Cambridge University Press. DOI: 10.1017/CBO9781139165044

Howell, K. E. (2015). *Empiricism, positivism and post-positivism*. Sage.

Hsu, A., & Schletz, M. (2024). Digital technologies–the missing link between climate action transparency and accountability? *Climate Policy*, 24(2), 193–210. DOI: 10.1080/14693062.2023.2237937

Huang, X., Wang, S., Yang, D., Hu, T., Chen, M., Zhang, M., Zhang, G., Biljecki, F., Lu, T., Zou, L., Wu, C. Y. H., Park, Y. M., Li, X., Liu, Y., Fan, H., Mitchell, J., Li, Z., & Hohl, A. (2024). Crowdsourcing Geospatial Data for Earth and Human Observations: A Review. In *Journal of Remote Sensing (United States)* (Vol. 4). DOI: 10.34133/remotesensing.0105

Hubacek, K., Baiocchi, G., Feng, K., Muñoz Castillo, R., Sun, L., & Xue, J. (2017). Global carbon inequality. *Energy, Ecology & Environment*, 2(6), 361–369. DOI: 10.1007/s40974-017-0072-9

Hughes, L., & Salinger, M. J. (2006). Climate Change and Heatwaves. *Australian Journal of Public Health*, 30(1), 1–8.

Hughes, T. P., Kerry, J. T., & Simpson, T. (2017). Global Warming and Reefs. *Science*, 359(6371), 158–160.

Hunter, L. (2018). *Carnivores of the world* (Vol. 117). Princeton University Press.

Hussain, S., Hussain, E., Saxena, P., Sharma, A., Thathola, P., & Sonwani, S. (2023). Navigating the impact of climate change in India: A perspective on climate action (SDG13) and sustainable cities and communities (SDG11). *Frontier in Sustainable Cities*, 5(1), 1–22. DOI: 10.3389/frsc.2023.1308684

Hutton, G. (2011). The economics of health and climate change: Key evidence for decision making. *Globalization and Health*, 7(1), 18. DOI: 10.1186/1744-8603-7-18 PMID: 21707990

IMF. (2024). *IMF data access to macroeconomic & financial data financial development indeks database.* https://data.imf.org/?sk=f8032e80-b36c-43b1-ac26-493c5b1cd33b

Im, K. S., Pesaran, M. H., & Shin, Y. (2003). Testing for unit roots in heterogeneous panels. *Journal of Econometrics*, 115(1), 53–74. DOI: 10.1016/S0304-4076(03)00092-7

Imperatives, S. (1987), Report of The World Commission on Environment and Development: Our Common Future.

Innocenti, M., Santarelli, G., Lombardi, G., Ciabini, L., Zjalic, D., Russo, M., & Cadeddu, C. (2023). How can climate change anxiety induce both pro-environmental behaviours and eco-paralysis? the mediating role of general self-efficacy. *International Journal of Environmental Research and Public Health*, 20(4), 3085. DOI: 10.3390/ijerph20043085 PMID: 36833780

Institute for Local Government. (2024). *Climate action.* Retrieved May 15, 2024, from https://www.ca-ilg.org/climate-action

Institute for Security Studies (ISS). (2010). *Climate change and natural resources. Conflicts in Africa.* Available from: https://issafrica.org/research/monographs/climate-change-and-natural-resources-conflicts-in-africa

Intergovernmental Panel on Climate Change. (2014). *Climate change report 2014: Impacts, adaptation and vulnerability. Fifth assessment report.* New York: Cambridge University Press.

Intergovernmental Panel on Climate Change. (2018). Global warming of 1.5°C. Retrieved from https://www.ipcc.ch/sr15/

IPCC. (1988). *Report of the Intergovernmental Panel on Climate Change.* United Nations Environment Programme (UNEP) and World Meteorological Organization (WMO).

IPCC. (1996). *Climate change 1995 - Impacts, adaptations and mitigation of climate change: Scientific-technical analyses contribution of working group ii to the second assessment report of the intergovernmental panel on climate change.* Cambridge University Press. Retrieved July 3, 2024, from https://www.ipcc.ch/site/assets/uploads/2018/03/ipcc_sar_wg_II_full_report.pdf

IPCC. (2007). Summary for Policymakers. A report of Working Group I of the Intergovernmental Panel on Climate Change, https://www.ipcc.ch/pdf/assessmentreport/ar4/wg1/ar4-wg1-spm.pdf

IPCC. (2014). *Climate change 2014 synthesis report, contribution of working groups I, II and III to the fifth assessment report of the ıntergovernmental panel on climate change*. IPCC, Geneva, Switzerland. Retrieved April 9, 2024, from https://www.ipcc.ch/site/assets/uploads/2018/02/SYR_AR5_FINAL_full.pdf

IPCC. (2014a). AR5 Working Group II. *Climatic Change*.

IPCC. (2014b). AR5 Working Group III. *Climatic Change*.

IPCC. (2021). *Climate Change 2021: The Physical Science Basis*. Intergovernmental Panel on Climate Change (IPCC).

IPCC. (2021). *The Physical Science Basis*. Intergovernmental Panel on Climate Change.

IPCC. (2022). *IPCC's WG II Sixth assessment report -WGII AR6, Climate change 2022: Impacts, adaptation and vulnerability working group II contribution to the sixth assessment report of the intergovernmental panel on climate change* Cambridge University Press, Cambridge, UK and New York, NY, USA. Retrieved April 22, 2024, from https://report.ipcc.ch/ar6/wg2/IPCC_AR6_WGII_FullReport.pdfDOI: 10.1017/9781009325844

IPCC. (2023). https://www.ipcc.ch/report/ar6/syr/downloads/report/IPCC_AR6_SYR_LongerReport.pdf

IPCC. (2024). *Climate change widespread, rapid, and intensifying*. https://www.ipcc.ch/2021/08/09/ar6-wg1-20210809-pr/

Iravani, A., Akbari, M. H., & Zohoori, M. (2017). Advantages and Disadvantages of Green Technology; Goals, Challenges and Strengths. *Int J Sci Eng Appl*, 6(9), 272–284. DOI: 10.7753/IJSEA0609.1005

Islam, N., & Winkel, J. (2017). Climate change and social inequality.

Islam, M. R., & Hasan, M. (2016). Climate-induced human displacement: A case study of Cyclone Aila in the south-west coastal region of Bangladesh. *Natural Hazards*, 81(2), 1051–1071. DOI: 10.1007/s11069-015-2119-6

Islam, M. R., & Shamsuddoha, M. (2017). Socioeconomic consequences of climate induced human displacement and migration in Bangladesh. *International Sociology*, 32(3), 277–298. DOI: 10.1177/0268580917693173

İzol, R., & Kaval, F. (2023). Kopenhag Okulu Bağlamında Türkiye'deki Siyasi Aktörlerin İklim Değişikliğine Yönelik Söylemlerinin Analizi. *Akdeniz İİBF Dergisi*, 23(1), 18–32.

Jacobs, M. (2013), Green Growth. *The Handbook of Global Climate and Environment Policy*, 197-214.

Jacobsen, K., & Landau, L. B. (2003). The dual imperative in refugee research: Some methodological and ethical considerations in social science research on forced migration. *Disasters*, 27(3), 185–206. DOI: 10.1111/1467-7717.00228 PMID: 14524045

Jahanger, A., Zaman, U., Hossain, M. R., & Awan, A. (2023). Articulating CO_2 emissions limiting roles of nuclear energy and ICT under the EKC hypothesis: An application of non-parametric MMQR approach. *Geoscience Frontiers*, 14(5), 101589. DOI: 10.1016/j.gsf.2023.101589

Jänicke, M. (2012). "Green Growth": From A Growing Eco-Industry to Economic Sustainability. *Energy Policy*, 48, 13–21. DOI: 10.1016/j.enpol.2012.04.045

Janicke, M., Binder, M., & Mönch, H. (1997). Dirty industries: Patterns of change in industrial countries. *Environmental and Resource Economics*, 9(4), 467–491. DOI: 10.1007/BF02441762

Jayawardhan, S. (2017). Vulnerability and climate change induced human displacement. *Consilience*, (17), 103–142.

Jeong, H. (2006). *Globalisation and the physical environment*. CHP.

Jeon, H. (2022). CO_2 emissions, renewable energy and economic growth in the US. *The Electricity Journal*, 35(7), 107170. DOI: 10.1016/j.tej.2022.107170

Jimoh, M. Y., Bikam, P., & Chikoore, H. (2021). The influence of socioeconomic factors on households' vulnerability to climate change in semiarid towns of Mopani, South Africa. *Climate (Basel)*, 9(1), 13. DOI: 10.3390/cli9010013

Jochem, E., & Madlener, R. (2003, July). The forgotten benefits of climate change mitigation: Innovation, technological leapfrogging, employment, and sustainable development. In *Workshop on the benefits of climate policy: Improving information for policy makers*. OECD Paris.

John, D., & Derakhshi, E. (2022). Low carbon mobility transitions and justice: A case of costa rica. *Development*, 65(1), 71–77. DOI: 10.1057/s41301-022-00331-6 PMID: 35250210

John, L. C. K., Mei, W. S., & Guang, Y. (2013). EE policies in three Chinese communities. In *International Handbook of 390 Research on Environmental Education* (pp. 178–188). American Educational Research Association. DOI: 10.4324/9780203813331-35

Johnson, B., & Christensen, L. (2008). *Educational research: Quantitative, qualitative, and mixed approaches*. Sage.

Joung, C. B., Carrell, J., Sarkar, P., & Feng, S. C. (2013). Categorisation of Indicators for Sustainable Manufacturing. *Ecological Indicators*, 24, 148–157. DOI: 10.1016/j.ecolind.2012.05.030

Jyoti, B., & Singh, A. K. (2020). Projected sugarcane yield in different climate change scenarios in Indian states: A state-wise panel data exploration. *International Journal of Food and Agricultural Economics*, 8(4), 343–365. https://www.foodandagriculturejournal.com/vol8.no4.pp343.pdf

Jyoti, B., Singh, A. K., & Ashraf, S. N. (2023). Appropriate technologies and their implications in the agricultural sector. In Alex Khang, P. H. (Ed.), *Advanced Technologies and AI-Equipped IoT Applications in High-Tech Agriculture* (pp. 65–87). IGI Global. DOI: 10.4018/978-1-6684-9231-4.ch004

Kabir, M., Habiba, U. E., Khan, W., Shah, A., Rahim, S., los Rios-Escalante, P. R. D., Farooqi, Z. U. R., & Ali, L. (2023). Climate change due to increasing concentration of carbon dioxide and its impacts on environment in 21st century; a mini review. In *Journal of King Saud University - Science* (Vol. 35, Issue 5). DOI: 10.1016/j.jksus.2023.102693

Kadioglu, I., & Farooq, S. (2017). Potential distribution of sterile oat (Avena sterilis L.) in Turkey under changing climate. *Turkish journal of weed science, 20*(2), 1-13.

Kahraman, S., & Şenol, P. (2018). İklim Değişikliği: Küresel, Bölgesel ve Kentsel Etkileri. *Akademia Sosyal Bilimler Dergisi*, 1, 353–370.

Kais, S., & Sami, H. (2016). An econometric study of the impact of economic growth and energy use on carbon emissions: Panel data evidence from fifty eight countries. *Renewable & Sustainable Energy Reviews*, 59, 1101–1110. DOI: 10.1016/j.rser.2016.01.054

Kälin, W. (2010). Conceptualising climate-induced displacement. *Climate change and displacement: Multidisciplinary perspectives, 81*, 102.

Kan, S., Chen, B., Meng, J., & Chen, G. (2019). An extended overview of natural gas use embodied in the world economy and supply chains: Policy implications from a time series analysis. Energy Policy, 137, 111068. DOI: 10.1016/j.enpol.2019.111068

Kandpal, A., Kumara, K., Sendhil, R., & Balaji, S. J. (2022). Technical efficiency in Indian wheat production: regional trends and way forward. In Kashyap, P. L., Gupta, V., Gupta, O. P., Sendhi, R., Gopalareddy, K., Jasrotia, P., & Singh, G. P. (Eds.), *New Horizons in Wheat and Barley* (pp. 475–490). Spinger Nature. DOI: 10.1007/978-981-16-4134-3_17

Kaplan, O. (2023). Türk Hukukunda İdarenin Yenilenebilir Enerji Kaynaklarından Elektrik Enerjisi Üretimi Yönünden İşlevlerinin İrdelenmesi. *Yaşar Hukuk Dergisi*, 5(2), 297–336.

Karaca, C. (2012). The Relationship Between Economic Development and Environmental Pollution: An Empirical Analysis on Developing Countries. *Çukurova University Journal of Social Sciences Institute, 21*(3), 139–156.

Karadeniz, C.B., Sari, S., & Çağlayan, A. B. (2018). İklim değişikliğinin Doğu Karadeniz turizmine olası etkileri. *Uluslararası Bilimsel Araştırmalar Dergisi (IBAD),* 170-179.

Karakaya, E., & Özçağ, M. (2001). Sürdürülebilir Kalkınma ve İklim Değişikliği: Uygulanabilecek İktisadi Araçların Analizi. First Conference in Fiscal Policy and Transition Economies, University of Manas, 1-7.

Karakaya, E. (2016). Paris İklim Anlaşması: İçeriği Ve Türkiye Üzerine Bir Değerlendirme. *Adnan Menderes Üniversitesi Sosyal Bilimler Enstitüsü Dergisi*, 3(1), 1–12. DOI: 10.30803/adusobed.188842

Kasztelan, A. (2017). Green Growth, Green Economy and Sustainable Development: Terminological and Relational Discourse. *Prague Economic Papers*, 26(4), 487–499. DOI: 10.18267/j.pep.626

Kathayat, B., Dixit, A. K., & Chandel, B. S. (2021). Inter-state variation in technical efficiency and total factor productivity of India's livestock sector. *Agricultural Economics Research Review*, 34(3), 59–72. DOI: 10.5958/0974-0279.2021.00015.X

Kaya, H. E. (2020). Kyoto'dan Paris'e Küresel İklim Politikaları. *Meriç Uluslararası Sosyal ve Stratejik Araştırmalar Dergisi*, 4(10), 165–191.

Kayigema, V., & Rugege, D. (2014). Women's perceptions of the girinka (one cow per poor family) programme, poverty alleviation and climate resilience in rwanda. *Agenda (Durban, South Africa)*, 28(3), 53–64. DOI: 10.1080/10130950.2014.939839

Keeling, C. D., Whorf, T. P., & Revelle, R. (1976). *Atmospheric Carbon Dioxide Concentrations and Their Changes: A Historical Overview*. Carbon Dioxide and Climate: A Scientific Assessment, 7-44.

Kellie-Smith, O., & Cox, P. (2011). Emergent dynamics of the climate–economy system in the Anthropocene. *Philosophical Transactions of the Royal Society A: Mathematical, Physical and Engineering Sciences, 369*(1938), 868-886. DOI: 10.1098/rsta.2010.0305

Kenner, D. (2019). *Carbon inequality: The role of the richest in climate change.* Routledge. DOI: 10.4324/9781351171328

Keskin, T. (2007). Enerji Verimliliği Kanunu ve Uygulama Süreci. *Mühendis ve Makina*, (569), 106–112.

Khalid, A. A., Mahmood, F., & Rukh, G. (2016). Impact of climate changes on economic and agricultural value added share in GDP. *Asian Management Research Journal*, 1(1), 35–48.

Khan, M., & Ozturk, I. (2021). Examining the direct and indirect effects of financial development on CO2 emissions for 88 developing countries. *Journal of Environmental Management*, 293, 112812. DOI: 10.1016/j.jenvman.2021.112812 PMID: 34058453

Khatibi, F. S., Dedekorkut-Howes, A., Howes, M., & Torabi, E. (2021). Can public awareness, knowledge and engagement improve climate change adaptation policies? *Discover Sustainability*, 2(1), 18. Retrieved April 22, 2023, from. DOI: 10.1007/s43621-021-00024-z

Khoshnood, K. (2018). Methodological and ethical challenges in research with forcibly displaced populations. *The Health of Refugees: Public Health Perspectives from Crisis to Settlement*, 209.

Kim, S. (2022). The effects of information and communication technology, economic growth, trade openness, and renewable energy on CO_2 emissions in OECD countries. *Energies*, 15(7), 2517. DOI: 10.3390/en15072517

King, R. G., & Levine, R. (1994, June). Capital Fundamentalism, Economic Development, and Economic Growth. In Carnegie-Rochester Conference Series on Public Policy. North-Holland. DOI: 10.1016/0167-2231(94)90011-6

Kjellstrom, T., Kovats, R. S., Lloyd, S. J., Holt, T., & Tol, R. S. J. (2009). The Direct Impact of Climate Change on Regional Labor Productivity. *Archives of Environmental & Occupational Health*, 64(4), 217–227. DOI: 10.1080/19338240903352776 PMID: 20007118

Klein, R. J., & Nicholls, R. J. (2004). *Coastal Vulnerability and Sea-Level Rise: The Role of Coastal Defenses.* Coastal Systems and Continental Margins, 123-134.

Klein, R. J., Schipper, E. L. F., & Dessai, S. (2005). Integrating mitigation and adaptation into climate and development policy: Three research questions. *Environmental Science & Policy*, 8(6), 579–588. DOI: 10.1016/j.envsci.2005.06.010

Kłopotek, M. (2017). The advantages and disadvantages of remote working from the perspective of young employees. *Organizacja i Zarządzanie: kwartalnik naukowy*, (4), 39-49.

Klusak, P., Agarwala, M., Burke, M., Kraemer, M., & Mohaddes, K. (2021). Rising temperatures, falling ratings: The effect of climate change on sovereign creditworthiness. SSRN *Electronic Journal*. DOI: 10.2139/ssrn.3811958

Koçak, E., Ulucak, R., & Ulucak, Z. Ş. (2020). The impact of tourism developments on CO_2 emissions: An advanced panel data estimation. *Tourism Management Perspectives*, 33, 100611. DOI: 10.1016/j.tmp.2019.100611

Koçaslan, G. (2014). Türkiye'nin Enerji Verimliliği Mevzuatı, Avrupa Birliği'ndeki Düzenlemeler ve Uluslararası-Ulusal Öneriler. C.Ü. İktisadi ve İdari Bilimler Dergisi, 15(2), 117-133.

Kocornik-Mina, A., & Rodriguez-Vega, L. (2018). Transport and climate change: A review. *Transport Policy*, 67, 101–114. DOI: 10.1016/j.tranpol.2017.05.009

Koenker, R., & Basset, G. S.Jr. (1978). Regression quantiles. *Econometrica*, 46(1), 33–50. DOI: 10.2307/1913643

Komolafe, A. A., Adegboyega, S. A., Anifowose, A. Y. B., Akinluyi, F. O., & Awoniran, D. R. (2014). Air pollution and climate change in Lagos, Nigeria: Needs for 391 proactive approaches to risk management and adaptation. *American Journal of Environmental Sciences*, 10(4), 412–423. DOI: 10.3844/ajessp.2014.412.423

Köse, İ. (2018). İklim Değişikliği Müzakereleri: Türkiye'nin Paris Anlaşmasını İmza Süreci. *Ege Stratejik Araştırmalar Dergisi*, 9(1), 55–81.

Koutsias, N., Arianoutsou, M., Kallimanis, A. S., Mallinis, G., Halley, J. M., & Dimopoulos, P. (2012). Where did the fires burn in Peloponnisos, Greece the summer of 2007? Evidence for a synergy of fuel and weather. *Agricultural and Forest Meteorology*, 156, 41–53. DOI: 10.1016/j.agrformet.2011.12.006

Krajcik, J., Czerniak, C., & Berger, C. (2014). *Teaching science in elementary and middle school: A project-based approach*. Routledge. DOI: 10.4324/9780203113660

Kristensen, H. S., & Mosgaard, M. A. (2020). A Review of Micro Level Indicators for A Circular Economy-Moving Away From The Three Dimensions of Sustainability? *Journal of Cleaner Production*, 243, 118531. DOI: 10.1016/j.jclepro.2019.118531

Kulkarni, S., Hof, A., Wijst, K., & Vuuren, D. (2022). Disutility of climate change damages warrants much stricter climate targets. *Preprint*. DOI: 10.21203/rs.3.rs-1788130/v1

Kumar, A. (2019). Globalization and Environmental Impacts: Challenges and Opportunities.

Kumar, R. A., & Paul, M. (2019). *Industry level analysis of productivity growth under market imperfections.* Working Paper, 207, Institute for Studies in Industrial Development, New Delhi.

Kumar, A., & Sharma, P. (2014). Climate change and sugarcane productivity in India: An econometric analysis. *Journal of Social and Development Sciences*, 5(2), 111–122. DOI: 10.22610/jsds.v5i2.811

Kumar, A., Sharma, P., & Ambrammal, S. K. (2015). Climatic effects on sugarcane productivity in India: A stochastic production function application. *International Journal of Economics and Business Research*, 10(2), 179–203. DOI: 10.1504/IJEBR.2015.070984

Kumar, P., & Balaji, N. (2019). Assessing the impact of variations in weather conditions on socio-economic status. *International Journal of Social Sciences & Economic Environment*, 4(2), 12–17. DOI: 10.53882/IJSSEE.2019.0402003

Kumar, P., Brander, L., Kumar, M., & Cuijpers, P. (2023). Planetary health and mental health nexus: Benefit of environmental management. *Annals of Global Health*, 89(1), 49. DOI: 10.5334/aogh.4079 PMID: 37521755

Kumar, S., & Arora, N. (2012). Evaluation of technical efficiency in Indian sugar industry: An application of full cumulative data envelopment analysis. *Eurasian Journal of Business and Economics*, 5(9), 57–78.

Kuznets, S. (1955). Economic growth and income inequality. *The American Economic Review*, 45(1), 1–28.

Laato, S., & Tregel, T. (2023). Into the Unown: Improving location-based gamified crowdsourcing solutions for geo data gathering. *Entertainment Computing*, 46, 100575. Advance online publication. DOI: 10.1016/j.entcom.2023.100575

Læsse, J. (2010). Education for Sustainable Development, Participation and Socio-Cultural Change. *Environmental Education Research*, 16(1), 39–57. DOI: 10.1080/13504620903504016

Larson, D. A., & Wilford, W. T. (1979). The Physical Quality of Life Index: A Useful Social Indicator? *World Development*, 7(6), 581–584. DOI: 10.1016/0305-750X(79)90094-9

Lawrence, S., López Ventura, J., Doody, L., & Peracio, P. (2017). Polisdigitocracy: Citizen engagement for climate action through digital technologies. *Field Actions Science Reports. The journal of field actions*, (Special Issue 16), 58-65.

Le Quéré, C., Jackson, R. B., Jones, M. W., Smith, A. J., Abernethy, S., Andrew, R. M., De-Gol, A. J., Willis, D. R., Shan, Y., Canadell, J. G., Friedlingstein, P., Creutzig, F., & Peters, G. P. (2020). Temporary reduction in daily global CO2 emissions during the COVID-19 forced confinement. *Nature Climate Change*, 10(7), 647–653. DOI: 10.1038/s41558-020-0797-x

Ledoux, L., Mertens, R., & Wolff, P. (2005, November). EU Sustainable Development Indicators: An Overview. *Natural Resources Forum*, 29(4), 392–403. DOI: 10.1111/j.1477-8947.2005.00149.x

Leedy, P. D., & Ormrod, J. E. (2013). *Practical Research: Planning and Design*. Prentice Hall.

Letcher, T. M. (2019). *Climate Change: Observed Impacts on Planet Earth*. Elsevier.

Levin, E., Beisekenov, N., Wilson, M., Sadenova, M., Nabaweesi, R., & Nguyen, L. (2023). Empowering Climate Resilience: Leveraging Cloud Computing and Big Data for Community Climate Change Impact Service (C3IS). *Remote Sensing (Basel)*, 15(21), 5160. Advance online publication. DOI: 10.3390/rs15215160

Lewis, W. A. (2013). *Theory of Economic Growth*. Routledge. DOI: 10.4324/9780203709665

Li, J., & Sheng, Y. (2012). An automated scheme for glacial lake dynamics mapping using Landsat imagery and digital elevation models: A case study in the Himalayas. *Int J Remote Sens*, 33: 5194–5213. DOI: 10.1080/01431161.2012.657370

Lin, H., Wang, X., Bao, G., & Xiao, H. (2022). Heterogeneous spatial effects of FDI on CO2 emissions in China. *Earth's Future, 10*(1), e2021EF002331. .DOI: 10.1029/2021EF002331

Linde, A., Bubeck, P., Dekkers, J., Moel, H., & Aerts, J. (2011). Future flood risk estimates along the river rhine. *Natural Hazards and Earth System Sciences*, 11(2), 459–473. DOI: 10.5194/nhess-11-459-2011

Lipper, L., Thornton, P., Campbell, B. M., Baedeker, T., Braimoh, A., Bwalya, M., Caron, P., Cattaneo, A., Garrity, D., Henry, K., Hottle, R., Jackson, L., Jarvis, A., Kossam, F., Mann, W., McCarthy, N., Meybeck, A., Neufeldt, H., Remington, T., & Torquebiau, E. F. (2014). Climate-smart agriculture for food security. *Nature Climate Change*, 4(12), 1068–1072. DOI: 10.1038/nclimate2437

Liu, K., & Baskaran, B. (2003). Thermal performance of green roofs through field evaluation Proceedings of 1st North American Green Roof Conference: Greening Rooftops for Sustainable Communities, Chicago, The Cardinal Group Toronto.

Liu, F., Shafique, M., & Luo, X. (2023). Literature review on life cycle assessment of transportation alternative fuels. *Environmental Technology & Innovation*, 32, 103343. DOI: 10.1016/j.eti.2023.103343

Liu, J., Potter, T., & Zahner, S. (2020). Policy brief on climate change and mental health/well-being. *Nursing Outlook*, 68(4), 517–522. DOI: 10.1016/j.outlook.2020.06.003 PMID: 32896304

Liu, Y., Ruiz-Menjivar, J., Zavala, M., & Zhang, J. (2023). Examining the effects of climate change adaptation on technical efficiency of rice production. *Mitigation and Adaptation Strategies for Global Change*, 28(55), 1–17. DOI: 10.1007/s11027-023-10092-3

Liu, Z. (2021). Identifying urban land use social functional units: A case study using OSM data. *International Journal of Digital Earth*, 14(12), 1798–1817. Advance online publication. DOI: 10.1080/17538947.2021.1988161

Local Governments for Sustainability. (2024). *Our pathways, our approach*. Retrieved May 21, 2024, from https://iclei.org/our_approach/

Lomborg, B. (2020). Welfare in the 21st century: Increasing development, reducing inequality, the impact of climate change, and the cost of climate policies. *Technological Forecasting and Social Change*, 156, 119981. DOI: 10.1016/j.techfore.2020.119981

Lorek, S., & Spangenberg, J. H. (2014). Sustainable Consumption Within A Sustainable Economy-Beyond Green Growth and Green Economies. *Journal of Cleaner Production*, 63, 33–44. DOI: 10.1016/j.jclepro.2013.08.045

Luo, B., Khan, A. A., Wu, X., & Li, H. (2023). Navigating carbon emissions in G-7 economies: A quantile regression analysis of environmental-economic interplay. *Environmental Science and Pollution Research International*, 30(47), 104697–104712. DOI: 10.1007/s11356-023-29722-z PMID: 37707736

Lv, Z., & Li, S. (2021). How financial development affects CO_2 emissions: A spatial econometric analysis. *Journal of Environmental Management*, 277, 111397. DOI: 10.1016/j.jenvman.2020.111397 PMID: 33039704

Lybbert, T., & Sumner, D. (2010). Agricultural technologies for climate change mitigation and adaptation in developing countries: policy options for innovation and technology diffusion.

Lyster, R., & Burkett, M. (2018). Climate-induced displacement and climate disaster law: Barriers and opportunities. In *Research handbook on climate disaster law* (pp. 97–114). Edward Elgar Publishing. DOI: 10.4337/9781786430038.00012

Lyytimäki, J., Antikainen, R., Hokkanen, J., Koskela, S., Kurppa, S., Känkänen, R., & Seppälä, J. (2018). Developing Key Indicators of Green Growth. *Sustainable Development (Bradford)*, 26(1), 51–64. DOI: 10.1002/sd.1690

Machado, J. A. F., & Santos Silva, J. M. C. (2019). Quantiles via moments. *Journal of Econometrics*, 213(1), 145–173. DOI: 10.1016/j.jeconom.2019.04.009

Ma, D., Sandberg, M., & Jiang, B. (2015). Characterizing the heterogeneity of the openstreetmap data and community. *ISPRS International Journal of Geo-Information*, 4(2), 535–550. Advance online publication. DOI: 10.3390/ijgi4020535

Mahajan, V., Nauriyal, D. K., & Singh, S. P. (2014). Efficiency and ranking of Indian pharmaceutical industry: Does type of ownership matter? *Eurasian Journal of Business and Economics*, 7(14), 29–50. DOI: 10.17015/ejbe.2014.014.02

Mahato, A. (2014). Climate Change and its Impact on Agriculture. *International Journal of Scientific and Research Publications*, 4(4), 1–6.

Mahmood, H., Maalel, N., & Zarrad, O. (2019). Trade openness and CO2 emissions: Evidence from Tunisia. *Sustainability (Basel)*, 11(12), 3295. DOI: 10.3390/su11123295

Maity, S., & Singh, K. (2021). Frontier production functions, technical efficiency and panel data: With application to tea gardens in India. *International Journal of Business and Globalisation*, 27(4), 571–591. DOI: 10.1504/IJBG.2021.113797

Malthus, T. R. (1986). *An essay on the principle of population (1798)*. The Works of Thomas Robert Malthus, London, Pickering & Chatto Publishers, 1, 1-139.

Margery, P., Read, S., & Lepoutre, J. (2017). TerraCycle: Outsmarting waste. In *Case Studies in Social Entrepreneurship* (pp. 114-133). Routledge.

Maroof, Z., Hussain, S., Jawad, M., & Naz, M. (2019). Determinants of industrial development: A panel analysis of South Asian economies. *Quality & Quantity*, 53(1), 391–1419. DOI: 10.1007/s11135-018-0820-8

Martinho, V. J. P. D. (2021). Insights Into Circular Economy Indicators: Emphasising Dimensions of Sustainability. *Environmental and Sustainability Indicators*, 10, 100119. DOI: 10.1016/j.indic.2021.100119

Matos, S., Viardot, E., Sovacool, B. K., Geels, F. W., & Xiong, Y. (2022). Innovation and climate change: A review and introduction to the special issue. *Technovation*, 117, 102612. DOI: 10.1016/j.technovation.2022.102612

Ma, W., Nasriddinov, F., Haseeb, M., Ray, S., Kamal, M., Khalid, N., & Ur Rehman, M. (2022). Revisiting the impact of energy consumption, foreign direct investment, and geopolitical risk on CO_2 emissions: Comparing developed and developing countries. *Frontiers in Environmental Science*, 10, 985384. DOI: 10.3389/fenvs.2022.985384

Mayer, B. (2011). The international legal challenges of climate-induced migration: Proposal for an international legal framework. *Colo. J. Int'l Envtl. L. & Pol'y*, 22, 357.

Mayer, R. E. (2009). *Multimedia learning*. Cambridge University Press. DOI: 10.1017/CBO9780511811678

Mazumdar, M., Rajeev, M., & Ray, S. C. (2009). Output and input efficiency of manufacturing firms in India: A case of the Indian pharmaceutical sector. Working Paper No. 219, The Institute for Social and Economic Change, Bangalore, India.

MC. C. (2020). *Perché l'inquinamento da Pm10 può agevolare la diffusione del virus*. https://www.ilsole24ore.com/art/l-inquinamentoparticolato-ha-agevolato-diffusionecoronavirus-ADCbb0D

McGough, A., Kavak, H., & Mahabir, R. (2024). Is more always better? Unveiling the impact of contributor dynamics on collaborative mapping. *Computational & Mathematical Organization Theory*, 30(2), 173–186. Advance online publication. DOI: 10.1007/s10588-023-09383-6

McMichael, A., McMichael, C., Berry, H., & Bowen, K. (2010). *Climate-related displacement: health risks and responses. Climate Change and Population Displacement: Multidisciplinary Perspectives*. Hart Publishing Ltd.

Meadows, D. H., Randers, J., & Meadows, D. L. (2013). The limits to growth (1972). In *The future of nature* (pp. 101–116). Yale University Press. DOI: 10.2307/j.ctt5vm5bn.15

Mehedi, H., Nag, A. K., & Farhana, S. (2010). *Climate Induced Displacement*. Case Study of Cyclone Aila in the Southwest Coastal Region of Bangladesh.

Mehta, Y., & Johan, R. A. (2017). Manufacturing sectors in India: Outlook and challenges. *Procedia Engineering*, 174(1), 90–104. DOI: 10.1016/j.proeng.2017.01.173

Mendelsohn, R., & Dinar, A. (2009). *Climate change and agriculture: An economic analysis of global impacts, adaptation and distributional effects*. Edward Elgar Publishing. DOI: 10.4337/9781849802239

Meşhur, H. F. A. (2010). A research on the attitudes of organizations towards telework. *Dokuz Eylül University Journal of Faculty of Economics and Administrative Sciences*, 25(1), 1–24.

MFA. (2023). *History of Türkiye- EU relations*, Retrieved July 03, 2024 from https://www.ab.gov.tr/brief-history_111_en.html

Mikayilov, J. I., Galeotti, M., & Hasanov, F. J. (2018). The impact of economic growth on CO2 emissions in Azerbaijan. *Journal of Cleaner Production*, 197, 1558–1572. DOI: 10.1016/j.jclepro.2018.06.269

Millward-Hopkins, J., & Oswald, Y. (2023). Reducing global inequality to secure human wellbeing and climate safety: A modelling study. *The Lancet. Planetary Health*, 7(2), e147–e154. DOI: 10.1016/S2542-5196(23)00004-9 PMID: 36754470

Ministry of Development. (2013). Tenth Development Plan (2013-2018).

Ministry of Foreign Affairs of the Republic of Turkey. (2024). *Kyoto protokolü*. Retrieved April 20, 2024, from https://www.mfa.gov.tr/kyoto-protokolu.tr.mfa

Ministry of Foreign Affairs of the Republic of Turkey. (2024). *Viyana sözleşmesi ve Montreal protokolü*. Retrieved April 20, 2024, from https://www.mfa.gov.tr/viyana-sozlesmesi-ve-montreal-protokolu.tr.mfa

Mitra, A., Sharma, C., & Véganzonès-Varoudakis, M. A. (2016). Infrastructure, ICT and firms' productivity and efficiency: An application to the Indian manufacturing. In De Beule, F., & Narayanan, K. (Eds.), *Globalization of Indian Industries*. India Studies in Business and Economics. DOI: 10.1007/978-981-10-0083-6_2

Monroe, M. C., Plate, R. R., Oxarart, A., Bowers, A., & Chaves, W. A. (2019). Identifying effective climate change education strategies: A systematic review of the research. *Environmental Education Research*, 25(6), 791–812. DOI: 10.1080/13504622.2017.1360842

Mooney, P., & Corcoran, P. (2014). Analysis of interaction and co-editing patterns amongst openstreetmap contributors. *Transactions in GIS*, 18(5), 633–659. Advance online publication. DOI: 10.1111/tgis.12051

Moon, J. A. (2004). *Reflection in learning and professional development: Theory and practice*. Routledge Falmer.

Moore, M., & Wesselbaum, D. (2023). Climatic factors as drivers of migration: A review. *Environment, Development and Sustainability*, 25(4), 2955–2975. DOI: 10.1007/s10668-022-02191-z

Moosa, I. A., & Smith, L. (2004). Economic Development Indicators as Determinants of Medal Winning At The Sydney Olympics: An Extreme Bounds Analysis. *Australian Economic Papers*, 43(3), 288–301. DOI: 10.1111/j.1467-8454.2004.00231.x

Moradi, M., Roche, S., & Mostafavi, M. A. (2023). Evaluating OSM Building Footprint Data Quality in Québec Province, Canada from 2018 to 2023: A Comparative Study. *Geomatics*, 3(4), 541–562. Advance online publication. DOI: 10.3390/geomatics3040029

Morris, M. D. (1978). A Physical Quality of Life Index. *Urban Ecology*, 3(3), 225–240. DOI: 10.1016/0304-4009(78)90015-3

Morshed, S. R., Fattah, M. A., Al Kafy, A., Alsulamy, S., Almulhim, A. I., Shohan, A. A. A., & Khedher, K. M. (2024). Decoding seasonal variability of air pollutants with climate factors: A geostatistical approach using multimodal regression models for informed climate change mitigation. *Environmental Pollution*, 345, 123463. Advance online publication. DOI: 10.1016/j.envpol.2024.123463 PMID: 38325513

Moshou, H., & Drinia, H. (2023). Climate Change Education and Preparedness of Future Teachers—A Review: The Case of Greece. *Sustainability (Basel)*, 15(2), 1177. DOI: 10.3390/su15021177

Mostaque, L. Y., & Hasan, S. (2015). Climate Change Induced Migration: Impact on Slumaisation. *Participatory Community Assessment for Priority Problem Diagnosis in Bajura District, Nepal: What Matters Most–Poverty or Climate Change?* 223.

Mühlhofer, E., Kropf, C. M., Riedel, L., Bresch, D. N., & Koks, E. E. (2024). OpenStreetMap for multi-faceted climate risk assessments. *Environmental Research Communications*, 6(1), 015005. Advance online publication. DOI: 10.1088/2515-7620/ad15ab

Mulenga, B. P., Wineman, A., & Sitko, N. J. (2017). Climate trends and farmers perceptions of climate change in Zambia. *Environmental Management*, 59(2), 291–306. DOI: 10.1007/s00267-016-0780-5 PMID: 27778064

Municipality Law No. 5393. (2005). Retrieved May 21, 2024, from https://www.mevzuat.gov.tr/mevzuatmetin/1.5.5393.pdf

Murphy, S. D. (2001). U.S. rejection of Kyoto Protocol process. *The American Journal of International Law*, 95(3), 647–650. DOI: 10.2307/2668508

Mutlu, M. Y., & Tezer, A. (2023). İklim Değişikliğine Mekânsal Uyum ve Azaltım Yaklaşımlarında Toprak Ekosistem Servislerinin Rolü. *Dirençlilik Dergisi*, 7(2), 305–324.

Myrdal, G. (1974). What is development? *Journal of Economic Issues*, 8(4), 729–736. DOI: 10.1080/00213624.1974.11503225

NASA. (2024). *Carbon dioxide*. Retrieved June 29, 2024, from https://climate.nasa.gov/vital-signs/carbon-dioxide/?intent=121

Naser, M. M., & Afroz, T. (2009). Human rights implications of climate change induced displacement. *Bond L. Rev.*, 21(3), i. DOI: 10.53300/001c.5543

Nath, S. K. (1962). The Theory of Balanced Growth. *Oxford Economic Papers*, 14(2), 138–153. DOI: 10.1093/oxfordjournals.oep.a040893

National Center for Science Education. (2019). Climate change in the classroom: A national survey of middle and high school science teachers. Available from: https://ncse.ngo/climate-change-classroom-national-survey-middle-and-high-school-science-teachers

National Research Council. (2000). Inquiry and the national science education standards: A guide for teaching and learning. National Academies Press.

National Wildlife Federation. (2021). National wildlife federation releases results of climate survey of American teenagers. Available from: https://www.nwf.org/News-and-Magazines/Media-Center/News-by-Topic/Global-Warming/2021/05-18-21-National-Wildlife-Federation-Releases-Results-of-Climate-Survey-of-American-Teenagers

Naudé, W. (2010). Climate change and industrial policy. *Sustainability (Basel)*, 3(7), 1003–1021. DOI: 10.3390/su3071003

Nguyen, T. T., Pham, T. A. T., & Tram, H. T. X. (2020). Role of information and communication technologies and innovation in driving carbon emissions and economic growth in selected G-20 countries. *Journal of Environmental Management*, 261, 110162. DOI: 10.1016/j.jenvman.2020.110162 PMID: 32148259

Ngxongo, N. A. (2021). The impact of climate change on visitor destination selection: A case study of the Central Drakensberg Region in Kwazulu-Natal. *Jàambá*, 13(1). Advance online publication. DOI: 10.4102/jamba.v13i1.1161 PMID: 34956552

Nhamo, G., & Muchuru, S. (2019). Climate adaptation in the public health sector in Africa: Evidence from United Nations Framework Convention on Climate Change National Communications. *Jàambá*, 11(1), 1–10. DOI: 10.4102/jamba.v11i1.644 PMID: 31049162

Nicholas, P., Breakey, S., White, B., Brown, M., Fanuele, J., Starodub, R., & Ros, A. (2020). Mental health impacts of climate change: Perspectives for the ed clinician. *Journal of Emergency Nursing: JEN*, 46(5), 590–599. DOI: 10.1016/j.jen.2020.05.014 PMID: 32828480

Nordhaus, W. D. (2021). Climate Club Futures: On the Effectiveness of Future Climate Clubs.

Nordhaus, W. D. (1993). Rolling the "DICE": An optimal transition path for controlling greenhouse gases. *Resource and Energy Economics*, 15(1), 27–50. DOI: 10.1016/0928-7655(93)90017-O

Nordhaus, W. D. (2010). Modeling induced innovation in climate-change policy. In *Technological change and the environment* (pp. 182–209). Routledge.

Nordhaus, W. D. (2019). Climate change: The ultimate challenge for economics. *The American Economic Review*, 109(6), 1991–2014. DOI: 10.1257/aer.109.6.1991

Novack, T., Vorbeck, L., & Zipf, A. (2024). An investigation of the temporality of OpenStreetMap data contribution activities. *Geo-Spatial Information Science*, 27(2), 259–275. Advance online publication. DOI: 10.1080/10095020.2022.2124127

Nunn, P.D., Aalbersberg, W., & Lata, S. (2014). Beyond the core: community governance for climate-change adaptation in peripheral parts of Pacific Island Countries. *Reg Environ Change,14*, 221–235. DOI: 10.1007/s10113-013-0486-7

Nunn, P. D. (2012). Understanding and adapting to sea-level rise. In Harris, F. (Ed.), *Global Environmental Issues* (2nd revised ed., pp. 87–104). Wiley. DOI: 10.1002/9781119950981.ch5

Nurudeen, A., & Usman, A. (2010). Government Expenditure and Economic Growth in Nigeria, 1970-2008: A disaggregated Analysis. *Business and Economics Journal*, 4(1), 1–11.

Nzuza, Z. W. (2021). Effect of climate change on the manufacturing sector. In Olarewaju, O. M., & Ganiyu, I. O. (Eds.), *Handbook of Research on Climate Change and the Sustainable Financial Sector* (pp. 463–476). IGI Global. DOI: 10.4018/978-1-7998-7967-1.ch028

O'leary, Z. (2014). Primary data: Surveys, interviews and observation. The essential guide to doing your research project, 201-216.

Obradovich, N., Migliorini, R., Paulus, M., & Rahwan, I. (2018). Empirical evidence of mental health risks posed by climate change. *Proceedings of the National Academy of Sciences of the United States of America*, 115(43), 10953–10958. DOI: 10.1073/pnas.1801528115 PMID: 30297424

Ockwell, D., Watson, J., Mallett, A., Haum, R., MacKerron, G., & Verbeken, A. M. (2010). Enhancing developing country access to eco-innovation: The case of technology transfer and climate change in a post-2012 policy framework.

Oluoch, S., Lal, P., Susaeta, A., Mugabo, R., Masozera, M., & Aridi, J. (2022). Public preferences for renewable energy options: A choice experiment in rwanda. *Frontiers in Climate*, 4, 874753. Advance online publication. DOI: 10.3389/fclim.2022.874753

Oluwaseyi, J., & Stilinski, D. (2024). *The Impact of Climate Change on International Migration: Analyzing the Social*. Political, and Economic Consequences of Climate-Induced Displacement.

Ömercioğlu, A. (2023). 5686 Sayılı Jeotermal Kaynaklar ve Doğal Mineralli Sular Kanunu'na Göre Tahsil Edilen İdare Payı ve Görevli Mahkeme Sorunu. *Euroasıa Journal Of Socıal Scıences & Humanıtıes*, 10(32), 159–173.

Opach, T., Navarra, C., Rød, J. K., Neset, T. S., Wilk, J., Cruz, S. S., & Joling, A. (2023). Identifying relevant volunteered geographic information about adverse weather events in Trondheim using the CitizenSensing participatory system. *Environment and Planning. B, Urban Analytics and City Science*, 50(7), 1806–1821. Advance online publication. DOI: 10.1177/23998083221136557

Opoku, S., Filho, W., Fudjumdjum, H., & Adejumo, O. (2021). Climate change and health preparedness in africa: Analysing trends in six african countries. *International Journal of Environmental Research and Public Health*, 18(9), 4672. DOI: 10.3390/ijerph18094672 PMID: 33925753

Orchard, S. E., Stringer, L. C., & Manyatsi, A. M. (2016). Farmer perceptions and responses to soil degradation in Swaziland. *Land Degradation & Development*, 28(1), 46–56. DOI: 10.1002/ldr.2595 PMID: 30393450

OSM. (2021). *OpenStreetMap Copyright and License*. https://www.openstreetmap.org/copyright

Our World in Data. (2020). CO_2 emissions by sector. Retrieved July 5, 2024, from https://ourworldindata.org/grapher/co-emissions-by-sector?time=latest&facet=none

Owen, D. A. L. (2018). *Smart water technologies and techniques: Data capture and analysis for sustainable water management*. John Wiley & Sons. DOI: 10.1002/9781119078678

Özcan, B. A. (2020). Ortak Mülkiyet Çerçevesinde İklim Değişikliği Sorununun Çözümünde Kyoto Protokolü'nün Etkisi. *Akdeniz İİBF Dergisi*, 20(2), 169–184.

Özekici, Y. K., & Silik, C. E. (2017). Türkiye'deki iklim değişikliği plan ve politikalarında turizm sektörünün yeri. *Gazi Üniversitesi Turizm Fakültesi Dergisi*, (2), 58–79.

Özmen, M. T. (2009). Sera Gazı- Küresel Isınma ve Kyoto Protokolü. İMO Dergisi, 453(1), 42-46.

Öztürkoğlu, Y. (2013). Tüm Yönleriyle Esnek Çalışma Modelleri. Beykoz Akademi Dergisi, 1(1), 109-129.

Paglialunga, E., Coveri, A., & Zanfei, A. (2022). Climate change and within-country inequality: New evidence from a global perspective. *World Development*, 159, 106030. DOI: 10.1016/j.worlddev.2022.106030

Pagoropoulos, A., Pigosso, D. C., & McAloone, T. C. (2017). The emergent role of digital technologies in the Circular Economy: A review. *Procedia CIRP*, 64, 19–24. DOI: 10.1016/j.procir.2017.02.047

Paksoy, S. (2019). Türkiye'nin İklim Aksiyonunun Bugünkü Durumu. Çukurova Üniversitesi Sosyal Bilimler Enstitüsü Dergisi, 28(3), 155-160.

Palinkas, L., O'Donnell, M., Lau, W., & Wong, M. (2020). Strategies for delivering mental health services in response to global climate change. DOI: 10.20944/preprints202010.0150.v1

Pall, P., Allen, M. R., & Stone, D. A. (2007). Human Contribution to the Length of the 2004 European Heatwave. *Nature*, 449(7164), 804–808. PMID: 17943116

Papava, V. (1994). The Role of The State in The Modern Economic System. *Problems of Economic Transition*, 37(5), 35–48. DOI: 10.2753/PET1061-1991370535

Paris Agreement. (2015). Retrieved June 30, 2024, from https://unfccc.int/sites/default/files/english_paris_agreement.pdf

Parks, B. C., & Roberts, J. T. (2013). Inequality and the global climate regime: breaking the north-south impasse. In *The Politics of Climate Change* (pp. 164–191). Routledge.

Parmesan, C., & Yohe, G. (2003). A Globally Coherent Fingerprint of Climate Change Impacts Across Natural Systems. *Nature*, 421(6918), 37–42. DOI: 10.1038/nature01286 PMID: 12511946

Parra, J. D., Said-Hung, E., & Montoya-Vargas, J. (2020). (Re) introducing critical realism as a paradigm to inform qualitative content analysis in *causal* educational research. *International Journal of Qualitative Studies in Education*. DOI: 10.1080/09518398.2020.1735555

Parsons, E. S., Jowell, A., Veidis, E., Barry, M., & Israni, S. T. (2024). Climate change and inequality. *Pediatric Research*, 1–8.

Pata, U. K., Yurtkuran, S., & Kalça, A. (2016). Energy Consumption and Economic Growth in Turkey: ARDL Bounds Testing Approach. *Marmara University Journal of Economic and Administrative Sciences*, 38(2), 255–271.

Patel, R., & Patel, A. (2024). Evaluating the impact of climate change on drought risk in semi-arid region using GIS technique. *Results in Engineering*, 21, 101957. Advance online publication. DOI: 10.1016/j.rineng.2024.101957

Pattberg, P., & Widerberg, O. (2015). Transnational multistakeholder partnerships for sustainable development: Conditions for success. *Ambio*, 45(1), 42–51. DOI: 10.1007/s13280-015-0684-2 PMID: 26202088

Pattnayak, S. S., & Chadha, A. (2013). Technical efficiency of Indian pharmaceutical firms: A stochastic frontier function approach. https://conference.iza.org/conference_files/pada2009/pattnayak_s5193.pdf

Paul, B. D. (2008). A history of the concept of sustainable development: Literature review. *The Annals of the University of Oradea. Economic Sciences Series*, 17(2), 576–580.

Pearce, D., & Atkinson, G. (1998). The concept of sustainable development: An evaluation of its usefulness ten years after Brundtland. *Environmental Economics and Policy Studies*, 134(2), 251–270. DOI: 10.1007/BF03353896

Pearson, N., & Niaufre, C. (2013). Desertification and drought related migrations in the Sahel–the cases of Mali and Burkina Faso. *The State of Environmental Migration, 3*.

Pearson, R. A., Smart, G., Wilkins, M., Lane, E., Harang, A., Bosserelle, C., Cattoën, C., & Measures, R. (2023). GeoFabrics 1.0. 0: An open-source Python package for automatic hydrological conditioning of digital elevation models for flood modelling. *Environmental Modelling & Software*, 170, 105842. DOI: 10.1016/j.envsoft.2023.105842

Pérez-Peña, M. D. C., Jiménez-García, M., Ruiz-Chico, J., & Peña-Sánchez, A. R. (2021). Analysis of research on the SDGs: The relationship between climate change, poverty and inequality. *Applied Sciences (Basel, Switzerland)*, 11(19), 8947. DOI: 10.3390/app11198947

Peri, G. (2004). Socio-Cultural Variables and Economic Success: Evidence From Italian Provinces 1951-1991. *Contributions in Macroeconomics*, 4(1), 20121025. DOI: 10.2202/1534-5998.1218

Peri, G., Rury, D., & Wiltshire, J. C. (2022). The economic impact of migrants from hurricane maria. *The Journal of Human Resources*, 0521-11655R1. DOI: 10.3368/jhr.0521-11655R1

Pesaran, M. H. (2004), General diagnostic tests for cross section dependence in panels. *IZA Discussion Paper Series No. 1240.*

Pesaran, M. H., Ullah, A., & Yamagata, T. (2008). A bias adjusted LM test of error cross section independence. *The Econometrics Journal*, 11(1), 105–127. DOI: 10.1111/j.1368-423X.2007.00227.x

Philippidis, G., Shutes, L., Robert, M., Ronzon, T., Tabeau, A., & Meijl, H. (2020). Snakes and ladders: World development pathways' synergies and trade-offs through the lens of the sustainable development goals. *Journal of Cleaner Production*, 267, 122147. DOI: 10.1016/j.jclepro.2020.122147 PMID: 32921933

Piętak, Ł. (2014). Review of Theories and Models of Economic Growth. *Comparative Economic Research.Central and Eastern Europe*, 17(1), 45–60.

Pinho, M. G. M., Flueckiger, B., Valentin, A., Kasdagli, M. I., Kyriakou, K., Lakerveld, J., Mackenbach, J. D., Beulens, J. W. J., & de Hoogh, K. (2023). The quality of OpenStreetMap food-related point-of-interest data for use in epidemiological research. *Health & Place*, 83, 103075. Advance online publication. DOI: 10.1016/j.healthplace.2023.103075 PMID: 37454481

Poddar, A. K. (2024). Climate Change and Migration: Developing Policies to Address the Growing Challenge of Climate-Induced Displacement. *The International Journal of Climate Change*, 16(1), 149–170. DOI: 10.18848/1835-7156/CGP/v16i01/149-170

Polenske, K. R. (2017), Growth Pole Theory and strategy Reconsidered: Domination, Linkages, and Distribution. *Regional Economic Development*, 91-111.

Pradhan, P., Costa, L., Rybski, D., Lucht, W., & Kropp, J. (2017). A systematic study of sustainable development goal (sdg) interactions. *Earth's Future*, 5(11), 1169–1179. DOI: 10.1002/2017EF000632

Priego, F. J., Rosselló, J., & Santana-Gallego, M. (2015). The Impact of climate change on domestic tourism: A gravity model for Spain. *Regional Environmental Change*, 15(2), 291–300. DOI: 10.1007/s10113-014-0645-5

Pronto, L., Prat-Guitart, N., & Caamaño, J. (2023). *Research for REGI Committee – Forest fires of summer 2022,* European Parliament, Policy Department for Structural and Cohesion Policies, Brussels.

Qin, Z. (2024). The relationship between climate change anxiety and pro-environmental behavior in adolescents: the mediating role of future self-continuity and the moderating role of green self-efficacy. DOI: 10.21203/rs.3.rs-3930493/v1

Queiroz, M., Lucas, F., & Sörensen, K. (2024). Instance generation tool for on-demand transportation problems. *European Journal of Operational Research*, 317(3), 696–717. Advance online publication. DOI: 10.1016/j.ejor.2024.03.006

Quinn, D. P., & Shapiro, R. Y. (1991). Economic Growth Strategies: The Effects of Ideological Partisanship on Interest Rates and Business Taxation in The United States. *American Journal of Political Science*, 35(3), 656–685. DOI: 10.2307/2111560

Rabe, B. G. (2002). Greenhouse & statehouse: The evolving state government role in climate change. *Pew Center on Global Climate Change.* Retrieved May 12, 2024, from https://www.c2es.org/wp-content/uploads/2002/11/states_greenhouse.pdf

Rabinovych, M., & Pintsch, A. (2023). Sustainable development: A common denominator for the EU's policy towards the Eastern partnership? *The International Spectator,* Vol. 58, No. 1, 38–57, Retrieved August 08, 2024 from https://Doi.Org/10.1080/03932729.2023.2165774

Raheem, I. D., Tiwari, A. K., & Balsalobre-Lorente, D. (2020). The role of ICT and financial development in CO2 emissions and economic growth. *Environmental Science and Pollution Research International*, 27(2), 1912–1922. DOI: 10.1007/s11356-019-06590-0 PMID: 31760620

Rahman, M. M., & Vu, X. B. (2020). The nexus between renewable energy, economic growth, trade, urbanisation and environmental quality: A comparative study for *Australia and Canada.Renewable Energy*, 155, 617–627. DOI: 10.1016/j.renene.2020.03.135

Rajesh, R. S. N. (2007). *Technical efficiency in the informal manufacturing enterprises: Firm level evidence from an Indian state*. MPRA Paper No. 7816. https://mpra.ub.uni-muenchen.de/7816/

Ramsey, J. D., Burford, C. L., Beshir, M. Y., & Jensen, R. C. (1983). Effects of Workplace Thermal Conditions on Safe Work Behavior. *Journal of Safety Research*, 4(3), 105–114. DOI: 10.1016/0022-4375(83)90021-X

Rao, N. D., & Min, J. (2018). Less global inequality can improve climate outcomes. *Wiley Interdisciplinary Reviews: Climate Change*, 9(2), e513. DOI: 10.1002/wcc.513

Redclift, M., & Sage, C. (1998). Global environmental change and global inequality: North/South perspectives. *International Sociology*, 13(4), 499–516. DOI: 10.1177/026858098013004005

Redish, A. (1984). Why Was Specie Scarce in Colonial Economies? An Analysis of The Canadian Currency, 1796-1830. *The Journal of Economic History*, 44(3), 713–728. DOI: 10.1017/S0022050700032332

Reiche, D. (2010). Renewable energy policies in the Gulf countries: A case study of the carbon-neutral "Masdar City" in Abu Dhabi. *Energy Policy*, 38(1), 378–382. DOI: 10.1016/j.enpol.2009.09.028

Reichman, D. R. (2022). Putting climate-induced migration in context: The case of Honduran migration to the USA. *Regional Environmental Change*, 22(3), 91. DOI: 10.1007/s10113-022-01946-8 PMID: 35814810

Reti, M. J. (2007). An Assessment of the Impact of Climate Change on Agriculture and Food Security in the Pacific: A Case Study in Vanuatu. FAO SAPA, ftp://ftp.fao.org/docrep/fao/011/i0530e/i0530e02.pdf

Reyes Cruz, C. A. (2023). *Causes of displacement: a look into the state of Puerto Rico's housing crisis before, during, and after Hurricane Maria* (Doctoral dissertation).

Reynolds, M. (2019). Coronavirus shows the enormous scale of the climate crisis. https://www.wired.co.uk/article/coronavirus-climate-change.

Richmond-Navarro, G., Madriz-Vargas, R., Ureña-Sandí, N., & Barrientos-Johansson, F. (2019). Research opportunities for renewable energy electrification in remote areas of costa rica. *Perspectives on Global Development and Technology*, 18(5-6), 553–563. DOI: 10.1163/15691497-12341530

Ritzema, H., & Loon-Steensma, J. (2017). Coping with climate change in a densely populated delta: A paradigm shift in flood and water management in the netherlands. *Irrigation and Drainage*, 67(S1), 52–65. DOI: 10.1002/ird.2128

Riyadh, Z., Rahman, M., Saha, S., Ahamed, T., & Current, D. (2021). Adaptation of agroforestry as a climate smart agriculture technology in bangladesh. *International Journal of Agricultural Research, Innovation and Technology*, 11(1), 49–59. DOI: 10.3329/ijarit.v11i1.54466

Roberts, J. T. (2001). Global inequality and climate change. *Society & Natural Resources*, 14(6), 501–509. DOI: 10.1080/08941920118490

Roberts, J. T., & Parks, B. (2006). *A climate of injustice: Global inequality, north-south politics, and climate policy*. MIT press.

Robine, J. M., Cheung, S. L. K., & Le Roy, S. (2008). Death Toll Exceeded 70,000 in Europe Heatwave. *Nature*, 455(7210), 43–44.

Rodima-Taylor, D., Olwig, M. F., & Chhetri, N. (2012). Adaptation as innovation, innovation as adaptation: An institutional approach to climate change. *Applied Geography (Sevenoaks, England)*, 33(0), 107–111. DOI: 10.1016/j.apgeog.2011.10.011

Rodrigues, A., Santiago, A., Laím, L., Viegas, D. X., & Zêzere, J. L. (2022). Rural fires—Causes of human losses in the 2017 fires in Portugal. *Applied Sciences (Basel, Switzerland)*, 12(24), 12561. DOI: 10.3390/app122412561

Romão, J., Palm, K., & Persson-Fischier, U. (2023). Open spaces for co-creation: A community-based approach to tourism product diversification. *Scandinavian Journal of Hospitality and Tourism*, 23(1), 94–113. Advance online publication. DOI: 10.1080/15022250.2023.2174183

Roos, M., Hoffart, F. M., Roos, M., & Hoffart, F. M. (2021). Climate change and responsibility. *Climate economics: A call for more pluralism and responsibility*, 121-155.

Ryan, S. J., Carlson, C. J., & Mordecai, E. A. (2019). Global Expansion and Redistribution of Aedes-borne Virus Transmission Risk. *Nature Microbiology*, 4(6), 1060–1067. PMID: 30921321

Safonov, G. (2019). *Social consequences of climate change - Building climate friendly and resilient communities via transition from planned to market economies*. Retrieved May 5, 2024, from https://library.fes.de/pdf-files/id-moe/15863.pdf

Şahin, D. (2018). APEC ülkelerinde turizm, ekonomik büyüme ve çevresel kalite ilişkisi: panel veri analizi. *İktisadi Yenilik Dergisi*, 5(2), 32-44.

Şahin, Ö. U. (2016). Kyoto Protokolü ve Kopenhag Mutabakatının Karşılaştırmalı Analizi. [JoA]. *Journal of Awareness*, 1(1), 5–10.

Sahu, P. K. (2015). Technical efficiency of domestic and foreign firms in Indian manufacturing: A firm level panel analysis. *Arthshastra Indian Journal of Economic & Research*, 4(2), 7–21. DOI: 10.17010/aijer/2015/v4i2/65534

Sahu, S. K., & Narayanan, K. (2011). Determinants of energy intensity in Indian manufacturing industries: A firm level analysis. *Eurasian Journal of Business and Economics*, 4(8), 13–30.

Sahu, S. K., & Narayanan, K. (2015). Environmental certification and technical efficiency: A study of manufacturing firms in India. *Journal of Industry, Competition and Trade*, 16(2), 1–17. DOI: 10.1007/s10842-015-0213-9

Salik, K. M., Shabbir, M., & Naeem, K. (2020). Climate-induced displacement and migration in Pakistan: Insights from Muzaffargarh and Tharparkar Districts.

Saliminezhad, A., & Lisaniler, F. G. (2018). Validity of Unbalanced Growth Theory and Sectoral Investment Priorities in Indonesia: Application of Feature Ranking Methods. *The Journal of International Trade & Economic Development*, 27(5), 521–540. DOI: 10.1080/09638199.2017.1398270

Samantarai, M., & Dutta, S. (2023). *Katherine Lucey and Solar Sister: empowering women in sub-Saharan Africa to create clean energy businesses*. The CASE Journal.

Samek Lodovici, M. (2021). The impact of teleworking and digital work on workers and society. .DOI: 10.2861/72272

Şanli, F. B., Bayrakdar, S., & İncekara, B. (2017). Küresel iklim değişikliğinin etkileri ve bu etkileri önlemeye yönelik uluslararası girişimler. *Süleyman Demirel Üniversitesi İktisadi ve İdari Bilimler Fakültesi Dergisi*, 22(1), 201–212.

San-Miguel-Ayanz, J., Durrant, T., Boca, R., Maianti, P., Libertá, G., Artés-Vivancos, T., Oom, D., Branco, A., de Rigo, D., Ferrari, D., Pfeiffer, H., Grecchi, R., Onida, M., & Löffler, P. (2022). *Forest Fires in Europe, Middle East and North Africa 2022*. EUR 31269 EN, Publications Office of the European Union. DOI: 10.2760/34094

Sarfo, I., Bortey, O., & Kumara, T. (2019). Effectiveness of adaptation strategies among coastal communities in ghana: the case of dansoman in the greater accra region. Current Journal of Applied Science and Technology, 1-12. DOI: 10.9734/cjast/2019/v35i630211

Sarkar, D., & Anderson, J. T. (2022). Corporate editors in OpenStreetMap: Investigating co-editing patterns. *Transactions in GIS*, 26(4), 1879–1897. Advance online publication. DOI: 10.1111/tgis.12910

Sarkodie, S. A., Strezov, V., Weldekidan, H., Asamoah, E. F., Owusu, P. A., & Doyı, I. N. Y. (2019). Environmental sustainability assessment using dynamic autoregressive-distributed lag simulations-nexus between greenhouse gas emissions, biomass energy, food and economic growth. *The Science of the Total Environment*, 668, 318–332. DOI: 10.1016/j.scitotenv.2019.02.432 PMID: 30852209

Şaşmaz, M. Ü., Sakar, E., Yayla, Y. E., & Akküçük, U. (2020). The relationship between renewable energy and human development in OECD countries: A panel data analysis. *Sustainability (Basel)*, 12(18), 7450. DOI: 10.3390/su12187450

Satpathy, L. D., Chatterjee, B., & Mahakud, J. (2017). Firm characteristics and total factor productivity: Evidence from Indian manufacturing firms. *Margin*, 11(1), 77–98. DOI: 10.1177/0973801016676013

Saura, J. R., Ribeiro-Soriano, D., & Zegarra Saldaña, P. (2022). Exploring the challenges of remote work on Twitter users' sentiments: From digital technology development to a post-pandemic era. *Journal of Business Research*, 142, 242–254. DOI: 10.1016/j.jbusres.2021.12.052

Scherr, S. J., Shames, S., & Friedman, R. (2012). From climate-smart agriculture to climate-smart landscapes. *Agriculture & Food Security*, 1(1), 1–15. DOI: 10.1186/2048-7010-1-12

Schneider, F. A., Ortiz, J. C., Vanos, J. K., Sailor, D. J., & Middel, A. (2023). Evidence-based guidance on reflective pavement for urban heat mitigation in Arizona. *Nature Communications*, 14(1), 1467. Advance online publication. DOI: 10.1038/s41467-023-36972-5 PMID: 36928319

Schreurs, M. A. (2004). The climate change divide: The European Union, The United States, and the future of the Kyoto Protocol. In Vig, N. J., & Faure, M. G. (Eds.), *Green giants? Environmental policies of the United States and the European Union* (pp. 207–230). MIT Press. DOI: 10.7551/mitpress/3363.003.0014

Schwan, S., & Yu, X. (2018). Social protection as a strategy to address climate-induced migration. *International Journal of Climate Change Strategies and Management*, 10(1), 43–64. DOI: 10.1108/IJCCSM-01-2017-0019

Schwartz, S., Benoit, L., Clayton, S., Parnes, M., Swenson, L., & Lowe, S. (2022). Climate change anxiety and mental health: Environmental activism as buffer. *Current Psychology (New Brunswick, N.J.)*, 42(20), 16708–16721. DOI: 10.1007/s12144-022-02735-6 PMID: 35250241

Scobie, M. (2019). Sustainable development and climate change adaptation: goal interlinkages and the case sids. DOI: 10.17875/gup2019-1213

Seetanah, B., & Fauzel, S. (2018). Investigating the impact of climate change on the tourism sector: Evidence from a sample of Island Economies. *Tourism Review*, 74(2), 194–203. DOI: 10.1108/TR-12-2017-0204

Selçuk, S. F. (2023). Uluslararası İklim Değişikliği Anlaşmaları ve Türkiye'nin Tutumu. *Ulusal Çevre Bilimleri Araştırma Dergisi*, 6(1), 9–19.

Semasinghe, W. M. (2020). Development, what does it really mean? *Acta Politica Polonica*, 49, 51–59. DOI: 10.18276/ap.2020.49-05

Sen, J., & Das, D. (2016). Technical efficiency in India's unorganized manufacturing sector: A non-parametric analysis. *International Journal of Business and Management*, 4(4), 92–101.

Sen, M. K., Dutta, S., Kabir, G., Pujari, N. N., & Laskar, S. A. (2021). An integrated approach for modelling and quantifying housing infrastructure resilience against flood hazard. *Journal of Cleaner Production*, 288, 125526. DOI: 10.1016/j.jclepro.2020.125526

Seritan, A., & Seritan, I. (2020). The time is now: Climate change and mental health. *Academic Psychiatry*, 44(3), 373–374. DOI: 10.1007/s40596-020-01212-1 PMID: 32162168

Sevim, B., & Ünlüönen, K. (2010). İklim değişikliğinin turizme etkileri: Konaklama işletmelerinde bir uygulama. *Erciyes Üniversitesi Sosyal Bilimler Enstitüsü Dergisi*, 1(28), 43–66.

Shahbaz, M., Hye, Q. M. A., Tiwari, A. K., & Leitão, N. C. (2013). Economic growth, energy consumption, financial development, international trade and CO_2 emissions in Indonesia. *Renewable & Sustainable Energy Reviews*, 25, 109–121. DOI: 10.1016/j.rser.2013.04.009

Shahbaz, M., Mahalik, M. K., Shah, S. H., & Sato, J. R. (2016). Time-varying analysis of CO2 emissions, energy consumption, and economic growth nexus: Statistical experience in the next 11 countries. *Energy Policy*, 98, 33–48. DOI: 10.1016/j.enpol.2016.08.011

Shahzadi, A., Yaseen, M. R., & Anwar, S. (2019). Relationship between Globalization and Environmental Degradation in Low Income Countries: An Application of Kuznets Curve. *Indian Journal of Science and Technology*, 12(19), 1–13. Advance online publication. DOI: 10.17485/ijst/2019/v12i19/143994

Shang, T., Samour, A., Abbas, J., Ali, M., & Tursoy, T. (2024). Impact of financial inclusion, economic growth, natural resource rents, and natural energy use on carbon emissions: The MMQR approach. *Environment, Development and Sustainability*, 1–31. DOI: 10.1007/s10668-024-04513-9

Sherman, M., Berrang-Ford, L., Lwasa, S., Ford, J., Namanya, D., Llanos-Cuentas, A., Maillet, M., & Harper, S.IHACC Research Team. (2016). Drawing the line between adaptation and development: A systematic literature review of planned adaptation in developing countries. *Wiley Interdisciplinary Reviews: Climate Change*, 7(5), 707–726. DOI: 10.1002/wcc.416

Shoib, S., Hussaini, S. S., Armiya'u, A. Y., Saeed, F., Őri, D., Roza, T. H., Gürcan, A., Agrawal, A., Solerdelcoll, M., Lucero-Prisno, D. E.III, Nahidi, M., Swed, S., Ahmed, S., & Chandradasa, M. (2023). Prevention of suicides associated with global warming: Perspectives from early career psychiatrists. *Frontiers in Psychiatry*, 14, 1251630. Advance online publication. DOI: 10.3389/fpsyt.2023.1251630 PMID: 38045615

Singer, H. W. (1949). Economic Progress in Underdeveloped Countries. *Social Research*, 1–11.

Singer, M. (2018). *Climate change and social inequality: The health and social costs of global warming*. Routledge. DOI: 10.4324/9781315103358

Singh, A. K., Ashraf, S. N., & Arya, A. (2019a). Estimating factors affecting technical efficiency in Indian manufacturing sector. *Eurasian Journal of Business and Economics*, 12(24), 65–86. DOI: 10.17015/ejbe.2019.024.04

Singh, A. K., & Jyoti, B. (2019). Measuring the climate variability impact on cash crops farming in India: An Empirical Investigation. *Agriculture and Food Sciences Research*, 6(2), 155–164. DOI: 10.20448/journal.512.2019.62.155.165

Singh, A. K., Jyoti, B., & Sankaranarayanan, K. G. (2024). A review for analyzing the impact of climate change on agriculture and food security in India. In Samanta, D., & Garg, M. (Eds.), *The Climate Change Crisis and Its Impact on Mental Health* (pp. 227–251). IGI Global. DOI: 10.4018/979-8-3693-3272-6.ch017

Singh, A. K., & Kumar, S. (2022a). Assessing the performance and factors affecting industrial development in Indian states: An empirical analysis. *Journal of Social Economic Research*, 8(2), 135–154. DOI: 10.18488/journal.35.2021.82.135.154

Singh, A. K., & Kumar, S. (2022b). Expert's perception on technology transfer and commercialization, and intellectual property rights in India: Evidence from selected research organizations. *Journal of Management, Economics, and Industrial Organization*, 6(1), 1–33. DOI: 10.31039/jomeino.2022.6.1.1

Singh, A. K., & Kumar, S. (2022c). Measuring the factors affecting annual turnover of the firms: A case study of selected manufacturing industries in India. *International Journal of Business Management and Finance Research*, 5(2), 33–45. DOI: 10.53935/26415313.v5i2.211

Singh, A. K., Narayanan, K. G. S., & Sharma, P. (2017). Effect of climatic factors on cash crop farming in India: An application of Cobb-Douglas production function model. *International Journal of Agricultural Resources, Governance and Ecology*, 13(2), 175–210. DOI: 10.1504/IJARGE.2017.086452

Singh, A. K., Narayanan, K. G. S., & Sharma, S. (2019b). Measurement of technical efficiency of climatic and non-climatic factors in sugarcane farming in Indian staes: Use of stochastic frontier production function approach. *Climatic Change*, 5(19), 150–166.

Singh, A. K., & Singh, B. J. (2021). Projected productivity of cash crops in different climate change scenarios in India: Use of marginal impact analysis technique. *Finance & Economics Review*, 3(1), 63–87. DOI: 10.38157/finance-economics-review.v3i1.281

Singh, A. K., Singh, B. J., & Ashraf, S. N. (2020). Implications of intellectual property protection, and science and technological development in the manufacturing sector in selected economies. *Journal of Advocacy. Research in Education*, 7(1), 16–35.

Singh, D., & Malik, G. (2018). Technical efficiency and its determinants: A panel data analysis of Indian public and private sector banks. *Asian Journal of Accounting Perspectives*, 11(1), 48–71. DOI: 10.22452/AJAP.vol11no1.3

Sırdaş, S. (2002).. . *Meteorolojik Kuraklık ve Türkiye Modellemesi, İstanbul Teknik Üniversitesi Dergisi*, 2(2), 95–103.

Sitarz, D. (1994). *Agenda 21. The earth strategy to save our planet*. Earth Press.

Slim, H. (1995). What is development? *Development in Practice*, 5(2), 143–148. DOI: 10.1080/0961452951000157114 PMID: 12288928

Slovic, A. D., Kanai, C., Sales, D. M., Rocha, S. C., de Souza Andrade, A. C., Martins, L. S., Coelho, D. M., Freitas, A., Moran, M., Mascolli, M. A., Caiaffa, W. T., & Gouveia, N. (2023). Spatial data collection and qualification methods for urban parks in Brazilian capitals: An innovative roadmap. *PLoS ONE, 18*(8 August). DOI: 10.1371/journal.pone.0288515

Smith, K. (2009). Climate change and radical energy innovation: the policy issues.

Smol, M. (2019). The Importance of Sustainable Phosphorus Management in The Circular Economy (CE) Model: The Polish Case Study. *Journal of Material Cycles and Waste Management*, 21(2), 227–238. DOI: 10.1007/s10163-018-0794-6

Sneddon, C., Howarth, R. B., & Norgaard, R. B. (2006). Sustainable development in a post-Brundtland world. *Ecological Economics*, 57(2), 253–268. DOI: 10.1016/j.ecolecon.2005.04.013

Sobh, R., & Perry, C. (2006). Research Design and Data Analysis in Realism Research. *European Journal of Marketing*, 40(11), 1194–1209. DOI: 10.1108/03090560610702777

Sobirov, Y., Makhmudov, S., Saibniyazov, M., Tukhtamurodov, A., Saidmamatov, O., & Marty, P. (2024). Investigating the Impact of Multiple Factors on CO_2 Emissions: Insights from Quantile Analysis. *Sustainability (Basel)*, 16(6), 2243. DOI: 10.3390/su16062243

Somuncu, M. (2018). İklim değişikliği Türkiye turizmi için bir tehdit mi, bir fırsat mı? *Tücaum 30. Yıl Uluslararası Coğrafya Sempozyumu*, 748-771.

Song, Y., Medda, F., & Wang, M. (2024). The value of resilience bond in financing flood resilient infrastructures: A case study of Towyn. *Journal of Sustainable Finance & Investment*, 14(4), 1–24. DOI: 10.1080/20430795.2024.2366200

Soni, A., Mittal, A., & Kapshe, M. (2017). Energy intensity analysis of Indian manufacturing industries. *Resource-Efficient Technologies*, 3(1), 353–357.

Southern Africa Environment Outlook. (2016). 2nd South Africa Environment Outlook. A report on the state of the environment. Executive Summary. Department of Environmental Affairs, Pretoria. https://www.environment.gov.za/sites/default/files/reports/environmentoutlook_execu tivesummary.pdf

Southern African Development Community. (2008). Southern Africa environmental outlook. Available from: https://www.environment.gov.za/sites/default/files/reports/environmentoutlook _chapter14.pdf

Steinfatt, K. (2020). Trade policies for a circular economy. DOI: 10.30875/4445c521-en

Stern, N. (2007). *The Economics of Climate Change: The Stern Review*. Cambridge University Press. DOI: 10.1017/CBO9780511817434

Steward, F. (2012). Transformative innovation policy to meet the challenge of climate change: Sociotechnical networks aligned with consumption and end-use as new transition arenas for a low-carbon society or green economy. *Technology Analysis and Strategic Management*, 24(4), 331–343. DOI: 10.1080/09537325.2012.663959

Sudarshan, A., Somanathan, E., Somanathan, R., & Tewari, M. (2021). Climate change may hurt Indian manufacturing due to heat stress on workers. https://epic.uchicago.in/climate-change-may-hurt-indian-manufacturing-due-to-heat-stress-on-workers/

Su, H. N., & Moaniba, I. M. (2017). Does innovation respond to climate change? Empirical evidence from patents and greenhouse gas emissions. *Technological Forecasting and Social Change*, 122, 49–62. DOI: 10.1016/j.techfore.2017.04.017

Sverdrup, H. U. (2019). The global sustainability challenges in the future: the energy use, materials supply, pollution, climate change and inequality nexus. In *What Next for Sustainable Development?* (pp. 49–75). Edward Elgar Publishing. DOI: 10.4337/9781788975209.00013

Talu, N., & Kocaman, H. (2018). Türkiye'de İklim Değişikliği ile Mücadelede Politikalar, Yasal ve Kurumsal Yapı. 21.

Tandoğan, D., & Genç, M. C. (2019). Türkiye'de turizm ve karbondioksit salımı arasındaki ilişki: Rals-engle ve granger eşbütünleşme yaklaşımı. *Anatolia: Turizm Araştırmaları Dergisi*, 30(3), 221–230.

Tarraga, J. M., Piles, M., & Camps-Valls, G. (2020). Learning drivers of climate-induced human migrations with Gaussian processes. *arXiv preprint arXiv:2011.08901*.

Teye, J. K., & Nikoi, E. G. (2022). Climate-induced migration in West Africa. In *Migration in West Africa: IMISCOE regional reader* (pp. 79–105). Springer International Publishing. DOI: 10.1007/978-3-030-97322-3_5

Thampy, A., & Tiwary, M. K. (2021). Local banking and manufacturing growth: Evidence from India. *IIMB Management Review, 33*(2), 95-104. DOI: 10.1016/j.iimb.2021.03.013

The World Bank. (2023a). https://databank.worldbank.org/reports.aspx?source=2&series=EN.ATM. CO2E.PC&country

The World Bank. (2023b). https://databank.worldbank.org/reports.aspx?source=2 &series=ST.INT.ARVL &country

Thomas, A., & Benjamin, L. (2018). Policies and mechanisms to address climate-induced migration and displacement in Pacific and Caribbean small island developing states. *International Journal of Climate Change Strategies and Management*, 10(1), 86–104. DOI: 10.1108/IJCCSM-03-2017-0055

Tol, R. S. (2005). The marginal damage costs of carbon dioxide emissions: An assessment of the uncertainties. *Energy Policy*, 33(16), 2064–2074. DOI: 10.1016/j.enpol.2004.04.002

Tomlinson, C. A. (2014). *The differentiated classroom: Responding to the needs of all learners.* ASCD.

Torres, P. H. C., Leonel, A. L., Pires de Araújo, G., & Jacobi, P. R. (2020). Is the brazilian national climate change adaptation plan addressing inequality? Climate and environmental justice in a global south perspective. *Environmental Justice*, 13(2), 42–46. DOI: 10.1089/env.2019.0043

Tremblay, M., & Trudel, M. È. S. O. (2013). The Climate-Induced Migration: What Protection for Displaced People? *The International Journal of Climate Change*, 4(4), 67–81. DOI: 10.18848/1835-7156/CGP/v04i04/57870

Trenberth, K. E., Dai, A., & van der Schrier, G. (2014). *Global Warming and Changes in Drought.* In: Climate Extremes and Society, 287-316. DOI: 10.1038/nclimate2067

Trott, C. D., Lam, S., Roncker, J., Gray, E. S., Courtney, R. H., & Even, T. L. (2023). Justice in climate change education: A systematic review. *Environmental Education Research*, 29(11), 1–38. DOI: 10.1080/13504622.2023.2181265

Truong, P. M., Le, N. H., Hoang, T. D. H., Nguyen, T. K. T., Nguyen, T. D., Kieu, T. K., Nguyen, T. N., Izuru, S., Le, V. H. T., Raghavan, V., Nguyen, V. L., & Tran, T. A. (2023). Climate Change Vulnerability Assessment Using GIS and Fuzzy AHP on an Indicator-Based Approach. *International Journal of Geoinformatics*, 19(2), 39–53. Advance online publication. DOI: 10.52939/ijg.v19i2.2565

Tuğaç, Ç. (2020). Dünyada ve Türkiye'de iklim değişikliği politikaları. In Sağır, H. (Ed.), *Ekolojik kriz ve küresel çevre politikaları* (pp. 221–264). Beta.

Türkeş, M. (1997). Hava ve iklim kavramları üzerine. *TÜBİTAK Bilim ve Teknik Dergisi*, 355, 36–37.

Türkeş, M. (2001). Global climate protection, climate change framework agreement and Turkey. *Plumbing Engineering, TMMOB Chamber of Mechanical Engineers, Periodical Technical Publication*, 61, 14–29.

Türkeş, M. (2006). Küresel İklimin Geleceği ve Kyoto Protokolü. *Jeopolitik*, (29), 99–107.

Türkeş, M., Sümer, U. M., & Çetiner, G. (2000). Kyoto Protokolü Esneklik Mekanizmaları. *Tesisat Dergisi*, (52), 84–100.

Türkiye İklim Stratejisi. (2010). Türkiye İklim Değişikliği Stratejisi 2010-2023. https://www.gmka.gov.tr/dokumanlar/yayinlar/Turkiye-Iklim-Degisikligi-Stratejisi.pdf adresinden alındı

Tyagi, S., & Nauryal, D. K. (2016). Profitability determinants in Indian drugs and pharmaceutical industry: An analysis of pre and post TRIPS period. *Eurasian Journal of Business and Economics*, 9(17), 1–21. DOI: 10.17015/ejbe.2016.017.01

Udeagha, M. C., & Breitenbach, M. C. (2023). The role of financial development in climate change mitigation: Fresh policy insights from South Africa. *Biophysical Economics and Sustainability*, 8(1), 134. DOI: 10.1007/s41247-023-00110-y

Ullah, T., Lautenbach, S., Herfort, B., Reinmuth, M., & Schorlemmer, D. (2023). Assessing Completeness of OpenStreetMap Building Footprints Using MapSwipe. *ISPRS International Journal of Geo-Information*, 12(4), 143. Advance online publication. DOI: 10.3390/ijgi12040143

Ulueren, M. (2001). Küresel ısınma, BM iklim değişikliği çerçeve sözleşmesi ve Kyoto Protokolü. *T.C. Dışişleri Bakanlığı Uluslararası Ekonomik Sorunlar Dergisi*, 3, Retrieved May 12, 2024, from https://www.mfa.gov.tr/kuresel-isinma-bm-iklim-degisikligi-cerceve-sozlesmesi-ve-kyto-protokolu.tr.mfa

UN. (2023). *Europe warming twice as fast as other continents, warns WMO*, Retrieved August 08, 2024 from https://news.un.org/en/story/2023/06/1137867

UN. (2024a). Conferences/Environment and sustainable development. https://www.un.org/en/conferences/environment/stockholm1972

UN. (2024b). Conferences/Environment and sustainable development. https://www.un.org/en/conferences/environment/rio1992

UN. (2024c). Conferences/Environment and sustainable development. https://www.un.org/en/conferences/environment/

UN. (2024d). *The 17 goals*. https://sdgs.un.org/goals

UNFCCC. (1992). *United nations framework convention on climate change*. UN. Retrieved April 20, 2024, from https://unfccc.int/resource/docs/convkp/conveng.pdf

UNFCCC. (1992). *United Nations Framework Convention on Climate Change*. United Nations.

UNFCCC. (1997). *Kyoto protocol to the united nations framework convention on climate change*. UN. Retrieved April 20, 2024, from https://unfccc.int/documents/2409

UNFCCC. (1997). *Kyoto Protocol to the United Nations Framework Convention on Climate Change*. United Nations.

UNFCCC. (2008). *UNFCCC resource guide module 4: for preparing the national communications of non-annex i parties module 4 measures to mitigate climate change*. United Nations Framework Convention on Climate Change.

UNFCCC. (2014). EU *Agrees 40% greenhouse gas cut by 2030,* Retrieved June 12, 2024 from https://unfccc.int/news/eu-agrees-40-greenhouse-gas-cut-by-2030

UNFCCC. (2015). *Paris Agreement*. United Nations.

UNFCCC. (2023). *History of the convention*, Retrieved July 03, 2024 from https://unfccc.int/process/the-convention/history-of-the-convention#Essential-background

UNFCCC. (2024a). *The Paris agreement. What is the Paris agreement?* UN. Retrieved April 27, 2024, from https://unfccc.int/process-and-meetings/the-paris-agreement

UNFCCC. (2024b). *What is the Kyoto protocol?* UN. Retrieved April 20, 2024, from https://unfccc.int/kyoto_protocol

United Nation Development Programme-UNDP. (1990). *Human development report 1990*. New York Oxford Oxford University Press.

United Nations Educational, Scientific and Cultural Organization. (2015). Not just hot air: putting climate change education into practice. https://unesdoc.unesco.org/images/0023/002330/233083e.pdf

United Nations Educational, Scientific and Cultural Organization. (2017). Sustainability Starts with Teachers: Environmental scan informing teacher education programming for ESD secondary teacher education. https://www.google.com/search?q=sustainability+starts+with+teachers&rlz=1C1GCE A_en-ZA996ZA996&oq=Sustainability+starts+&aqs=chrome.1.69i57j0i512l5j0i15i22 i30j0i10i22i30l3.7975j0j15&sourceid=chrome&ie=UTF-8

United Nations Environment Programme. (2008a). *Africa atlas of our changing environment*. Available from: https://wedocs.unep.org/handle/20.500.11822/7717

United Nations Environment Programme. (2016). *Summary of the sixth global environmental outlook GEO-6 regional assessments: Key findings and policy messages UNEP/EA.2/INF/17.* Available from: http://www.scpclearinghouse.org/sites/default/files/summary_of_the_sixth_glo bal_environment_outlook_geo6_regional_assessments_key_findings_and_po licy_messages_unep_ea2_inf_17-2016 geo-6_summary_en.p.pdf

United Nations Framework Convention on Climate Change. (1992). FCCC/INFORMAL/84,GE.05- 62220(E), 200705. Available online: https://unfccc.int/resource/docs/convkp/conveng.pdf

United Nations Türkiye. (2024). *Our work on the sustainable development goals in Türkiye.* https://turkiye.un.org/en/sdgs

United Nations. (1992). *United Nations Framework Convention on Climate Change.* https://unfccc.int/files/essential_background/background_publications_htmlpdf/application/pdf/conveng.pdf

United States Environmental Protection Agency-EPA. (2010). *Climate Change Indicators in the United States.* 1200 Pennsylvania Avenue, N.W. (6207J) Washington, DC 20460.

UNU WIDER. (2024). *World income inequality database (WIID).* https://www4.wider.unu.edu/?ind=1&type=ChoroplethSeq&year=70&byCountry=false&slider=buttons

Upadhyay, R. K. (2020). Markers for Global Climate Change and Its Impact on Social, Biological and Ecological Systems: A Review. *American Journal of Climate Change*, 09(03), 159–203. Advance online publication. DOI: 10.4236/ajcc.2020.93012

Upreti, P. (2015). Factors Affecting Economic Growth in Developing Countries. *Major Themes in Economics*, 17(1), 37–54.

Uysal Oğuz, C. (2010). İklim değişikliği ile mücadelede yerel yönetimlerin rolü: Seattle örneği [The role of local governments in fighting with climate change: The example of Seattle]. *Yönetim ve Ekonomi Dergisi*, 17(2), 25–41.

Uzmen, R. (2007). *Global Warming and Climate Change: Is Humanity Facing a Great Disaster?* (2nd ed.). Bilge Kültür Sanat.

Van Den Berg, H. (2016). *Economic Growth and Development.* World Scientific Publishing Company. DOI: 10.1142/9058 van Vuuren, D. P., Edmonds, J., Kainuma, M., Riahi, K., & Weyant, J. (2011). A special issue on the RCPs. *Climatic Change*, 109(1-2), 1–4. DOI: 10.1007/s10584-011-0157-y

Vazquez-Brust, D. A., & Sarkis, J. (2012). *Green Growth: Managing The Transition to Sustainable Economies*. Springer Netherlands. DOI: 10.1007/978-94-007-4417-2

Vera, I., & Langlois, L. (2007). Energy Indicators for Sustainable Development. *Energy*, 32(6), 875–882. DOI: 10.1016/j.energy.2006.08.006

Vezirgiannidou, S. E. (2013). Climate and energy policy in the United States: The battle of ideas. *Environmental Politics*, 4(22), 593–609. DOI: 10.1080/09644016.2013.806632

Vushe, A. (2021). Proposed research, science, technology, and innovation to address current and future challenges of climate change and water resource management in Africa. *Climate Change and Water Resources in Africa: Perspectives and Solutions Towards an Imminent Water Crisis*, 489-518.

Walker, G., & King, S. D. (2009). *Our Planet is Heating Up: How Can We Deal with Global Warming?* (Akpınar, Ö., Trans.). Boğaziçi University Press.

Wang, J., Xing, J., Mathur, R., Pleim, J., Wang, S., Hogrefe, C., Gan, C., Wong, D. C., & Hao, J. (2017). Historical trends in PM2.5 Related premature mortality during 1990-2010 across the Northern Hemisphere. *Environmental Health Perspectives*, 125(3), 400–408. DOI: 10.1289/EHP298 PMID: 27539607

Wang, Q., & Zhang, F. (2021). The effects of trade openness on decoupling carbon emissions from economic growth–evidence from 182 countries. *Journal of Cleaner Production*, 279, 123838. DOI: 10.1016/j.jclepro.2020.123838 PMID: 32863606

Warsame, A., Sheik-Ali, I., Hussein, H., & Barre, G. (2023). Assessing the long- and short-run effects of climate change and institutional quality on economic growth in Somalia. *Environmental Research Communications*, 5(5), 055010. DOI: 10.1088/2515-7620/accf03

Webb, N. M., Troper, J. D., & Fall, R. (2008). Constructive activity and learning in collaborative small groups. *Journal of Educational Psychology*, 100(1), 1–13. PMID: 19578558

Wei, Y., Wang, K., Liao, H., & Tatano, H. (2016). Economics of climate change and risk of disasters in Asia-Pacific region. *Natural Hazards*, 84(S1), S1–S5. DOI: 10.1007/s11069-016-2590-8

Westerlund, J. (2007). Testing for error correction in panel data. *Oxford Bulletin of Economics and Statistics*, 69(6), 709–748. DOI: 10.1111/j.1468-0084.2007.00477.x

Weyant, J. P. (2017). Some contributions of integrated assessment models of global climate change. *Review of Environmental Economics and Policy*, 11(1), 115–137. DOI: 10.1093/reep/rew018

WHO World Health Organization. (2020). *WHO*. https://www.who.int/reproductivehealth/en/

WHO. (2020). *Coronavirus disease (COVID-19) Pandemic*. WHO. https://www.who.int/emergencies/diseases/novel-coronavirus-2019

WHO. (2021). *Heatwaves and Health: Guidance for the Public*. World Health Organization (WHO)._https://data.worldbank.org/indicator/NY.GDP.PCAP.CD?locations=ET

World Bank. (2022). Country Climate and Development Report: Türkiye. The World Bank Group.

World Bank. (2024). *World development indicators*. https://databank.worldbank.org/source/world-development-indicators

World Economic Forum. (2022). *BiodiverCities by 2030: Transforming cities' relationship with nature*. Retrieved April 30, 2024, from https://www3.weforum.org/docs/WEF_BiodiverCities_by_2030_2022.pdf

World Health Organization. (2009). *Global health risks - Mortality and burden of disease attributable to selected major risks* (Technical report).

World Tourism Organization. (2008). *Climate Change and Tourism Responding to Global Challenges*. e-unwto.org/doi/epdf/10.18111/9789284412341

Xue, S., Massazza, A., Akhter-Khan, S. C., Wray, B., Husain, M. I., & Lawrance, E. L. (2024). Mental health and psychosocial interventions in the context of climate change: A scoping review. *Npj Mental Health Research*, 3(1), 10. Advance online publication. DOI: 10.1038/s44184-024-00054-1 PMID: 38609540

Xu, G., & Zhang, W. (2024). Abating carbon emissions at negative costs: Optimal energy reallocation in China's industry. *Environmental Impact Assessment Review*, 105(1), 1–13. DOI: 10.1016/j.eiar.2023.107388

Yale Program on Climate Change Communication. (2021). Climate opinions by state. Available from: https://climatecommunication.yale.edu/visualizations-data/ycom-us/

Yamashita, J., Seto, T., Iwasaki, N., & Nishimura, Y. (2023). Quality assessment of volunteered geographic information for outdoor activities: An analysis of OpenStreetMap data for names of peaks in Japan. *Geo-Spatial Information Science*, 26(3), 333–345. Advance online publication. DOI: 10.1080/10095020.2022.2085188

Yáñez-Arancibia, A., Dávalos-Sotelo, R., & Day, J. W. (2014). *Ecological Dimensions for Sustainable Socio Economic Development* (Vol. 64). WIT Press.

Yang, Z., & Zhao, Y. (2014). Energy consumption, carbon emissions, and economic growth in India: Evidence from directed acyclic graphs. *Economic Modelling*, 38, 533–540. DOI: 10.1016/j.econmod.2014.01.030

Yellowlees, P. (2022). Climate change impacts on mental health will lead to increased digitization of mental health care. *Current Psychiatry Reports*, 24(11), 723–730. DOI: 10.1007/s11920-022-01377-6 PMID: 36214930

Yıldız, T. D. (2012). *3213 Sayılı Maden Kanunu Öncesinde ve Sonrasında Türkiye'de Maden Mevzuatında Yapılan Değişikliklerin İncelenmesi*. İstanbul Teknik Üniversitesi.

Yin, R. K. (2011). *Qualitative research from start to finish*. The Guilford Press.

Young, M. (2008). *Bringing knowledge back in from social constructivism to social realism in the sociology of education*. Taylor and Francis.

Yuhas, A. (2019). Cyclone Idai may be 'one of the worst' disasters in the Southern Hemisphere. *New York Times*, 19 March 2019. Available from: https://www.nytimes.com/2019/03/19/world/africa/cyclone-idai-mozambique.html

Yurtkuran, S. (2022). Gelen turist sayısının en fazla olduğu 10 ülkede turizm ile CO_2 Salımı arasındaki ilişki: Panel Fourier Toda-Yamamoto nedensellik analizi. *Erciyes Üniversitesi İktisadi ve İdari Bilimler Fakültesi Dergisi*, (61), 281–303. DOI: 10.18070/erciyesiibd.988886

Zahid, K. B., & Shah, H. (2024). Impact of climate change on farm-level technical efficiency in Punjab, Pakistan. *Climate Research*, 92(1), 65–77. DOI: 10.3354/cr01727

Zalasiewicz, J., & Williams, M. (2021). Climate change through Earth history. In *Climate Change* (3rd ed.). Observed Impacts on Planet Earth., DOI: 10.1016/B978-0-12-821575-3.00003-7

Zedler, J. B. (2000). Progress in wetland restoration ecology. *Trends in Ecology & Evolution*, 15(10), 402–407. DOI: 10.1016/S0169-5347(00)01959-5 PMID: 10998517

Zhai, F., & Zhuang, J. (2012). Agricultural impact of climate change: A general equilibrium analysis with special reference to Southeast Asia. *Climate Change Economics (Singapore)*, 3(2). Advance online publication. DOI: 10.1142/S2010007812500088

Zhang, X., An, J., Zhou, Y., Yang, M., & Zhao, X. (2024). How sustainable is OpenStreetMap? Tracking individual trajectories of editing behavior. *International Journal of Digital Earth*, 17(1), 2311320. Advance online publication. DOI: 10.1080/17538947.2024.2311320

Zhao, Y., & Liu, S. (2023). Effects of climate change on economic growth: A perspective of the heterogeneous climate regions in Africa. *Sustainability (Basel)*, 15(9), 7136. DOI: 10.3390/su15097136

Zhou, Y., Shao, H., Cao, Y., & Lui, E. M. (2021). Application of buckling-restrained braces to earthquake-resistant design of buildings: A review. *Engineering Structures*, 246, 112991. DOI: 10.1016/j.engstruct.2021.112991

Zikirya, B., Wang, J., & Zhou, C. (2021). The relationship between CO_2 emissions, air pollution, and tourism flows in China: A panel data analysis of Chinese Provinces. *Sustainability (Basel)*, 13(20), 11408. DOI: 10.3390/su132011408

Zilberman, D., Lipper, L., McCarthy, N., & Gordon, B. (2018). Innovation in response to climate change. *Climate smart agriculture: building resilience to climate change*, 49-74.

About the Contributors

Berfin Göksoy Sevinçli completed her undergraduate education in the Department of Public Administration at Selçuk University in 2014 and her master's degree in the Department of Public Administration at Yüzüncü Yıl University in 2017 with a thesis examining the effects of internal migration on urbanization through the example of Van. Afterwards, she participated in field research on rural development, migration, and various aspects of poverty by being included in the Poverty program determined as a priority area within the scope of the 100/2000 project of the Council of Higher Education. She was appointed as a research assistant at Bitlis Eren University in 2018. In 2022, she completed his doctorate with a thesis examining the transformation of Islamic cities at the Department of Political Science and Public Administration at Necmettin Erbakan University. As of 2022, she continues to serve as an assistant professor at the Department of Urbanization and Environmental Problems at Bitlis Eren University in the Department of Political Science and Public Administration. Göksoy Sevinçli has academic studies in the fields of urbanization policies, urban area management, and social policy.

Dilek Alma Savaş completed her undergraduate education at Atatürk University, Department of Economics. She completed her master's degree in Economic Theory at Manisa Celal Bayar University, Institute of Social Sciences and her doctorate in Economics at Bitlis Eren University, Faculty of Economics and Administrative Sciences. She currently works as an assistant professor at Bitlis Eren University, Faculty of Economics and Administrative Sciences. As of the end of 2023, she has 6 international articles, 1 of which is SSCI, 4 books/book editorships, 6 book chapters, 11 national and international congress/symposium proceedings, and 1 scientific research project. Alma Savaş, who continues her studies in the field of microeconomics, is married and has 1 child.

* * *

Munir Ahmad is a seasoned professional in the realm of Spatial Data Infrastructure (SDI), Geo-Information Productions, Information Systems, and Information Governance, boasting over 25 years of dedicated experience in the field. With a PhD in Computer Science, Dr. Ahmad's expertise spans Spatial Data Production, Management, Processing, Analysis, Visualization, and Quality Control. Throughout his career, Dr. Ahmad has been deeply involved in the development and deployment of SDI systems specially in the context of Pakistan, leveraging his proficiency in Spatial Database Design, Web, Mobile & Desktop GIS, and Geo Web Services Architecture. His contributions to Volunteered Geographic Information (VGI) and Open Source Geoportal & Metadata Portal have significantly enriched the geospatial community. As a trainer and researcher, Dr. Ahmad has authored over 50 publications, advancing the industry's knowledge base and fostering innovation in Geo-Tech, Data Governance, and Information Infrastructure, and Emerging Technologies. His commitment to Research and Development (R&D) is evident in his role as a dedicated educator and mentor in the field.

Fatma Fehime Aydın is a faculty member at Van Yüzüncü Yıl University, Faculty of Economics and Administrative Sciences. She specializes in macroeconomics, economic growth, development economics, regional development, sustainable economy, energy, environment, income distribution, employment and poverty. Assoc. Prof. Aydın pursues her academic work with a multidisciplinary approach, combining economic theory and practice with other disciplines such as sociology, political science, and environmental science. Her numerous scientific research articles have been published in national and international peer-reviewed journals and have attracted wide academic attention. In addition, she actively participates in various conferences and symposiums organized in the field of economics, sharing her academic knowledge and research findings. As part of her teaching activities, Assoc. Prof. Aydın teaches economics courses at undergraduate and graduate levels and provides students with in-depth knowledge on topics such as macroeconomics, regional economics, economic development, and energy economics. She has also supervised many theses and dissertations at the master's and doctoral levels.

Eda Bozkurt received her Bachelor's degree in Economy in 2006, her Ph.D. degree in Economy from Ataturk University, Turkey, in 2014. She is interested in development economics and economic growth, sustainable development. Now, she is an Assoc. Prof. at Faculty of Open and Distance Education, Ataturk University, Turkey.

William Estuardo Carrillo Barahona is an Engineer in Environmental Biotechnology, Master's Degree in Global Change, Natural Resources and

Sustainability, Doctoral Student in Agricultural, Agri-Environmental and Food Resources and Technologies. He has experience in national and international research centers in the environmental area. He has collaborated in linkage and research projects, active researcher of the IITMS research group. Teaching Member of the Higher Polytechnic School of Chimborazo in the Environmental Engineering program of the Morona Santiago Campus, of the Higher Polytechnic School of Chimborazo.

Ajay Chandel is working as an Associate Professor at Mittal School of Business, Lovely Professional University, Punjab. He has 14 years of teaching and research experience. He has published papers in SCOPUS, WOS, and UGC listed Journals in areas like Social Media Marketing, E-Commerce, and Consumer Behaviour. He has published cases on SMEs and Social Entrepreneurship in The Case Centre, UK. He also reviews The Case Journal, Emerald Group Publishing, and International Journal of Business and Globalisation, Inderscience. He has authored and developed MOOCs on Tourism and Hospitality Marketing under Epg-Pathshala- A gateway to all postgraduate courses (a UGC-MHRD project under its National Mission on Education Through ICT (NME-ICT).

Krishan Gopal is currently working as Associate Professor at Mittal School of Business (MHRD NIRF India Rank 34; ACBSP USA, Accredited), Lovely Professional University, Phagwara, Punjab (India). He has received his Ph.D degree from Lovely Professional University Punjab in the year 2020. His area of interest in teaching and research includes Business Research Methods, Strategic Management and Consumer Behaviour. Dr. Krishan Gopal has published more than 12 research papers in refereed national and international journals, 15 book chapters and cases, edited one book, attended various national and international seminars and acted as resource person in refresher course on mixed research methods. He has written 4 chapters of "Tourism and Hospitality Marketing" available on "e-PG-Pathshala"- a Project of UGC (MHRD).

Güven Güney received his Bachelor's degree in business administration in 2005, his Ph.D. degree in Business Administration from Ataturk University, Turkey, in 2018. He is interested in finance. Now, he is an Asst. Prof. at Faculty of Open and Distance Education, Ataturk University, Turkey.

Bhim Jyoti has been working with V.C.S.G., UUHF, College of Forestry, Ranichauri, Tehri Garhwal, Uttarakhand, India as Assistant Professor of Seed and Technology. She received MSc in Seed and Technology from CCS University Meerut and PhD in Seed and Technology from G. B. Pant University of Agriculture, Pantnagar, Uttarakhand. She has published several research articles in the seed and technology, plant growth, seed vigour, genetic variability, Biotechnological approaches, climate change and agricultural productivity.

Preet Kanwal completed her Post-Doctoral Fellowship at the ESG, University of Quebec in Montreal, Canada, furthering research expertise in her field. During her Post Doctoral Fellowship, she worked on exploring demographic sensitivity and its impact on discrimination in HEIs while considering insights from Canada and India. Dr. Kanwal is an academician and researcher with over 20 years of experience in Human Resource Management. She holds a PhD in Human Resource Management from I.K. Gujral Punjab Technical University, Punjab, India, where she conducted an empirical study on the quality of work life in the textile industry of Punjab, India. She also has international experience teaching MBA students from Victoria University (Online), Australia, and Sunway University (Online), Malaysia. Dr. Kanwal's research interests revolve around the quality of work-life, work-life balance, industrial relations, labour laws, social security and labour welfare, organizational stress, and behaviour, particularly in the context of the textile industry. She has presented her research at numerous national and international conferences and published her work in Scopus-indexed and UGC-listed Journals. Additionally, she has authored several book chapters on related topics, underscoring her commitment to advancing knowledge in her field. Her contributions extend beyond teaching and research; she has been actively involved in organizing and leading training sessions, workshops, and conferences, both as a participant and as a session chair.

Ezgi Kovancı completed her undergraduate degree in 2009 in the Department of International Relations at Kocaeli University and also studied in the same department at Marburg University through the Erasmus student exchange program during her university years. She pursued her master's degree in Integrated Environmental Studies at the University of Southampton and completed her Ph.D. in Urbanization and Environmental Issues at Ankara University's Faculty of Political Science. In 2013, she began working as a research assistant and conducted thesis research at Trier University during her doctoral studies. She completed her Ph.D. with a dissertation titled "Gender Discussion in the Context of the Climate Crisis." Since 2021, she has been working as an Assistant Professor at Adıyaman University. In March 2024, she was promoted to Associate Professor and currently teaches courses in environmental policy, urbanization policy, and gender studies.

Shashank Mittal has done his FPM in Organizational behavior and Human Resource from Indian Institute of Management Raipur. He holds B. Tech from I.E.T. Lucknow. His post FPM work experience includes almost five years of industry and academics exposures in multiple roles. Prior to joining FPM, he has over two years of industrial experience and three years of teaching experience in various organizations. He has published multiple papers in ABDC ranked and SSCI indexed journals of international repute such as Journal of Knowledge Management, Journal of Behavioral and Experimental Finance, International Journal of Conflict Management, Journal of Management and Organization and Current Psychology. His current research interest includes Social identity, Knowledge exchanges, Status, Proactive helping, Employer branding, Organizational justice, and Humanitarian relief management. He enjoys badminton and cycling during leisure time.

Amara Nisar holds an MPhil in Environmental Science from the University of the Punjab, Lahore. Her research spans various critical areas of environmental science, including toxicology, climate change, Geographic Information Systems (GIS), environmental contaminants, ecosystem health, environmental monitoring, policy development, and environmental degradation. Amara is dedicated to leveraging open-source data for environmental monitoring and policy development, aiming to address the pressing challenges of climate change and environmental sustainability..

Miguel Angel Osorio Rivera, Environmental Engineer, Master's Degree in Engineering for the Environment and Territory. Coordinator of the IITMS research group. Teaching Member of the Higher Polytechnic School of Chimborazo. He has scientific articles in high-impact databases such as WEB OF SCIENCE and SCOPUS. He has led several outreach and research projects. He currently serves as Academic Coordinator of ESPOCH-Morona Santiago Headquarters.

Noor ul Safa is an environmental scientist with a strong background in research and laboratory expertise. She holds a BS and MS in Environmental Sciences from GC Women University Sialkot. Her research interests span across climate change, toxicology, GIS, biodiversity conservation, and health risk assessment. Notably, she has conducted research on microplastics in indoor air and their associated health risks, demonstrating her proficiency in laboratory equipment and techniques. Additionally, she has contributed to various pilot-based research projects, showcasing her ability to collaborate and innovate. With a solid academic foundation and hands-on experience, Noor Ul Safa is a promising researcher in her field.

Ajay K. Singh is working as an Associate Professor (Research) in the Department of Humanities and Social Science, Graphic Era (Deemed to be University) Dehradun. He has worked as Assistant Professor (Economics) in the School of Liberal Arts & Management, DIT University Dehradun for 6 years. He did Post-Doctorate Research with EDI of India, Ahmedabad, Gujarat (India). He received MPhil (Economics) from DAVV Indore (India), and PhD (Economics) from IIT Indore (India. He has published several research papers in the diversified area such as climate change, agricultural productivity, assessment of food security; estimation of GFSI, development of environmental sustainability index (ESI) and its association with socio-economic indicators; measurement and determinants of entrepreneurship ecosystem, and dimension of sustainable development and its interlinkages with economic development.

Anugamini P. Srivastava is Assistant Professor in HRM at Symbiosis Institute of Business Management Pune, Symbiosis International (Deemed) University. She is Guest Editor for the upcoming special issue of European Journal of Training and Development, (Emerald, UK). She has book chapters and several research papers to indexed, abstracted and ABDC ranked journals of international repute to her credit. She is Recognized reviewer of Elsevier Journals and has reviewed articles for international journals. Her research interests are Human resource practices, Training, leadership, and other relevant topics of organisational behaviour.

Selvi Vural completed her bachelor's degree education at Hatay Mustafa Kemal University, Department of Civil Air Transportation Management, her master's degree education at Akdeniz University, and her PhD at Bursa Uludağ University, Department of Business Administration/Management and Organization. In her master's thesis, she examined the issue of disability in the context of work culture in accommodation establishments, and in his PhD thesis, she examined the effects of stereotypes on inclusive behaviors towards disabled employees in the air transportation sector. The author is interested in the fields of organizational psychology, organizational behavior, human resources, and aviation management. She continues to work at Gümüşhane University, Department of Aviation Management.

Mohit Yadav is an Associate Professor in the area of Human Resource Management at Jindal Global Business School (JGBS). He has a rich blend of work experience from both Academics as well as Industry. Prof. Mohit holds a Ph.D. from Department of Management Studies, Indian Institute of Technology Roorkee (IIT Roorkee) and has completed Master of Human Resource and Organizational Development (MHROD) from prestigious Delhi School of Economics, University of Delhi. He also holds a B.Com (Hons.) degree from University of Delhi and UGC-JRF scholarship. He has published various research papers and book chapters with reputed publishers like Springer, Sage, Emerald, Elsevier, Inderscience etc. and presented research papers in national and International conferences both in India and abroad. He has many best paper awards on his credit too. He is reviewer of various international journals like Computers in Human Behavior, Policing etc.

His areas of interest are Organizational Behavior, HRM, Recruitment and Selection, Organizational Citizenship Behavior, Quality of work life and role.

Enes Yalçın graduated from Bornova Anatolian High School in 2004. In 2010, he graduated from Ankara University Faculty of Veterinary Medicine. From 2011 to 2015, he worked as a Veterinarian at the Bornova Veterinary Control Institute. He later pursued studies in public administration, earning a Bachelor's degree from Dokuz Eylül University in 2014, a Master's degree from Selçuk University in 2016, and a Ph.D. from Dokuz Eylül University in 2021. He is also a student at Ege University Faculty of Letters, Department of English Translation and Interpreting. Between 2015 and 2022, he served as a Research Assistant in the Department of Political Science and Public Administration at Selçuk University Faculty of Economics and Administrative Sciences. Currently, he is working as an Assistant Professor in the Department of Political Science and Public Administration at İzmir Kâtip Çelebi University Faculty of Economics Administrative Sciences. He is married with two children.

Index

A

Action Plan 39, 266, 287, 406
Adaptation 21, 25, 26, 34, 35, 36, 37, 38, 40, 41, 42, 43, 44, 48, 49, 52, 55, 59, 60, 61, 63, 64, 65, 66, 68, 69, 70, 71, 72, 73, 74, 75, 77, 78, 84, 90, 91, 92, 93, 95, 96, 98, 99, 100, 102, 103, 107, 111, 127, 130, 132, 138, 139, 141, 148, 164, 165, 166, 167, 168, 170, 172, 173, 174, 178, 180, 181, 182, 184, 185, 186, 187, 191, 193, 195, 196, 197, 241, 256, 263, 264, 265, 266, 270, 271, 274, 284, 288, 289, 302, 313, 324, 329, 330, 331, 334, 342, 354, 355, 358, 360, 361, 362, 363, 368, 371, 375, 378, 379, 380, 381, 387, 390, 394, 395, 396, 397, 398, 403, 405, 406, 407, 409, 413, 416, 421, 424, 428, 430, 431, 435, 436, 438, 439, 440, 441, 445, 446, 449, 450, 454, 456, 457, 459, 460, 461, 462
Adaptation Strategies 21, 49, 52, 63, 68, 69, 72, 74, 78, 90, 99, 100, 107, 111, 130, 132, 139, 165, 167, 168, 170, 173, 174, 180, 186, 187, 256, 284, 289, 302, 330, 368, 375, 378, 379, 380, 387, 394, 395, 396, 397, 413, 421, 424, 445, 456
Agreements 35, 38, 42, 65, 72, 80, 142, 181, 183, 184, 193, 216, 255, 267, 278, 286, 287, 288, 313, 319
Approach 3, 9, 10, 14, 23, 26, 41, 44, 57, 61, 68, 72, 73, 85, 90, 92, 95, 99, 100, 129, 131, 133, 142, 160, 168, 175, 178, 189, 192, 194, 195, 211, 215, 234, 237, 241, 250, 256, 262, 264, 266, 267, 271, 311, 314, 315, 316, 319, 320, 324, 329, 332, 337, 340, 357, 360, 362, 368, 369, 370, 383, 384, 385, 388, 391, 392, 395, 399, 401, 402, 407, 408, 425, 430, 434, 450, 453, 458, 462

B

Biodiversity Conservation 11, 274, 389, 391, 396, 413

C

Capital Formation 114, 115, 120, 121, 122, 126
Carbon Dioxide 138, 141, 150, 151, 152, 154, 155, 160, 177, 178, 197, 203, 220, 230, 238, 241, 247, 260, 261, 262, 263, 264, 266, 272, 274, 275, 278, 279, 280, 286, 289, 290, 294, 295, 310, 311, 322, 333, 339, 383, 425, 429, 445
Carbon Footprint 11, 179, 185, 248, 249, 266, 274, 330, 430, 442, 453
Carbon Pricing 165, 172, 176, 179, 195, 274, 294, 317, 399, 430, 447
Case Studies 21, 27, 38, 41, 44, 49, 60, 63, 83, 89, 90, 100, 340, 346, 357, 359, 405, 408, 409, 423, 425, 450, 453, 461
Circular Economy 18, 19, 20, 185, 191, 406, 407, 421, 423, 429, 435, 452, 455, 457, 460, 462
Clean Development Mechanism 287, 288, 291
Climate 1, 10, 11, 13, 14, 18, 21, 22, 23, 24, 25, 26, 27, 28, 29, 30, 31, 32, 33, 34, 35, 36, 37, 38, 39, 40, 41, 42, 43, 44, 45, 46, 47, 48, 49, 50, 51, 52, 53, 54, 55, 56, 57, 58, 59, 60, 61, 62, 63, 64, 65, 66, 67, 68, 69, 70, 71, 72, 73, 74, 75, 76, 77, 78, 79, 80, 81, 82, 83, 84, 85, 86, 87, 88, 89, 90, 91, 92, 93, 94, 95, 96, 97, 98, 99, 100, 101, 102, 103, 104, 105, 106, 107, 108, 110, 111, 116, 126, 127, 128, 129, 130, 131, 132, 133, 134, 137, 138, 139, 141, 142, 143, 144, 147, 148, 149, 150, 151, 152, 153, 154, 155, 156, 157, 158, 159, 160, 161, 163, 164, 165, 166, 167, 168, 169, 170, 171, 172, 173, 174, 175, 176, 177, 178, 179, 180, 181, 182, 183, 184, 185, 186, 187, 188, 189, 190, 191, 192,

193, 194, 195, 196, 197, 199, 200, 201, 202, 203, 204, 205, 206, 209, 211, 212, 213, 215, 216, 217, 219, 220, 222, 226, 230, 231, 232, 233, 234, 237, 238, 239, 240, 241, 242, 243, 244, 245, 246, 247, 248, 249, 250, 251, 255, 256, 257, 258, 259, 260, 261, 262, 263, 264, 265, 266, 267, 268, 269, 270, 271, 272, 273, 274, 275, 277, 278, 279, 280, 281, 282, 283, 284, 285, 286, 287, 288, 289, 290, 291, 292, 293, 294, 295, 296, 301, 302, 303, 304, 305, 306, 307, 308, 309, 310, 311, 312, 313, 314, 316, 317, 318, 319, 320, 321, 322, 323, 324, 326, 327, 329, 330, 331, 332, 333, 334, 335, 336, 339, 340, 341, 342, 343, 345, 347, 348, 349, 351, 352, 353, 354, 355, 356, 357, 358, 359, 360, 361, 362, 363, 364, 365, 367, 368, 369, 370, 371, 372, 373, 374, 375, 376, 377, 378, 379, 380, 381, 382, 383, 384, 385, 387, 388, 389, 390, 391, 392, 393, 394, 395, 396, 397, 398, 399, 400, 401, 402, 403, 404, 405, 406, 407, 408, 409, 410, 411, 412, 413, 414, 415, 416, 417, 418, 419, 420, 421, 423, 424, 425, 426, 427, 428, 429, 430, 431, 432, 434, 435, 436, 437, 438, 439, 440, 441, 442, 443, 444, 445, 446, 447, 448, 449, 450, 451, 453, 454, 455, 456, 457, 458, 459, 460, 461, 462, 463

Climate Change 1, 10, 11, 13, 14, 21, 22, 23, 24, 25, 26, 27, 28, 29, 30, 31, 32, 33, 34, 35, 36, 37, 38, 39, 40, 41, 42, 43, 44, 45, 46, 47, 48, 49, 50, 51, 52, 54, 60, 61, 62, 63, 64, 65, 66, 67, 68, 69, 70, 71, 72, 73, 74, 75, 76, 77, 78, 79, 80, 81, 82, 83, 84, 85, 86, 87, 88, 89, 90, 91, 92, 93, 94, 95, 96, 97, 98, 99, 100, 102, 103, 104, 105, 106, 107, 108, 110, 111, 116, 126, 127, 128, 129, 130, 131, 132, 133, 134, 137, 138, 139, 141, 142, 143, 144, 148, 149, 150, 151, 152, 153, 154, 155, 156, 157, 158, 159, 160, 161, 163, 164, 165, 166, 167, 168, 169, 170, 171, 172, 173, 174, 175, 176, 177, 178, 179, 180, 181, 182, 183, 184, 185, 186, 188, 189, 190, 191, 192, 193, 194, 195, 196, 197, 199, 200, 201, 202, 203, 204, 205, 206, 209, 211, 212, 213, 215, 216, 217, 219, 220, 222, 226, 230, 231, 232, 233, 234, 237, 238, 239, 240, 241, 242, 243, 244, 245, 246, 247, 248, 249, 250, 251, 255, 256, 257, 258, 259, 260, 261, 262, 263, 264, 265, 266, 267, 268, 269, 270, 271, 272, 273, 274, 275, 277, 278, 279, 280, 281, 282, 283, 284, 285, 286, 287, 288, 289, 290, 291, 292, 293, 294, 295, 296, 301, 302, 303, 304, 305, 306, 307, 308, 310, 311, 312, 313, 314, 316, 317, 318, 319, 322, 323, 324, 326, 327, 329, 330, 331, 332, 333, 334, 335, 336, 339, 340, 341, 342, 343, 345, 347, 348, 349, 351, 352, 353, 354, 355, 356, 357, 358, 359, 360, 361, 362, 363, 364, 365, 367, 368, 369, 371, 373, 374, 376, 378, 379, 380, 382, 383, 384, 385, 387, 388, 389, 390, 391, 393, 394, 395, 396, 397, 398, 399, 401, 402, 403, 404, 405, 406, 407, 408, 409, 410, 412, 413, 414, 415, 416, 417, 418, 419, 420, 421, 423, 424, 425, 426, 427, 428, 430, 431, 432, 434, 435, 436, 438, 439, 440, 441, 442, 443, 444, 445, 446, 450, 451, 453, 454, 456, 457, 458, 459, 460, 461, 462, 463

Climate Change Education 329, 330, 343, 345, 348, 352, 353, 354, 355, 356, 357, 358, 359, 360, 363, 364

Climate Crisis 22, 23, 24, 25, 26, 189, 193, 230, 242, 243, 244, 245, 251, 273, 282, 301, 302, 318

Climate Finance 26, 40, 142, 178, 183, 274, 288, 397, 398, 399, 410, 430, 455

Climate-Induced Migration 29, 30, 31, 32, 42, 45, 47, 48, 49, 50, 51, 52, 53, 54, 55, 56, 57, 58, 59, 60, 62, 63, 64,

65, 66, 67, 68, 72, 73, 75, 76, 77, 78, 80, 81, 82
Climate Resilience 28, 29, 63, 71, 72, 77, 93, 127, 142, 163, 182, 183, 185, 186, 189, 192, 193, 255, 274, 375, 379, 383, 388, 389, 394, 398, 400, 405, 407, 408, 413, 414, 420, 423, 430, 456, 457
CO2 Emissions 106, 110, 151, 152, 154, 161, 200, 202, 203, 204, 205, 206, 208, 209, 211, 212, 214, 220, 222, 223, 224, 225, 227, 233, 234, 235, 236, 237, 260, 265, 269, 270, 271, 272, 310, 311, 317, 416
Community Resilience 192, 380, 397
Concepts 1, 2, 3, 8, 9, 13, 14, 143, 217, 247, 264, 268, 342, 345, 346, 348, 349, 350, 353, 354

D

Decarbonization 185, 274, 310, 414
Desertification 27, 30, 39, 40, 47, 49, 50, 52, 61, 62, 74, 76, 77, 138, 151, 201, 257, 280, 281, 333, 394
Digital Technologies 127, 187, 423, 424, 435, 441, 445, 455, 457, 461, 462
Disaster Response 83, 90, 92, 93, 94, 95, 96, 98, 99, 100, 166, 190, 370, 381
Disasters 10, 26, 28, 33, 34, 38, 42, 46, 50, 52, 60, 62, 63, 64, 70, 75, 84, 85, 86, 87, 89, 91, 93, 94, 108, 137, 138, 157, 163, 186, 190, 216, 239, 248, 249, 255, 258, 259, 266, 268, 286, 306, 308, 365, 388, 389, 395, 427, 436, 437, 440, 441
Displacement 25, 29, 30, 32, 34, 36, 40, 43, 47, 48, 49, 50, 51, 52, 53, 54, 55, 57, 60, 61, 62, 63, 64, 65, 66, 67, 68, 69, 70, 71, 72, 73, 74, 75, 76, 77, 78, 80, 81, 82, 86, 88, 89, 90, 93, 95, 166, 192

E

Economic and Financial Factors 222
Economic Costs 84, 163, 164, 165, 166, 167, 168, 169, 170, 173, 175, 177, 178, 179, 180, 184, 186, 187, 188, 189, 190, 193, 194
Economic Development 2, 5, 6, 7, 8, 9, 14, 16, 17, 18, 19, 20, 42, 56, 59, 60, 92, 108, 127, 140, 144, 146, 159, 167, 175, 218, 234, 238, 283, 326, 407
Economic Growth 2, 3, 4, 5, 6, 7, 8, 9, 10, 11, 12, 13, 14, 15, 16, 17, 18, 19, 20, 23, 42, 59, 60, 71, 85, 92, 137, 138, 140, 141, 144, 145, 146, 147, 158, 160, 161, 164, 165, 170, 175, 176, 180, 195, 196, 197, 200, 202, 204, 211, 217, 218, 219, 222, 223, 224, 225, 230, 231, 232, 233, 234, 235, 236, 237, 238, 244, 294, 295, 316, 336, 389, 390, 391, 392, 407, 409, 411, 420
Economic Impacts of Climate Change 137, 148, 167, 168, 195, 243
Economic Models 1, 2, 163, 170, 175, 193, 242
Economy 2, 3, 4, 5, 7, 8, 9, 10, 11, 12, 13, 14, 15, 16, 17, 18, 19, 20, 31, 43, 51, 53, 56, 58, 59, 91, 92, 137, 138, 140, 146, 147, 149, 152, 159, 163, 166, 167, 169, 171, 172, 173, 175, 176, 177, 178, 184, 185, 188, 190, 191, 193, 195, 196, 200, 203, 216, 217, 231, 238, 242, 243, 245, 249, 262, 265, 279, 283, 293, 295, 305, 316, 317, 323, 388, 393, 399, 401, 406, 407, 416, 419, 421, 423, 424, 429, 430, 431, 435, 441, 444, 445, 452, 455, 457, 460, 462, 463
Education for Sustainable Development 18, 329, 336, 337, 341, 349, 350, 354
Emission Reduction Targets 142, 274, 288, 291, 295, 312, 317, 324
Emissions 10, 11, 22, 25, 32, 35, 37, 74, 75, 78, 106, 110, 134, 137, 138, 142, 149, 150, 151, 152, 154, 155, 156, 157, 161, 165, 171, 172, 175, 176, 177, 178, 179, 180, 181, 182, 183, 184, 185, 187, 190, 191, 197, 199, 200, 202, 203, 204, 205, 206, 208, 209, 211, 212, 214, 216, 220, 222, 223, 224, 225, 227, 230, 231, 233, 234, 235, 236, 237, 238, 239, 240, 241, 242, 243, 247, 248, 249, 255,

256, 259, 260, 261, 262, 263, 264, 265, 266, 267, 268, 269, 270, 271, 272, 274, 277, 278, 279, 280, 282, 283, 284, 286, 287, 288, 289, 290, 291, 292, 293, 294, 295, 307, 309, 310, 311, 312, 313, 314, 316, 317, 318, 319, 320, 322, 323, 326, 327, 332, 333, 335, 360, 391, 392, 393, 394, 395, 397, 398, 399, 400, 402, 403, 404, 405, 406, 407, 412, 414, 415, 416, 418, 424, 428, 429, 430, 431, 434, 435, 441, 442, 443, 444, 447, 455, 463

Environment 9, 10, 11, 12, 13, 17, 18, 30, 46, 53, 61, 62, 68, 72, 86, 87, 97, 98, 99, 110, 111, 137, 140, 141, 146, 158, 159, 161, 164, 179, 185, 195, 196, 200, 201, 202, 204, 205, 211, 215, 216, 217, 218, 219, 230, 231, 232, 237, 240, 246, 248, 256, 261, 271, 278, 280, 281, 283, 284, 285, 286, 294, 301, 302, 303, 311, 316, 336, 343, 344, 345, 346, 347, 348, 349, 350, 351, 352, 353, 354, 355, 358, 360, 362, 364, 365, 371, 382, 383, 384, 399, 409, 411, 412, 424, 425, 431, 441, 442, 443, 446, 447, 450, 452, 453, 462

Environmental Education 18, 336, 337, 339, 340, 341, 344, 347, 349, 350, 351, 354, 360, 361, 362, 363, 364

European Commission 266, 270, 303, 316

European Green Deal 301, 316, 317, 323

European Union 37, 156, 182, 245, 261, 262, 263, 272, 301, 304, 309, 312, 313, 315, 318, 319, 320, 322, 325, 326, 403

Extreme Weather 22, 24, 28, 29, 33, 34, 40, 49, 69, 73, 76, 77, 78, 83, 85, 86, 88, 90, 93, 143, 144, 147, 149, 152, 164, 167, 168, 172, 180, 182, 190, 193, 246, 248, 255, 257, 259, 278, 324, 331, 332, 352, 368, 375, 378, 382, 395, 419, 424, 427, 435, 437, 438, 440, 441, 445, 456

F

Finance 8, 16, 26, 35, 40, 59, 122, 133, 142, 167, 178, 183, 269, 274, 288, 387, 396, 397, 398, 399, 400, 401, 410, 412, 413, 415, 430, 446, 455, 459, 462

Fit for 55 Package 317, 318

Fossil Fuels 11, 138, 151, 157, 175, 185, 205, 216, 220, 230, 232, 259, 261, 269, 278, 279, 280, 285, 287, 293, 294, 295, 302, 303, 309, 388, 392, 399, 404, 409, 432, 442, 453

Future 9, 10, 11, 12, 13, 14, 15, 16, 17, 22, 23, 26, 32, 35, 38, 41, 42, 43, 44, 48, 49, 50, 51, 73, 74, 75, 76, 77, 78, 79, 84, 85, 86, 87, 89, 90, 95, 96, 97, 100, 104, 142, 143, 144, 148, 154, 163, 170, 171, 173, 174, 179, 180, 184, 185, 186, 187, 188, 189, 192, 193, 194, 199, 210, 215, 216, 217, 218, 219, 231, 235, 236, 238, 242, 251, 255, 258, 266, 272, 274, 277, 278, 281, 285, 293, 294, 296, 302, 304, 305, 314, 315, 316, 317, 318, 319, 323, 333, 351, 363, 368, 380, 381, 388, 389, 391, 394, 397, 400, 401, 402, 409, 411, 413, 414, 416, 417, 418, 419, 420, 421, 423, 424, 425, 428, 431, 437, 441, 445, 452, 454, 455, 456, 457, 458, 459, 460, 463

G

Global Climate Change 49, 94, 103, 156, 158, 197, 258, 272, 283, 311, 333, 385, 432

Global Inequality 21, 22, 23, 24, 25, 26, 32, 35, 38, 41, 45, 46, 47

Global Warming 47, 104, 137, 138, 139, 141, 142, 143, 148, 149, 151, 152, 159, 161, 168, 175, 177, 178, 181, 186, 188, 200, 201, 205, 219, 220, 238, 240, 241, 242, 243, 246, 249, 256, 257, 258, 260, 267, 268, 279, 282, 287, 290, 292, 295, 296, 301, 302, 303, 305, 335, 339, 347, 352, 354, 356, 361, 388, 414, 418, 425

Green Growth 2, 3, 12, 13, 14, 15, 16, 18, 19, 20, 127, 407, 460

Greenhouse 10, 11, 22, 35, 36, 37, 137, 138, 139, 141, 142, 148, 149, 150, 151, 154, 155, 156, 157, 161, 164, 165, 169, 171, 175, 176, 177, 180, 183, 184, 191, 195, 197, 199, 200, 204, 205, 216, 220, 225, 231, 238, 239, 240, 241, 242, 243, 248, 249, 255, 256, 257, 259, 260, 261, 262, 263, 264, 265, 268, 269, 272, 274, 277, 278, 279, 280, 281, 282, 283, 284, 287, 288, 289, 290, 291, 292, 293, 294, 295, 296, 309, 310, 311, 316, 317, 319, 320, 322, 325, 327, 332, 333, 335, 352, 389, 391, 392, 399, 402, 403, 407, 418, 424, 425, 428, 429, 432, 441, 442, 443, 463

Greenhouse Gas 10, 11, 22, 35, 36, 37, 137, 138, 142, 149, 150, 154, 155, 156, 157, 161, 165, 171, 176, 177, 180, 183, 184, 191, 195, 199, 200, 204, 216, 220, 225, 231, 238, 239, 240, 241, 242, 243, 248, 249, 255, 256, 257, 259, 260, 261, 262, 263, 264, 265, 268, 269, 274, 277, 278, 279, 280, 282, 283, 284, 287, 288, 289, 290, 291, 292, 293, 295, 296, 309, 311, 316, 317, 319, 320, 322, 325, 327, 332, 335, 391, 392, 399, 402, 403, 407, 418, 424, 428, 441, 442, 443, 463

Greenhouse Gas Emissions 10, 11, 22, 35, 37, 137, 138, 142, 149, 150, 154, 155, 156, 157, 161, 165, 171, 176, 177, 180, 183, 184, 191, 199, 200, 204, 216, 225, 231, 239, 240, 241, 242, 243, 248, 249, 255, 256, 259, 260, 261, 262, 263, 264, 265, 268, 269, 274, 277, 278, 279, 280, 282, 283, 284, 287, 288, 289, 290, 291, 292, 293, 316, 319, 320, 322, 327, 332, 335, 391, 392, 399, 402, 403, 407, 418, 424, 428, 441, 442, 443, 463

I

Importance 2, 3, 6, 8, 9, 11, 13, 14, 20, 37, 65, 90, 108, 127, 128, 152, 157, 163, 186, 190, 219, 240, 255, 261, 278, 280, 285, 286, 288, 289, 293, 294, 309, 315, 388, 399, 424, 431, 441

Industrial Development 106, 107, 108, 110, 111, 115, 127, 128, 130, 131, 132, 409

Innovation 4, 6, 10, 14, 16, 43, 58, 59, 92, 95, 97, 106, 108, 109, 127, 142, 165, 175, 176, 177, 178, 179, 184, 185, 187, 189, 190, 191, 193, 224, 236, 310, 316, 317, 319, 325, 337, 369, 370, 387, 389, 390, 392, 393, 398, 400, 401, 407, 408, 410, 412, 414, 415, 417, 418, 421, 423, 424, 425, 428, 429, 430, 431, 434, 436, 439, 444, 446, 447, 450, 451, 452, 453, 454, 455, 456, 457, 458, 459, 460, 461, 462, 463

International Cooperation 22, 26, 35, 36, 37, 38, 42, 43, 44, 49, 60, 63, 66, 72, 73, 76, 78, 139, 142, 163, 165, 175, 178, 179, 180, 181, 182, 183, 184, 188, 192, 193, 280, 287, 291, 293, 397, 400, 401, 403, 405, 415, 418

International Tourism 203, 205, 208, 209

Interventions 34, 40, 54, 68, 85, 86, 88, 90, 93, 94, 95, 96, 97, 99, 100, 102, 104, 171, 175, 185, 264, 310, 368, 377, 439

K

Kyoto Protocol 141, 142, 161, 219, 261, 262, 263, 264, 272, 273, 278, 282, 283, 287, 288, 290, 291, 292, 295, 310, 311, 312, 313

M

Mental Health 83, 84, 85, 86, 87, 88, 89, 90, 91, 92, 93, 94, 95, 96, 97, 98, 99, 100, 101, 102, 103, 104, 132, 144, 158, 306

Millennium Development Goals 219, 350

Mitigation 25, 32, 35, 37, 60, 63, 127, 130,

137, 141, 163, 165, 167, 168, 170, 171, 172, 174, 175, 176, 178, 179, 180, 181, 182, 184, 186, 187, 189, 193, 216, 237, 256, 264, 265, 266, 271, 274, 283, 288, 302, 313, 322, 329, 330, 342, 355, 358, 359, 360, 368, 375, 377, 378, 383, 385, 387, 389, 390, 391, 392, 393, 394, 398, 409, 417, 424, 428, 429, 431, 434, 436, 439, 445, 446, 450, 454, 457, 459, 461

Mitigation Strategies 63, 141, 163, 172, 175, 179, 274, 375, 377, 390, 391, 392, 393, 394, 429

N

Nature-Based Solutions 186, 388, 417, 419, 456

O

OpenStreetMap 367, 369, 370, 371, 372, 373, 374, 378, 379, 380, 382, 383, 384, 385

OSM 367, 368, 369, 370, 371, 372, 373, 374, 375, 376, 377, 378, 379, 380, 381, 382, 383, 384

P

Panel Data Analysis 107, 133, 199, 200, 202, 203, 204, 205, 207, 209, 213, 214

Paris Agreement 21, 34, 35, 36, 38, 42, 66, 72, 76, 78, 141, 142, 161, 178, 181, 184, 188, 255, 263, 264, 272, 273, 278, 279, 282, 283, 287, 288, 292, 296, 301, 313, 314, 318, 320, 327, 403, 405, 412, 414, 415, 417, 418, 420

Policies 3, 4, 5, 6, 7, 8, 9, 15, 17, 23, 25, 26, 29, 31, 32, 34, 36, 37, 38, 41, 42, 43, 44, 45, 46, 48, 53, 54, 59, 60, 64, 65, 71, 82, 85, 88, 93, 94, 95, 96, 97, 98, 99, 100, 126, 127, 128, 137, 139, 140, 142, 145, 147, 153, 155, 156, 157, 169, 178, 181, 184, 186, 187, 188, 189, 190, 192, 196, 200, 210, 225, 231, 232, 241, 242, 244, 245, 246, 249, 256, 263, 264, 265, 266, 267, 269, 272, 274, 277, 278, 281, 283, 286, 287, 288, 292, 293, 294, 301, 309, 310, 311, 312, 313, 314, 315, 317, 318, 319, 320, 321, 322, 323, 324, 325, 341, 362, 368, 369, 376, 389, 401, 402, 404, 405, 406, 407, 408, 409, 410, 411, 413, 414, 416, 417, 418, 419, 421, 425, 429, 430, 447, 449, 450, 458, 462

Policy Formulation 93, 367, 369, 380, 411

Policy Frameworks 35, 37, 38, 44, 65, 66, 79, 178, 183, 188, 406, 408, 432, 447

Policy Integration 71, 73, 78, 387, 401, 402, 404, 405

Policy Responses 34, 49, 50, 51, 64, 66, 67, 68, 75, 85

Psychological Distress 83, 85, 86, 88, 89, 90, 100

R

Realist Social Theory 329, 339

Renewable Energy 10, 11, 13, 16, 25, 37, 42, 43, 70, 71, 140, 155, 156, 157, 169, 175, 178, 179, 181, 183, 185, 187, 189, 190, 204, 211, 213, 223, 224, 225, 227, 230, 231, 232, 233, 234, 235, 236, 245, 256, 268, 277, 278, 280, 284, 285, 287, 293, 296, 301, 304, 309, 311, 313, 316, 317, 318, 319, 320, 388, 390, 392, 393, 398, 399, 402, 404, 406, 407, 408, 409, 410, 412, 416, 420, 421, 423, 424, 428, 429, 430, 431, 432, 433, 434, 435, 444, 445, 446, 447, 448, 449, 451, 452, 453, 454, 455, 457, 462

Resilience 10, 28, 29, 34, 35, 36, 39, 40, 42, 43, 44, 45, 49, 56, 58, 60, 63, 64, 65, 66, 68, 69, 70, 71, 72, 73, 74, 76, 77, 78, 84, 92, 93, 95, 96, 97, 98, 99, 100, 101, 127, 138, 142, 157, 163, 167, 177, 178, 179, 180, 182, 183, 184, 185, 186, 189, 191, 192, 193, 194, 219, 249, 255, 265, 266, 274, 280, 289, 302, 307, 329, 330, 331, 358, 359, 368, 375, 378, 379, 380,

383, 388, 389, 391, 394, 395, 396, 397, 398, 399, 400, 402, 403, 404, 405, 406, 407, 408, 410, 413, 414, 416, 417, 419, 420, 423, 428, 430, 431, 433, 434, 437, 438, 440, 441, 442, 443, 444, 445, 450, 451, 456, 457, 462, 463

S

Sea-Level Rise 27, 34, 38, 40, 44, 49, 50, 60, 61, 70, 75, 77, 90, 160, 164, 166, 179, 363, 407, 416
Social Development 176, 177, 401
Social Justice 7, 9, 10, 11, 12, 13, 15, 21, 22, 23, 32, 34, 35, 36, 37, 38, 41, 43, 44, 269
Sustainability 3, 6, 7, 8, 9, 10, 11, 12, 13, 14, 16, 17, 18, 19, 23, 45, 46, 48, 56, 68, 101, 109, 127, 129, 131, 137, 140, 141, 152, 157, 161, 163, 167, 179, 187, 191, 192, 193, 194, 195, 196, 197, 213, 214, 215, 216, 217, 235, 237, 239, 240, 241, 245, 247, 248, 249, 267, 271, 278, 281, 284, 285, 286, 287, 293, 296, 310, 315, 316, 320, 338, 344, 345, 347, 348, 350, 351, 353, 354, 358, 362, 363, 364, 387, 388, 389, 392, 394, 406, 408, 413, 414, 415, 416, 417, 420, 443, 450, 451, 452, 453, 459
Sustainable 2, 3, 7, 8, 9, 10, 11, 12, 13, 14, 15, 16, 17, 18, 19, 20, 21, 23, 26, 31, 32, 36, 40, 42, 43, 44, 48, 51, 54, 60, 62, 63, 64, 66, 68, 69, 70, 71, 72, 73, 76, 77, 78, 79, 80, 97, 107, 127, 129, 131, 137, 156, 157, 165, 166, 167, 168, 170, 175, 177, 178, 179, 180, 182, 184, 185, 186, 187, 189, 190, 191, 193, 201, 215, 216, 218, 219, 220, 231, 232, 233, 235, 236, 237, 238, 240, 243, 248, 249, 255, 256, 261, 264, 266, 267, 274, 280, 281, 283, 285, 286, 288, 291, 293, 294, 295, 296, 301, 310, 313, 314, 315, 316, 317, 324, 325, 329, 330, 331, 332, 336, 337, 341, 344, 347, 348, 349, 350, 351, 353, 354, 358, 359, 361, 368, 371, 377, 385, 387, 388, 389, 390, 391, 392, 393, 394, 395, 396, 397, 398, 399, 400, 401, 403, 404, 405, 407, 408, 409, 410, 411, 412, 413, 414, 415, 416, 417, 418, 419, 420, 421, 424, 425, 428, 429, 431, 432, 434, 435, 441, 442, 443, 445, 449, 450, 451, 452, 453, 454, 455, 456, 457, 458, 459, 460, 461, 462
Sustainable Development 11, 16, 17, 18, 19, 20, 21, 26, 36, 42, 43, 44, 48, 51, 60, 63, 69, 72, 76, 77, 80, 107, 127, 166, 168, 182, 184, 215, 216, 218, 219, 220, 231, 233, 236, 237, 238, 255, 264, 274, 283, 286, 288, 296, 314, 315, 324, 325, 329, 331, 332, 336, 337, 341, 344, 347, 348, 349, 350, 353, 354, 359, 361, 368, 371, 387, 388, 389, 390, 391, 392, 394, 397, 398, 399, 401, 403, 404, 405, 407, 408, 409, 410, 411, 412, 413, 414, 415, 416, 417, 418, 419, 420, 421, 435, 451, 461
Sustainable Development Goals 11, 127, 216, 219, 237, 314, 315, 324, 331, 350, 387, 389, 391, 409, 420, 421
Sustainable Economy 2, 3, 8, 9, 10, 11, 12, 13, 14, 15, 17, 18, 185, 416, 445

T

Technical Efficiency 105, 107, 113, 114, 116, 117, 127, 129, 130, 131, 132, 133, 134
Technology 3, 4, 5, 18, 26, 35, 42, 43, 53, 72, 78, 96, 97, 98, 99, 105, 106, 109, 128, 129, 132, 146, 147, 149, 161, 167, 171, 175, 178, 182, 183, 187, 192, 211, 224, 225, 232, 234, 235, 247, 250, 251, 258, 283, 287, 288, 291, 292, 325, 340, 343, 344, 345, 346, 347, 348, 349, 353, 357, 359, 361, 387, 396, 397, 398, 399, 400, 401, 404, 410, 412, 413, 414, 415, 417, 418, 421, 423, 424, 425, 428, 429, 432, 433, 435, 437, 438, 439,

442, 443, 444, 446, 447, 448, 449, 450, 452, 453, 454, 455, 457, 458, 459, 460, 461, 462, 463
Tourism 27, 70, 166, 168, 169, 199, 200, 201, 202, 203, 204, 205, 206, 208, 209, 210, 211, 212, 213, 214, 239, 244, 245, 249, 258, 269, 384, 407, 427
Tourism Sector 199, 200, 201, 202, 203, 204, 209, 213, 244

U

United Nations 5, 17, 66, 74, 141, 142, 153, 159, 161, 181, 200, 202, 216, 217, 218, 219, 237, 257, 261, 272, 273, 278, 279, 280, 281, 282, 288, 289, 291, 292, 296, 302, 310, 311, 314, 326, 330, 334, 350, 352, 364, 365, 388, 403, 413, 415, 418, 462

W

Water Scarcity 49, 70, 74, 76, 169, 201, 390, 395, 403